Circuits, Signals, and Systems

The MIT Electrical Engineering and Computer Science Series

Harold Abelson and Gerald Jay Sussman with Julie Sussman, *Structure and Interpretation of Computer Programs*, 1985

William McC. Siebert, *Circuits, Signals, and Systems*, 1986

Circuits, Signals, and Systems

William McC. Siebert

The MIT Press
Cambridge, Massachusetts   London, England

McGraw-Hill Book Company
New York   St. Louis   San Francisco   Montreal   Toronto

This book is one of a series of texts written by faculty of the Electrical Engineering and Computer Science Department at the Massachusetts Institute of Technology. It was edited and produced by The MIT Press under a joint production-distribution arrangement with the McGraw-Hill Book Company.

Ordering Information:

North America
Text orders should be addressed to the McGraw-Hill Book Company.
All other orders should be addressed to The MIT Press.

Outside North America
All orders should be addressed to The MIT Press or its local distributor.

Third printing, 1986

This book was set under the direction of the author using the TEX typesetting system and was printed and bound by Halliday Lithograph in the United States of America.

Library of Congress Cataloging in Publication Data
Siebert, William McC.
  Circuits, signals, and systems.
  (The MIT electrical engineering and computer science series)
  Includes index.
  1. Electric circuits. 2. Discrete-time systems.
3. Linear time invariant systems. I. Title. III. Series.
TK454.S57   1985   621.319'2   85–4302
ISBN 0–262–19229–2 (MIT Press)
ISBN 0–07–057290–9 (McGraw-Hill)

We must be grateful to God that He created the world in such a way that everything simple is true and everything complicated is untrue.

–Gregory Skovoroda
(18th-century Ukrainian
philosopher)

Our life is frittered away by detail ··· simplify, simplify!
–Henry David Thoreau

La simplicité c'est la plus grande sagesse.
–unknown French author

One of the principal objects of theoretical research in any department of knowledge is to find the point of view from which the subject appears in its greatest simplicity.
–Josiah Willard Gibbs

Seek simplicity and distrust it.
–Alfred North Whitehead

Contents

9   The Unit Sample Response and Discrete-Time Convolution

10   Convolutional Representations of Continuous-Time Systems

11   Impulses and the Superposition Integral

12   Frequency-Domain Methods for General LTI Systems

The core curriculum taken by all undergraduates in the Department of Electrical Engineering and Computer Science at MIT consists of four courses. Until a few years ago, two of these were relatively traditional introductory electrical engineering courses in circuit theory and linear system theory, and two were introductory computer science courses in languages and architectures. By 1978, however, it had become clear that the needs and interests of the students in the department are diverse. For some, the core electrical engineering courses are the first step on a path leading to professional careers in electronic circuit and device design, control and communication system design, or engineering applications of electromagnetic fields and waves. For other students, these core courses are the last they will take with an engineering or physical science flavor. In 1979—facing up to this diversity—the department redesigned the core electrical engineering subjects in both content and style. Topics such as elementary electronic devices and circuits were added, the mathematical emphasis was broadened to include more applications, and modest laboratory exercises were incorporated to provide further experience with engineering reality. Complementary changes were introduced in the core computer science subjects, as described in other books in this series.

This book has evolved from a set of lecture notes for the second of the electrical engineering core subjects. The background assumed is an appreciation of the constitutive relations for common electrical circuit elements (including simple semiconductor devices and operational amplifiers), some skill at exploiting Kirchhoff's Laws to write dynamic equations for simple circuits in either node or state form, and an ability to solve such dynamic equations in simple cases when the drives are either zero or (possibly complex) exponentials. Chapter 1 reviews and somewhat extends this background material. A textbook for the prerequisite course is in preparation.

Chapters 2–4 introduce operational methods (the unilateral Laplace transform), system functions, and the complex frequency domain in a circuit context, leading to an input-output (functional or "black-box") characterization of linear time-invariant (LTI) circuit behavior. Interconnections of LTI systems are explored in Chapters 5 and 6, with particular emphasis on the practical and conceptual consequences of feedback. The next two chapters use the precise parallels between lumped continuous-time systems (described by differential equations and Laplace transforms) and discrete-time systems (described by difference equations and Z-transforms) both to review the mathematical structure of LTI circuits as presented in the earlier chapters and to introduce important applications of that structure in a broader context.

Discrete-time systems also provide a convenient vehicle for introducing the input-output characterization of LTI systems directly in the time domain through the unit sample response and convolution. This is done in Chapter 9. In Chapter 10, these ideas are extended to continuous-time systems. The mathematical subtleties of continuous-time impulses are carefully explored in Chapter 11 and resolved through an operational or generalized-function approach.

Examination of the general convolutional characterization of the black-box behavior of LTI systems reveals two overlapping categories of systems—those that are *causal* (although not necessarily stable) with inputs specified for $t > t_0$ and with the effects of past inputs implied by a state at $t = t_0$, and those that are *stable* (although not necessarily causal) with inputs typically specified for all time, $-\infty < t < \infty$. The first category—causal systems—may be loosely identified as systems of *control* type; appropriate analytical tools include the ones described in the first half of the book. The second category—stable systems—may be loosely described as systems of *communication* type; appropriate analytical tools are based on bilateral transforms—particularly (in an introductory treatment that does not require complex function theory as a prerequisite) the Fourier transform as described in the second half of the book.

The eigenfunction property of complex exponentials for LTI systems and the significant role thereby conferred on sums or integrals of exponentials as signal representations are explored in Chapters 12 and 13, and the fundamental properties of Fourier series and transforms are derived. The implications of these properties for such important applications as sampling, filtering, and frequency shifting (modulation) are discussed in the next four chapters. Throughout (but especially in Chapter 16, where duration-bandwidth relationships and the uncertainty principle are derived) emphasis is placed on the insights that can be gained from looking at system behavior simultaneously in the time domain and the frequency domain. To facilitate this process, time-frequency (duality) relationships are developed in as symmetric a way as possible. The power of this approach is particularly evident in the detailed examination of communication system engineering principles in Chapter 17.

Chapter 18 develops an application of a different kind—digital signal processing. The discrete-time Fourier transform (derived in Chapter 14 as a dual interpretation of the Fourier series formulas) is employed to probe various approaches to the use of digital hardware for carrying out operations such as low-pass filtering of analog waveforms.

In the next to last chapter, it is shown that knowledge of such averages of the input waveform as correlation functions and power density spectra is sufficient to determine corresponding averages of the output waveforms of LTI systems. The relationship of characterizing signals in terms of averages to the idea of a random process is discussed. Chapter 20 applies these results to explain the performance advantages of such wideband communication systems as PCM and FM.

Throughout, I have sought to balance two somewhat conflicting requirements. On the one hand, as befits an introductory course designed for sophomores and juniors, I have tried to hold the mathematical level to the absolute minimum necessary for an explication of the basic principles being discussed. On the other hand, as befits the breadth and maturity of our students' interests, I have chosen to talk about those topics (independent of difficulty or subtlety) that seem most exciting in terms of their mathematical, philosophical, or applicational interest. The first requirement, for example, has led me to resist the temptation to formulate the circuit dynamic equations in matrix form. The second has induced me to introduce certain aspects of generalized functions and random processes even though the discussion may be dangerously oversimplified. The balance is particularly tricky in connection with proofs of mathematical theorems; in general, I have ignored rigorous formalisms, trying to follow Heaviside's advice: "The best of all proofs is to set out a fact descriptively so that it can be seen to be a fact." I have also been sensitive to Bertrand Russell's aphorism: "A book should have either intelligibility or correctness; to combine the two is impossible." This is, perhaps, a bit bleak, but where I have had to choose, I have opted for intelligibility.

The problems at the end of each chapter (except the last) are an integral part of the text. Some are intended to provide practice in the topics of the chapter; the simplest of these are separately identified as exercises and usually include answers. Many of the problems, however, are designed to extend or amplify the text material. Reading through all the problems, to discover at least the kinds of topics discussed there, should be considered an essential part of studying each chapter.

At MIT, the material in this book is the basis for a 14-week course meeting about 5 hours a week in groups of various sizes (2 large lectures, 2 smaller recitations, and 1 very small tutorial). The lectures and accompanying demonstrations attempt to convey broad insights and perspectives that are difficult to communicate in other ways; they are also used to extend topics such as computational methods and the bilateral Laplace transform that are only briefly discussed in the text. Four laboratory assignments are part of the course; presently these cover the design of active filters, a comparison of numerical integration techniques for differential equations, a study of systems constructed from tapped delay lines, and the properties of FM modulators and phase-locked loops.

Not all topics in every chapter of the book are covered each term; not all topics that are covered are mastered by every student. Nevertheless, after five years of experience with the course in this form, we believe that most students come away with a broad understanding of complex system behavior as well as a set of elementary skills that can serve as adequate preparation for more advanced courses on more specialized topics. Although the book has been designed as an introductory text rather than a comprehensive treatise or a handbook of useful formulas and algorithms for solving problems, it could, I believe, be used in other ways. In particular, complemented with further notes and practice material of the instructor's choosing, the book would make an appropriate basis for a year-long senior-level preprofessional course in many curricula.

Despite what it may say on the title page, no book of this kind is the work of one person. Both the content and style have been enormously influenced by my early associations with that great teacher Ernst Adolph Guillemin. Over the more than 30 years that I've been teaching some of this material, numerous colleagues and students have also had an enormous impact. I am particularly happy to acknowledge the significant contributions of Hal Abelson, Rob Buckley, Mike Dawson, Bob Gallager, Lennie Gould, Bart Johnson, Bob Kennedy, Marvin Keshner, Jae Lim, Peter Mathys, Bill Schreiber, Cam Searle, Steve Senturia, Gerry Sussman, Art Smith, Dick Thornton, George Verghese, Stuart Wagner, Alan Willsky, John Wyatt, Mark Zahn, and Victor Zue. I also deeply appreciate the many hours devoted to this book and the notes that preceded it by the MIT Press editor, Larry Cohen; by a number of secretaries over the years, including in particular Barbara Ricker and Sylvia Nelson; by Pat McDowell, who drew the figures; and by Amy Hendrickson, who assisted me in the final typesetting in TeX on computer facilities generously made available by Lou Braida. My wife Sandy and my children, who gave up numerous opportunities so that Daddy could work on "the book," deserve my most heartfelt thanks. Finally, I am grateful to a succession of Department Heads and Deans (and most recently to the Bernard M. Gordon Engineering Curriculum Development Fund) for the financial support that made this book possible, and for the faith—in the absence of any compelling evidence—that it might someday be completed.

Circuits, Signals, and Systems

1

DYNAMIC EQUATIONS AND THEIR
SOLUTIONS FOR SIMPLE CIRCUITS

1.0 Introduction

The goal of this first chapter is twofold: to remind the reader of the basic principles of electrical circuit analysis, and to formulate these principles in appropriate ways so that we can develop them further in the chapters to come. Circuits (or networks) are, of course, arrangements of *interconnected elements*. But the word "circuit" can refer either to a *real* reticulated structure that we build in the laboratory out of elements such as resistors, capacitors, and transistors, interconnected by wires or printed-circuit busses, or it can refer to a *model* that we develop abstractly. For the most part in this book, we shall be discussing circuits in this latter sense (although we should never forget for long that, as engineers, we are interested in circuit models primarily as aids to the design and understanding of real systems). Our first task is therefore to define what we choose the words "interconnected" and "elements" to imply as abstractions. The circuit model then becomes a graphic way of specifying a set of *dynamic equations* that describe the behavior of the circuit. But such a description is usually only implicit; in the latter part of this chapter, we shall explore how simple dynamic equations can be solved to yield an explicit specification of the circuit response to simple stimuli. A goal of later chapters will be to extend and refine the ideas of this chapter into a collection of powerful tools for the analysis and design of the complex systems that characterize modern engineering practice.

1.1 Constitutive Relations for Elements

In models of electrical circuits, the elements or branches are characterized by equations (called *constitutive relations*) relating branch voltages and currents.*
The simplest abstract electrical elements are the linear resistors, capacitors,

---

*It is perhaps useful to point out that most of the ideas to be studied in this book also apply to a variety of other situations in which the important dynamic variables are *efforts* and *flows* (e.g., mechanical forces and velocities, temperatures and heat flows, chemical potentials and reaction rates). In addition, many models proposed in the social and biological sciences are described by equations similar to those we shall be investigating. Some texts go to elaborate lengths to formalize these *analogies*. We are not convinced such efforts are worthwhile, since most students make the necessary translations easily. Examples from non-electrical applications are scattered throughout the problems in this book.

inductors, and ideal sources described in Figure 1.1–1. Note that the reference directions for current, $i(t)$, and voltage, $v(t)$, in the constitutive relations are always *associated* as shown; that is, the positive direction for $i(t)$ is selected to be through the element from the positive reference terminal for $v(t)$ towards the negative terminal. The units of $i(t)$ and $v(t)$ are *amperes* and *volts* respectively.

Circuit elements may have more than two terminals. Perhaps the most important abstract multiterminal element is the *ideal controlled* (or *dependent*) *source*, of which there are four basic types as shown in Figure 1.1–2. Ideal controlled sources arise most commonly as idealizations for such active elements as transistors and op-amps in their linear regions. The *ideal op-amp*, for example, is an important special case of an ideal voltage-controlled voltage source obtained

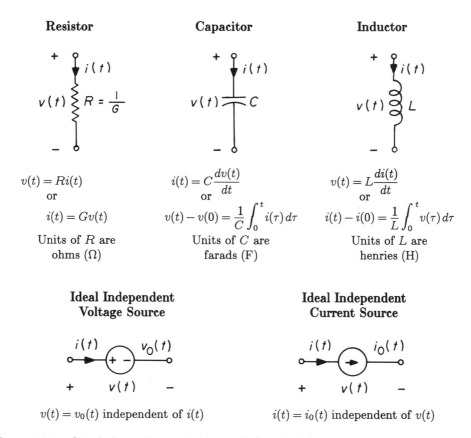

Figure 1.1–1. Simple linear 2-terminal lumped electrical elements and their constitutive relations. Note how current sources are distinguished from voltage sources; the orientation of the arrow or of the $+$ and $-$ signs inside the source symbol identifies the positive reference direction for the source quantity.

in the limit as the gain, α, becomes very large. It has its own special symbol as shown in Figure 1.1–3. The ideal op-amp is always used in a feedback circuit that achieves a finite output voltage by driving the input voltage difference, $\Delta v(t)$, nearly to zero. Other examples of multiterminal elements, such as coupled coils and transformers, transducers, and gyrators, are discussed in the problems at the end of Chapter 3.

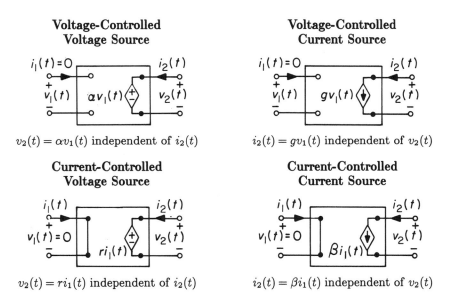

Figure 1.1–2. Ideal controlled (dependent) sources and their constitutive relations. Note that diamonds are used to identify dependent sources and circles to identify independent sources.

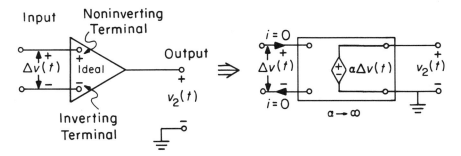

Figure 1.1–3. Ideal op-amp.

The ideal 2-terminal elements (excluding the independent sources) shown in Figures 1.1–1, 2, 3 are *linear*; that is, their dynamic variables satisfy the

SUPERPOSITION (LINEARITY) PRINCIPLE:*

If $i'(t)$ and $v'(t)$ are *any* pair of functions that satisfy the constitutive relation of an element, and if $i''(t)$ and $v''(t)$ are any other pair satisfying the same constitutive relation, then the element is said to obey the *superposition principle* (or equivalently to be *linear*) if the pair of functions $i(t) = ai'(t) + bi''(t)$ and $v(t) = av'(t) + bv''(t)$ also satisfy the constitutive relation for any choices of the constants a and b.

The 2-terminal elements described in Figures 1.1–1, 2, 3 (again excluding independent sources) also satisfy the

TIME-INVARIANCE PRINCIPLE:*

If $i(t)$ and $v(t)$ are *any* pair of functions that satisfy the constitutive relation of the element, then the element is *time-invariant* if $i(t - T)$ and $v(t - T)$ also satisfy the constitutive relation for *any* value of T.

Circuits composed entirely (except for independent sources) of linear time-invariant elements are examples of *linear time-invariant (LTI) systems*. The concept of an LTI system is more general, however, as we shall see in later chapters.

The most common *non*-linear element is probably the *diode*, whose idealized constitutive relation is shown in Figure 1.1–4. Also shown in this figure is the constitutive relation for what is surely the most important time-varying element—the *switch*. Circuits containing non-linear or time-varying elements are extremely useful. (See Problems 1.13–1.15 for some examples.) But the analysis of such circuits is often difficult. There are relatively few general principles or techniques for studying the behavior of non-linear circuits; each new circuit is likely to present a new analytical problem. In contrast, the theory of LTI systems consists of a rich collection of theorems, concepts, and methods providing powerful tools for understanding and design. As a result, the necessary non-linearities in practical electronic circuits are often restricted to isolated locations interconnected by LTI systems. Such an arrangement may vastly simplify the analysis while providing enough design freedom to achieve the desired dynamic effects. When such isolation and localization are impossible, as for example in some high-speed integrated circuits, the design process may reduce to employing numerical methods to study the performance of the device as various parameters are systematically varied. Computerized circuit simulation programs intended for this purpose have been developed, but the wide availability of such simulation

---

*The extension of these definitions to multiterminal elements is straightforward. For other examples of 2-terminal elements that are or are not linear and/or time-invariant, see Exercise 1.1.

programs has not eliminated the need to understand the mathematics of LTI systems, which remains a powerful *language* in terms of which complex system behavior can be discussed.

Ideal Diode

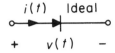

$i(t) = 0$ for $v(t) \leq 0$
$v(t) = 0$ for $i(t) \geq 0$

Ideal Switch

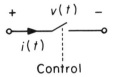

Control

Control is an independent function of time with two states—"open" or "closed"
$i(t) = 0$ in open state
$v(t) = 0$ in closed state

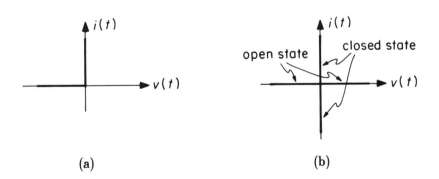

(a)　　　　　　　　　　　　　　　　(b)

Figure 1.1–4. The ideal diode (a) and the ideal switch (b).

1.2 Interconnection Constraints: Kirchhoff's Laws

In addition to the constraints imposed by the constitutive relations of the branch elements, the branch voltages and currents in electric circuits are further constrained by the two fundamental laws portrayed in Figure 1.2–1.

> *KIRCHHOFF'S CURRENT LAW (KCL):*
> The algebraic sum of the currents entering any circuit node is zero. (More generally, the sum of the currents passing inward through any network cut set must equal zero. A *cut set* is any set of branches which, if cut, would divide the circuit into two parts.)

> *KIRCHHOFF'S VOLTAGE LAW (KVL):*
> The algebraic sum of the directed voltage drops around any circuit mesh is zero. (More generally, the sum of the voltage drops around any closed path in the circuit must equal zero.)

Both laws follow from Maxwell's equations* provided that the circuit is so designed and the variables are changed sufficiently slowly that all significant electromagnetic energy is stored inside the "elements" rather than in the spaces between the elements; the energy storage can then be described as *lumped*. For circuits of bench-top size composed of typical lumped elements, Kirchhoff's Laws are a good approximation for signal frequencies less than a few tens of megahertz.

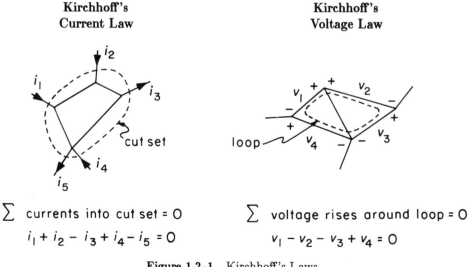

Kirchhoff's Current Law

$\sum$ currents into cut set = O

$i_1 + i_2 - i_3 + i_4 - i_5 = 0$

Kirchhoff's Voltage Law

$\sum$ voltage rises around loop = O

$v_1 - v_2 - v_3 + v_4 = 0$

Figure 1.2–1. Kirchhoff's Laws.

---

*For a careful investigation of the conditions under which Kirchhoff's Laws are valid, see, for example, R. M. Fano, L. J. Chu, and R. B. Adler, *Electromagnetic Fields, Energy, and Forces* (New York, NY: John Wiley, 1960).

1.3 Dynamic Equations in Node and State Form

Together, Kirchhoff's Laws and the constitutive relations provide a set of $2N$ independent equations for the N voltages and N currents associated with the N branches of a circuit. The formulation and solution of these dynamic equations* for a circuit is, however, usually much simplified by employing one or another of several special procedures that substantially reduce the number of unknowns. Two such special procedures—leading to what are called *node equations* or *state equations*—are particularly important in applications. The first special procedure is the

NODE EQUATIONS PROCEDURE:

1. Pick a *reference node*. The resulting equations will usually be simplest if the chosen node is the one that is common to the largest number of voltage sources and/or the largest number of branches.

2. Assign a *node voltage variable* to every other node, except that only one of two nodes connected by an ideal voltage source (whether independent or dependent) need be assigned a node voltage variable. (In particular, we do not need to assign a node voltage variable to any node connected to the reference node by a chain of one or more ideal voltage sources.) The number of assigned node voltage variables is thus one less than the number of nodes minus the number of ideal voltage sources.[†] Each node voltage variable measures the voltage of the corresponding node with respect to the reference node.

3. Write a KCL equation in terms of node voltage variables at each node to which such a variable is assigned. (If one or more ideal voltage sources are connected to the node, write the KCL equation for a cut set enclosing the desired node and the voltage sources, as shown in Example 1.3–1.)

The node equations procedure thus leads to as many equations and unknowns as there are assigned node voltage variables. In general, this number is very much less than twice the number of branches. Once the node voltages are known, any desired branch voltages or currents can usually be found quite easily.

---

*In much of the electrical engineering literature these are called *equilibrium equations*, but this seems a rather inappropriate label since a circuit is rarely in equilibrium in a mechanical or thermodynamical sense. In physics the analogous equations are called *dynamic equations* or *equations of motion*, and we shall adopt the former term.

[†]If the network contains a loop of voltage sources (or a cut set of current sources), Kirchhoff's Laws imply that the values of these sources are not independent; one source can thus be deleted without modifying the behavior of the circuit. This rule for the number of independent node variables assumes that such deletion has been carried out.

Example 1.3–1

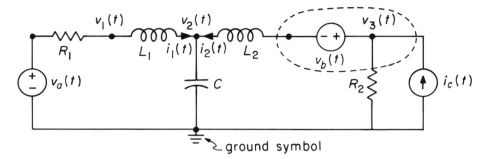

Figure 1.3–1. Circuit to illustrate the node equations procedure.

Following the procedure above for the circuit of Figure 1.3–1, we write the node equations as follows:

1. As our reference node we pick the one marked by the ground symbol. This node is chosen because four branches including one voltage source join there.
2. We assign three node voltage variables as shown. The circuit has six nodes; note that 6 nodes − 2 voltage sources − 1 = 3 independent node voltages.
3. We then write KCL equations at these three labelled nodes:

   i) For the currents leaving the node labelled $v_1(t)$:

$$\frac{v_1(t) - v_a(t)}{R_1} + \frac{1}{L_1} \int_0^t [v_1(\tau) - v_2(\tau)]\, d\tau + i_1(0) = 0. \qquad (1.3\text{–}1)$$

   ii) For the currents leaving the node labelled $v_2(t)$:

$$\frac{1}{L_1} \int_0^t [v_2(\tau) - v_1(\tau)]\, d\tau - i_1(0) + C\frac{dv_2(t)}{dt} + \frac{1}{L_2} \int_0^t [v_2(\tau) - v_3(\tau) + v_b(\tau)]\, d\tau - i_2(0) = 0.$$
$$(1.3\text{–}2)$$

   iii) For the currents leaving the cut set defined by the dotted loop in Figure 1.3–1, enclosing the node labelled $v_3(t)$ and the voltage source $v_b(t)$:

$$\frac{1}{L_2} \int_0^t [v_3(\tau) - v_b(\tau) - v_2(\tau)]\, d\tau + i_2(0) + \frac{v_3(t)}{R_2} = i_c(t). \qquad (1.3\text{–}3)$$

The result is three simultaneous integro-differential equations in the three unknown node voltages, $v_1(t)$, $v_2(t)$, and $v_3(t)$.

▶ ▶ ▶

Example 1.3–2

As a second example of the node equations procedure, consider the circuit of Figure 1.3–2.

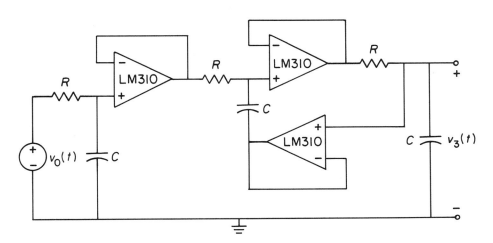

Figure 1.3–2. Another circuit to illustrate the node equations procedure.

If the op-amps are ideal, each voltage follower in Figure 1.3–2 can be replaced by the unit-gain controlled source shown in Figure 1.3–3.

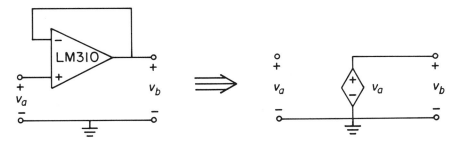

Figure 1.3–3. Controlled source equivalent of a voltage follower.

The result is the equivalent circuit of Figure 1.3–4, which contains three dependent voltage sources. (Note that the input current of an ideal voltage follower is zero. A voltage follower thus acts as a *buffer* or *isolator*. Despite the fact that its output voltage is equal to its input voltage, a voltage follower cannot in general be replaced by a wire between input and output without altering circuit behavior.)

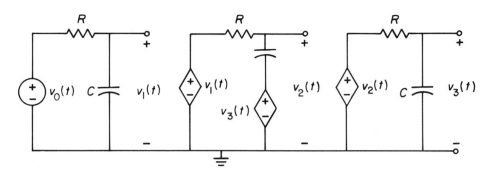

Figure 1.3–4. Equivalent circuit of Figure 1.3–2.

The circuit of Figure 1.3–4 has 8 nodes and 4 voltage sources; hence $8-4-1=3$ node voltage variables, $v_1(t)$, $v_2(t)$, and $v_3(t)$, are required. The most appropriate choice of reference node is, again, the real circuit ground, since it is a junction of 6 branches and is common to all the voltage sources. Writing KCL equations at each variable node yields three equations in three unknowns:

$$C\frac{dv_1(t)}{dt} + \frac{1}{R}[v_1(t) - v_0(t)] = 0 \qquad (1.3-4)$$

$$C\frac{d}{dt}[v_2(t) - v_3(t)] + \frac{1}{R}[v_2(t) - v_1(t)] = 0 \qquad (1.3-5)$$

$$C\frac{dv_3(t)}{dt} + \frac{1}{R}[v_3(t) - v_2(t)] = 0. \qquad (1.3-6)$$

▶ ▶ ▶

The second special procedure for writing dynamic equations is the

STATE EQUATIONS PROCEDURE:

1. Replace each inductor L_j temporarily by an ideal current source of value $i_j(t)$, and each capacitor C_k by an ideal voltage source of value $v_k(t)$.

2. Solve the resulting circuit, consisting of resistors and sources only, for the voltages $v_j(t)$ (across the current sources replacing the inductors) and the currents $i_k(t)$ (through the voltage sources replacing the capacitors). For an LTI circuit, this will yield a set of equations of the form $v_j(t)$ (or $i_k(t)$) equals a weighted sum of inductor currents, $i_j(t)$, capacitor voltages, $v_k(t)$, and independent source quantities.

3. Replace $v_j(t) = L_j\dfrac{di_j(t)}{dt}$ and $i_k(t) = C_k\dfrac{dv_k(t)}{dt}$ on the left in these equations to give a set of *first-order* differential equations in the set of state variables—the inductor currents and the capacitor voltages.

Example 1.3–3

To help make these formal steps more concrete, let's reconsider the circuit of Example 1.3–1. Assigning branch voltage and current variables to the inductors and capacitors leads to the situation shown in Figure 1.3–5.

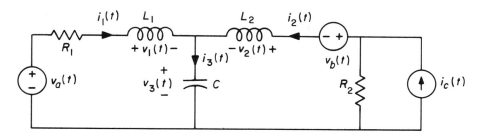

Figure 1.3–5. Circuit to illustrate the state equations procedure.

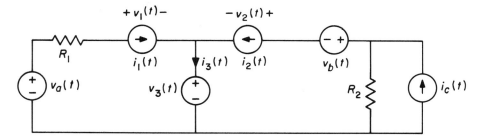

Figure 1.3–6. Figure 1.3–5 with energy-storage elements replaced by sources.

Replacing inductors and capacitors by ideal current and voltage sources respectively yields the circuit of Figure 1.3–6. Elementary resistive circuit analysis then gives

$$v_1(t) = -R_1 i_1(t) - v_3(t) + v_a(t) \tag{1.3-7}$$

$$v_2(t) = -R_2 i_2(t) - v_3(t) - v_b(t) + R_2 i_c(t) \tag{1.3-8}$$

$$i_3(t) = i_1(t) + i_2(t). \tag{1.3-9}$$

Since $v_1(t) = L_1 \dfrac{di_1(t)}{dt}$, $v_2(t) = L_2 \dfrac{di_2(t)}{dt}$, and $i_3(t) = C \dfrac{dv_3(t)}{dt}$, we obtain the dynamic equations in state form:

$$\frac{di_1(t)}{dt} = -\frac{R_1}{L_1} i_1(t) - \frac{1}{L_1} v_3(t) + \frac{1}{L_1} v_a(t) \tag{1.3-10}$$

$$\frac{di_2(t)}{dt} = -\frac{R_2}{L_2} i_2(t) - \frac{1}{L_2} v_3(t) - \frac{1}{L_2} v_b(t) + \frac{R_2}{L_2} i_c(t) \tag{1.3-11}$$

$$\frac{dv_3(t)}{dt} = \frac{1}{C} i_1(t) + \frac{1}{C} i_2(t). \tag{1.3-12}$$

Note that the left-hand side of each of these equations is the first derivative of a state variable $\big(i_1(t),\ i_2(t),\ \text{or}\ v_3(t)\big)$ and the right-hand side is a function of state variables and independent sources $\big(v_a(t),\ v_b(t),\ \text{and}\ i_c(t)\big)$ only.

▶ ▶ ▶

The inductor currents and capacitor voltages are called *state* variables because their present values summarize the accumulated effects of past experiences insofar as these may influence future behavior. This follows because the inductor currents and capacitor voltages determine the present distribution of stored energy in the circuit.* The choice of inductor currents and capacitor voltages as state variables is not, however, unique; for example, any equal number of independent linear combinations of the inductor currents and capacitor voltages could also serve as a set of state variables because the capacitor voltages and inductor currents can be uniquely derived from them. (See Problem 1.1.)

The number of independent state variables is called the *order* (or *degree*) of the circuit. The procedure described above suggests that the order is equal to the number of capacitors and inductors in the circuit. If, however, the circuit has loops of capacitors and voltage sources, or cut sets of inductors and current sources, the number of independent state variables (and the number of state equations) is reduced by one for each such loop or cut set. This happens because KVL or KCL equations constrain the values of the ideal sources in the first step of the state equations procedure. (See Problem 1.2.)

The state equations describe the local evolution of the state. Their form is important: The rate of change of the state is a function of the present state and the present inputs. As a result of this orderly structure, the state form of the dynamic equations has certain advantages over the node form, particularly for proving theorems or for describing general properties of circuits. However, different procedures (such as node or state) for writing dynamic equations will in general require different numbers of variables and equations, and will yield equations of different complexity. Selecting the "best" procedure is an art, not a science, and depends on the particular circuit and the objectives of the analysis. Both the node and state procedures can be extended to networks of arbitrary complexity, including time-varying and non-linear elements. Formal circuit analysis algorithms exist for either procedure that will automatically produce the dynamic equations once the network topology and constitutive relations for the elements are specified. Moreover, for LTI circuits at least, the effort required to obtain explicit analytical solutions of the dynamic equations is roughly independent of the procedure used to formulate the equations—it is primarily dependent on the *order* of the system, as we shall see.

---

*Recall that the product of a branch voltage and the associated current is the instantaneous power input to that branch element. For an LTI capacitor C, the integral of the instantaneous power, which is the stored energy at time t, is

$$\int_{-\infty}^{t} v_C(\tau) i_C(\tau) \, d\tau = \int_{-\infty}^{t} v_C(\tau) C \frac{dv_C(\tau)}{d\tau} \, d\tau = \tfrac{1}{2} C v_C^2(t).$$

Similarly, for an LTI inductor L, the stored energy at time t is $\tfrac{1}{2} L i_L^2(t)$. In both of these formulas we have tacitly assumed that at $t = -\infty$ the element is in the *zero state*, $v_C(-\infty) = i_L(-\infty) = 0$.

1.4 Block Diagrams

Once the dynamic equations have been written in state form for any system—electrical circuit or not—it is easy to devise a *block diagram*, and from this an electronic circuit, that behaves analogously. By "analogously" we mean that if the inputs to the electronic circuit have the same waveform or time shape as the inputs to the actual system, then the waveforms of the state variables (or any combinations of state variables and inputs that may be selected as outputs) will also be the same in the actual system and in the electronic analog. Of course, in the process of translation from the actual system to the electronic analog, we are free to choose the units of the electronic circuit variables corresponding to the units of the actual system variables in any convenient way. We can also choose the time scale of the analog to speed up the representation of some slow action (such as a geological process) or slow down a fast action (such as the effects of an explosion). For many years, electronic *analog computers* of this kind have been effective design tools for complex, expensive, hard-to-modify systems such as aircraft or missile control systems. Today, to be sure, the dynamic analysis of such systems is usually done digitally, but block diagrams remain useful conceptually, and electronic circuits designed on the basis of such diagrams still have many applications in real-time situations such as audio signal processing.

Some of the more common elements appearing in simple block diagrams are shown in Figure 1.4–1. Each block defines a relationship between one or more *outputs* (the labels on arrows leaving the block) and one or more *inputs* (the labels on arrows entering the block). When blocks are connected together, the output of one becomes an input to another. The following example illustrates how interconnections of such blocks can describe a given set of state equations.

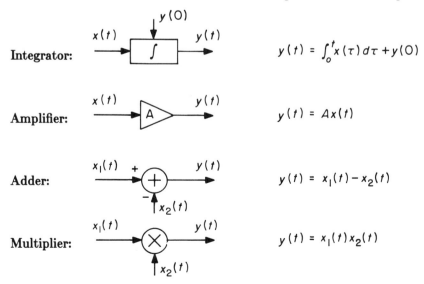

Integrator: $y(t) = \int_{o}^{t} x(\tau)\,d\tau + y(0)$

Amplifier: $y(t) = Ax(t)$

Adder: $y(t) = x_1(t) - x_2(t)$

Multiplier: $y(t) = x_1(t)x_2(t)$

Figure 1.4–1. Simple block diagram elements.

Example 1.4–1

The circuit of Figure 1.3–5 led to three state equations:

$$\frac{di_1(t)}{dt} = -\frac{R_1}{L_1}i_1(t) - \frac{1}{L_1}v_3(t) + \frac{1}{L_1}v_a(t) \tag{1.4-1}$$

$$\frac{di_2(t)}{dt} = -\frac{R_2}{L_2}i_2(t) - \frac{1}{L_2}v_3(t) - \frac{1}{L_2}v_b(t) + \frac{R_2}{L_2}i_c(t) \tag{1.4-2}$$

$$\frac{dv_3(t)}{dt} = \frac{1}{C}i_1(t) + \frac{1}{C}i_2(t). \tag{1.4-3}$$

These equations can be simulated by the block diagram of Figure 1.4–2, composed of integrators, adders, and amplifiers. Because the system is 3rd-order, three integrators are needed whose outputs represent the state variables, $i_1(t)$, $i_2(t)$, and $v_3(t)$. The key to devising or analyzing such diagrams is to focus on the inputs to the integrators, that is, on the derivatives of the state variables. Each derivative is composed of a sum of weighted inputs and state variables in accordance with the corresponding state equation. The way in which the diagram is drawn and the dashed boxes in Figure 1.4–2 should make this structure readily apparent. If we identify, say, the voltage $v_c(t)$ across R_2 in the circuit of Figure 1.3–5 as the output of interest, then we may add several blocks to the diagram to realize $v_c(t) = R_2(i_c(t) - i_2(t))$.

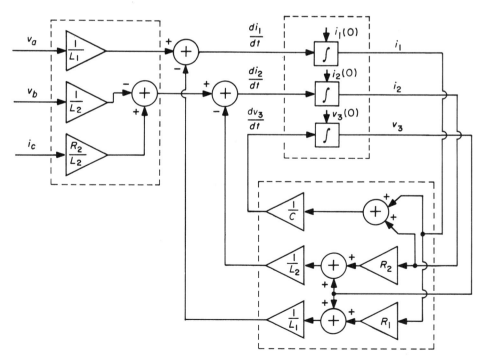

Figure 1.4–2. Block diagram simulation of equations (1.4–1, 2, 3).

▶ ▶ ▶

Example 1.4–2

To synthesize an electronic circuit analogous to the block diagram of Figure 1.4–2, we can interconnect the op-amp circuits of Figure 1.4–3 with appropriate element values.* Each of the circuits in Figure 1.4–3 actually combines several of the functions of the basic blocks of Figure 1.4–1. An appropriate interconnection is shown in Figure 1.4–4. Note that all the variables, independent of symbol, are in fact voltages. The labels next to the resistors are resistances but, since only ratios of resistances influence behavior, they can all be scaled by a common factor to any convenient range of values. The value C_0 of the integrator capacitors determines the time scale of the analog and may be chosen as desired.

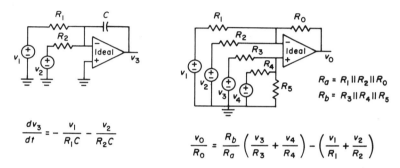

$$\frac{dv_3}{dt} = -\frac{v_1}{R_1 C} - \frac{v_2}{R_2 C}$$

$R_a = R_1 \| R_2 \| R_0$
$R_b = R_3 \| R_4 \| R_5$

$$\frac{v_0}{R_0} = \frac{R_b}{R_a}\left(\frac{v_3}{R_3} + \frac{v_4}{R_4}\right) - \left(\frac{v_1}{R_1} + \frac{v_2}{R_2}\right)$$

Figure 1.4–3. Op-amp realizations of integrators and adders.

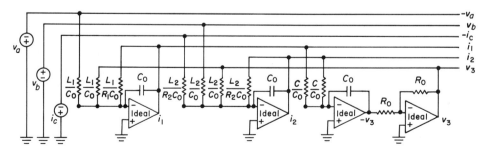

Figure 1.4–4. Electronic circuit implementation of (1.4–1, 2, 3).

▶ ▶ ▶

---

*The name *operational amplifier* was originally assigned because these units were first used to realize such operations as integrator and adder in analog computers.

1.5 Solutions of the Dynamic Equations

However derived—via the node or the state procedure described in Section 1.3 or some other procedure—the dynamic equations for a lumped circuit typically have the form of a set of differential equations in several variables.* Such equations describe the unknown responses to known stimuli only implicitly; they have the form

$$\text{(operations on responses)} = \text{(operations on stimuli)}.$$

On the other hand, what is frequently sought is an explicit (or operational) description in the form

$$\text{response} = \text{(operations on stimuli)}.$$

To achieve an operational description, the dynamic equations must be solved (integrated) rather than simply evaluated. Moreover, the solution may not be unique (because the results of operating on two different responses may be the same, so that both satisfy the dynamic equations). To obtain a unique solution, we must have additional auxiliary information, such as initial conditions or initial state.

If the circuit is LTI, an explicit closed-form solution of the dynamic equations can nearly always be achieved (at least in principle and for a wide class of input functions). The simplest such solutions are composed by combining solutions to two special situations:

  a) The drives or inputs are zero.

  b) The drives or inputs are exponentials in time.

The remainder of this chapter will review these two special solution situations for LTI circuits. And one of the purposes of the chapters that follow is to show that these situations are not in fact as "special" as they seem to be.

1.6 Solutions of the Dynamic Equations When the Inputs Are Zero

If all the independent sources are zero in some finite $(t_0 < t < t_1)$ or semi-infinite $(t > t_0)$ time interval, then the branch voltages and currents may be zero in that interval, but they do not have to be zero—energy stored in the circuit by inputs during $t < t_0$ may act as an effective drive for what is variously called the *natural* or *homogeneous* or *zero-input response* (ZIR) of the circuit. It is easy to show that the voltages and currents in a lumped LTI circuit during a zero-input interval may have nonzero values if and only if their waveforms are sums of particular appropriate exponential time functions (called *normal modes*) with specific (in general, complex) *time constants* whose reciprocals are the *natural frequencies* of the circuit. The basic issues are most readily explained through an example.

---

*Node equations for circuits containing inductors will also include integrals of the node voltages (see Example 1.3–1). But such integrals can be "cleared" by differentiating the equations.

Example 1.6–1

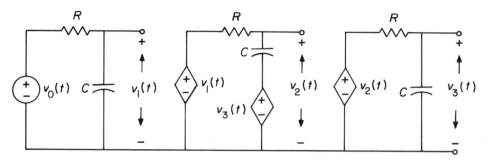

Figure 1.6–1. Equivalent circuit of Figure 1.3–4.

The voltage-follower circuit of Example 1.3–2 led to the equivalent circuit of Figure 1.6–1. In that example, we derived the following dynamic equations for this circuit by the node method:

$$C\frac{dv_1(t)}{dt} + \frac{1}{R}v_1(t) = \frac{1}{R}v_0(t) \tag{1.6–1}$$

$$C\frac{d}{dt}[v_2(t) - v_3(t)] + \frac{1}{R}[v_2(t) - v_1(t)] = 0 \tag{1.6–2}$$

$$C\frac{dv_3(t)}{dt} + \frac{1}{R}[v_3(t) - v_2(t)] = 0. \tag{1.6–3}$$

Suppose that $v_0(t) = 0$ in some interval $t_0 < t < t_1$. Try solutions during that interval of the form

$$v_1(t) = V_1 e^{st}, \quad v_2(t) = V_2 e^{st}, \quad v_3(t) = V_3 e^{st} \tag{1.6–4}$$

to obtain

$$\left(Cs + \frac{1}{R}\right)V_1 e^{st} \qquad\qquad\qquad\qquad = 0 \tag{1.6–5}$$

$$-\frac{1}{R}V_1 e^{st} + \left(Cs + \frac{1}{R}\right)V_2 e^{st} \qquad - CsV_3 e^{st} = 0 \tag{1.6–6}$$

$$-\frac{1}{R}V_2 e^{st} + \left(Cs + \frac{1}{R}\right)V_3 e^{st} = 0. \tag{1.6–7}$$

The common factor e^{st} can be cancelled since it is nonzero for any finite s and t. The resulting set of three linear algebraic equations in V_1, V_2, and V_3 is consistent with nonzero values of V_1, V_2, and V_3 if and only if the determinant of the coefficients vanishes, that is, if

$$\begin{vmatrix} Cs+\dfrac{1}{R} & 0 & 0 \\ -\dfrac{1}{R} & Cs+\dfrac{1}{R} & -Cs \\ 0 & -\dfrac{1}{R} & Cs+\dfrac{1}{R} \end{vmatrix} = \dfrac{(RCs)^3 + 2(RCs)^2 + 2RCs + 1}{R^3} = 0. \qquad (1.6\text{--}8)$$

The roots of this *characteristic equation* are the *characteristic* (or *natural*) *frequencies*[*] of the circuit:

$$s = -\dfrac{1}{RC}, \quad \dfrac{-1 \pm j\sqrt{3}}{2RC}. \qquad (1.6\text{--}9)$$

Thus $v_1(t)$, $v_2(t)$, and $v_3(t)$ may have the form (1.6–4) with nonzero values of (at least some of) the amplitudes V_1, V_2, and V_3, provided that s has one of the values in (1.6–9). For each allowed value of s, however, constraints are imposed on V_1, V_2, and V_3 by (1.6–5, 6, 7). Thus, if $s = -1/RC$, it is immediately apparent from (1.6–7) that V_2 must be zero and from (1.6–6) that $V_1 = V_3$. Hence, one nonzero solution to our zero-input problem is

$$v_1(t) = v_3(t) = Ae^{-t/RC}, \quad v_2(t) = 0 \qquad (1.6\text{--}10)$$

where A is an arbitrary constant. Given these node voltages, one can readily compute all the branch voltages and currents. These, too, will be proportional to the arbitrary constant factor A.

A similar result holds separately for each of the natural frequencies. For each, the node voltages (and hence all the branch currents and voltages) have the form of constants times exponential factors in the corresponding frequency; the constants are determined by (1.6–5, 6, 7) up to a single common arbitrary (in general, complex) factor, independent for each natural frequency. Thus, for $s = \left(-1 \pm j\sqrt{3}\right)/2RC$, it is apparent from (1.6–5) that $V_1 = 0$, and from (1.6–7) that $V_2 = V_3 e^{\pm j\pi/3}$, so that two other solutions to our zero-input problem are

$$v_1(t) = 0 \qquad (1.6\text{--}11)$$

$$v_2(t) = B_1 e^{-t/2RC} e^{j\sqrt{3}t/2RC} \qquad (1.6\text{--}12)$$

$$v_3(t) = B_1 e^{-j\pi/3} e^{-t/2RC} e^{j\sqrt{3}t/2RC} \qquad (1.6\text{--}13)$$

and

$$v_1(t) = 0 \qquad (1.6\text{--}14)$$

$$v_2(t) = B_2 e^{-t/2RC} e^{-j\sqrt{3}t/2RC} \qquad (1.6\text{--}15)$$

$$v_3(t) = B_2 e^{j\pi/3} e^{-t/2RC} e^{-j\sqrt{3}t/2RC} \qquad (1.6\text{--}16)$$

If several sets of node voltages (and the associated branch voltages and currents) independently satisfy Kirchhoff's Laws and the branch constitutive relations for a circuit of linear elements under zero-input conditions, then it is easy to argue that node and branch variables composed by *superimposing* (adding) the corresponding variables from each set also satisfy Kirchhoff's Laws and the branch constitutive relations,

---

[*]Also called the *singularities* of the circuit because for these values of s the node equations are *singular* and have non-unique solutions.

that is, the combination obeys the dynamic equations for the circuit under zero-input conditions. Moreover, if the actual voltages and currents in the circuit are real, then the coefficients of the terms corresponding to complex conjugate natural frequencies must be complex conjugates. Hence, the most general form for the zero-input-response node voltages of the circuit of Figure 1.6–1 is

$$v_1(t) = Ae^{-t/RC} \tag{1.6-17}$$

$$v_2(t) = Be^{-t/2RC}e^{j\sqrt{3}t/2RC} + B^*e^{-t/RC}e^{-j\sqrt{3}t/2RC}$$

$$= 2|B|e^{-t/2RC}\cos\left(\frac{\sqrt{3}t}{2RC} + \angle B\right) \tag{1.6-18}$$

$$v_3(t) = Ae^{-t/RC} + Be^{-j\pi/3}e^{-t/2RC}e^{j\sqrt{3}t/2RC}$$

$$\qquad\qquad + B^*e^{j\pi/3}e^{-t/2RC}e^{-j\sqrt{3}t/2RC}$$

$$= Ae^{-t/RC} + 2|B|e^{-t/2RC}\cos\left(\frac{\sqrt{3}t}{2RC} + \angle B - \frac{\pi}{3}\right) \tag{1.6-19}$$

where the asterisk (*) indicates complex conjugation. To find values for the real constant A and the complex constant $B = |B|e^{j\angle B}$, we must be given separately information equivalent to knowing the state of the system at some moment, typically the beginning of the zero-input interval (that is, initial conditions).

Note that the number of arbitrary (real) constants that must be determined from auxiliary information is equal to the *degree* of the characteristic equation, which is also the same as the number of normal modes, the number of natural frequencies, the number of state variables, or in general the *order* of the system. If the roots of the characteristic equation are not distinct or simple (that is, if the characteristic equation contains a repeated factor $(s - s_k)^n$), then the normal modes include not only the term $e^{s_k t}$ but also $te^{s_k t}, t^2 e^{s_k t}, \ldots, t^{n-1}e^{s_k t}$; s_k is said to be a natural frequency of *order* or *multiplicity* n. The number of arbitrary constants must be expanded correspondingly.

▶ ▶ ▶

1.7 Solutions of the Dynamic Equations for Exponential Inputs

If all of the independent inputs to an LTI circuit during some finite or semi-infinite time interval are proportional to the same exponential time function e^{st} (where s may be complex), then all of the voltages and currents in the circuit during that interval may have the same form—proportional to the same time function e^{st}— and the proportionality factors may readily be found. When we say "may have the same form" we mean simply that such functions satisfy the dynamic differential equations. But we have already pointed out that such equations specify the circuit behavior only implicitly. The solutions of exponential form are thus not the only solutions when the inputs are exponentials. The *complete solution* during an interval in which the inputs are exponentials consists of a *particular solution* of the form e^{st} plus the natural response to zero input with the constants chosen so that the complete solution matches a given initial state or equivalent information—as we shall illustrate in the following examples.

Example 1.7–1

Constant inputs can be considered a special case of exponential inputs with $s = 0$. A particular solution for a circuit with constant inputs, then, corresponds to all of the branch variables having appropriate constant values; this is the "d-c" (for "direct-current") or steady-state solution that the complete response will approach asymptotically if the constant inputs are maintained indefinitely.

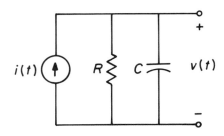

Figure 1.7–1. Example 1.7–1 circuit.

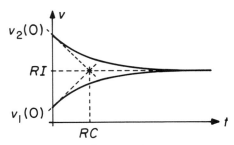

Figure 1.7–2. Constant input response.

For first-order systems the complete solution to a constant input can be written down by inspection. Thus the node equation (or state equation—they are essentially the same) for the circuit of Figure 1.7–1 is

$$i(t) = C\frac{dv(t)}{dt} + \frac{1}{R}v(t). \tag{1.7–1}$$

If $i(t) = I = $ constant, $t > 0$, then a particular solution is $v(t) = V = $ constant. The $dv(t)/dt$ term in (1.7–1) is then zero so that

$$I = \frac{V}{R}. \tag{1.7–2}$$

The zero-input response is readily shown (by substitution into (1.7–1)) to have the form $Ae^{-t/RC}$ so that the complete solution is

$$v(t) = RI + Ae^{-t/RC}. \tag{1.7–3}$$

If the value of $v(t)$ at $t = 0$ is known to be $v(0)$, then it follows at once from (1.7–3) that

$$A = v(0) - RI \tag{1.7–4}$$

so that

$$v(t) = RI + (v(0) - RI)e^{-t/RC}, \quad t > 0. \tag{1.7–5}$$

The form of $v(t)$ is shown in Figure 1.7–2 for several values of $v(0)$. Note that the time $t = RC$ corresponds to the moment at which the response has progressed approximately 2/3 of the way from the initial to the final value. (The exact amount is $1 - \frac{1}{e} \approx 0.632$). Note also that a straight line with the initial slope of the response intercepts the final value at $t = RC$.

An ability to write down the complete response (1.7−5) for a first-order system with constant input, and to sketch it accurately as in Figure 1.7−2—directly from the problem statement without any intermediate steps—should be part of the skills of every engineer or scientist. Some examples of how this skill is useful in analyzing practical electronic circuits are given in Problems 1.13 and 1.14.

▶ ▶ ▶

Example 1.7−2

To illustrate the effect of exponential drives in a more complicated case, let's continue the analysis of the circuit of Examples 1.3−2 and 1.6−1. Let the drive be $v_0(t) = V_0 e^{st}$, and assume that the node voltages have the form $v_i(t) = V_i e^{st}$. The dynamic node equations become

$$CsV_1 e^{st} + \frac{1}{R}V_1 e^{st} = \frac{1}{R}V_0 e^{st} \qquad (1.7-6)$$

$$Cs\left(V_2 e^{st} - V_3 e^{st}\right) + \frac{1}{R}\left(V_2 e^{st} - V_1 e^{st}\right) = 0 \qquad (1.7-7)$$

$$CsV_3 e^{st} + \frac{1}{R}\left(V_3 e^{st} - V_2 e^{st}\right) = 0. \qquad (1.7-8)$$

Cancelling the common factor e^{st} and solving the simultaneous equations, we obtain

$$V_1 = \frac{V_0}{RCs + 1} \qquad (1.7-9)$$

$$V_2 = \frac{V_0(RCs + 1)}{(RCs)^3 + 2(RCs)^2 + 2(RCs) + 1} \qquad (1.7-10)$$

$$V_3 = \frac{V_0}{(RCs)^3 + 2(RCs)^2 + 2(RCs) + 1}. \qquad (1.7-11)$$

We consider two special cases:

a) The *unit step response* of the circuit is the response when the input is the *unit step function* $u(t)$:

$$v_0(t) = u(t) = \begin{cases} 1, & \text{if } t \geq 0 \\ 0, & \text{if } t < 0. \end{cases} \qquad (1.7-12)$$

By convention,* we take the fact that we are asking for the unit step response to imply that the circuit has been at rest for $t < 0$, that is,

$$v_1(t) = v_2(t) = v_3(t) = 0, \quad t < 0. \qquad (1.7-13)$$

Since the voltages on the capacitors cannot change instantaneously, the state at $t = 0+$ is described by

$$v_1(0+) = v_2(0+) = v_3(0+) = 0. \qquad (1.7-14)$$

---

*Unfortunately, this convention is not used by all workers in the field. If the circuit is not at rest at $t = 0$, we shall talk about the response to a *constant*, such as $v_0(t) = 1$, $t > 0$, rather than the unit step response.

For $t \geq 0$, $v_0(t)$ has the form

$$v_0(t) = V_0 e^{st} \tag{1.7-15}$$

with $V_0 = 1, s = 0$. Thus particular solutions are, from (1.7-9, 10, 11),

$$v_1(t) = V_1 e^{st}\Big|_{V_0=1, s=0} = 1 \tag{1.7-16}$$

$$v_2(t) = V_2 e^{st}\Big|_{V_0=1, s=0} = 1 \tag{1.7-17}$$

$$v_3(t) = V_3 e^{st}\Big|_{V_0=1, s=0} = 1 . \tag{1.7-18}$$

Using the results of Example 1.6–1, we have as complete solutions

$$v_1(t) = 1 + Ae^{-t/RC} \tag{1.7-19}$$

$$v_2(t) = 1 + Be^{-t/2RC} e^{j\sqrt{3}t/2RC} + B^* e^{-t/2RC} e^{-j\sqrt{3}t/2RC} \tag{1.7-20}$$

$$v_3(t) = 1 + Ae^{-t/RC} + Be^{-j\pi/3}e^{-t/2RC}e^{j\sqrt{3}t/2RC} \\ + B^* e^{j\pi/3}e^{-t/2RC}e^{-j\sqrt{3}t/2RC} . \tag{1.7-21}$$

Matching the conditions at $t = 0+$ gives $A = -1$, $B = \dfrac{-e^{-j\pi/6}}{\sqrt{3}}$, $B^* = \dfrac{-e^{j\pi/6}}{\sqrt{3}}$, and

$$v_1(t) = \left[1 - e^{-t/RC}\right]u(t) \tag{1.7-22}$$

$$v_2(t) = \left[1 - \frac{2}{\sqrt{3}}e^{-t/2RC} \cos\left(\frac{\sqrt{3}t}{2RC} - \frac{\pi}{6}\right)\right]u(t) \tag{1.7-23}$$

$$v_3(t) = \left[1 - e^{-t/RC} - \frac{2}{\sqrt{3}}e^{-t/2RC} \sin\left(\frac{\sqrt{3}t}{2RC}\right)\right]u(t) . \tag{1.7-24}$$

These waveforms are sketched in Figure 1.7–3.

b) The *sinusoidal steady-state response* of the circuit is the response to the input

$$v_0(t) = V_0 \cos\omega t, \quad -\infty < t < \infty . \tag{1.7-25}$$

This input is not of the form $V_0 e^{st}$, but it is closely related to such a form since the cosine function can be written in either of two ways:

$$1. \quad \cos\omega t = \frac{e^{j\omega t} + e^{-j\omega t}}{2} \tag{1.7-26}$$

$$2. \quad \cos\omega t = \Re\left[e^{j\omega t}\right] \tag{1.7-27}$$

(where the symbol "$\Re$" stands for "take real part of"). Because the circuit is LTI and composed of real elements, we can find the desired response by first finding the response to

$$v_0(t) = V_0 e^{j\omega t}, \quad -\infty < t < \infty \tag{1.7-28}$$

and then doing either of the following:

    1. Adding the responses for ω positive and ω negative and dividing by 2,

or

    2. Taking the real part of the response.

Usually the second procedure is simpler. Since we are interested in the sinusoidal steady state, the desired complete responses are just the components proportional to the drive exponential $e^{j\omega t}$. Any finite natural response terms that might have been initiated when the sinusoidal drive was applied "at" $t = -\infty$ have presumably long since died out. Taking $v_3(t)$ as typical, the steady-state response to $v_0(t) = V_0 e^{j\omega t}$ is

$$v_3(t) = \frac{V_0 e^{j\omega t}}{(j\omega RC)^3 + 2(j\omega RC)^2 + 2(j\omega RC) + 1} = H(j\omega)V_0 e^{j\omega t} \tag{1.7-29}$$

where

$$H(j\omega) = |H(j\omega)|e^{j\angle H(j\omega)} = \frac{1}{(j\omega RC)^3 + 2(j\omega RC)^2 + 2(j\omega RC) + 1} \tag{1.7-30}$$

is called the *frequency response* of the circuit. Hence the steady-state response to $v_o(t) = V_0 \cos\omega t = \Re[V_0 e^{j\omega t}]$ is

$$v_3(t) = \Re[H(j\omega)V_0 e^{j\omega t}] = V_0|H(j\omega)| \cos[\omega t + \angle H(j\omega)], \quad -\infty < t < \infty. \tag{1.7-31}$$

The steady-state response to a sinusoidal waveform of arbitrary frequency can thus be described by plotting the magnitude and phase of the frequency response vs. ω as shown in Figure 1.7–4. It is easy to show from (1.7–30) that

$$|H(j\omega)|^2 = H(j\omega)H(-j\omega) = \frac{1}{1 + (\omega RC)^6}. \tag{1.7-32}$$

The frequency response has a magnitude ≈ 1 for $\omega \ll 1/RC$, and it is small compared to 1 for $\omega \gg 1/RC$. Thus, this circuit with $v_0(t)$ as its input and $v_3(t)$ as its output *passes* low frequencies and (to a degree) *stops* or *rejects* high frequencies. It is thus called a *lowpass filter*. Since the transition between the pass and rejection *bands* occurs near $\omega = 1/RC$, this is called the *band-edge* or *cutoff frequency* of the filter. As we shall later discuss, there are many different forms of lowpass filters; filters for which

$$|H(j\omega)|^2 = \frac{1}{1 + (\omega/\omega_0)^{2n}} \tag{1.7-33}$$

are called *Butterworth filters* of order n and cutoff frequency ω_0.

▶ ▶ ▶

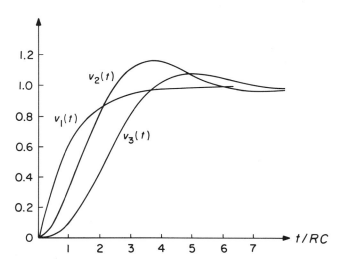

Figure 1.7–3. Step responses of third-order Butterworth lowpass filter.

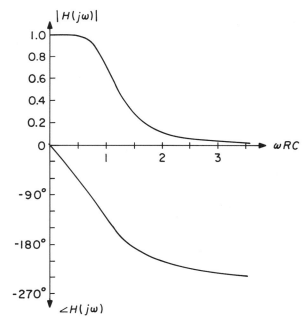

Figure 1.7–4. Frequency response of third-order Butterworth lowpass filter.

1.8 Summary

Using either the state or node methods, it is straightforward to derive a set of dynamic differential equations for any lumped circuit. If the circuit is LTI, the solution of these equations to find either the zero-input response or a particular response to an exponential drive begins by assuming exponential solutions of the form Ae^{st}. As a result, the differential equations are converted into a set of algebraic equations in the exponential amplitudes. In the case of the zero-input response, these equations are homogeneous and thus permit nonzero amplitudes only for certain characteristic frequencies. The general form of the zero-input response, then, is a sum of exponential terms at the characteristic frequencies with the amplitudes of the terms derived from knowledge of the circuit state at the beginning of the zero-input interval (or equivalent). In the case of an exponential drive proportional to e^{st}, the algebraic equations derived from assuming solutions of the form Ae^{st} are non-homogeneous and thus yield in general specific amplitudes for a particular solution to the problem. The complete solution is obtained by adding to this particular solution terms corresponding to a zero-input response, with appropriate amplitudes so that the complete solution matches the known state of the system at some moment (or equivalent information).

These "classical" methods for analyzing circuit behavior can be stream-lined and extended in a variety of ways. For example, the algebraic equations that result from substituting trial solutions of the form Ae^{st} into the dynamic differential equations can be written directly from the circuit diagram (i.e., without first writing the differential equations) by using impedance methods. Probably you have already had some experience with impedance techniques in earlier courses. Developing impedance ideas through a new tool—the Laplace transform—will be one goal of the next chapter. The Laplace transform also provides a formal procedure for finding the response of an LTI circuit to an arbitrary input—not just the exponentials or sums of exponentials to which the "classical" methods of this chapter would seem to be limited. As we shall ultimately come to understand, however, the Laplace transformation has this power because virtually any waveform can be represented as a "sum" of exponentials; this is the essential idea behind what is called *Fourier analysis*. Thus the usefulness of frequency-domain methods for studying LTI system behavior, including impedance ideas and their generalization to the notion of a system function, is not restricted to exponential drives but is applicable in general.

The landscape of LTI analysis procedures is extraordinarily varied and rich. This wealth of perspectives, in turn, makes the language of LTI models an extraordinarily valuable tool for both the understanding and design of complex dynamical structures such as those employed in signal processing, communications and control, and measurements. Helping you learn to "speak" this language—not just to "solve" problems—is the real goal of this book.

EXERCISES FOR CHAPTER 1

Exercise 1.1

In the following assume that all waveforms are defined in the interval $-\infty < t < \infty$.

a) Demonstrate that each of the following constitutive relations describes a 2-terminal element that is both linear and time-invariant.

i) $\dfrac{dv(t)}{dt} + v(t) = i(t)$

ii) $v(t) = i(t-1)$

iii) $i(t) = \displaystyle\int_{-\infty}^{t} v(\tau)\cos(t-\tau)\,d\tau$

iv) $i(t) = \displaystyle\int_{0}^{t} v(\tau)\,d\tau + i(0)$

b) Demonstrate that each of the following constitutive relations describes an element that is not linear and/or is not time-invariant.

i) $v(t)i(t) = 1$

ii) $i(t) = \displaystyle\int_{0}^{t} v(\tau)\,d\tau$

iii) $v(t) = i(t)\cos t$

iv) $v(t) = i(t) + 1$

Exercise 1.2

a) Pick a reference node and an appropriate (set of) node voltage(s) for each of the circuits below, and write the corresponding dynamic equation(s) in node form.

b) Pick an appropriate (set of) state variable(s) for each of the circuits below and write the corresponding dynamic equation(s) in state form.

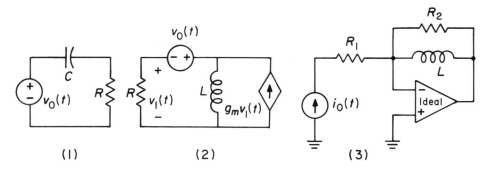

(1)   (2)   (3)

Answers:

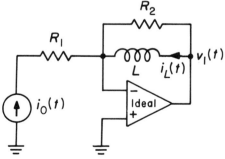

1a)

$$C\frac{d[v_1(t) - v_0(t)]}{dt} + \frac{v_1(t)}{R} = 0$$

1b)

$$C\frac{dv_C(t)}{dt} = -\frac{v_0(t) + v_C(t)}{R}$$

2a)

$$\frac{1}{L}\int_0^t v_2(\tau)\,d\tau + i_L(0) + \frac{v_2(t) - v_0(t)}{R} - g_m(v_2(t) - v_0(t)) = 0$$

2b)

$$L\frac{di_L(t)}{dt} = v_0(t) + \frac{i_L(t)}{g_m - \dfrac{1}{R}}$$

3a)

$$\frac{1}{L}\int_0^t v_1(\tau)\,d\tau + i_L(0) + \frac{v_1(t)}{R_2} = -i_0(t)$$

3b)

$$L\frac{di_L(t)}{dt} = -[i_0(t) + i_L(t)]R_2$$

Exercise 1.3

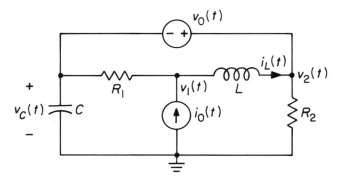

a) Pick the reference node as shown by the ground symbol (this is *not* necessarily the best reference node to pick in this circuit). Show that the node equations in the variables $v_1(t)$ and $v_2(t)$ are

$$\frac{1}{L}\int_0^t [v_1(\tau) - v_2(\tau)]\, d\tau + i_L(0) + \frac{1}{R_1}[v_1(t) - v_2(t) + v_0(t)] = i_0(t)$$

$$\frac{1}{R_2}v_2(t) + \frac{1}{R_1}[v_2(t) - v_0(t) - v_1(t)] + \frac{1}{L}\int_0^t [v_2(\tau) - v_1(\tau)]\, d\tau$$

$$- i_L(0) + C\frac{d}{dt}[v_2(t) - v_0(t)] = 0.$$

b) Show that state equations in the state variables $v_C(t)$ and $i_L(t)$ are

$$C\frac{dv_C(t)}{dt} = -\frac{1}{R_2}v_C(t) - \frac{1}{R_2}v_0(t) + i_0(t)$$

$$L\frac{di_L(t)}{dt} = -R_1 i_L(t) - v_0(t) + R_1 i_0(t).$$

c) Show that the equations derived in (a) and (b) are consistent.

Exercise 1.4

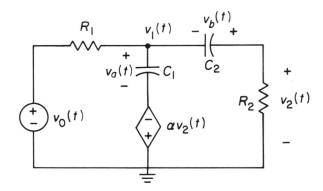

a) Show that the node equations in the variables $v_1(t)$ and $v_2(t)$ for the circuit above are

$$\frac{v_2(t)}{R_2} + C_2 \frac{d}{dt}[v_2(t) - v_1(t)] = 0$$

$$C_2 \frac{d}{dt}[v_1(t) - v_2(t)] + C_1 \frac{d}{dt}[v_1(t) + \alpha v_2(t)] + \frac{1}{R_1}[v_1(t) - v_0(t)] = 0.$$

b) Show that state equations in the variables $v_a(t)$ and $v_b(t)$ are

$$C_1 \frac{dv_a(t)}{dt} = -\frac{R_1 + R_2}{R_1 R_2 (1 + \alpha)} v_a(t) - \frac{R_1 - \alpha R_2}{R_1 R_2 (1 + \alpha)} v_b(t) + \frac{v_0(t)}{R_1}$$

$$C_2 \frac{dv_b(t)}{dt} = -\frac{1}{(1 + \alpha) R_2} v_a(t) - \frac{1}{(1 + \alpha) R_2} v_b(t).$$

c) Draw an actual circuit diagram employing an op-amp and appropriate other elements which would have the incremental equivalent circuit shown above.

Exercise 1.5

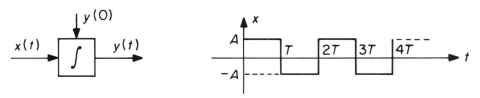

Sketch the output of the ideal integrator above if the input is as shown. Illustrate the effect of various choices for $y(0)$.

Exercise 1.6

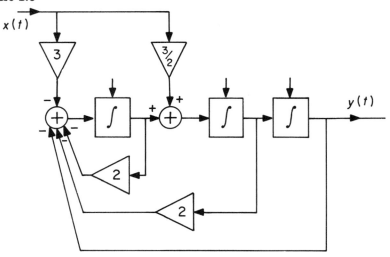

a) Taking the outputs of the integrators as state variables, write the dynamic equations for this system in state form.

b) Eliminate intermediate state variables to obtain the input-output differential equation

$$\frac{d^3y(t)}{dt^3} + 2\frac{d^2y(t)}{dt^2} + 2\frac{dy(t)}{dt} + y(t) = \frac{3}{2}\frac{dx(t)}{dt}.$$

Exercise 1.7

Find the natural frequencies and the form of the ZIR for each of the following circuits.

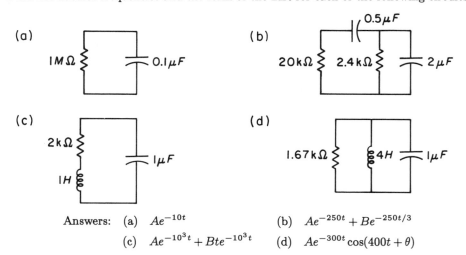

(a) $1\,M\Omega$ $0.1\,\mu F$

(b) $0.5\,\mu F$ $20\,k\Omega$ $2.4\,k\Omega$ $2\,\mu F$

(c) $2\,k\Omega$ $1\,H$ $1\,\mu F$

(d) $1.67\,k\Omega$ $4\,H$ $1\,\mu F$

Answers: (a) Ae^{-10t} (b) $Ae^{-250t} + Be^{-250t/3}$

(c) $Ae^{-10^3 t} + Bte^{-10^3 t}$ (d) $Ae^{-300t}\cos(400t + \theta)$

PROBLEMS FOR CHAPTER 1

Problem 1.1

From the physical point of view, the voltages on capacitors and currents in inductors are obviously a satisfactory set of state variables since their present values describe the distribution of stored energies in the circuit and thus specify the entire effect that past stimuli can have on future behavior. But equally obviously they are not unique in this respect; any other set of quantities from which the capacitor voltages and inductor currents could be algebraically derived would also be an appropriate alternative set of state variables.

a) By manipulating the node equations of Example 1.3–2 into state form, prove that the node voltages for that circuit are an appropriate set of state variables.

b) Show by means of some simple examples that node voltages are not always an appropriate set of state variables.

Problem 1.2

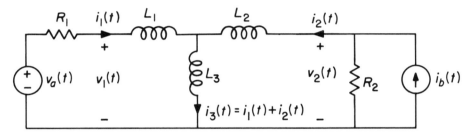

The three inductors in the circuit above may represent three independent coils or an equivalent circuit for two coupled coils (see Problem 3.7). In either case, KCL at the central node allows just two of the inductor currents to be independent, so that the circuit has only two independent state variables and is hence of second rather than third order.

a) Choosing the independent state variables to be $i_1(t)$ and $i_2(t)$, write expressions for $v_1(t)$ and $v_2(t)$ entirely in terms of $i_1(t)$, $i_2(t)$, $v_a(t)$, and $i_b(t)$.

b) Considering the T of inductors as a 2-port (see appendix to Lecture 3), find equations characterizing it in the form

$$\frac{di_1(t)}{dt} = \Gamma_{11}v_1(t) + \Gamma_{12}v_2(t)$$

$$\frac{di_2(t)}{dt} = \Gamma_{21}v_1(t) + \Gamma_{22}v_2(t)$$

and determine $\{\Gamma_{ij}\}$ in terms of $\{L_k\}$.

c) Combine the results of (a) and (b) to obtain state equations in normal form for the circuit above.

Problem 1.3

With appropriate parameter values, the circuit shown below functions as a *non-inverting integrator*.

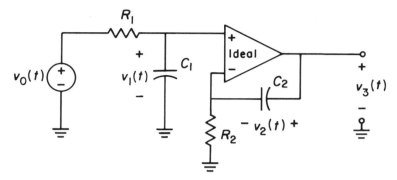

a) Write state equations for this circuit in terms of the capacitor voltages $v_1(t)$ and $v_2(t)$.

b) Show that the input-output differential equation relating $v_3(t)$ and $v_0(t)$ has the form

$$\frac{dv_3(t)}{dt} = K v_0(t)$$

provided that $R_1 C_1 = R_2 C_2$. Find the magnitude and sign of K.

Problem 1.4

From the standpoint of input-output behavior, many different block diagrams may be *equivalent* in the sense of being described by the same input-output differential equation. Show that each of the following block diagrams is equivalent in this sense to the diagram of Exercise 1.6.

(a)

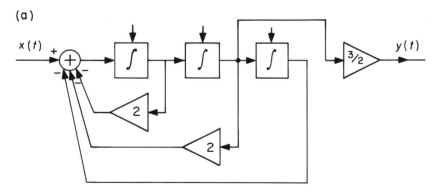

(b)

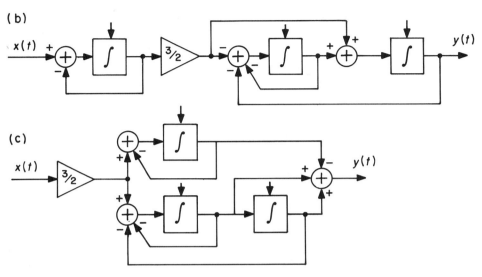

(c)

Problem 1.5

The input-output behavior of *any* 3rd-order system can be simulated by the following block diagram:

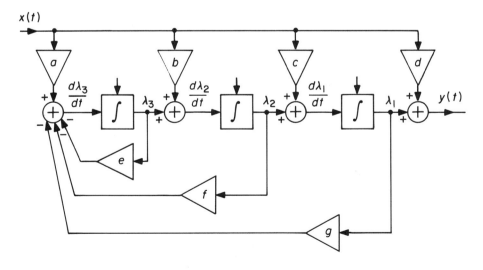

a) Write the dynamical equations for this system in state form. That is, express $d\lambda_1(t)/dt$, $d\lambda_2(t)/dt$, $d\lambda_3(t)/dt$, and the output $y(t)$ in terms of the state variables $\lambda_1(t)$, $\lambda_2(t)$, $\lambda_3(t)$, and the input $x(t)$. (HINT: focus on the relationships among the variables imposed by the adder blocks.)

b) Eliminate the state variables and their derivatives to obtain a single differential equation of the form

$$\frac{d^3y(t)}{dt^3} + \alpha_2\frac{d^2y(t)}{dt^2} + \alpha_1\frac{dy(t)}{dt} + \alpha_0 y(t) = \beta_3\frac{d^3x(t)}{dt^3} + \beta_2\frac{d^2x(t)}{dt^2} + \beta_1\frac{dx(t)}{dt} + \beta_0 x(t)$$

relating the input and the output.

c) Derive formulas for the gains a, b, $\ldots$, g in terms of the coefficients $\{\alpha_i\}$ and $\{\beta_i\}$ and thus show that any input-output 3rd-order differential equation can be realized in this way.

The procedure illustrated by this problem generalizes to equations of arbitrary order.

Problem 1.6

This problem discusses another form of block diagram that can simulate the input-output behavior of any 3rd-order LTI system. Extension to LTI systems of arbitrary order should be obvious.

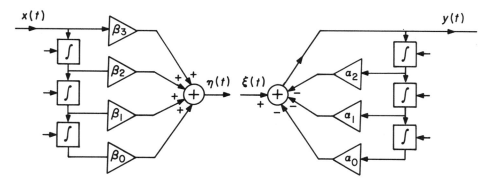

a) Show that the block diagram to the left above is described by the input-output differential equation

$$\frac{d^3\eta(t)}{dt^3} = \beta_3\frac{d^3x(t)}{dt^3} + \beta_2\frac{d^2x(t)}{dt^2} + \beta_1\frac{dx(t)}{dt} + \beta_0 x(t).$$

b) Show that the block diagram to the right above is described by the input-output differential equation

$$\frac{d^3y(t)}{dt^3} + \alpha_2\frac{d^2y(t)}{dt^2} + \alpha_1\frac{dy(t)}{dt} + \alpha_0 y(t) = \frac{d^3\xi(t)}{dt^3}.$$

c) If the two block diagrams above are connected together in *cascade* so that $\xi(t) = \eta(t)$, the overall structure is described by

$$\frac{d^3y(t)}{dt^3} + \alpha_2\frac{d^2y(t)}{dt^2} + \alpha_1\frac{dy(t)}{dt} + \alpha_0y(t) = \beta_3\frac{d^3x(t)}{dt^3} + \beta_2\frac{d^2x(t)}{dt^2} + \beta_1\frac{dx(t)}{dt} + \beta_0x(t)$$

and thus with appropriate choices of gains can simulate the input-output behavior of any 3rd-order system as claimed. However, a simpler system results if the structures above are cascaded in *reversed order* as shown below. (In later chapters we shall show that the overall input-output result of cascading LTI systems is independent of the order in which they are cascaded.)

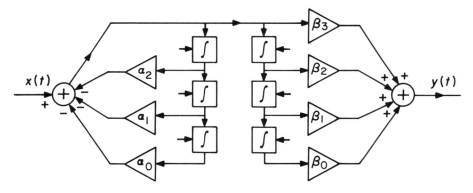

Since the inputs to the integrators at the top of each chain of integrators are the same, their outputs are the same, and so on down the chain. Hence only a single chain of integrators is really necessary, as shown below. Show directly for the system below that the input-output differential equation is the same as that given above.

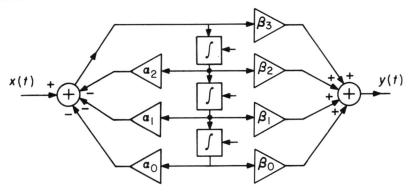

Problem 1.7

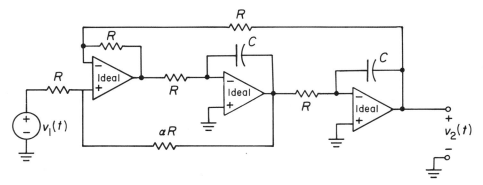

a) Pick an appropriate set of variables and write differential equations describing the system above in normal state form.

b) Derive from these equations a single input-output differential equation relating $v_1(t)$ and $v_2(t)$.

c) If R, C, and α may take on any positive values, what range of natural frequencies may be realized with this circuit?

Problem 1.8

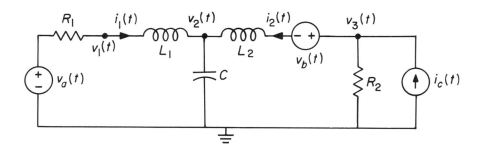

In Example 1.3–1, node equations were written for the above circuit. With $v_a(t) = v_b(t) = i_c(t) = 0$ these equations become:

$$\frac{v_1(t)}{R_1} + \frac{1}{L_1} \int_0^t [v_1(\tau) - v_2(\tau)]\, d\tau + i_1(0) = 0$$

$$\frac{1}{L_1} \int_0^t [v_2(\tau) - v_1(\tau)]\, d\tau - i_1(0) + C\frac{dv_2(t)}{dt} + \frac{1}{L_2} \int_0^t [v_2(\tau) - v_3(\tau)]\, d\tau - i_2(0) = 0$$

$$\frac{1}{L_2} \int_0^t [v_3(\tau) - v_2(\tau)]\, d\tau + i_2(0) + \frac{v_3(t)}{R_2} = 0 .$$

Assuming $R_1 = R_2 = 1$ kΩ, $L_1 = L_2 = 0.1$ H, $C = 0.2$ μF, parallel the development of Example 1.6–1 to find the natural frequencies and the normal modes of this circuit. (HINT: The analysis will be simplified if you first differentiate all three equations to clear out intergrals and constants. As a check you should find that the natural frequencies are $10^{-4}s = -1, -1/2 \pm j\sqrt{3}/2$).

Problem 1.9

The dynamic state equations for an LTI system under ZIR conditions may be written in matrix form as

$$\dot{\lambda}] = [A] \times \lambda] .$$

Here $\lambda]$ is a column matrix of the state variables $\lambda_i(t)$, $\dot{\lambda}]$ is a column matrix of the time derivatives $d\lambda_i(t)/dt$, and $[A]$ is a square matrix of coefficients a_{ij}. To find the natural frequencies, as in Example 1.6–1, set

$$\lambda_i(t) = \Lambda_i e^{st}$$

and substitute to obtain

$$s\Lambda]e^{st} = [A] \times \Lambda]e^{st}$$

where $\Lambda]$ is the column matrix of amplitudes Λ_i. Cancelling the nonzero factor e^{st} leads to

$$([A] - s[I]) \times \Lambda] = 0$$

where $[I]$ is the identity matrix. The natural frequencies are thus the s-roots of the equation

$$\det([A] - s[I]) = 0$$

which are called the *eigenvalues* of the matrix $[A]$.

For the circuit of Examples 1.3–1 and 1.3–3, as well as Problem 1.8, the $[A]$ matrix is (Example 1.3–3)

$$[A] = \begin{bmatrix} \dfrac{-R_1}{L_1} & 0 & \dfrac{-1}{L_1} \\ 0 & \dfrac{-R_2}{L_2} & \dfrac{-1}{L_2} \\ \dfrac{1}{C} & \dfrac{1}{C} & 0 \end{bmatrix} .$$

For the element values of Problem 1.8, find the eigenvalues of $[A]$ and show that they are the same as the natural frequencies found in Problem 1.8. (The eigenvalues of a matrix with real elements are generally easier to find numerically than the roots of a polynomial of equivalent degree.)

Problem 1.10

Continuing the analysis of the circuit of Problems 1.8 and 1.9, suppose we seek the response $v_3(t)$ to the input $v_0(t) = 2u(t)$. (Recall that a step drive implies that the circuit is at rest for $t < 0$.)

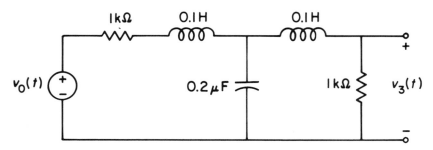

a) From the fact that the current in the right-hand inductor is zero at $t = 0$, conclude that $v_3(0+) = 0$.

b) From the fact that the voltage on the capacitor is zero at $t = 0$ conclude that
$$\dot{v}_3(0+) = \frac{dv_3(t)}{dt}\bigg|_{t=0+} = 0.$$

c) From the fact that the current in the left-hand inductor is zero at $t = 0$, conclude that $\ddot{v}_3(0+) = \frac{d^2 v_3(t)}{dt^2}\bigg|_{t=0+} = 0.$

d) From the results of problem 1.8 and a consideration of the steady-state value of $v_3(t)$ conclude that

$$v_3(t) = A + Be^{-10^4 t} + Ce^{-10^4 t/2}e^{j\sqrt{3}\times10^4 t/2} + C^* e^{-10^4 t/2}e^{-j\sqrt{3}\times10^4 t/2}, \quad t > 0$$

and find the (real) values of A and B and the (complex) value of C or C^*. (HINT: compare Example 1.7–2.)

Problem 1.11

In the circuit of Exercise 1.4, let $C_1 = 1 \ \mu F$, $C_2 = 1.5 \ \mu F$, $R_1 = 1 \ k\Omega$, $R_2 = 333 \ \Omega$, $\alpha = 3$.

a) Find the natural frequencies and the form of the ZIR.

b) Suppose $v_0(t) = 3e^{-400t}$ volts in the interval $t > 0$. Find the component of the response $v_2(t)$ that has the form $V_2 e^{-400t}$.

c) Find the complete solution for $v_2(t)$ in the interval $t > 0$ if it is known that
$$v_2(0+) = -15 \text{ volts}, \quad \dot{v}_2(0+) = \frac{dv_2(t)}{dt}\bigg|_{t=0+} = 5 \times 10^3 \text{ volts/sec}.$$

d) What values of the initial capacitor voltages $v_a(t)$ and $v_b(t)$ at $t = 0+$ correspond to the values of $v_2(0+)$ and $\dot{v}_2(0+)$ given in (c)?

Problem 1.12

The equation

$$\frac{d^2 y_1(t)}{dt^2} + \epsilon \left[y_1^2(t) - 1 \right] \frac{dy_1(t)}{dt} + y_1(t) = x(t)$$

is an example of a non-linear inhomogeneous second-order differential equation of the *Van der Pol* type. Such equations have been extensively studied, and (for $\epsilon \ll 1$) are often proposed as approximate descriptions of a variety of slightly non-linear systems with known input $x(t)$ and output $y_1(t)$.

a) Define another variable $y_2(t)$ by the equation

$$y_2(t) = \frac{dy_1(t)}{dt}.$$

Using $y_1(t)$ and $y_2(t)$ as state variables, write a pair of first-order equations in state form that are equivalent to Van der Pol's equation.

b) For $\epsilon = 0$, the Van der Pol equation describes an LTI system. Find the general (real) form of the zero-input response.

c) For $\epsilon \ll 1$, and starting at rest so that initially $|y_1(t)| \ll 1$, the equation describes an approximately LTI 2nd-order system with negative damping:

$$\frac{d^2 y_1(t)}{dt^2} - \epsilon \frac{dy_1(t)}{dt} + y_1(t) = x(t).$$

Find the general (real) form of the zero-input response under these conditions. Sketch $y_1(t)$.

The results of (c) should be an exponentially increasing sinusoid of frequency about 1 rad/sec and time constant $2/\epsilon$. As $|y_1(t)|$ increases, it obviously becomes less reasonable to neglect the $y_1^2(t)$ term compared with 1. Very roughly, we may expect the "average damping" to be less negative as $|y_1(t)|$ grows; eventually the oscillation stabilizes near the point where the average of $y_1^2(t)$ is 1, so that the "average damping" is zero. For small ϵ, the oscillation is nearly sinusoidal with period 2π; for larger ϵ, the waveform becomes markedly non-sinusoidal with an increased period.

Problem 1.13

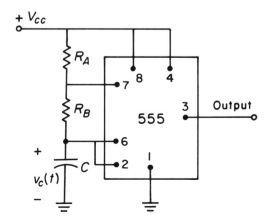

The *555-type timer* is a classic integrated circuit with many uses. The figure on page 39 shows the 555 connected to form a free-running relaxation-oscillator or multivibrator. Under these conditions the 555 behaves as a voltage-controlled switch. When power is applied with C discharged, current flows from the supply V_{CC} through R_A and R_B to charge C. During this period the output voltage is near V_{CC}. When $v_C(t)$, the voltage across C, reaches $(2/3)V_{CC}$, the 555 effectively switches terminals 7 and 3 to ground. C thus discharges through R_B, and the output voltage drops to near zero. When $v_C(t)$ reaches $(1/3)V_{CC}$, the 555 switches again—disconnecting terminal 7 so that C recharges through R_A and R_B from V_{CC}, and connecting the output to V_{CC}. When $v_C(t)$ reaches $(2/3)V_{CC}$, the cycle repeats.

a) Sketch the output voltage and the voltage $v_C(t)$ across C more or less to scale in accordance with the description above, assuming that C is discharged at the moment power is applied.

b) Show that the period of the oscillation is given by

$$T = 0.693(R_A + 2R_B)C.$$

Problem 1.14

An oscillator whose frequency is a function of the value of some voltage (or current) is usually called a *voltage- (or current-) controlled oscillator* (abbreviated VCO). VCO's have a variety of uses in intrumentation, modulation, phase-locked loops, etc., and are often included as subcomponents of integrated circuits. The LM1800, for example, is widely used as an FM stereo demodulator. It contains as part of a phased-locked loop the VCO described by the following diagram:

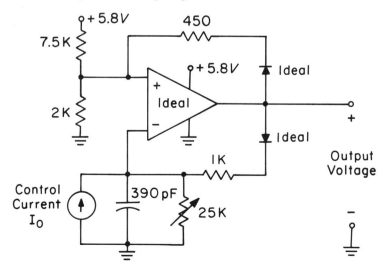

a) Assume a control current $I_0 = 0$. If the device is turned on from rest, the capacitor holds the inverting op-amp input momentarily at ground while the non-inverting input is above ground. The output voltage thus goes to the supply voltage,

5.8 volts, and both diodes are forward-biased. The non-inverting op-amp input remains at a voltage $V_+(max)$ as the capacitor starts to charge. What is $V_+(max)$?

b) The output voltage and the non-inverting op-amp input voltage will stay at 5.8 volts and $V_+(max)$, respectively, until the capacitor voltage and the inverting op-amp input voltage exceed $V_+(max)$, at which point the output voltage will suddenly fall to zero, both diodes will become reverse-biased, the non-inverting op-amp input voltage will fall to $V_+(min)$, and the capacitor will start to discharge through the 25 kΩ resistor (which is variable in the actual circuit but which we will assume fixed for this problem). What is the value of $V_+(min)$?

c) The output voltage and the non-inverting op-amp input voltage will stay at 0 volts and $V_+(min)$, respectively, until the capacitor voltage and the inverting op-amp input voltage fall below $V_+(min)$, at which point the output suddenly rises to 5.8 volts and the cycle repeats. How long does it take the capacitor voltage to fall from $V_+(max)$ to $V_+(min)$?

d) How long does it take the capacitor voltage to rise from $V_+(min)$ to $V_+(max)$?

e) Sketch approximately to scale the output voltage and the inverting and non-inverting op-amp input voltages.

f) What is the period of the output voltage?

g) Describe with simple sketches the qualitative effect of the control current if it is small $(-40\,\mu\text{amp} < I_0 < 40\,\mu\text{amp})$. (Note that for this range the effect of I_0 during the charging interval can be ignored.)

h) Compute the period of the output voltage for 4–5 points in the range $-40\,\mu\text{amp} < I_0 < 40\,\mu\text{amp}$ and sketch the output frequency (= 1/period) vs. I_0. This is the control characteristic for the VCO (although in this case the control is assumed to be a current, so it might be more appropriate to call it a CCO).

Problem 1.15

An electrical analog of the heart and circulatory system provides an interesting example of a system that is both non-linear and time-varying, and yet can be analyzed entirely using the simple techniques described in Example 1.7–1.

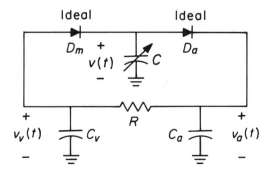

In the diagram above, C_a and C_v describe the elastic capacities of the arteries and veins, respectively (voltage is analogous to pressure, charge is analogous to volume), and R represents the resistance of the peripheral circulation (current is analogous to volume flow). The heart is modelled by the time-varying ventricular capacitance C

and two diodes D_m and D_a, representing the mitral and aortic valves. (Only the left side of the mammalian heart is considered here; the right heart and the pulmonary circulation, which should be part of this loop, have been omitted for simplicity.)

a) During *diastole*, the heart fills with returned blood. To model this phase, assume D_m is conducting, D_a is open, and C has the constant value C_d. At the beginning of diastole ($t = 0$), assume the charge on C_a is $q_a(0)$ and the combined charge on C_d and C_v together is $Q - q_a(0)$, where $Q =$ constant is the total charge (blood volume) in the system. Find $q_a(t)$ and $q_v(t)$. For the assumptions about the diodes to be consistent, $q_a(0)$ must satisfy some condition. What is that condition?

b) At $t = T/2$, the muscles of the heart wall contract. (Electrical correlates of the signals initiating contraction can be picked up on the external chest surface and constitute the largest peaks in the *electrocardiogram*.) During the succeeding interval, called *systole*, the mitral and aortic valves are respectively closed and opened by the build-up of pressure in the heart, and blood is squeezed out into the arteries. To model this phase, assume that the value of C is suddenly reduced from C_d to $C_s < C_d$. (Imagine suddenly pulling the plates of the capacitor further apart.) The voltage across C will quickly rise, turning D_m off and D_a on. The charge on C_v at the start of the period $T/2 < t < T$ will be the same as the previously found value $q_v(T/2)$, whereas the combined charge on C (now equal to C_s) and C_a together will be $Q - q_v(T/2)$. Find $q_a(t)$ and $q_v(t)$ for $T/2 < t < T$.

c) At $t = T$, the value of C is suddenly increased to $C = C_d$, and the cycle repeats. After a number of cycles the charge pattern will approach periodicity. How can one calculate this final periodic behavior?

d) Show that the voltages (pressures) shown below correctly describe the periodic condition for $Q = 1$, $C_a = C_s = 1$, $C_v = C_d = 20$, $T = 2RC_1$, $q_a(0) = 0.1768$, and $C_1 = C_a(C_v + C_d)/(C_a + C_v + C_d)$.

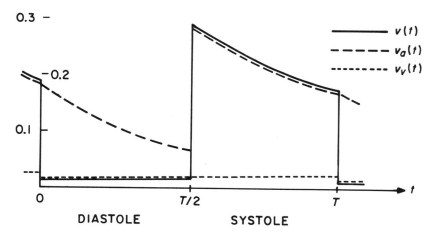

e) Note that $v_a(t) > v_v(t)$ at all times in the example of (d), corresponding to a continuous flow of blood from the arteries into the veins. The total charge in the system is, however, $Q =$ constant at all times; no charge is lost. Energy, nevertheless, is continuously "lost" to the system—dissipated as heat in the resistor R. Where does this energy come from?

2

THE UNILATERAL LAPLACE TRANSFORM

2.0 Introduction

An engineered dynamic system is typically characterized by the operation it is intended to perform—amplifier, motor controller, radio direction finder, etc. For a physical system observed in nature (for example, the solar system), it is enough to ask "How does it work?" For an engineered system, it is necessary also to ask "What is it supposed to do?" A similar interest in overall objectives arises in studies of biological and social systems, which are generally thought to have evolved under constraints that make them appear purposeful.

The study of engineered systems thus involves a continual interplay of *functional* (input-output, "black box") and *structural* (circuit, state, block diagram) system descriptions. To proceed from a structural to a functional description, we must *analyze* the given structure to determine its input-output behavior. Alternately, we may wish to *synthesize* a structure having a desired functional response.

Mathematically, a functional description of a system has the form of an *operator** such as

$$y(t) = f[x(t), \lambda_1(0), \lambda_2(0), \ldots, \lambda_n(0)] \qquad (2.0-1)$$

which explicitly assigns a unique output waveform $y(t)$, $t > 0$, to each input waveform $x(t)$, $t > 0$, and each initial state described by $\lambda_1(0), \lambda_2(0), \ldots, \lambda_n(0)$. A particular system may have several equivalent† functional descriptions, just as it may have several equivalent structural descriptions (for example, a circuit diagram, a set of node equations, a set of state equations). Our principal goal in

---

*In the sequel, when we wish to talk generally about the behavior of systems without implying any particular dimensions (voltage, current, displacement, temperature, etc.) for the input and output time functions, we shall usually designate the input by $x(t)$, the output by $y(t)$, and state variables by $\lambda_i(t)$. Systems can, of course, have more than one input or output; if so, appropriate formal changes in (2.0−1) would be required. Note that a system characterized by a (point) *function*, such as $y(t) = x^2(t)$, is more restricted than a system characterized by an *operator*, such as $y(t) = \int_0^t x(\tau)\,d\tau + y(0)$; in the first case the present value of the output depends only on the present value of the input, whereas in the second it depends on input values at many times—for example, the entire interval from $t = 0$ to the present, and even earlier times as reflected in $y(0)$.

†Two operators are *equivalent* if they give the same response to the same input and state for every input and state.

this chapter is to use Laplace transform techniques to derive what is called the frequency-domain form of functional description of an LTI system. In a later chapter, we shall study an equivalent time-domain form.

2.1 The Unilateral Laplace Transform

The *unilateral Laplace transform* (abbreviated L-transform) is itself an operator that maps a function of time into a function of a complex variable $s = \sigma + j\omega$ according to the formula

$$X(s) = L[x(t)] = \int_0^\infty x(t)e^{-st}\,dt \ . \qquad (2.1\text{--}1)$$

It must be emphasized that it is the fact that this integral maps $x(t)$ into a function of s that is important. We are interested in the values of $X(s)$ throughout some region of the complex s-plane, not just at a single point, because (as we shall discuss) given $X(s)$ in a region of the s-plane we can in general recover $x(t)$ uniquely for $t > 0$, that is, the Laplace transform as an operator is *biunique*.

This property of biuniqueness is critical to the usefulness of the Laplace transform.* Thus suppose we are interested in describing the operator $y(t) = f[x(t), \{\lambda_i(0)\}]$ characterizing some system. Instead of doing this directly, it often turns out to be simpler and more illuminating to describe the operator $Y(s) = F[X(s), \{\lambda_i(0)\}]$ relating the Laplace transforms of the input and output; because of biuniqueness this is equivalent to the description sought. For reasons to be described, s is called the *complex frequency*. The operator $Y(s) = F[X(s), \{\lambda_i(0)\}]$ is thus said to characterize the system in the *frequency domain*, whereas $y(t) = f[x(t), \{\lambda_i(0)\}]$ characterizes it in the *time domain*:

$$\begin{array}{ccc} & \text{input} & \text{output} \\ & \downarrow & \downarrow \\ \text{time domain} \quad \rightarrow & x(t) \ \Rightarrow & y(t) = f[x(t), \{\lambda_i(0)\}] \\ & \updownarrow & \updownarrow \\ \text{frequency domain} \quad \rightarrow & X(s) \ \Rightarrow & Y(s) = F[X(s), \{\lambda_i(0)\}] \end{array}$$

As an analytical tool for computing the response to specific simple inputs, the Laplace transform is primarily useful for systems that are equivalent functionally

---

*As one might expect from the name, the L-transform was one of the many contributions to mathematics and physics of the Marquis Pierre Simon de Laplace (1749–1827), who pointed out the biunique relationship between the two functions and applied the results to the solution of differential equations in a paper published in 1779 with the rather cryptic title "On what follows." The real value of the L-transform in applications seems not to have been appreciated, however, for over a century, until it was essentially rediscovered and popularized by the eccentric British engineer Oliver Heaviside (1850–1925), whose studies had a major impact on many aspects of modern electrical engineering.

to a lumped LTI circuit of intermediate complexity— say 3rd-order. For non-linear or time-varying systems $F[\cdot]$ is usually no easier to describe or evaluate than $f[\cdot]$. Very small LTI systems (1st-order) with simple inputs can generally be solved most efficiently by direct methods—as in Example 1.7–1. Very large systems require mechanical aids in any event; Laplace transform techniques require the manipulation of matrices with algebraic elements (functions of s) and the extraction of the roots of high-order polynomials—both of which are clumsy operations for a computer. But within its limited and important domain, the Laplace transform is an astonishingly effective tool for solving circuit problems, and the insights gained into general system properties and behavior are even more important.

2.2 Examples of $\mathcal{L}$-Transforms and Theorems

Example 2.2–1

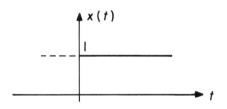

$$x(t) = 1, \quad t > 0$$

$$X(s) = \int_0^\infty x(t)e^{-st}\, dt = \int_0^\infty 1e^{-st}\, dt = -\frac{1}{s}e^{-st}\Big|_0^\infty = \frac{1}{s}$$

provided that $\Re e[s] > 0$ so that $e^{-st} \to 0$ as $t \to \infty$.
▶ ▶ ▶

Example 2.2–2

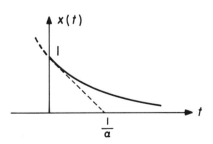

$$x(t) = e^{-\alpha t}, \quad t > 0$$

$$X(s) = \int_0^\infty e^{-\alpha t}e^{-st}\, dt = \frac{-1}{s+\alpha}e^{-(s+\alpha)t}\Big|_0^\infty = \frac{1}{s+\alpha}$$

provided that $\Re e[s] > -\alpha$ so that $e^{-(s+\alpha)t} \to 0$ as $t \to \infty$.
▶ ▶ ▶

Note that the values of $x(t)$ for $t < 0$ do not influence $X(s)$ (and thus obviously cannot be recovered from $X(s)$). Consequently

a) $L[1] = L[u(t)] = \dfrac{1}{s}, \quad \Re e[s] > 0;$

b) $L[e^{-\alpha t}] = L[e^{-\alpha t}u(t)] = L[u(-t) + e^{-\alpha t}u(t)] = \dfrac{1}{s + \alpha}, \quad \Re e[s] > -\alpha.$

(The *bilateral* Laplace transform, defined as $\int_{-\infty}^{\infty} x(t)e^{-st}\,dt$, is affected by $x(t)$, $t < 0$, and has a number of other properties that will be briefly explored in an appendix to Chapter 14.)

It is also important to note that the integral defining $X(s)$ often exists only in a limited region of the s-plane called the *domain of (absolute) convergence*. It should be obvious that, provided $x(t)$ grows no faster than $e^{\sigma_0 t}$ for some finite value of σ_0 and is otherwise well-behaved, the product $e^{-st}x(t)$ will be absolutely integrable in the right half-plane, $\Re e[s] > \sigma_0$. The smallest (real) number σ_0 such that $e^{-\sigma t}x(t)$ is absolutely integrable for all $\sigma > \sigma_0$ is called the *abscissa of (absolute) convergence*. The domain of absolute convergence may be the entire s-plane, so that $\sigma_0 = -\infty$; this will be true, for example, if $x(t)$ is a *pulse*, that is, if $x(t)$ is nonzero only for a finite time interval. On the other hand, if $x(t)$ grows faster than any exponential, for example, $x(t) = e^{t^2}$, then there will be no domain of convergence and Laplace transform methods cannot be applied.

Most of the $X(s)$ of interest to us will be *rational functions* (ratios of polynomials in s). The roots of the denominator polynomial are values of s for which the function $X(s) \to \infty$, and these are called *poles*. The roots of the numerator are called *zeros*. By definition, there can be no poles in the region of convergence; for a rational function $X(s) = L[x(t)]$, the domain of convergence is the region to the right of the rightmost pole. Non-rational transforms will often appear in our studies, however, as the following examples illustrate.

Example 2.2–3

$x(t) = u(t - T), \quad t > 0, T > 0$

$$X(s) = \int_0^{\infty} x(t)e^{-st}\,dt = \int_T^{\infty} 1e^{-st}\,dt = -\frac{1}{s}e^{-st}\Big|_T^{\infty} = \frac{1}{s}e^{-sT}$$

provided that $\Re e[s] > 0$ so that $e^{-st} \to 0$ as $t \to \infty$.

▶ ▶ ▶

Together, Examples 2.2–3 and 2.2–1 illustrate a general theorem:

> *DELAY THEOREM:*
>
>   Let $L[x(t)] = X(s)$. Then for $T > 0$,
>
> $$L[x(t-T)u(t-T)] = X(s)e^{-sT}.$$

(The proof of the Delay Theorem follows immediately from a simple change of variable in (2.1–1). Note that the Delay Theorem is not in general true for $T < 0$, as can be seen by considering the L-transforms of $x(t) = u(t)$ and of $x(t+1)u(t+1) = u(t+1)$. Can you give a statement of the conditions under which it will be true for $T < 0$?)

Example 2.2–4

$$x(t) = \begin{cases} 1, & 0 < t < T \\ 0, & t > T. \end{cases}$$

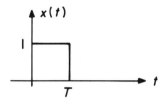

$$X(s) = \int_0^\infty x(t)e^{-st}\,dt = \int_0^T 1e^{-st}\,dt = -\frac{1}{s}e^{-st}\bigg|_0^T = \frac{1}{s}\left(1 - e^{-sT}\right), \quad \text{all } s.$$

▶ ▶ ▶

Note that the pulse $x(t)$ in Example 2.2–4 can be written as a difference of two time functions:

$$x(t) = 1 - u(t-T), \quad t > 0, T > 0.$$

The resulting $X(s)$ is the difference of the transforms of the individual time functions. This, too, is an example of a general theorem:

> *LINEARITY THEOREM:*
>
>   Let $L[x_1(t)] = X_1(s)$ and $L[x_2(t)] = X_2(s)$. Then
>
> $$L[ax_1(t) + bx_2(t)] = aX_1(s) + bX_2(s).$$

(Again, the proof of the Linearity Theorem follows immediately from (2.1–1).)

   Of course, neither of the transforms obtained in Examples 2.2–3 and 2.2–4 is a rational function because of the presence of the e^{-sT} terms. Nevertheless, $(1/s)e^{-sT}$ behaves very much like $1/s$ near $s = 0$; that is, it has a pole at $s = 0$. Since e^{-sT} is well-behaved for all finite s, the domain of convergence in Example 2.2–3 is $\Re[s] > 0$. Despite appearances, the transform of the pulse in Example 2.2–4, $(1/s)\left(1 - e^{-sT}\right)$, is well-behaved for all s, including $s = 0$; indeed, no constraints are necessary on s for the L-transform of the pulse waveform in Example 2.2–4 to exist, so the domain of convergence is the entire plane.

The Delay and Linearity Theorems are useful for finding, without integrating, the $\mathcal{L}$-transforms of many interesting functions. Some examples are:

Example 2.2–5

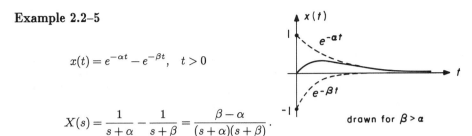

$$x(t) = e^{-\alpha t} - e^{-\beta t}, \quad t > 0$$

$$X(s) = \frac{1}{s + \alpha} - \frac{1}{s + \beta} = \frac{\beta - \alpha}{(s + \alpha)(s + \beta)}.$$

drawn for $\beta > \alpha$

The domain of convergence is $\Re e[s] > -\alpha$ or $\Re e[s] > -\beta$, whichever is further to the right.

▶ ▶ ▶

Example 2.2–6

$$x(t) = \sin \omega_0 t = \frac{1}{2j}\left[e^{j\omega_0 t} - e^{-j\omega_0 t}\right], \quad t > 0$$

$$X(s) = \frac{1}{2j}\frac{1}{s - j\omega_0} - \frac{1}{2j}\frac{1}{s + j\omega_0} = \frac{\omega_0}{s^2 + \omega_0^2}, \quad \Re e[s] > 0.$$

▶ ▶ ▶

Example 2.2–7

$$x(t) = \begin{cases} \sin \omega_0 t, & 0 < t < \pi/\omega_0 \\ 0, & \text{elsewhere} \end{cases}$$

$$= \sin \omega_0 t + \sin \omega_0 \left[t - \frac{\pi}{\omega_0}\right] u\left[t - \frac{\pi}{\omega_0}\right]$$

$$X(s) = \frac{\omega_0}{s^2 + \omega_0^2} + \frac{\omega_0}{s^2 + \omega_0^2}e^{-s\pi/\omega_0} = \frac{\omega_0}{s^2 + \omega_0^2}\left(1 + e^{-s\pi/\omega_0}\right), \quad \text{all } s.$$

This $X(s)$ has no poles anywhere in the finite s-plane. (The apparent poles at $s = \pm j\omega_0$ are cancelled by zeros of $\left(1 + e^{-s\pi/\omega_0}\right)$ at those points.)

▶ ▶ ▶

Henceforth we shall usually drop any explicit indication of domains of convergence, relying implicitly on the fact that there is *some* value of σ_0 such that for $\Re e[s] > \sigma_0$, *all* transforms arising in a particular problem are well-defined.

2.3 The Inverse Laplace Transform

The scheme proposed in Section 2.1 for exploiting Laplace transforms in system characterization requires that we be able to reverse the process described above and recover $x(t)$ given $X(s)$. It can be shown (it is not at all obvious) that this can be accomplished through the *inverse Laplace transform*

$$x(t) = \mathcal{L}^{-1}[X(s)] = \frac{1}{2\pi j} \int_C X(s)e^{st}\, ds \qquad (2.3\text{--}1)$$

where the integral is a line integral along an appropriate contour C in the complex plane. This integral can be remarkably effective in many cases, but its significance and efficient handling depend on an appreciation of the theory of functions of a complex variable, which puts it beyond our scope. (We shall, however, have a bit more to say about this topic after introducing the Fourier transform in Chapter 14.)

For our purposes it will be adequate to carry out the inverse transformation by manipulating $X(s)$ into a sum of terms, each of which we can recognize as the direct transform of a simple time function. Exploiting the Linearity Theorem and the Uniqueness Theorem (one form of which we now state) gives the desired inversion.

> *UNIQUENESS THEOREM:*
>
> If $X_1(s) = \mathcal{L}[x_1(t)]$ and $X_2(s) = \mathcal{L}[x_2(t)]$ exist and are equal in any small region of the s-plane, then $X_1(s) = X_2(s)$ throughout their common region of convergence and $x_1(t) = x_2(t)$ for almost all $t > 0$.

This theorem implies that for practical purposes the inverse transformation is unique.*

The inverse transform technique that we are going to describe will work in any case in which $X(s)$ is a rational function, and in some other cases as well. The key is to appreciate that every rational function can be expanded in *partial fractions*. Specifically, if the rational function is *proper* (degree of numerator less than the degree of the denominator—we shall remove this restriction in Chapter 11) and if the roots of the denominator— the poles of $X(s)$—are *simple*

---

*Of course, nothing can be learned from $X(s)$ about $x(t)$ for $t < 0$. "Almost all" means that $x_1(t)$ and $x_2(t)$ might differ at isolated points, for example, $x_1(t) = 1$, $x_2(t) = \begin{Bmatrix} 1, & t \neq 1 \\ 0, & t = 1 \end{Bmatrix}$; but such differences generally have no practical consequences for reasons we shall explore in Chapter 11. For a proof of the uniqueness theorem, see, e.g., D. V. Widder, *The Laplace Transform* (Princeton, NJ: Princeton Univ. Press, 1946) p. 63. Uniqueness is closely related to the fact that $X(s)$ is an *analytic function* throughout its domain of convergence—that is, $X(s)$ can be expanded in a convergent Taylor's series about any point in its domain.

or *distinct* (we shall remove this restriction shortly), then we may always write

$$X(s) = \frac{a_n s^n + a_{n-1} s^{n-1} + \cdots + a_0}{s^m + b_{m-1} s^{m-1} + \cdots + b_0} = \frac{a_n s^n + a_{n-1} s^{n-1} + \cdots + a_0}{(s - s_{p_1})(s - s_{p_2}) \ldots (s - s_{p_m})}$$

$$= \frac{k_1}{s - s_{p_1}} + \frac{k_2}{s - s_{p_2}} + \cdots + \frac{k_m}{s - s_{p_m}} \tag{2.3-2}$$

where $k_1, k_2, \ldots, k_m$ are an appropriate set of constants called *residues*.* Once the k_i are found, we may write the corresponding $x(t)$ immediately as

$$x(t) = k_1 e^{s_{p_1} t} + k_2 e^{s_{p_2} t} + \cdots + k_m e^{s_{p_m} t}, \quad t > 0. \tag{2.3-3}$$

This, or its generalization to include multiple-order poles, is often called the *Heaviside Expansion Theorem*.

It should be evident from (2.3–2) that the residues for $X(s)$ with simple poles can be easily found from the formula

$$k_i = [X(s)(s - s_{p_i})]_{s = s_{p_i}}. \tag{2.3-4}$$

The following examples illustrate the partial-fraction procedure for finding inverse Laplace transforms for $X(s)$ with simple poles.

Example 2.3–1

$$X(s) = \frac{s+3}{s^2 + s} = \frac{s+3}{s(s+1)} = \frac{k_0}{s} + \frac{k_1}{s+1}$$

where

$$k_0 = [X(s)s]_{s=0} = \left.\frac{s+3}{s+1}\right|_{s=0} = 3$$

$$k_1 = [X(s)(s+1)]_{s=-1} = \left.\frac{s+3}{s}\right|_{s=-1} = -2.$$

Thus

$$X(s) = \frac{3}{s} - \frac{2}{s+1} \implies x(t) = 3 - 2e^{-t}, \quad t > 0.$$

Since the partial fraction expansion is an algebraic identity, it can *and should* always be checked by multiplying the terms back together again:

$$X(s) = \frac{3(s+1) - 2s}{s(s+1)} = \frac{s+3}{s^2 + s}.$$

---

*Note carefully the way in which $X(s)$ in (2.3–2) is normalized so that the highest power of s in the denominator has coefficient unity. Failure to carry out this normalization is a common source of error in evaluating residues.

Example 2.3-2

$$X(s) = \frac{s+1}{s^2+1} = \frac{s+1}{(s+j)(s-j)} = \frac{k_+}{s+j} + \frac{k_-}{s-j}$$

where

$$k_+ = [X(s)(s+j)]_{s=-j} = \frac{s+1}{s-j}\Big|_{s=-j} = \frac{1-j}{-2j} = \frac{1}{\sqrt{2}}e^{j\pi/4}$$

$$k_- = [X(s)(s-j)]_{s=j} = \frac{s+1}{s+j}\Big|_{s=j} = \frac{1+j}{2j} = \frac{1}{\sqrt{2}}e^{-j\pi/4}.$$

Thus

$$x(t) = \frac{1}{\sqrt{2}}e^{j\pi/4}e^{-jt} + \frac{1}{\sqrt{2}}e^{-j\pi/4}e^{jt} = \sqrt{2}\cos(t-\pi/4), \quad t>0.$$

Note that because the coefficients of the numerator and denominator polynomials are real, the poles occur in conjugate complex pairs and have conjugate complex residues.

▶ ▶ ▶

Example 2.3-3

$$X(s) = \frac{1 - e^{-(s+\alpha)T}}{s+\alpha}.$$

This is not a rational function, but it can be written as a sum of products of rational functions and e^{-sT} factors:

$$X(s) = \frac{1}{s+\alpha} - \frac{e^{-\alpha T}}{s+\alpha}e^{-sT}.$$

Hence, from the Delay Theorem,

$$x(t) = e^{-\alpha t} - e^{-\alpha T}\left[e^{-\alpha(t-T)}u(t-T)\right] = e^{-\alpha t} - e^{-\alpha t}u(t-T), \quad t>0.$$

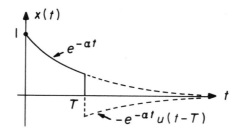

This is a pulse-type waveform, and we note that $X(s)$ does not in fact have a pole at $s = -\alpha$ or anywhere else in the finite s-plane; the domain of convergence is the entire plane.

▶ ▶ ▶

2.4 Multiple-Order Poles

The partial-fraction expansion procedure becomes slightly more complex if any roots of the denominator polynomial are repeated or multiple-order. Extra terms must then be added to the expansion corresponding to all powers of the repeated factor up to the order of the pole. The following example should make the ideas clear.

Example 2.4–1

$$X(s) = \frac{1}{(s+1)^3(s+2)} = \frac{k_1''}{(s+1)^3} + \frac{k_1'}{(s+1)^2} + \frac{k_1}{s+1} + \frac{k_2}{s+2}.$$

The residue k_2 and the coefficient k_1'' can easily be found by an obvious extension of the previous method:

$$k_2 = [X(s)(s+2)]_{s=-2} = \frac{1}{(s+1)^3}\bigg|_{s=-2} = -1$$

$$k_1'' = \left[X(s)(s+1)^3\right]_{s=-1} = \frac{1}{s+2}\bigg|_{s=-1} = 1.$$

These two terms by themselves are not enough to represent $X(s)$ (as can readily be seen by multiplying the expansion out with $k_1' = k_1 = 0$). To find k_1 and k_1' several methods are possible:

a) Multiply back together and match coefficients with the original function:

$$\frac{1}{(s+1)^3} + \frac{k_1'}{(s+1)^2} + \frac{k_1}{s+1} - \frac{1}{s+2} = \frac{1}{(s+1)^3(s+2)}$$

$$= \frac{(s+2) + k_1'(s+1)(s+2) + k_1(s+1)^2(s+2) - (s+3)^3}{(s+1)^3(s+2)}.$$

Matching highest powers gives $k_1 s^3 - s^3 = 0$ or $k_1 = 1$. Matching next highest powers gives $k_1' s^2 + k_1(2s^2 + 2s^2) - 3s^2 = 0$ or $k_1' = -1$.

b) Expand $X(s)(s+1)^3$ in a power series in $(s+1)$:

$$X(s)(s+1)^3 = \frac{1}{s+2} = \frac{1}{(s+1)+1} = 1 - (s+1) + (s+1)^2 - \cdots$$

(where we have used the expansion $\frac{1}{1+x} = 1 - x + x^2 - x^3 + \cdots$ which is valid for $|x| < 1$, that is, for s near -1). Then

$$X(s) = \frac{1 - (s+1) + (s+1)^2 - \cdots}{(s+1)^3} = \frac{1}{(s+1)^3} - \frac{1}{(s+1)^2} + \frac{1}{s+1} - \cdots.$$

The coefficients of the negative powers are the desired terms, that is, $k_1'' = 1$, $k_1' = -1$, and $k_1 = 1$. (Since the coefficients of the Taylor's-series expansion of

$X(s)(s+1)^3$ are related to the derivatives of this expression, formulas are often given in mathematics books for k_1 and k_1' in terms of derivatives, but these are rarely the most effective way to compute these coefficients.)

c) Subtract away the $k_1''/(s+1)^3$ term, leaving a function with only a second-order pole. Repeating gives a function with only a first-order pole, which can be expanded as before. Thus

$$X(s) - \frac{k_1''}{(s+1)^3} = \frac{1}{(s+1)^3(s+2)} - \frac{1}{(s+1)^3} = \frac{1-(s+2)}{(s+1)^3(s+2)}$$

$$= \frac{-1}{(s+1)^2(s+2)} = \frac{k_1'}{(s+1)^2} + \frac{k_1}{s+1} - \frac{1}{s+2}.$$

Then

$$k_1' = \left[\frac{-1}{(s+1)^2(s+2)} (s+1)^2 \right]_{s=-1} = -1$$

etc.

▶ ▶ ▶

To evaluate the time functions corresponding to $\mathcal{L}$-transforms with multiple-order poles, we need to know the time function whose transform is $1/(s+\alpha)^n$. This can be found from repeated application of the following theorem.

MULTIPLY-BY-t THEOREM:

Let $\mathcal{L}[x(t)] = X(s)$. Then

$$\mathcal{L}[t\,x(t)] = -\frac{dX(s)}{ds}. \tag{2.4--1}$$

The proof follows at once on differentiating the basic defining formula

$$X(s) = \int_0^\infty x(t)e^{-st}\,dt.$$

Thus we conclude that

$$\mathcal{L}[te^{-\alpha t}] = -\frac{d}{ds}\frac{1}{s+\alpha} = \frac{1}{(s+\alpha)^2} \tag{2.4--2}$$

so that, by induction,

$$\mathcal{L}[t^n e^{-\alpha t}] = \frac{n!}{(s+\alpha)^{n+1}}. \tag{2.4--3}$$

In particular, completing Example 2.4–1, if

$$X(s) = \frac{1}{(s+1)^3(s+2)} = \frac{1}{(s+1)^3} - \frac{1}{(s+1)^2} + \frac{1}{s+1} - \frac{1}{s+2}$$

then

$$x(t) = \frac{1}{2}t^2 e^{-t} - te^{-t} + e^{-t} - e^{-2t}, \quad t > 0.$$

The appendix to this chapter contains a brief table of $\mathcal{L}$-transform pairs and a list of important properties. More extensive tables are widely available.

2.5 Circuit Analysis with the Laplace Transform

In the preceding chapter, we explained how the dynamic equations characterizing the behavior of a circuit are derived from two kinds of information—the constitutive relations between branch voltages and currents that describe the elements, and the constraints among the same variables that arise from their interconnections and Kirchhoff's Laws. Since the constitutive relations in general involve derivatives and/or integrals, the dynamic equations are in general differential equations. Alternatively, as suggested in Section 2.1, we could describe both the elements and the interconnection constraints in terms of relations between the unilateral Laplace transforms of the branch voltages and currents. The result for LTI circuits is a set of algebraic equations that are much easier than differential equations to manipulate and interpret.

Let's begin by replacing the constitutive relations for the simple 2-terminal lumped electrical elements of Figure 1.1–1 by equivalent relations in the frequency domain. Thus, a linear resistor is equally adequately described by the instantaneous version of Ohm's Law, $v(t) = Ri(t)$, or by the relation

$$V(s) = RI(s) \tag{2.5-1}$$

between the $\mathcal{L}$-transforms of the voltage and the current that follows from the Linearity Theorem. Mathematically, (2.5–1) has the same form as Ohm's Law for the "voltage" $V(s)$ across a "resistor" R carrying a "current" $I(s)$. Of course, $V(s)$ and $I(s)$ are not a voltage and a current, but rather the transforms of a voltage and a current. Nevertheless, it is often very convenient to draw a circuit diagram "in the frequency domain" in which the branch variables are the transforms of the actual branch time functions and the elements are described by constitutive relations between these transforms.

To derive similar frequency-domain constitutive relations for linear capacitors and inductors, we need the following theorem.

DIFFERENTIATION THEOREM:
Let $\mathcal{L}[x(t)] = X(s)$. Then

$$\boxed{\mathcal{L}\left[\frac{dx(t)}{dt}\right] = sX(s) - x(0).} \tag{2.5-2}$$

The proof of this theorem proceeds as follows. By definition*

$$\mathcal{L}\left[\frac{dx(t)}{dt}\right] = \int_0^\infty \frac{dx(t)}{dt} e^{-st}\, dt.$$

Integrate by parts, that is, use the formula $\int u\, dv = uv - \int v\, du$ with $u = e^{-st}$, $du = -se^{-st}\, dt$, $dv = \dfrac{dx(t)}{dt} dt$, $v = x(t)$, to give

---

*We assume $dx(t)/dt$ is well-behaved. It is interesting to note, however, that the right-hand side of (2.5–2) is apparently meaningful even when $x(t)$ has discontinuities, suggesting that some sort of significance can be attached to $dx(t)/dt$ even in such cases—as we shall explore further in Chapter 11.

$$\int_0^\infty \frac{dx(t)}{dt} e^{-st} \, dt = x(t)e^{-st}\big|_0^\infty + s\int_0^\infty x(t)e^{-st} \, dt$$
$$= -x(0) + sX(s)$$

where we have assumed $x(t)e^{-st}\big|^\infty = 0$, as it generally must if s is in the domain of convergence of $X(s) = \mathcal{L}[x(t)]$. This completes the proof.

Using this theorem to transform the equation $i(t) = C\,dv(t)/dt$ characterizing a capacitor C yields

$$I(s) = CsV(s) - Cv(0) \tag{2.5--3}$$

or

$$V(s) = \frac{1}{Cs}I(s) + \frac{v(0)}{s}. \tag{2.5--4}$$

Mathematically, (2.5–4) has the same form as Ohm's Law for the "voltage" $V(s)$ across a "resistor" $1/Cs$ carrying a "current" $I(s)$ in series with a "voltage source" $v(0)/s$. In the "frequency-domain" circuit, then, a capacitor will be represented by an *impedance*, $Z(s) = 1/Cs$, in series with a "voltage" source $v(0)/s$ reflecting the initial state. Similarly, transforming the equation $v(t) = L\,di(t)/dt$ characterizing an inductor L yields

$$V(s) = LsI(s) - Li(0) \tag{2.5--5}$$

or

$$I(s) = \frac{1}{Ls}V(s) + \frac{i(0)}{s}. \tag{2.5--6}$$

Hence in the "frequency-domain" circuit, an inductor is represented by an impedance $Z(s) = Ls$ in parallel with a "current source" $i(0)/s$ describing the initial state.

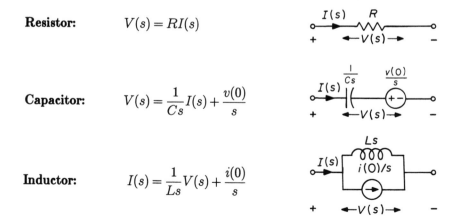

Resistor:	$V(s) = RI(s)$	
Capacitor:	$V(s) = \dfrac{1}{Cs}I(s) + \dfrac{v(0)}{s}$	
Inductor:	$I(s) = \dfrac{1}{Ls}V(s) + \dfrac{i(0)}{s}$	

Figure 2.5–1. Simple LTI elements in the frequency domain.

The frequency-domain element descriptions derived above are summarized in Figure 2.5–1. Note that the "branch variables" in these elements are the transforms of the voltages and currents that are actually present. The "element values" are the impedances associated with these elements, and the "sources" appear only in the frequency-domain descriptions. Note also that impedance is analogous to resistance, that is, it is the ratio of a voltage (transform) to a current (transform); the reciprocal of impedance—analogous to conductance—is called *admittance*. In general, the symbol $Z(s)$ is used to represent the impedance of an arbitrary element; $Y(s)$ is used to represent the admittance of an arbitrary element.

As a result of the Linearity Theorem, the constraints imposed on the time-domain branch voltages and currents by Kirchhoff's Laws carry over without alteration into the frequency domain:

$$time\ domain \qquad\qquad frequency\ domain$$

$$\sum_{cut\ set} i_j(t) = 0 \quad\Longleftrightarrow\quad \sum_{cut\ set} I_j(s) = 0$$

$$\sum_{loop} v_j(t) = 0 \quad\Longleftrightarrow\quad \sum_{loop} V_j(s) = 0. \qquad (2.5-7)$$

These constraints can be represented schematically in the frequency-domain circuit by connecting the elements in exactly the same way as they are connected in the time-domain circuit. How impedance methods and the L-transform are combined to yield efficient solution procedures for LTI circuit problems is best explained through several examples.

Example 2.5–1

In Example 1.7–1 we computed by "classical" methods the response of the simple first-order circuit shown in Figure 2.5–2 to a constant input. The result, including an arbitrary initial voltage $v(0)$ on the capacitor, was

$$v(t) = RI + (v(0) - RI)e^{-t/RC}, \quad t > 0. \qquad (2.5-8)$$

This is plotted in Figure 2.5–3.

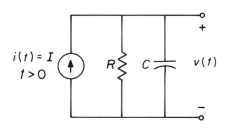

Figure 2.5–2. Example 2.5–1 circuit.

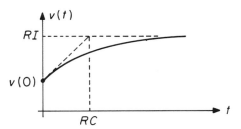

Figure 2.5–3. Constant input response.

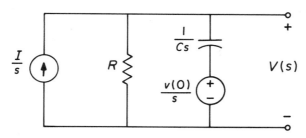

Figure 2.5–4. Frequency-domain form of circuit of Figure 2.5–2.

The frequency-domain form of this circuit is shown in Figure 2.5–4. The resistor and the capacitor have been replaced by their impedances, R and $1/Cs$, respectively. A source $v(0)/s$ has been inserted in series with the capacitor to describe the effects of the initial state. The constant current source I in the time-domain circuit has been replaced in the frequency-domain circuit by its transform I/s. We seek to determine $V(s)$, the transform of the voltage $v(t)$. Treating $V(s)$ as a node voltage, we can use elementary resistive circuit theory to derive the node equation

$$\frac{V(s)}{R} + \left(V(s) - \frac{v(0)}{s}\right)Cs - \frac{I}{s} = 0. \tag{2.5–9}$$

Solving (2.5–9) for $V(s)$ and expanding in partial fractions gives

$$V(s) = \frac{(I/C) + v(0)s}{s(s + (1/RC))} = \frac{RI}{s} + \frac{v(0) - RI}{s + (1/RC)} \tag{2.5–10}$$

which we recognize immediately from Examples 2.2–1 and 2.2–2 as the transform of

$$v(t) = RI + (v(0) - RI)e^{-t/RC}, \quad t > 0 \tag{2.5–11}$$

which is identical with the result obtained in Example 1.7–1. Of course, this is such a simple problem that the answer can be written down directly—which is obviously preferable to using transforms.

▶ ▶ ▶

Example 2.5–2

The kind of situation in which the Laplace transform is most useful is illustrated by the problem described in Figure 2.5–5. To solve it, write node equations for the frequency-domain circuit as if it were a resistive circuit with branch resistances equal to the impedances. Selecting $V_C(s)$ and $V_1(s)$ as node voltage variables, we obtain by summing currents away from the nodes

$$\frac{V_C(s) - V_0(s)}{R_1} + \left(V_C(s) - \frac{v_C(0)}{s}\right)Cs + \frac{V_C(s) - V_1(s)}{Ls} + \frac{i_L(0)}{s} = 0 \tag{2.5–12}$$

$$\frac{V_1(s) - V_C(s)}{Ls} + \frac{V_1(s)}{R_2} - \frac{i_L(0)}{s} - I_0(s) = 0. \tag{2.5–13}$$

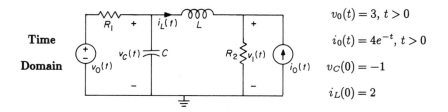

$$v_0(t) = 3,\ t > 0$$

$$i_0(t) = 4e^{-t},\ t > 0$$

$$v_C(0) = -1$$

$$i_L(0) = 2$$

$$R_1 = 0.5\,\Omega, \quad R_2 = 1\,\Omega, \quad L = 1\,\mathrm{H}, \quad C = 0.5\,\mathrm{F}$$

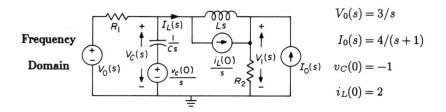

$$V_0(s) = 3/s$$

$$I_0(s) = 4/(s+1)$$

$$v_C(0) = -1$$

$$i_L(0) = 2$$

Figure 2.5–5. Time- and frequency-domain circuits for Example 2.5–2.

Solving (2.5–12, 13) for $V_1(s)$ gives (after some algebra)

$$
\begin{aligned}
V_1(s) \;=\; & \frac{R_2\!\left(s^2 + s\frac{1}{R_1 C} + \frac{1}{LC}\right)}{s^2 + s\!\left(\frac{1}{R_1 C} + \frac{R_2}{L}\right) + \frac{1}{LC}\!\left(1 + \frac{R_2}{R_1}\right)}\, I_0(s) \\[2mm]
& + \frac{\frac{R_2}{LCR_1}}{s^2 + s\!\left(\frac{1}{R_1 C} + \frac{R_2}{L}\right) + \frac{1}{LC}\!\left(1 + \frac{R_2}{R_1}\right)}\, V_0(s) \\[2mm]
& + \frac{s\frac{R_2}{L}}{s^2 + s\!\left(\frac{1}{R_1 C} + \frac{R_2}{L}\right) + \frac{1}{LC}\!\left(1 + \frac{R_2}{R_1}\right)}\, \frac{v_C(0)}{s} \\[2mm]
& + \frac{s^2 R_2 + s\frac{R_2}{R_1 C}}{s^2 + s\!\left(\frac{1}{R_1 C} + \frac{R_2}{L}\right) + \frac{1}{LC}\!\left(1 + \frac{R_2}{R_1}\right)}\, \frac{i_L(0)}{s}.
\end{aligned}
\tag{2.5–14}
$$

To catch algebraic mistakes, it is good practice to carry circuit analyses about this far in literal form so that the dimensions of terms being added can be checked for consistency (using the facts that s has dimensions t^{-1}, RC and L/R have dimensions t, and LC has dimensions t^2). But to proceed further, it is helpful to substitute the element values given in Figure 2.5–5 to obtain

$$
\begin{aligned}
V_1(s) \;=\; & \frac{s^2 + 4s + 2}{s^2 + 5s + 6}\, I_0(s) \;+\; \frac{4}{s^2 + 5s + 6}\, V_0(s) \\[2mm]
& + \frac{s}{s^2 + 5s + 6}\, \frac{v_C(0)}{s} \;+\; \frac{s(s+4)}{s^2 + 5s + 6}\, \frac{i_L(0)}{s}.
\end{aligned}
\tag{2.5–15}
$$

Substituting the appropriate functions for the transforms of the sources and the initial conditions as given in Figure 2.5–5, and factoring $s^2 + 5s + 6 = (s+2)(s+3)$, yields

$$
\begin{aligned}
V_1(s) &= \frac{4(s^2 + 4s + 2)}{(s+1)(s+2)(s+3)} + \frac{12}{s(s+2)(s+3)} - \frac{1}{(s+2)(s+3)} + \frac{2(s+4)}{(s+2)(s+3)} \\
&= \frac{6s^3 + 25s^2 + 27s + 12}{s(s+1)(s+2)(s+3)} = \frac{2}{s} - \frac{2}{s+1} + \frac{5}{s+2} + \frac{1}{s+3}.
\end{aligned}
$$

$$(2.5\text{–}16)$$

Inverse transforming yields

$$
v_1(t) = 2 - 2e^{-t} + 5e^{-2t} + e^{-3t}, \quad t > 0 \tag{2.5–17}
$$

which is the final answer sought.

▶ ▶ ▶

     Clearly the $\mathcal{L}$-transform is a powerful and efficient solution procedure for problems of the type illustrated in Example 2.5–2. But the algebra (which we did not display above) can be much reduced by exploiting one or more of the simplification techniques studied in earlier courses for linear resistive circuits—equivalent series and parallel resistances, voltage- and current-divider formulas, superposition, Thévenin and Norton theorems, ladder method, etc. These apply just as well to impedances in frequency-domain circuits, as shown in the following example.

Example 2.5–3

Suppose we replace the left-hand part of the circuit of Figure 2.5–5 by its Thévenin equivalent as shown in Figure 2.5–6.

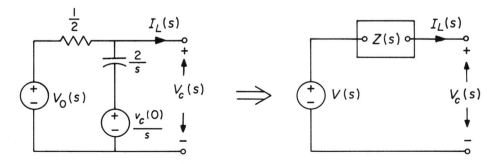

Figure 2.5–6. Thévenin equivalent for part of the circuit of Figure 2.5–5.

If we assume an open circuit (that is, $I_L(s) = 0$), we have by superposition and use of the voltage-divider formula,

$$V_C(s) = \frac{2/s}{(1/2)+(2/s)}V_0(s) + \frac{1/2}{(1/2)+(2/s)}\frac{v_C(0)}{s}$$

$$= \frac{4V_0(s)}{s+4} + \frac{v_C(0)}{s+4} = V(s)$$

where $V(s)$ is the Thévenin voltage source. To find the Thévenin impedance $Z(s)$, set the sources $V_0(s)$ and $v_C(0)/s$ to zero and compute the parallel combination of the two impedances $1/2$ and $2/s$:

$$Z(s) = \frac{(1/2)\cdot(2/s)}{(1/2)+(2/s)} = \frac{2}{s+4}.$$

The equivalent circuit now appears as in Figure 2.5–7.

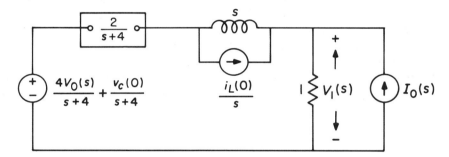

Figure 2.5–7. Equivalent circuit after Thévenin replacement of left part.

Again using superposition, divider formulas, etc., we can write $V_1(s)$ by inspection:

$V_1(s) =$

$$\underbrace{\frac{1}{1+s+\frac{2}{s+4}}\left[\frac{v_C(0)}{s+4} + \frac{4V_0(s)}{s+4}\right]}_{\substack{\text{voltage}\\\text{divider}}} + \underbrace{\frac{s\cdot 1}{1+s+\frac{2}{s+4}}\frac{i_L(0)}{s}}_{\substack{\text{current}\\\text{divider}}} + \underbrace{\frac{\left(s+\frac{2}{s+4}\right)\cdot 1}{1+s+\frac{2}{s+4}}I_0(s)}_{\substack{\text{parallel}\\\text{impedances}}}$$

or

$$V_1(s) = \frac{s^2+4s+2}{s^2+5s+6}I_0(s) + \frac{4}{s^2+5s+6}V_0(s)$$

$$+ \frac{s}{s^2+5s+6}\frac{v_C(0)}{s} + \frac{s(s+4)}{s^2+5s+6}\frac{i_L(0)}{s}$$

which is exactly the same as before.

▶ ▶ ▶

2.6 Summary

Because of the biunique relationship between a time function and its Laplace transform, one can characterize the input-output behavior of a system either in the time domain—by describing how to find the output time function given the input time function and appropriate information about the initial state— or in the frequency domain—by describing how to find the transform of the output time function given the transform of the input time function and appropriate state information. For LTI circuits, the time-domain description is in terms of differential equations, but the frequency-domain description leads to linear algebraic equations that are much easier to manipulate. Moreover, applying Laplace transforms to the constitutive relations for LTI elements and to Kirchhoff's Laws leads to the idea of replacing the time-domain circuit with a frequency-domain circuit in which the elements are replaced by their impedances. Solving for desired branch variables then reduces to an elementary problem in resistive circuit theory.

The Laplace transform is a very powerful "crank" for "turning out" in this way solutions to transient circuit problems, as shown in Examples 2.5–1, 2, 3. But the value of impedance methods and the frequency domain goes well beyond their usefulness in solving transient circuit problems, as we shall begin to see in the next chapter.

APPENDIX TO CHAPTER 2
Table II.1—Short Table of Unilateral L-Transforms

$$X(s) = \int_0^\infty x(t)e^{-st}\,dt$$

$x(t) = L^{-1}[X(s)]$		$X(s) = L[x(t)]$
$\delta(t)$*	$\Longleftrightarrow$	1
$u(t) = 1$	$\Longleftrightarrow$	$\dfrac{1}{s}$
$e^{-\alpha t}$	$\Longleftrightarrow$	$\dfrac{1}{s+\alpha}$
t^n	$\Longleftrightarrow$	$\dfrac{n!}{s^{n+1}}$
$t^n e^{-\alpha t}$	$\Longleftrightarrow$	$\dfrac{n!}{(s+\alpha)^{n+1}}$
$\sin\omega_0 t$	$\Longleftrightarrow$	$\dfrac{\omega_0}{s^2 + \omega_0^2}$
$\cos\omega_0 t$	$\Longleftrightarrow$	$\dfrac{s}{s^2 + \omega_0^2}$
$e^{-\alpha t}\cos\omega_0 t$	$\Longleftrightarrow$	$\dfrac{s+\alpha}{(s+\alpha)^2 + \omega_0^2}$

Note: $x(t)$ is defined by $L^{-1}[X(s)]$ for $t \ge 0$ only.

Table II.2—Important Unilateral L-Transform Theorems

Linearity	$ax_1(t) + bx_2(t)$	$\Longleftrightarrow$	$aX_1(s) + bX_2(s)$
Delay	$x(t-T)u(t-T)$	$\Longleftrightarrow$	$X(s)e^{-sT}, \quad T > 0$
Time Multiplication	$tx(t)$	$\Longleftrightarrow$	$-\dfrac{dX(s)}{ds}$
$e^{-\alpha t}$ Multiplication	$e^{-\alpha t}x(t)$	$\Longleftrightarrow$	$X(s+\alpha)$
Scaling	$x(at)$	$\Longleftrightarrow$	$\dfrac{1}{a}X\left[\dfrac{s}{a}\right], \quad a > 0$
Differentiation	$\dfrac{dx(t)}{dt}$	$\Longleftrightarrow$	$sX(s) - x(0)$
Integration	$\displaystyle\int_0^t x(\tau)\,d\tau$	$\Longleftrightarrow$	$\dfrac{X(s)}{s}$
Initial-Value	$x(0)$	$=$	$\displaystyle\lim_{s\to\infty} sX(s)$
Final-Value[†]	$x(\infty)$	$=$	$\displaystyle\lim_{s\to 0} sX(s)$
Convolution[‡]	$\displaystyle\int_0^t x_1(\tau)x_2(t-\tau)\,d\tau$	$\Longleftrightarrow$	$X_1(s)X_2(s)$

*See Chapter 11   †Provided $sX(s)$ has no poles in $\Re e[s] \ge 0$ ‡See Chapter 10

EXERCISES FOR CHAPTER 2

Exercise 2.1

Show that the $\mathcal{L}$-transforms of each of the following time functions are as given. Sketch each time function and show on a diagram the locations of all finite poles and zeros of the transform. Indicate by shading on this diagram the region of absolute convergence. Throughout, assume that α and ω_0 are positive quantities.

$$x(t),\ t > 0 \qquad\qquad X(s)$$

a) $\quad 1 - e^{-\alpha t} \quad\Longleftrightarrow\quad \dfrac{\alpha}{s(s+\alpha)}, \quad \Re e[s] > 0$

b) $\quad e^{+\alpha t}\sin\omega_0 t \quad\Longleftrightarrow\quad \dfrac{\omega_0}{(s-\alpha)^2 + \omega_0^2}, \quad \Re e[s] > \alpha$

c) $\quad e^{-\alpha t}\cos(\omega_0 t + \pi/4) \quad\Longleftrightarrow\quad \dfrac{s + \alpha - \omega_0}{\sqrt{2}[(s+\alpha)^2 + \omega_0^2]}, \quad \Re e[s] > -\alpha$

d) $\quad te^{-\alpha t}\cos\omega_0 t \quad\Longleftrightarrow\quad \dfrac{(s+\alpha)^2 - \omega_0^2}{[(s+\alpha)^2 + \omega_0^2]^2}, \quad \Re e[s] > -\alpha$

e) $\quad \begin{cases} 1, & 1 < t < 2 \\ 0, & \text{elsewhere} \end{cases} \quad\Longleftrightarrow\quad \dfrac{1}{s}\left[e^{-s} - e^{-2s} \right], \quad \text{all } s$

f) $\quad \begin{cases} t, & 0 < t < 1 \\ 2 - t, & 1 < t < 2 \\ 0, & \text{elsewhere} \end{cases} \quad\Longleftrightarrow\quad \left(\dfrac{1 - e^{-s}}{s} \right)^2, \quad \text{all } s$

g) $\quad \begin{cases} 1, & -1 < t < 0 \\ 0, & \text{elsewhere} \end{cases} \quad\Longleftrightarrow\quad 0$

h) $\quad \cosh\alpha t \quad\Longleftrightarrow\quad \dfrac{s}{s^2 - \alpha^2}, \quad \Re e[s] > \alpha$

i) $\quad e^{-\alpha t}u(t - 1) \quad\Longleftrightarrow\quad \dfrac{e^{-(s+\alpha)}}{s + \alpha}, \quad \Re e[s] > -\alpha$

j) $\quad 1 - (1 + \alpha t)e^{-\alpha t} \quad\Longleftrightarrow\quad \dfrac{\alpha^2}{s(s+\alpha)^2}, \quad \Re e[s] > 0$

Exercise 2.2

Show that the inverse unilateral $\mathcal{L}$-transforms of the following functions of s are as given. Sketch each time function and show on a diagram the locations of all finite poles and zeros of the transform. Indicate by shading on this diagram the region of absolute convergence.

$$X(s) \qquad\qquad x(t),\ t > 0$$

a) $\quad \dfrac{1}{(s+1)(s+2)} \quad\Longleftrightarrow\quad e^{-t} - e^{-2t}$

b) $\quad \dfrac{s+3}{(s+1)(s+2)} \quad\Longleftrightarrow\quad 2e^{-t} - e^{-2t}$

(Exercise 2.2 continued on the next page)

Exercise 2.2 (cont.)

$$\underline{X(s)} \qquad \underline{x(t), \quad t > 0}$$

c) $\quad \dfrac{1}{s^2 - \alpha^2} \quad \Longleftrightarrow \quad \dfrac{1}{\alpha}\sinh \alpha t$

d) $\quad \dfrac{1}{(s+\alpha)^2 + \omega_0^2} \quad \Longleftrightarrow \quad \dfrac{1}{\omega_0}e^{-\alpha t}\sin \omega_0 t$

e) $\quad \dfrac{1}{s}(1 - e^{-s})^2 \quad \Longleftrightarrow \quad \begin{cases} 1, & 0 < t < 1 \\ -1, & 1 < t < 2 \\ 0, & \text{elsewhere} \end{cases}$

f) $\quad \dfrac{1}{s^2 + 1}(1 - e^{-2\pi s}) \quad \Longleftrightarrow \quad \begin{cases} \sin t, & 0 < t < 2\pi \\ 0, & \text{elsewhere} \end{cases}$

g) $\quad \dfrac{s+2}{(s+1)^2} \quad \Longleftrightarrow \quad (1+t)e^{-t}$

h) $\quad \dfrac{1}{s^2(s-1)} \quad \Longleftrightarrow \quad e^t - (1+t)$

i) $\quad \dfrac{s}{s^2 + 2s + 2} \quad \Longleftrightarrow \quad \sqrt{2}e^{-t}\cos(t + \pi/4)$

j) $\quad \dfrac{1}{s^3 + 2s^2 + 2s + 1} \quad \Longleftrightarrow \quad e^{-t} + \dfrac{2}{\sqrt{3}}e^{-t/2}\cos\left(\dfrac{\sqrt{3}t}{2} - \dfrac{5\pi}{6}\right)$

Exercise 2.3

The state equation describing a system is

$$\frac{d\lambda(t)}{dt} = -2\lambda(t) + x(t).$$

Show that the response of this system to an input $x(t) = u(t-1)$ is described by

$$\lambda(t) = \frac{1}{2}\left(1 - e^{-2(t-1)}\right)u(t-1).$$

Exercise 2.4

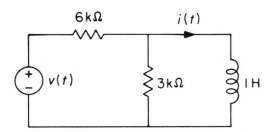

Show that the current in the circuit above is

$$i(t) = 2e^{-10^3 t} - e^{-2 \times 10^3 t}\,\text{milliamp}, \quad t \ge 0$$

if it is known that $v(t) = 6e^{-10^3 t}$ volt, $t > 0$, and $i(0) = 1\,\text{mamp}$. Sketch $i(t)$.

PROBLEMS FOR CHAPTER 2

Problem 2.1

a) Use the basic definition of the unilateral $\mathcal{L}$-transform and integration by parts to derive the Integration Theorem of Table II.2:

$$\mathcal{L}\left[\int_0^t x(\tau)\,d\tau\right] = \frac{1}{s}\mathcal{L}[x(t)].$$

b) Derive the same result from the Differentiation Theorem of Section 2.5 by setting

$$y(t) = \frac{dx(t)}{dt}, \quad x(t) = \int_0^t y(\tau)\,d\tau + x(0).$$

c) Use the Integration Theorem to devise an impedance representation for a capacitor described by the constitutive relation

$$v_C(t) = \frac{1}{C}\int_0^t i_C(\tau)\,d\tau + v_C(0)$$

and show that it is the same as that given in Figure 2.5–1.

d) Use the Differentiation or Integration Theorem to show that an integrator has the frequency-domain representation shown below.

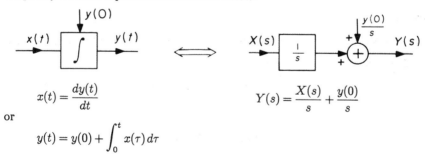

$$x(t) = \frac{dy(t)}{dt}$$

$$Y(s) = \frac{X(s)}{s} + \frac{y(0)}{s}$$

or

$$y(t) = y(0) + \int_0^t x(\tau)\,d\tau$$

What are the corresponding frequency-domain representations for adders and gain elements as used in block diagrams such as Problem 1.5? Draw a frequency-domain transformation of the block diagram in Problem 1.5.

Problem 2.2

Inverse transforming the frequency-domain representation for the capacitor given in Figure 2.5–1 suggests that a capacitor initially charged to a voltage $v(0)$ at $t = 0$ should be indistinguishable in any circuit for $t > 0$ from an initially uncharged capacitor in series with a battery $v(0)$. Do you believe this? Analyze several simple circuit situations to convince yourself that this equivalence is reasonable.

Problem 2.3

a) Argue that the frequency-domain descriptions of capacitors and inductors correspond to the equivalent circuits below.

$$I(s) = CsV(s) - \underbrace{Cv(0)}_{I_0(s)} \quad \Longrightarrow$$

$$I_0(s) = Cv(0)$$

$$V(s) = LsI(s) - \underbrace{Li(0)}_{V_0(s)} \quad \Longrightarrow$$

$$V_0(s) = Li(0)$$

In these circuits note that $I_0(s) = Cv(0)$ (= constant) must not be thought of as a constant current source—it corresponds to a current source whose transform is a constant, and that is a very different thing. (A time function whose transform is a constant is called an *impulse*; we shall return to this topic in much detail in Chapter 11.)

b) Show that the circuits given in (a) are in impedance terms the Thévenin or Norton equivalents of the circuits given in Figure 2.5–1.

Problem 2.4

Two interesting theorems about L-transforms are the Initial and Final Value Theorems:

INITIAL VALUE THEOREM:

If both $x(t)$ and $\dfrac{dx(t)}{dt}$ are L-transformable and if $\lim\limits_{s\to\infty} sL[x(t)]$ exists, then

$$\lim_{s\to\infty} sL[x(t)] = x(0).$$

FINAL VALUE THEOREM:

If both $x(t)$ and $\dfrac{dx(t)}{dt}$ are L-transformable and if $sL[x(t)]$ has no poles on the $j\omega$-axis or in the right half-plane, then

$$\lim_{s\to 0} sL[x(t)] = \lim_{t\to\infty} x(t).$$

These theorems are easily proved from the Differentiation Theorem,

$$sL[x(t)] = L\left[\frac{dx(t)}{dt}\right] + x(0).$$

Thus as $s \to \infty$ with $\Re[s] > 0$ (in the region of absolute convergence)

$$\lim_{s \to \infty} \mathcal{L}\left[\frac{dx(t)}{dt}\right] = \lim_{s \to \infty} \int_0^\infty \frac{dx(t)}{dt}e^{-st}\, dt = \int_0^\infty \frac{dx(t)}{dt} \lim_{s \to \infty} e^{-st}\, dt = 0$$

which gives the first theorem. On the other hand as $s \to 0$ with $\Re[s] \geq 0$ (in the region of absolute convergence)

$$\lim_{s \to 0} \mathcal{L}\left[\frac{dx(t)}{dt}\right] = \lim_{s \to 0} \int_0^\infty \frac{dx(t)}{dt}e^{-st}\, dt = \int_0^\infty \frac{dx(t)}{dt} \lim_{s \to 0} e^{-st}\, dt$$

$$= \int_0^\infty \frac{dx(t)}{dt}\, dt = x(\infty) - x(0)$$

which gives the second. (For a further discussion of the Initial Value Theorem see Problem 11.15.)

a) Apply these theorems to each of the following functions $X(s)$ to find (where possible) $x(0)$ and $\lim_{t \to \infty} x(t)$:

1. $\dfrac{1 - e^{-sT}}{s}$ 4. $\dfrac{1}{s^2(s+1)}$

2. $\dfrac{1}{s}e^{-sT}$ 5. $\dfrac{s}{s^2 + 1}$

3. $\dfrac{1}{s(s+1)^2}$ 6. $\dfrac{(s+1)^2 - 1}{[(s+1)^2 + 1]^2}$

b) In each case find $x(t)$ and show that the results of the theorems agree.

Problem 2.5

a) Find the Laplace transform of each of the time functions shown below. (HINT: It may sometimes be easier to find the transform of the derivative of the given function and use the Integration Theorem to find the desired transform.)

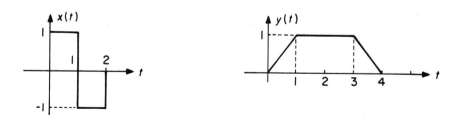

b) Show that your results are consistent with the Initial and Final Value Theorems of Problem 2.4.

Problem 2.6

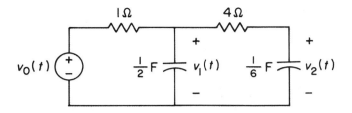

a) Use impedance methods to find $V_2(s) = \mathcal{L}[v_2(t)]$ in terms of $V_0(s) = \mathcal{L}[v_0(t)]$, $v_1(0)$, and $v_2(0)$.

b) Use the Differentiation and Initial Value Theorems to evaluate $\dot{v}_2(0) = \dfrac{dv_2(t)}{dt}\bigg|_{t=0}$ from the answer to (a). Express your answer in terms of $v_1(0)$ and $v_2(0)$. Show that the result does not depend on $v_0(t)$ as long as $v_0(0)$ is finite.

c) Check your answer to (b) directly from the circuit. (HINT: How is the current in the 4 Ω resistor related to $\dot{v}_2(t)$?)

d) Find by any means the input-output differential equation relating $v_2(t)$ and $v_0(t)$. Transform this equation to check the result obtained in (a). (HINT: To transform the second-order equation you may find it useful to derive first the general relation

$$\mathcal{L}\left[\frac{d^2 x(t)}{dt^2}\right] = s^2 X(s) - sx(0) - \dot{x}(0). \ )$$

e) Find $v_2(t)$, $t > 0$, if $v_0(t) = e^{-2t}$ volt, $v_2(0) = 1$ volt, $\dot{v}_2(0) = -2$ volts/sec.

Problem 2.7

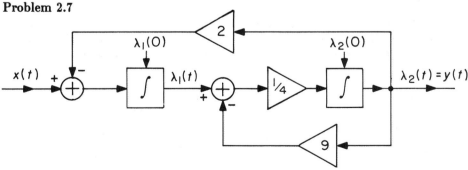

The block diagram above describes the setup of an analog computer for solving a particular problem.

a) Write the dynamic equations for this system in state form.

b) Transform and solve these equations to obtain an expression for $Y(s)$ in terms of $X(s)$, $\lambda_1(0)$, and $\lambda_2(0)$.

c) Find $y(t)$, $t > 0$, for $x(t) = 2$, $t > 0$, $\lambda_1(0) = 4$, and $\lambda_2(0) = -0.5$.

d) Repeat (c) for the same initial state and the input $x(t) = 7e^{-2t}$, $t > 0$.

Problem 2.8

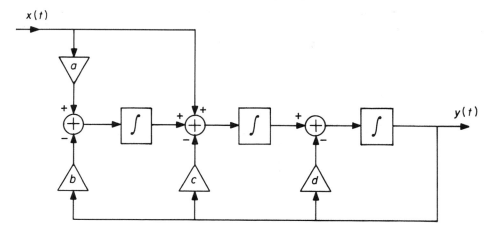

Determine how the transform $Y(s)$ of $y(t)$ is related to the transform $X(s)$ of $x(t)$ if the system above is in the zero state at $t = 0$.

Problem 2.9

Stated loosely, the *Maximum Principle of Pontryagin* says that to take a system from one state to another in the shortest possible time, subject to constraints on the magnitude of certain variables, one should operate continuously at the extremes, shifting from one extreme condition to another in a systematic way dependent on the initial and final states. For example, to take your car from rest at one place to rest at another in the shortest possible time you should (obviously) apply maximum acceleration up to the last possible moment such that maximum braking will just bring you to a screeching halt at the desired spot. Control systems of this sort are often rather picturesquely called "bang-bang" systems. As an example, consider a second-order system characterized by state variables $\lambda_1(t)$ and $\lambda_2(t)$ satisfying the state equations

$$\frac{d\lambda_1(t)}{dt} = \lambda_2(t)$$

$$\frac{d\lambda_2(t)}{dt} = -3\lambda_2(t) - 2\lambda_1(t) + x(t)$$

$$y(t) = \lambda_1(t)$$

where $x(t)$ and $y(t)$ are the input and output, respectively.

a) Using $\mathcal{L}$-transforms, solve for $Y(s) = \Lambda_1(s) = \mathcal{L}[\lambda_1(t)]$ and $\Lambda_2(s) = \mathcal{L}[\lambda_2(t)]$ in terms of $X(s) = \mathcal{L}[x(t)]$, $\lambda_1(0)$, and $\lambda_2(0)$.

b) Consider the particular initial state $\lambda_1(0) = 1$, $\lambda_2(0) = 2$. Solve for $y(t)$ if $x(t) = 0$, $t > 0$. Under these conditions, how long would it take the system to come essentially to rest? For example, how long would it take for both the state variables to become less than 10% of their initial values?

c) Show that the system can be brought to rest at time T_2 (that is, $y(t) \equiv 0$, all $t > T_2$) by the input shown in the figure if the values of T_1 and T_2 are properly chosen. Find T_1 and T_2, and compare the resulting value of T_2 with the time to come essentially to rest when $x(t) = 0$, $t > 0$.

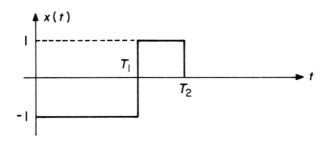

Problem 2.10

The useful application domain for $\mathcal{L}$-transforms is normally considered to be restricted to LTI systems. But the solution of some time-varying differential equations is also simpler in the frequency domain. An example is *Bessel's equation*

$$t\frac{d^2x(t)}{dt^2} + \frac{dx(t)}{dt} + tx(t) = 0.$$

a) Use the Differentiation and Multiply-by-t Theorems to transform this equation into a first-order differential equation in $X(s) = \mathcal{L}[x(t)]$.

b) The equation in $X(s)$ can be integrated by separation of variables. Show (by substitution) that a solution is

$$X(s) = \frac{C}{\sqrt{s^2 + 1}}$$

where C is an arbitrary constant.

c) The zero-order Bessel function $x(t) = J_0(t)$ is a solution of the original equation. Indeed, the irrational function obtained in (b)—with an appropriate choice of C— is $\mathcal{L}[J_0(t)]$. $J_0(0)$ is normally defined to be 1; what is the appropriate choice of C?

Problem 2.11

This problem explores features of a model for freeway traffic. Experimentation indicates that the acceleration of a car in heavy traffic depends mostly on its velocity relative to the car immediately in front of it. Specifically, assume that

$$\frac{dv_1(t)}{dt} = k[v_0(t) - v_1(t)]$$

where $v_0(t)$ = velocity of leading car

$v_1(t)$ = velocity of following car

k = constant measuring the sensitivity of the driver of the following car. (Typically $0.2 < k < 0.8 \text{ sec}^{-1}$.)

a) Determine the $\mathcal{L}$-transform ratio $\dfrac{V_1(s)}{V_0(s)}$ assuming $v_1(0) = 0$.

b) Assume that both cars are initally at rest. Let $v_0(t) = t$, $t > 0$. Find $v_1(t)$. Sketch $v_1(t)$ for $k = 1$.

c) Consider now a string of similar cars following one another in single file with velocities $v_0(t)$, $v_1(t)$, $v_2(t)$, etc. Assume also a more realistic relationship between the velocities of the nth and $(n-1)^{\text{st}}$ cars:

$$\frac{dv_n(t)}{dt} = k[v_{n-1}(t - T) - v_n(t - T)]$$

where the reaction time, T, typically is 1-2 seconds in practice. Let the velocity of the lead car be

$$v_0(t) = \cos 2\pi f_0 t + V$$

where V is the common average velocity with which the whole file moves. Derive an expression for the sinusoidal steady-state velocity component of the nth car.

d) For the conditions of (c), show that your result implies

$$|v_n(t) - V| \le \frac{1}{\left|1 + \dfrac{j2\pi f_0}{k} e^{j2\pi f_0 T}\right|^n}.$$

Show from this that the amplitude of the velocity variations of the nth car decreases with increasing n for all f_0 if and only if $kT < 0.5$ (a condition that requires the drivers to react reasonably fast but not to be unduly sensitive). What would be the consequences of $kT > 0.5$?

3

SYSTEM FUNCTIONS

3.0 Introduction

The examples studied in the preceding chapter suggest that the frequency-domain description of the response of any lumped LTI circuit has a certain simple structure. In this chapter and the next we shall study this structure in some detail. Our study will not necessarily make it easier for us to determine the voltages and currents in any specific circuit problem. But our goals in this book go well beyond merely presenting efficient problem-solving techniques. Specifically, we hope to develop the insight and understanding necessary for the design of complex systems. For this, a full appreciation of the general properties of circuit behavior is even more important than skill at detailed circuit analysis.

3.1 A Superposition Formula for LTI Circuits

Example 2.5–2 led to a formula of the form

$$V_1(s) = H_1(s)I_0(s) + H_2(s)V_0(s) + H_3(s)\frac{v_C(0)}{s} + H_4(s)\frac{i_L(0)}{s} \qquad (3.1-1)$$

for the transform $V_1(s)$ of the voltage across a particular pair of terminals in terms of the transforms $I_0(s)$ and $V_0(s)$ of two external sources and the transforms $v_C(0)/s$ and $i_L(0)/s$ of sources replacing the initial capacitor voltage and inductor current. The four functions $H_1(s)$, $H_2(s)$, $H_3(s)$, and $H_4(s)$, which relate the sources to $V_1(s)$, were derived by impedance methods.

It should be immediately evident from the Superposition Theorem of linear resistive circuit theory that the form of this result is general. That is, for any lumped LTI circuit we can always write a generalized *superposition formula*

$$Y(s) = \sum_{m=1}^{M} H_{em}(s)X_m(s) + \sum_{n=1}^{N} H_{in}(s)\frac{\lambda_n(0)}{s} \qquad (3.1-2)$$

where

$Y(s)$ = $\mathcal{L}$-transform of the circuit voltage or current that is the object of our analysis, the one that we choose to designate as the *output* or *response* of the circuit;

$X_m(s)$ = $\mathcal{L}$-transform of the mth independent external voltage or current source considered as an *input* or *stimulus* to the circuit; M is the number of external sources;

$\dfrac{\lambda_n(0)}{s}$ = $\mathcal{L}$-transform of the source describing the effect of the value $\lambda_n(0)$ of the nth state variable at $t = 0$ (typically, a capacitor voltage or inductor current); N is the *order* of the system;

$H_{em}(s), H_{in}(s)$ = functions of s relating each external source or initial condition source (respectively) to the output.

Equation (3.1−2) is the general form of the functional or operator description of LTI systems in the frequency domain that was promised in Section 2.1.

The two summation terms on the right in (3.1−2) are given separate names:

$$\sum_{n=1}^{N} H_{in}(s)\frac{\lambda_n(0)}{s} = \text{Zero Input Response (ZIR)} \qquad (3.1{-}3)$$

$$\sum_{m=1}^{M} H_{em}(s)X_m(s) = \text{Zero State Response (ZSR).} \qquad (3.1{-}4)$$

The ZIR term (loosely, the "free" or "natural" response) is not a function of inputs in the interval $t \geq 0$; it is determined by the initial state at $t = 0$, which in turn depends on inputs for $t < 0$. If all the external input sources are zero for $t \geq 0$, that is, if $X_m(s) = 0$, all m, then the ZIR is the total response for $t \geq 0$. On the other hand, the ZSR term (loosely, the "forced" or "driven" response) is not a function of the initial state. In particular, if the initial state is the *zero state*,* that is, if $\lambda_n(0) = 0$, all n, then the ZSR is the total response for all $t \geq 0$. The generalized superposition formula thus states that the total output at any time $t_0 \geq 0$ is a sum of the ZIR—the continuing effects at t_0 of inputs for $t < 0$—and the ZSR—the effects at t_0 of the inputs in the interval $0 \leq t < t_0$.

Note that we shall use the words "input" and "output" to refer either to the time functions, $x(t)$ and $y(t)$, or to their transforms, $X(s)$ and $Y(s)$; by the Uniqueness Theorem, these are alternative ways of describing the same things— the input or drive and the output or response. Note also that the ZIR and ZSR terms are not at all the same things as the "transient response" and the "steady-state response." Even if the input is a constant or a continuing sinusoid, so that "transient" and "steady-state" have unambiguous meanings, the ZSR term in general contains "transient" as well as "steady-state" components. We shall explore the relationships among these various response components more carefully in Section 3.3.

---

*It is perhaps worth pointing out that the zero state is a unique state for LTI systems in that it is independent of the choice of state variables (which is not unique).

3.2 System Functions

The separate elements making up the summations on the right in (3.1–2) all have the same structure—a product of the transform of a source $(X_m(s)$ or $\lambda_n(0)/s)$ and a function of s derived from the network $(H_{em}(s)$ or $H_{in}(s))$. Thus we can interpret the factor $H_{em}(s)$ or $H_{in}(s)$ in each elementary term as the ratio of the L-transform of the response component to the L-transform of the source producing that component. Such a ratio of the transform of a response to the transform of a source is called a *system function*. System functions in electrical systems are classified as *driving-point* or *transfer* functions, accordingly as they relate voltages and currents at, respectively, the same or different *ports*.* They may be dimensionless *ratios* (voltage/voltage or current/current) or may have the dimensions of *impedance* (voltage/current) or *admittance* (current/voltage). Of course, since system functions may be used to relate drives and responses that are not voltages and currents—and not even electrical quantities—the range of possible dimensions is limitless.

Note, however, that system functions are always defined as output divided by input. To see why it is important to emphasize this point, consider the following example.

Example 3.2–1

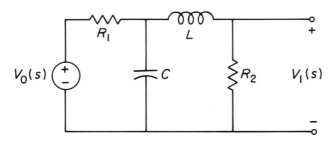

Figure 3.2–1. Circuit for Example 3.2–1.

The circuit shown in Figure 3.2–1 is part of the situation considered in Example 2.5–2 where we concluded that the ZSR term relating the output $V_1(s)$ and the input $V_0(s)$ is

$$V_1(s) = \frac{\dfrac{R_2}{R_1LC}}{s^2 + s\left(\dfrac{1}{R_1C} + \dfrac{R_2}{L}\right) + \dfrac{1}{LC}\left(1 + \dfrac{R_2}{R_1}\right)} V_0(s). \tag{3.2–1}$$

The system function is then given by

$$H_1(s) = \frac{\text{output}}{\text{input}} = \frac{V_1(s)}{V_0(s)} = \frac{\dfrac{R_2}{R_1LC}}{s^2 + s\left(\dfrac{1}{R_1C} + \dfrac{R_2}{L}\right) + \dfrac{1}{LC}\left(1 + \dfrac{R_2}{R_1}\right)}. \tag{3.2–2}$$

---

*For a discussion of the concept of a *port* (loosely, a pair of terminals) see the appendix to this chapter.

$H_1(s)$ is a (dimensionless) transfer function since $V_1(s)$ and $V_0(s)$ are defined at different ports.

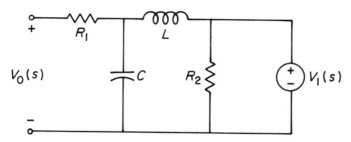

Figure 3.2–2. Circuit of Figure 3.2–1 with source at right.

Suppose, however, we were to drive this circuit with a voltage source at the right-hand port and measure the voltage response at the left as shown in Figure 3.2–2. We readily compute that in this case (which was not considered in Example 2.5–2) the ZSR term in $V_0(s)$ (which now represents the output) determined by $V_1(s)$ (which now represents the input) is

$$V_0(s) = \frac{\dfrac{1}{Cs}}{\dfrac{1}{Cs} + Ls} V_1(s) = \frac{\dfrac{1}{LC}}{s^2 + \dfrac{1}{LC}} V_1(s) \tag{3.2–3}$$

which corresponds to the system function

$$H_2(s) = \frac{\text{output}}{\text{input}} = \frac{V_0(s)}{V_1(s)} = \frac{\dfrac{1}{LC}}{s^2 + \dfrac{1}{LC}}. \tag{3.2–4}$$

Note that $H_2(s)$ is a ratio of the voltage across the left-hand pair of terminals to the voltage across the right-hand pair; so is $1/H_1(s)$ as defined above. But $1/H_1(s)$ and $H_2(s)$ are totally different!

▶ ▶ ▶

The point of this example is to show that it is generally necessary to identify not only what pair of variables in a circuit are related by a system function, but also which variable is the source and which the response. The simplest way to make this evident is to adopt the convention that system functions are always defined as output divided by input. Consequently, the reciprocal of a system function is not necessarily a system function. There is, however, one very important case in which a system function correctly describes the relationship between two variables irrespective of which is the drive and which the response, as illustrated in the next example.

Example 3.2–2

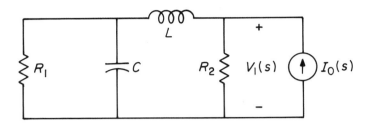

Figure 3.2–3. Circuit for Example 3.2–2.

The circuit shown in Figure 3.2–3 is also part of the situation considered in Example 2.5–2, where we concluded that the ZSR term relating the output $V_1(s)$ and the input $I_0(s)$ is

$$V_1(s) = \frac{R_2\left(s^2 + \dfrac{s}{R_1 C} + \dfrac{1}{LC}\right)}{s^2 + s\left(\dfrac{1}{R_1 C} + \dfrac{R_2}{L}\right) + \dfrac{1}{LC}\left(1 + \dfrac{R_2}{R_1}\right)} I_0(s). \tag{3.2–5}$$

The system function is then given by

$$H_3(s) = \frac{\text{output}}{\text{input}} = \frac{V_1(s)}{I_0(s)} = \frac{R_2\left(s^2 + \dfrac{s}{R_1 C} + \dfrac{1}{LC}\right)}{s^2 + s\left(\dfrac{1}{R_1 C} + \dfrac{R_2}{L}\right) + \dfrac{1}{LC}\left(1 + \dfrac{R_2}{R_1}\right)}. \tag{3.2–6}$$

$H_3(s)$ is a driving-point impedance since $V_1(s)$ and $I_0(s)$ are the voltage and current at the same port.

Suppose, however, we were to drive this circuit with a voltage source instead of a current source and measure the current response instead of the voltage, as shown in Figure 3.2–4.

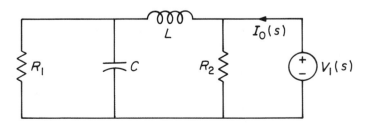

Figure 3.2–4. Circuit of Figure 3.2–3 driven by a voltage source.

We readily compute

$$I_0(s) = \frac{V_1(s)}{R_2} + \frac{V_1(s)}{Ls + \dfrac{R_1/Cs}{R_1 + 1/Cs}}$$

$$= \frac{s^2 + s\left(\dfrac{1}{R_1C} + \dfrac{R_2}{L}\right) + \dfrac{1}{LC}\left(1 + \dfrac{R_2}{R_1}\right)}{R_2\left(s^2 + \dfrac{s}{R_1C} + \dfrac{1}{LC}\right)} V_1(s). \tag{3.2--7}$$

The system function is then given by

$$H_4(s) = \frac{\text{output}}{\text{input}} = \frac{I_0(s)}{V_1(s)} = \frac{s^2 + s\left(\dfrac{1}{R_1C} + \dfrac{R_2}{L}\right) + \dfrac{1}{LC}\left(1 + \dfrac{R_2}{R_1}\right)}{R_2\left(s^2 + \dfrac{s}{R_1C} + \dfrac{1}{LC}\right)} \tag{3.2--8}$$

and is a driving-point admittance.

▶ ▶ ▶

Note that in this example, unlike the situation with transfer system functions, interchanging which variable is considered the input and which the output simply inverts the system function. This is evidently a general property of driving-point impedances and admittances. Can you explain why driving-point and transfer system functions are different in this respect? The driving-point impedance or admittance characterizes the behavior of a 2-terminal (single-port) LTI network no matter how it is driven or connected externally. Similar comprehensive descriptions are possible for multiterminal LTI networks designed for their transfer properties, but such descriptions require more than the specification of a single transfer function (see the appendix to this chapter).

3.3 System Functions as Response Amplitudes to Exponential Drives

As explained in Chapter 2, system functions are readily computed by impedance methods with L and C replaced by impedances Ls and $1/Cs$, and the network then solved as if it were a resistive circuit. This same procedure—with s replaced by $j\omega$—is used (as you know from earlier studies) to find the *sinusoidal steady-state frequency response* of a network. Thus, for $s = j\omega$, the system function $H(s)$ has the following interpretation:

> If the input to an LTI network is the complex exponential $Xe^{j\omega t}$ and the steady-state output is the complex exponential $Ye^{j\omega t}$, then
>
> $$Y/X = H(j\omega)$$
>
> where $H(s)$ is the system function relating the input and output.

In other words, $e^{j\omega t}$ as a drive gives $H(j\omega)e^{j\omega t}$ as the "steady-state" response after the "transients" have died away. Indeed, this result is commonly used as the basis for the experimental measurement of $H(j\omega)$ for an LTI system whose internal structure is unknown, concealed inside a "black box." We now show with an example that $H(s)$ has a similar interpretation even when $s \neq j\omega$, so that $e^{st} = e^{\sigma t}e^{j\omega t}$ cannot correspond to any straightforward notion of "steady-state" (since $e^{\sigma t}$, $\sigma \neq 0$, either grows or decays with time).

Example 3.3–1

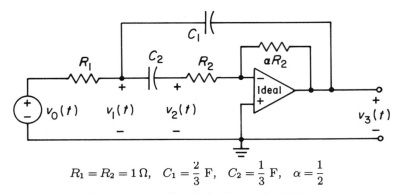

$$R_1 = R_2 = 1\,\Omega, \quad C_1 = \frac{2}{3}\,\text{F}, \quad C_2 = \frac{1}{3}\,\text{F}, \quad \alpha = \frac{1}{2}$$

Figure 3.3–1. Circuit for Example 3.3–1.

In Problem 3.1, you will show that $H(s)$ for the circuit in Figure 3.3–1 is

$$H(s) = \frac{V_3(s)}{V_0(s)} = \frac{\dfrac{-\alpha}{(1+\alpha)R_1 C_1}s}{s^2 + \dfrac{s}{1+\alpha}\left(\dfrac{1}{R_1 C_1} + \dfrac{1}{R_2 C_2} + \dfrac{1}{R_2 C_1}\right) + \dfrac{1}{(1+\alpha)R_1 C_1 R_2 C_2}} \qquad (3.3\text{–}1)$$

which for the given values becomes

$$H(s) = \frac{-s/2}{s^2 + 4s + 3} = \frac{-s/2}{(s+1)(s+3)}. \qquad (3.3\text{–}2)$$

Now let $v_0(t) = e^{s_0 t}$, $t > 0$, where s_0 is an arbitrary complex number. Then $\mathcal{L}[v_0(t)] = V_0(s) = 1/(s - s_0)$, and under ZSR conditions

$$V_3(s) = H(s)V_0(s) = \frac{-s/2}{(s+1)(s+3)(s - s_0)}. \qquad (3.3\text{–}3)$$

If $s_0 \neq -1, -3$, we may expand

$$V_3(s) = \frac{\dfrac{-1}{4(1 + s_0)}}{s+1} + \frac{\dfrac{3}{4(3 + s_0)}}{s+3} + \frac{\dfrac{-s_0/2}{(s_0 + 1)(s_0 + 3)}}{s - s_0} \qquad (3.3\text{–}4)$$

where we recognize the residue in the $(s - s_0)$ term as $H(s_0)$ so that the ZSR is

$$v_3(t) = -\frac{1}{4(1+s_0)}e^{-t} + \frac{3}{4(3+s_0)}e^{-3t} + H(s_0)e^{s_0 t}, \quad t > 0. \qquad (3.3\text{--}5)$$

If s_0 is purely imaginary, $s_0 = j\omega_0$, this result describes the usual sinusoidal-drive situation—the first two terms are "transients" resulting from the sudden application to a resting network of the drive $e^{j\omega_0 t}$ at $t = 0$; in time they die away, leaving the "steady-state" response $H(j\omega_0)e^{j\omega_0 t}$. But even if $s_0 \neq j\omega_0$, the last term—the "driven" term—in the expression for $v_3(t)$ will come to dominate after a while provided that $\Re e[s_0] > -1$. (If $\Re e[s_0] < -1$, the "driven" term vanishes more rapidly than the "transient" terms; the latter thus ultimately become relatively more important, although all three terms may be decaying. Note also that the word "driven" here in quotes means simply that term having the same form as the drive—that is, $H(s_0)e^{s_0 t}$. The total driven (zero-state) response contains both the "driven" term and "transient" terms.)

▶ ▶ ▶

The result obtained in Example 3.3–1 is general:

> If the input to an LTI system has the form $e^{s_0 t}$, then the output will become predominantly $H(s_0)e^{s_0 t}$ as time passes, provided that s_0 lies in that part of the s-plane to the right of the rightmost pole of $H(s)$. This region is called the *domain of convergence* for $H(s)$.

This important observation illustrates the complete way in which $H(s)$ is a generalization of $H(j\omega)$, and justifies calling s the *complex frequency*.

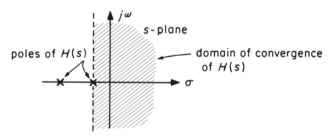

Figure 3.3–2. Domain of convergence.

3.4 System Functions and the Input-Output Differential Equation

There is a complete and close relationship between the system function $H(s)$ and the input-output differential equation obtained from the node or state differential equations by eliminating all of the unknown variables except the output. Several examples will make the general relationship evident.

Example 3.4–1

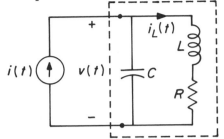
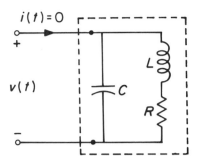

Figure 3.4–1. Circuit for Example 3.4–1.

Differential state equations for the circuit to the left in Figure 3.4–1 in terms of $v(t)$ and $i_L(t)$, the capacitor voltage and inductor current, are

$$L\frac{di_L(t)}{dt} = -Ri_L(t) + v(t) \tag{3.4–1}$$

$$C\frac{dv(t)}{dt} = i(t) - i_L(t). \tag{3.4–2}$$

Solving (3.4–2) for $i_L(t)$ and substituting into (3.4–1) give the input-output equation in terms of the drive $i(t)$ and the response $v(t)$:

$$\frac{d^2v(t)}{dt^2} + \frac{R}{L}\frac{dv(t)}{dt} + \frac{1}{LC}v(t) = \frac{1}{C}\left(\frac{di(t)}{dt} + \frac{R}{L}i(t)\right). \tag{3.4–3}$$

The natural frequencies of this circuit—that is, the frequencies present in $v(t)$ when $i(t) = 0$ and the terminals are open-circuit as shown to the right in Figure 3.4–1—are the roots of the characteristic equation

$$s^2 + \frac{R}{L}s + \frac{1}{LC} = 0 \tag{3.4–4}$$

derived from the left-hand side of the input-output equation.

The system function relating the drive and the response is the driving-point impedance

$$Z(s) = \frac{V(s)}{I(s)} = \frac{\text{output}}{\text{input}} \tag{3.4–5}$$

which is readily found by series and parallel impedance arguments to be

$$Z(s) = \frac{\frac{1}{Cs}(Ls + R)}{\frac{1}{Cs} + Ls + R} = \frac{\frac{1}{C}\left(s + \frac{R}{L}\right)}{s^2 + \frac{R}{L}s + \frac{1}{LC}} = \frac{V(s)}{I(s)}. \tag{3.4–6}$$

Cross-multiplying gives

$$\left(s^2 + \frac{R}{L}s + \frac{1}{LC}\right)V(s) = \frac{1}{C}\left(s + \frac{R}{L}\right)I(s). \tag{3.4–7}$$

Comparison with the input-output differential equation shows that here, and in general, to derive the input-output differential equation from the system function it is only necessary to cross-multiply and identify $d/dt \leftrightarrow s$. Often, indeed, the easiest way to derive the input-output differential equation is to use impedance methods first to find the system function. The tricky process of eliminating intermediate variables and their derivatives is thus much simplified. Algebraic equations are easier to manipulate than differential equations.

The relationship between the system function and the input-output differential equation can also be deduced in the opposite direction by exploiting the result derived in Section 3.3—that e^{st} as an input gives ultimately the output $H(s)e^{st}$ for s in the domain of convergence. Thus if we substitute $i(t) = e^{st}$ and $v(t) = H(s)e^{st}$ into the input-output differential equation above, we obtain

$$H(s)s^2 e^{st} + H(s)\frac{R}{L}se^{st} + H(s)\frac{1}{LC}e^{st} = \frac{1}{C}\left(se^{st} + \frac{R}{L}e^{st}\right). \qquad (3.4\text{--}8)$$

Solving for $H(s)$ gives the same result $H(s) = Z(s)$ as derived above by impedance methods.*

▶ ▶ ▶

An important conclusion to be derived from the relationship between the system function and the input-output differential equation is that the natural frequencies of the circuit are the roots of the denominator polynomial, that is, the poles of $H(s)$.[†] Loosely, the output $Y(s) = H(s)X(s)$ can be finite when the input $X(s)$ is zero only if $H(s)$ is infinite, and this happens only for the values of s that are the poles of $H(s)$.

Example 3.4–2

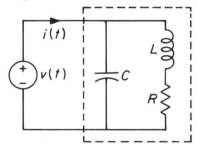
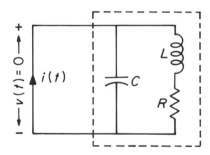

Figure 3.4–2. Circuit for Example 3.4–2.

*Attempts to derive the form of the ZSR and $H(s)$ from the input-output differential equation by direct application of the $\mathcal{L}$-transform Differentiation Theorem can lead to difficulties. See Problem 3.3.

[†]If the numerator and denominator polynomials of $H(s)$ contain a common factor, that is, if a zero of $H(s)$ cancels a pole, the relationship between $H(s)$ and the ZIR is less close. Specifically, the ZIR may contain a term whose complex frequency is not a pole of $H(s)$. However, there is no signal $x(t)$, $t < 0$, that when applied to the normal input corresponding to $H(s)$, will generate such a ZIR term for $t > 0$. On the other hand, if such a ZIR term is excited, for example, by driving the network at some other input, then there is no input that can be applied at the normal input that will cancel the effects of this ZIR term in a finite time; such a system is said to be *uncontrollable*. For an example, see Exercise 4.5.

Note that the poles of $H(s)$ are the natural frequencies under the condition that the input is zero. If the input is a current source as in Example 3.4–1, then zero input implies that the input terminals are open-circuit. But if the input is a voltage source, zero input implies that the input terminals are shorted. Thus consider (as shown in Figure 3.4–2) the same circuit as before but driven by a voltage source $v(t)$; the response will now be taken to be the current $i(t)$. The input-output roles of $v(t)$ and $i(t)$ are interchanged. Hence the characteristic equation is

$$\frac{1}{C}\left(s + \frac{R}{L}\right) = 0. \tag{3.4–9}$$

The root of the characteristic equation, $s = -R/L$, defines the functional form of the current, $i(t) = Ie^{-Rt/L}$, that can flow under short-circuit conditions, that is, $v(t) = 0$, as shown to the right in Figure 3.4–2. The system function in this case is the driving-point admittance*

$$Y(s) = \frac{I(s)}{V(s)}. \tag{3.4–10}$$

Of course,

$$Y(s) = \frac{1}{Z(s)} = \frac{C\left(s^2 + \dfrac{R}{L}s + \dfrac{1}{LC}\right)}{s + \dfrac{R}{L}}. \tag{3.4–11}$$

The natural frequency is the pole of $Y(s)$, or the zero of $Z(s)$, under short-circuit conditions.

▶ ▶ ▶

The observation that the poles of the system function $H(s)$ are the natural frequencies of the circuit is, of course, consistent with the fact that

$$Y(s) = H(s)X(s) \tag{3.4–12}$$

is the $\mathcal{L}$-transform of the ZSR output $y(t)$ to the input $x(t)$. If $Y(s) = \mathcal{L}[y(t)]$ is expanded in partial fractions, the poles of $H(s)$ yield terms describing that part of the response whose form is determined by the circuit rather than the drive.

3.5 Summary

The response of a linear time-invariant circuit can always be interpreted as the sum of the zero-state response and the zero-input response. Each of these components is in turn a superposition of terms describing the separate effects of each of the external sources and each of the internal sources reflecting the initial state. Each term has, in the frequency domain, the form of a product of the $\mathcal{L}$-transform of the nth source times a function of s, $H_n(s)$, called the system

---

*The word "admittance" was coined by Heaviside to describe the ratio of a current to a voltage. Regrettably, the symbol $Y(s)$ for admittance is well-established. Note the distinction in this book between $Y(s)$ as the admittance in the electric-circuit example leading to (3.4–11) and $Y(s)$ as the $\mathcal{L}$-transform of $y(t)$ in a general formula such as (3.4–12).

function, which relates the response to that particular source or drive. System functions are usually easy to determine by impedance methods from a structural description of the circuit. If the particular drive has the form $e^{s_0 t}$, then the response component will approach $H_n(s_0)e^{s_0 t}$ after a while if s_0 is in the domain of convergence for $H_n(s)$, that is, in the region to the right of the rightmost pole of $H_n(s)$. Since the poles of $H_n(s)$ are the natural frequencies of the system, $e^{s_0 t}$ under these conditions decays more slowly (or grows more rapidly) than the natural response terms, and this explains its eventual dominance. Finally, $H_n(s)$ and the input-output differential equation relating the response to that particular source contain essentially identical information. Thus, in general, the system function summarizes everything there is to know about the input-output behavior of an LTI system.

APPENDIX TO CHAPTER 3

System Function Characterization of LTI 2-Ports

As illustrated in Example 3.2–1, a system function describes the behavior of a network only under certain specified source and termination conditions. Complete "black-box" descriptions of multiterminal LTI networks adequate to characterize the behavior of the network under any conditions of drive and load are possible, however, and are often extremely useful. An important special case is discussed in this appendix.

Consider a network composed of linear time-invariant R's, L's, and C's, ideal transformers, controlled sources, etc., enclosed in a box so that it is accessible only through four terminals. Suppose further that the external connections to the 4-terminal network constrain the terminal currents to be *paired*—the currents in each pair being equal in magnitude and opposite in sign. Such a 4-terminal network is called a 2-*port*.* One external arrangement that ensures 2-port behavior is shown in Figure 3.A–1. Obviously many other possibilities exist; a sufficient condition is that there be no connection between the external networks connected to ports 1 and 2 except through the 2-port. The description of a 4-terminal network as a 2-port obviously does not characterize the behavior of the network under all external conditions, but it is adequate for many purposes.

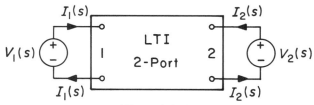

Figure 3.A–1.

Even if we do not know what the detailed circuit arrangements may be inside an LTI 2-port (that is, if we are forced to consider the 2-port as a "black box" accessible only through electrical measurements made at its terminal pairs), we can still conclude

---

*Extending this terminology, a network connected to the rest of the world through n paired terminals is called an *n-port*. A 2-terminal network is always a 1-port, since no matter how it is connected, KCL guarantees that the currents at the two terminals are paired. A 3-terminal network may always be represented as a 2-port without loss of generality, as shown in Figure 3.A–2.

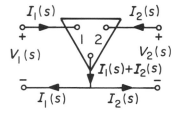
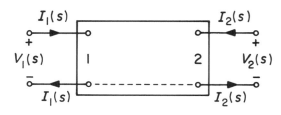

Figure 3.A–2.

from superposition and impedance arguments such as discussed in this chapter that the ZSR currents $I_1(s)$ and $I_2(s)$ must be given by equations of the form

$$I_1(s) = Y_{11}(s)V_1(s) + Y_{12}(s)V_2(s)$$
$$I_2(s) = Y_{21}(s)V_1(s) + Y_{22}(s)V_2(s)$$

in terms of the external voltage sources $V_1(s)$ and $V_2(s)$. The various system functions $Y_{ij}(s)$ are called the *short-circuit driving-point* and *transfer admittances* of the 2-port because they can, at least in principle, be inferred from measurements made on the circuit with one or the other of the terminal pairs shorted, as shown in Figure 3.A–3.

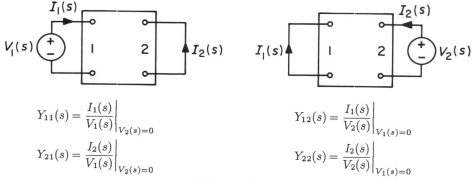

$$Y_{11}(s) = \left.\frac{I_1(s)}{V_1(s)}\right|_{V_2(s)=0} \qquad\qquad Y_{12}(s) = \left.\frac{I_1(s)}{V_2(s)}\right|_{V_1(s)=0}$$

$$Y_{21}(s) = \left.\frac{I_2(s)}{V_1(s)}\right|_{V_2(s)=0} \qquad\qquad Y_{22}(s) = \left.\frac{I_2(s)}{V_2(s)}\right|_{V_1(s)=0}$$

Figure 3.A–3.

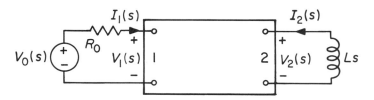

Figure 3.A–4.

The short-circuit admittances characterize the ZSR behavior of the 2-port under any external conditions that satisfy the paired-current condition. For example, if the 2-port is loaded with an inductor at port 2 and driven from a circuit with Thévenin parameters as shown in Figure 3.A–4, then the source and load impose the conditions

$$V_1(s) = V_0(s) - I_1(s)R_0$$
$$V_2(s) = -LsI_2(s).$$

Combining with the two short-circuit admittance equations, we may eliminate $V_1(s)$, $I_1(s)$, and $I_2(s)$ to obtain the overall transfer ratio under these conditions

$$\frac{V_2(s)}{V_0(s)} = \frac{-\dfrac{1}{R_0}Y_{21}(s)}{\left(\dfrac{1}{R_0}+Y_{11}(s)\right)\left(\dfrac{1}{Ls}+Y_{22}(s)\right) \;-\; Y_{12}(s)Y_{21}(s)}.$$

(See Problem 3.6 for another way of deriving this result. If the 2-port is reciprocal, the derivation is even simpler; see Problem 3.5. Many alternate methods of characterizing a 2-port are possible, one or another of which may be simpler in a particular situation. See Problems 3.7 and 3.8 for examples.)

If limitations are imposed on the kinds of elements out of which the 2-port is constructed, then the $Y_{ij}(s)$ will in general have to satisfy certain conditions—some of which will be discussed in Chapter 4. One of the more interesting conditions is *reciprocity*,

$$Y_{12}(s) = Y_{21}(s)$$

which is guaranteed if the 2-port contains LTI R's, L's, C's, and transformers, but no controlled sources. For a further discussion of reciprocity, see Problem 3.5.

EXERCISES FOR CHAPTER 3

Exercise 3.1

Consider the circuit below in which the current source is the input and the voltage across the resistor R is the output:

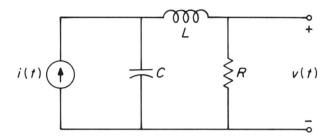

a) Show that the system function relating the input and output is

$$H(s) = \frac{\dfrac{R}{LC}}{s^2 + \dfrac{R}{L}s + \dfrac{1}{LC}}.$$

What are the dimensions of the separate terms in the denominator and numerator? What are the dimensions of $H(s)$?

b) Determine the input-output differential equation.

c) Show that the form of the ZIR is

$$v(t) = Ae^{-t/2}\cos\left(\frac{\sqrt{3}t}{2} + \theta\right)$$

if $R = 1\,\Omega$, $L = 1$ H, $C = 1$ F.

Exercise 3.2

Use impedance methods to derive the input-output differential equation for the circuit below and thus show that it behaves as a *double integrator*, that is, $v_1(t) \sim d^2 v_2(t)/dt^2$.

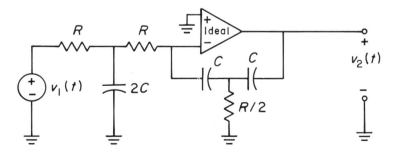

Exercise 3.3

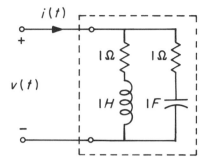 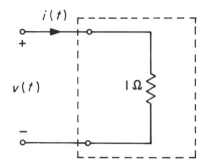

a) Show that the driving-point impedance of the network on the left is identical for all s to that on the right; that is,

$$Z(s) = \frac{V(s)}{I(s)} = 1.$$

b) Many years ago Joseph Slepian, writing in the *Transactions* of the old American Institute of Electrical Engineers, proposed as a puzzle the description of a test to be performed solely at the electrical terminals that would permit you to discover which of the two "black boxes" above you had been handed. A flood of letters resulted and the argument went on for months. Some writers tried to prove that no successful test was possible; others maintained that under certain excitation conditions the circuits would behave differently. What is your position?

Exercise 3.4

Experiments on an LTI system lead to the following conclusions:

a) Independent of the state of the system at $t = 0$, an input $x(t) = e^{-2t}$, $t > 0$, yields an output of the form $y(t) = 3e^{-2t} + (k_0 + k_1 t)e^{-t}$, $t > 0$;

b) Independent of the state of the system at $t = 0$, an input $x(t) = e^{-3t}$, $t > 0$, yields an output of the form $y(t) = (k_2 + k_3 t)e^{-t}$, $t > 0$.

Argue that, if the system function $H(s)$ is a proper fraction ($H(s) \to 0$ as $s \to \infty$), it must be

$$H(s) = \frac{3(s+3)}{(s+1)^2}.$$

PROBLEMS FOR CHAPTER 3

Problem 3.1

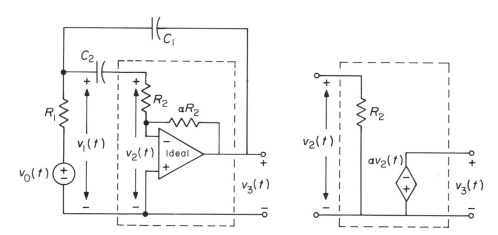

a) Argue that the dashed section in the circuit above is equivalent to the circuit shown to the right, so that an overall equivalent circuit takes the form shown below.

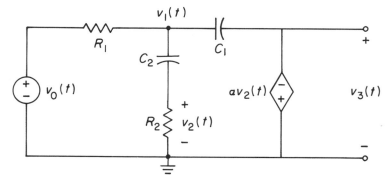

b) Using impedance methods, write ZSR node equations for the nodes whose voltages are labelled $v_1(t)$ and $v_2(t)$. Show that your results are consistent with the node equations in differential form given in Exercise 1.4.

c) Solve these node equations for the system function $H(s) = V_3(s)/V_0(s)$ and check your result with the formula given in Example 3.3–1.

d) Take the $\mathcal{L}$-transform of the differential state equations given in Exercise 1.4 under zero-state conditions. Solve these equations for $H(s) = V_3(s)/V_0(s)$, where $V_3(s) = -\alpha V_2(s)$, and compare your result with that derived in (c).

Problem 3.2

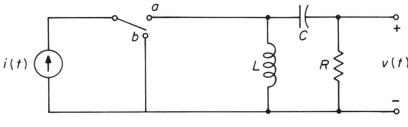

$$C = 0.01 \ \mu\text{F}, \ R = 3 \ \text{k}\Omega, \ L = 20 \ \text{mH}$$

a) With the switch in position a, find the system function $H(s) = V(s)/I(s)$, where $v(t)$ is the ZSR to the input $i(t)$, $V(s) = \mathcal{L}[v(t)]$, etc.
b) Suppose the current $i(t)$ has been held at a constant value $i(t) = 1$ mamp with the switch in position a for a very long time prior to $t = 0$. At $t = 0$ the switch is moved to position b. Find the values of $v(t)$ and $dv(t)/dt$ just after the switch is changed.
c) Find the natural frequencies of the circuit.
d) Find $v(t)$ for all time after the switch is switched to b.

Problem 3.3

Lynn Iyar, one of the more mathematically inclined students in the class, was not satisfied with the rather informal ways described in Section 3.4 for relating the input-output differential equation and the system function for an LTI circuit. Why not, she thought, simply apply the $\mathcal{L}$-transform Differentiation Theorem to obtain the desired relationship directly? Thus the RC circuit shown to the right corresponds to the input-output differential equation

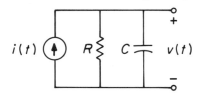

$$C\frac{dv(t)}{dt} + \frac{1}{R}v(t) = i(t). \tag{1}$$

Applying the $\mathcal{L}$-transform Differentiation Theorem,

$$\mathcal{L}\left[\frac{dx(t)}{dt}\right] = s\mathcal{L}[x(t)] - x(0) \tag{2}$$

yields

$$CsV(s) - v(0) + \frac{1}{R}V(s) = I(s). \tag{3}$$

Since the system function describes the ZSR, Lynn argued we should set $v(0) = 0$ and solve for

$$H(s) = \frac{V(s)}{I(s)} = \frac{\dfrac{1}{C}}{s + \dfrac{1}{RC}} \tag{4}$$

which is the same as the result obtained by impedance methods.

Lynn then tried to apply this method to more complex situations. She readily derived the formula

$$L\left[\frac{d^2x(t)}{dt}\right] = sL\left[\frac{dx(t)}{dt}\right] - \frac{dx(t)}{dt}\bigg|_{t=0}$$

$$= s^2 L[x(t)] - sx(0) - \frac{dx(t)}{dt}\bigg|_{t=0} \tag{5}$$

which can obviously be extended to derivatives of any order. However, when she tried to use (2) and (5) to transform the second-order equation of Example 3.4–1, relating the input $i(t)$ and the output $v(t)$ for the circuit shown to the right,

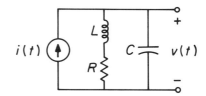

$$\frac{d^2v(t)}{dt^2} + \frac{R}{L}\frac{dv(t)}{dt} + \frac{1}{LC}v(t) = \frac{1}{C}\left(\frac{di(t)}{dt} + \frac{R}{L}i(t)\right) \tag{6}$$

she obtained

$$\left(s^2 + \frac{R}{L}s + \frac{1}{LC}\right)V(s) - \left(s + \frac{R}{L}\right)v(0) - \frac{dv(t)}{dt}\bigg|_{t=0} = \frac{1}{C}\left(s + \frac{R}{L}\right)I(s) - \frac{1}{C}i(0). \tag{7}$$

Now setting $v(0) = dv(t)/dt|_{t=0} = 0$ does not appear to give the same result as impedance methods for the ZSR system function $H(s) = V(s)/I(s)$ unless $i(0) = 0$.

Lynn was baffled and consulted her roommate Anna Logg. Anna, whose approach to problems was more physical than mathematical, took one look at the circuits and said "Oh, the trouble is that setting $v(t)$ and $dv(t)/dt$ to zero in the second circuit does not necessarily imply that the circuit is in the zero state." Show that Anna is right and explain how this accounts for Lynn's difficulties in interpreting equation (7). Show in general that a necessary and sufficient condition such that setting the output and its first $N-1$ derivatives to zero forces an Nth-order system to be in the zero state is that the input-output system function have no zeros for finite values of s.

Problem 3.4

Most systems, linear or not, eventually yield a periodic response to a suddenly applied periodic input. Unless the system happens to be in exactly the right state at the time of input application, however, there will be a nonzero interval at the start during which "transients" die out. A classical method of calculating the ultimate periodic response in such cases is to treat the state at some moment after the periodic response has been established as an algebraic unknown, compute the response of the system to the next complete period of the stimulus in terms of this unknown initial state, equate the state at the end of this period to the initial state, and solve the resulting equations for the required initial state. An alternative scheme for LTI systems uses the $\mathcal{L}$-transform and insight into the system response structure as illustrated below.

a) Let $x(t)$ be constructed by repeating periodically with period T a pulse waveform $x_p(t)$ of duration $\leq T$ as indicated in the figure to the right. Show that

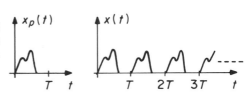

$$\mathcal{L}[x(t)] = X(s) = \frac{X_p(s)}{1 - e^{-sT}}, \quad \text{where} \quad X_p(s) = \mathcal{L}[x_p(t)].$$

(HINT: Argue that, for $\Re e[s] > 0$,

$$\frac{1}{1 - e^{-sT}} = 1 + e^{-sT} + e^{-2sT} + e^{-3sT} + \cdots.)$$

b) As a specific example, find the $\mathcal{L}$-transform of the waveform $x_0(t)$ shown to the right.

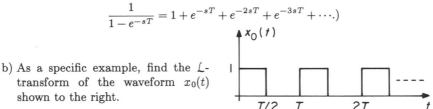

c) Suppose that $x_0(t)$ from (b) is the input to an LTI system with system function $H(s) = 1/(s + \alpha)$, $\alpha > 0$. Sketch the pole locations in the s-plane of the ZSR, $Y(s) = H(s)X(s)$. (Pay particular attention to the possibility of poles located along $s = j\omega$.)

d) The poles in (c) can be divided into two classes—a finite number of poles located inside the half-plane $\Re e[s] < 0$, and an infinite number of poles located along the $j\omega$-axis. The former correspond to transient terms in $y(t)$ that decay with time; the latter correspond to continuing sinusoids that superimpose to comprise the periodic part of the response. Find the transform of the periodic part of the ZSR in (c) by subtracting away from $Y(s)$ those terms in the partial-fraction expansion of $Y(s)$ that correspond to the left-half-plane poles.

e) Rearrange the result in (d) so it has the form derived in (a) and thus identify the transform $Y_p(s)$ of one period of $y(t)$. Do an inverse transform to obtain $y_p(t)$ and sketch your result.

Problem 3.5

Physically, an LTI 2-port is said to be *reciprocal** if the ZSR current in a short across one pair of terminals in response to an arbitrary voltage source across the other pair of terminals is independent of which end is shorted and which driven—as implied in the figure below. The 2-port is reciprocal if $i_a(t) = i_b(t)$ (ZSR) for all $v(t)$.

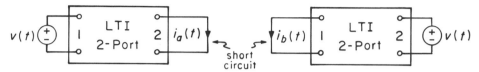

a) If the 2-port is described by the short-circuit admittance equations of the appendix to this chapter, show that a necessary and sufficient condition for reciprocity is

$$Y_{12}(s) = Y_{21}(s).$$

b) Show that an equivalent test for reciprocity is that the voltages $v_a(t)$ and $v_b(t)$ be the same under the two conditions illustrated below for an arbitrary current $i(t)$.

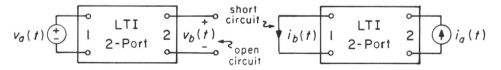

c) Show that another equivalent test for reciprocity is that

$$\frac{V_b(s)}{V_a(s)} = \frac{I_b(s)}{I_a(s)}$$

where the voltages and currents are defined by the circuits illustrated below.

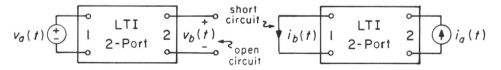

---

*It can be shown that a sufficient condition for a 2-port to be reciprocal is that it contain only LTI R's, L's, C's, and coupled coils (e.g., ideal transformers). This is sometimes called the *Network Reciprocity Theorem*. It is straightforward to show that this Reciprocity Theorem is a consequence of the symmetry of the node equations as they are usually written for such circuits; that is, the term in the KCL equation at node i proportional to the voltage at node j is identical to the term in the KCL equation at node j proportional to the voltage at node i (see, e.g., E. A. Guillemin, *Introductory Circuit Theory* (New York, NY: John Wiley, 1953) p. 148ff). A more elegant but rather less transparent argument follows from Tellegen's Theorem—see Problem 4.4 and, e.g., C. A. Desoer and E. S. Kuh, *Basic Circuit Theory* (New York, NY: McGraw-Hill, 1969) p. 681ff. The reciprocity concept is readily extended to n-ports and even to non-linear circuits. A key theorem due to Brayton is that a network composed of interconnected reciprocal subnetworks is reciprocal (see, e.g., G. F. Oster, A. S. Perelson, and A. Katchalsky, *Quart. Rev. Biophysics, 6* (1973): 1–138).

Problem 3.6

a) Show that any reciprocal LTI 2-port can be represented by the *equivalent* Π-*circuit** shown below where the 1-port admittances $Y_a(s)$, $Y_b(s)$, and $Y_c(s)$ are defined in terms of the short-circuit admittances of the 2-port by

$$Y_a(s) = Y_{11}(s) + Y_{12}(s)$$
$$Y_b(s) = -Y_{12}(s) = -Y_{21}(s)$$
$$Y_c(s) = Y_{22} + Y_{21}(s).$$

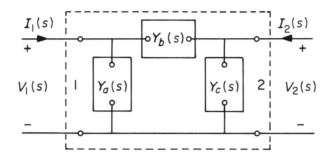

b) Use the equivalent circuit above and simple techniques of resistive network theory (e.g., Thévenin-Norton equivalences, voltage-divider formulas, etc.) to derive the transfer ratio $V_2(s)/V_0(s)$ in the following diagram and show that the result agrees with the formula derived in the appendix to this chapter.

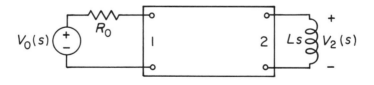

---

*It must be understood that this is only a mathematically equivalent circuit; there is no guarantee that three 1-port circuits having the admittances $Y_a(s)$, $Y_b(s)$, and $Y_c(s)$ could necessarily actually be constructed out of positive R's, L's, and C's, etc. Moreover, the equivalence holds only for terminal-pair behavior (2-port behavior); for example, the bottom terminals of each pair are shorted together (that is, they are at the same potential) in the equivalent circuit but may not be so connected in the actual 4-terminal network.

Problem 3.7

If the currents rather than the voltages at the ports are taken as the independent variables, then the LTI 2-port shown in the figure below can be characterized by the equations

$$V_1(s) = Z_{11}(s)I_1(s) + Z_{12}(s)I_2(s)$$
$$V_2(s) = Z_{21}(s)I_1(s) + Z_{22}(s)I_2(s)$$

where the $Z_{ij}(s)$ are called the *open-circuit driving-point* and *transfer impedances* of the 2-port.

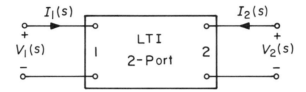

a) Devise experiments to measure the open-circuit impedances analogous to those described in the appendix to this chapter for the short-circuit admittances.

b) Derive the following expressions for the open-circuit impedances in terms of the short-circuit admittances, and vice versa:

$$Z_{11}(s) = \frac{Y_{22}(s)}{Y_{11}(s)Y_{22}(s) - Y_{12}(s)Y_{21}(s)}; \qquad Y_{11}(s) = \frac{Z_{22}(s)}{Z_{11}(s)Z_{22}(s) - Z_{12}(s)Z_{21}(s)}$$

$$Z_{12}(s) = \frac{-Y_{21}(s)}{Y_{11}(s)Y_{22}(s) - Y_{12}(s)Y_{21}(s)}; \qquad Y_{12}(s) = \frac{-Z_{21}(s)}{Z_{11}(s)Z_{22}(s) - Z_{12}(s)Z_{21}(s)}$$

$$Z_{21}(s) = \frac{-Y_{12}(s)}{Y_{11}(s)Y_{22}(s) - Y_{12}(s)Y_{21}(s)}; \qquad Y_{21}(s) = \frac{-Z_{12}(s)}{Z_{11}(s)Z_{22}(s) - Z_{12}(s)Z_{21}(s)}$$

$$Z_{22}(s) = \frac{Y_{11}(s)}{Y_{11}(s)Y_{22}(s) - Y_{12}(s)Y_{21}(s)}; \qquad Y_{22}(s) = \frac{Z_{11}(s)}{Z_{11}(s)Z_{22}(s) - Z_{12}(s)Z_{21}(s)}$$

c) Show that the tests for reciprocity described in Problem 3.5 imply and are implied by

$$Z_{12}(s) = Z_{21}(s).$$

d) Show that any reciprocal LTI 2-port can be represented by the equivalent T-circuit shown below, where the 1-port impedances $Z_a(s)$, $Z_b(s)$, and $Z_c(s)$ are defined in terms of the open-circuit impedances of the 2-port by

$$Z_a(s) = Z_{11}(s) - Z_{12}(s)$$
$$Z_b(s) = Z_{12}(s) = Z_{21}(s)$$
$$Z_c(s) = Z_{22}(s) - Z_{21}(s).$$

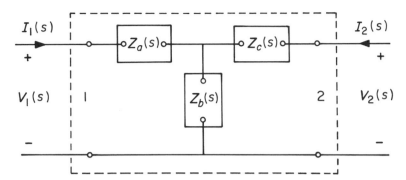

e) Open-circuit 2-port impedances are very useful for describing the properties of a pair of *coupled coils*—a set of two coils or inductors arranged physically (e.g., wound on a common core) so that the changing flux generated by a changing current in one coil links both coils and thus induces voltages in both coils. The symbol for a pair of coupled coils is shown below.

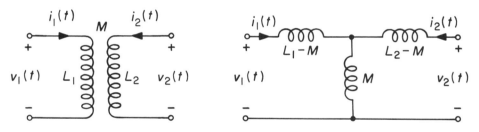

The corresponding equations are

$$v_1(t) = L_1\frac{di_1(t)}{dt} + M\frac{di_2(t)}{dt}$$
$$v_2(t) = M\frac{di_1(t)}{dt} + L_2\frac{di_2(t)}{dt}.$$

The *mutual inductance* M may have either sign (depending on the choice of reference directions for the currents) and is constrained in magnitude by the fact that physically the *coupling coefficient*, $k = |M|/\sqrt{L_1 L_2}$, must be less than unity. Argue that the arrangement of three inductors in a T-circuit as shown to the right above is mathematically equivalent as a 2-port to the pair of coupled coils shown to the left. Could any pair of coupled coils therefore physically be replaced by three uncoupled inductors of appropriate values without altering the behavior of the circuit? Explain.

f) Devise an alternative equivalent circuit for a pair of coupled coils in the form of a Π-circuit of inductors, based on the short-circuit admittance equivalent circuit of Problem 3.6. Give values for the elements in your circuit in terms of L_1, L_2, and M.

Problem 3.8

In addition to the open-circuit impedance or short-circuit admittance representations for LTI 2-ports discussed in Problem 3.7 and the appendix to this chapter, four other ways of characterizing LTI 2-port behavior are possible—corresponding to the four remaining ways of picking two independent variables from the four terminal quantities $V_1(s)$, $V_2(s)$, $I_1(s)$, and $I_2(s)$. The preferred choice among the 6 possibilities in a particular practical case usually involves a balance of two factors:

1. For which representation is the description of a particular 2-port simplest?
2. For which representation are the constraints imposed by the external circuits most readily expressed?

This problem explores these factors with several examples.

a) The *ideal transformer* shown to the right is characterized by the equations

$$v_2(t) = nv_1(t)$$
$$ni_2(t) = -i_1(t).$$

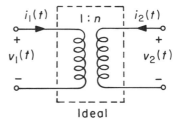

Ideal

It may be considered the limit of a pair of coupled coils (see Problem 3.7) as $k \to 1$ and $L_1 \to \infty$, $L_2 \to \infty$ with $L_2/L_1 = n^2$. (n is approximately the ratio of the number of turns in the secondary winding (L_2) to the number of turns in the primary winding (L_1)). Show that both the open-circuit impedances and the short-circuit admittances of Problem 3.7 and the appendix to this chapter are infinite for the ideal transformer, but that the so-called *ABCD representation* of a 2-port,

$$V_2(s) = A(s)V_1(s) + B(s)I_1(s)$$
$$I_2(s) = C(s)V_1(s) + D(s)I_1(s)$$

exists. Find $A(s)$, $B(s)$, $C(s)$, and $D(s)$ for the ideal transformer.

b) A simplified incremental circuit for a transistor is shown to the right. Find the parameter values corresponding to this circuit for the *hybrid representation* of a 2-port,

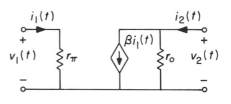

$$V_1(s) = H_{11}(s)I_1(s) + H_{12}(s)V_2(s)$$
$$I_2(s) = H_{21}(s)I_1(s) + H_{22}(s)V_2(s).$$

c) Determine the condition that must be satisfied by the $H_{ij}(s)$ parameters of (b) if the 2-port described by this representation is reciprocal. Is the incremental circuit for the transistor reciprocal?

d) Suppose two 2-ports are connected in parallel to form the single 2-port represented by the dashed box in the figure below. (The purpose of the 1:1 ideal transformer is to ensure that the 2-port conditions remain satisfied for each 2-port in the parallel combination. The transformer is unnecessary if, for example, both constituent 2-ports have common grounds connecting their lower terminals as shown by the dashed lines.) Find the short-circuit admittances characterizing the parallel connection in terms of the short-circuit admittances of each constituent 2-port.

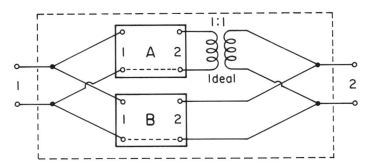

e) Draw a diagram showing how two 2-ports should be interconnected to form a new 2-port such that the open-circuit driving-point and transfer impedances of the interconnection are the sums of the corresponding open-circuit driving-point and transfer impedances of the constituent 2-ports.

Problem 3.9

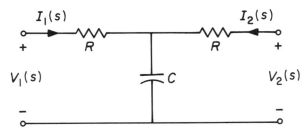

a) For the circuit above, find the short-circuit driving-point and transfer admittances in the representation

$$I_1(s) = Y_{11}(s)V_1(s) + Y_{12}(s)V_2(s)$$
$$I_2(s) = Y_{21}(s)V_1(s) + Y_{22}(s)V_2(s).$$

b) Show that the circuit on the next page is equivalent to that above, in the sense that it has the same 2-port representation.

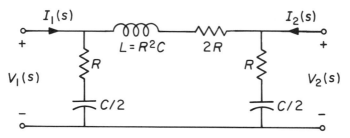

c) Show that the circuit below is also equivalent to those above as a 2-port.

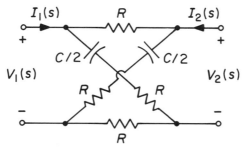

d) Is there any set of measurements that could be made at the ports of the three circuits above to distinguish one from the others?

e) Is there any set of measurements that could be made at the terminals of the three circuits above to distinguish one from the others?

Problem 3.10

a) A *negative-impedance converter* is a 2-port device that converts an impedance at its output to appear at the input as the *negative* of that impedance, as shown below.

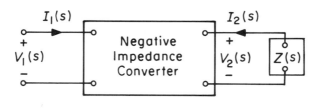

$$Z_{in}(s) = \frac{V_1(s)}{I_1(s)} = -Z(s)$$

A *gyrator* is a 2-port device that converts an impedance at its output to appear at the input as the *reciprocal* of that impedance, as shown below.

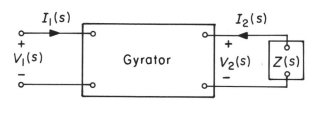

$$Z_{in}(s) = \frac{V_1(s)}{I_1(s)} = \frac{R^2}{Z(s)}$$

Describe the similarities and differences in the input impedances to these two devices if $Z(s)$ is a pure capacitor, $Z(s) = 1/Cs$. (Consider in particular the behavior for $s = j\omega$.) Which input impedance, if either, is indistinguishable from the impedance of an inductor?

b) Both the negative-impedance converter and the gyrator can be described by appropriate choices of the parameters in the $ABCD$ representation of a 2-port described in Problem 3.8:

$$V_2(s) = A(s)V_1(s) + B(s)I_1(s)$$
$$I_2(s) = C(s)V_1(s) + D(s)I_1(s).$$

Determine the values of the parameters that describe each device (the answers may not be unique). Are these devices reciprocal (see Problem 3.5)?

c) Show that the circuit below behaves as a negative-impedance converter.

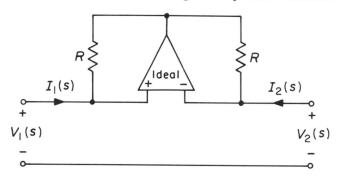

d) Show that the circuit on the next page behaves as a gyrator. "NIC" is a negative-impedance converter as in (c).

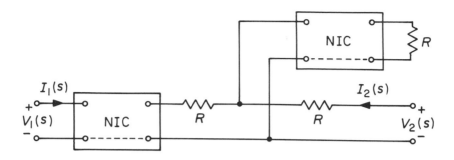

Problem 3.11

a) Another circuit that can be used to realize a gyrator (see Problem 3.10) is as shown below. Argue that

$$Z(s) = \frac{V(s)}{I(s)} = \frac{Z_1(s)Z_3(s)Z_5(s)}{Z_2(s)Z_4(s)}.$$

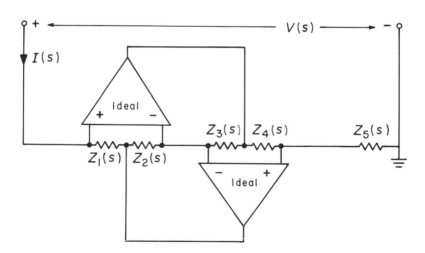

b) If $Z_4(s)$ is a capacitor and the remaining impedances are resistors, the circuit above behaves at its terminals as an inductor of value $L = \frac{R_1 R_3 R_5 C_4}{R_2}$. The circuit can thus be used to replace an inductor in a filter design, provided that the inductor has one terminal grounded. Unfortunately, lowpass filters, such as the Butterworth filter shown on the next page, have inductors with both terminals above ground.

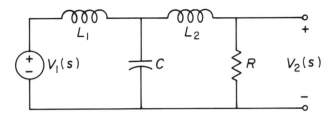

Butterworth lowpass filter, cutoff frequency $= \omega_0$ rad/sec.

$$L_1 = \frac{3R}{2\omega_0}, \quad L_2 = \frac{R}{2\omega_0}, \quad C = \frac{4}{3R\omega_0}$$

One way of using the circuit in (a) to realize such a filter is to observe that if all of the impedances in any circuit are divided by ks, then all voltage ratios in that circuit remain unchanged. (Convince yourself that this statement is true.) However, inductors get converted into resistors of value L/k, resistors get converted into capacitors of value k/R, and capacitors get converted into "double capacitors" with impedances $1/kCs^2$. A "double capacitor" can be realized with the circuit in (a) if $Z_1(s)$ and $Z_3(s)$ are capacitors. Use these ideas to find values for the elements in the circuit below that will realize a Butterworth lowpass filter with cutoff frequency equal to 1000 Hz. Try to keep all resistor values in the range 10 kΩ to 100 kΩ.

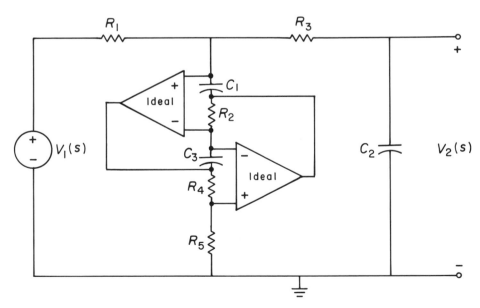

Problem 3.12

A *transducer* is a system for transforming energy from one form to another. Frequently, one of the forms is electrical; examples are an almost endless variety of mechano-electric devices such as motors, generators, loudspeakers, phonograph pickups, and accelerometers, as well as a host of thermo-electric, chemo-electric, optico-electric and other devices. The important dynamic behavior of a transducer often corresponds to a situation in which the perturbations in the variables about some steady condition are small enough that the system can be described incrementally as a linear 2-port. In addition, the energy transduction efficiency of many transducers is high enough that the device can be modelled as nearly lossless. In these cases, the linear 2-port must satisfy an interesting reciprocity condition, as this problem will illustrate.

a) The diagram below shows a simple mechano-electric transducer constructed from a massless plate of area A supported a variable distance $x(t)$ away from a fixed ground plane by electrically insulated springs. A mechanical force $f(t)$ can be applied to the plate. The plate can also be charged electrically through a flexible wire, forming a capacitor with the ground plane. This arrangement describes the essential features of a variety of useful devices such as condenser microphones and force transducers.

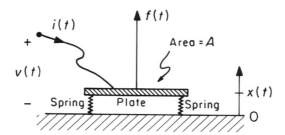

The stored energy in this system is

$$E[q(t), x(t)] = \frac{q^2(t)x(t)}{2\epsilon_0 A} + \frac{K(x(t) - x_0)^2}{2}$$

where

$q(t) = $ the electrical charge on the movable plate,

$\epsilon_0 = $ the permittivity of free space,

$x_0 = $ the resting length of the springs,

$K = $ the effective combined stiffness of the springs.

Since the system is lossless, electrical or mechanical work done on the system yields a corresponding increase in stored energy. Incrementally,

$$\Delta E[q(t), x(t)] = v(t)\Delta q(t) + f(t)\Delta x(t).$$

Thus, it must follow that

$$v(t) = \frac{\partial E[q(t), x(t)]}{\partial q(t)}, \qquad f(t) = \frac{\partial E[q(t), x(t)]}{\partial x(t)}.$$

Find formulas for $v(t)$ and $f(t)$ in terms of $q(t)$ and $x(t)$.*

b) Assume that each of the variables can be described as a large constant (quiescent) value plus a small perturbation:

$$f(t) = f_{00} + f_i(t), \quad |f_{00}| \gg |f_i(t)|$$
$$v(t) = v_{00} + v_i(t), \quad |v_{00}| \gg |v_i(t)|$$
$$x(t) = x_{00} + x_i(t), \quad |x_{00}| \gg |x_i(t)|$$
$$q(t) = q_{00} + q_i(t), \quad |q_{00}| \gg |q_i(t)|.$$

Let $F_i(s)$, $V_i(s)$, $U_i(s)$, and $I_i(s)$ stand respectively for the $\mathcal{L}$-transforms of $f_i(t)$, $v_i(t)$, $dx_i(t)/dt$, and $dq_i(t)/dt$. Show that these quantities are related by the following 2-port equations:

$$F_i(s) = \frac{K}{s} U_i(s) + \frac{q_{00}}{\epsilon_0 A s} I_i(s)$$
$$V_i(s) = \frac{q_{00}}{\epsilon_0 A s} U_i(s) + \frac{x_{00}}{\epsilon_0 A s} I_i(s).$$

c) Note that the *cross* or *coupling* terms in the 2-port equations in (b) (i.e., the term in the force equation proportional to the current, and the term in the voltage equation proportional to the velocity) have identical coefficients. This is a reciprocity condition of exactly the same sort discussed in Problem 3.5 for purely electrical circuits. Show that such a reciprocity condition holds for any system in which the stored energy can be written as a function of the displacement $x(t)$ at the mechanical terminal and the charge $q(t)$ at the electrical terminal.† HINT: Make use of the mathematical fact that

$$\frac{\partial}{\partial q(t)} \left(\frac{\partial E[q(t), x(t)]}{\partial x(t)} \right) = \frac{\partial}{\partial x(t)} \left(\frac{\partial E[q(t), x(t)]}{\partial q(t)} \right).$$

---

*The internal energy written in terms of generalized "displacements," such as $x(t)$ and $q(t)$, is called in physics the *Hamiltonian* of the system. The partial derivative of the Hamiltonian with respect to a particular "displacement" yields the associated generalized "force." If the system contains stored magnetic and kinetic energy, then the corresponding "displacements" are the magnetic flux and the mechanical momentum, and the "forces" are the electric current and the mechanical velocity, respectively.

†A reciprocity condition of this kind is called a *Maxwell relation* in thermodynamics.

4

POLES AND ZEROS

4.0 Introduction

The system function $H(s)$ for any lumped LTI circuit always has the form of a ratio of polynomials in s, that is, a rational function. By the Fundamental Theorem of Algebra, any polynomial can be factored in terms of its roots,

$$a_n s^n + a_{n-1} s^{n-1} + a_{n-2} s^{n-2} + \cdots + a_1 s + a_0$$
$$= a_n(s - s_1)(s - s_2)\cdots(s - s_n).$$

Thus a rational system function $H(s)$ can always be written in the form

$$H(s) = K \frac{(s - s_{z1})(s - s_{z2})\cdots(s - s_{zN})}{(s - s_{p1})(s - s_{p2})\cdots(s - s_{pM})}$$

and is completely specified (except for a real multiplicative constant) by the roots of the numerator and denominator polynomials—the *zeros*, s_{zi}, and the *poles*, s_{pi}, of $H(s)$. The type of system function (e.g., driving-point impedance or transfer ratio), the types of elements that compose the circuit, and the topology of the network all place constraints on zero and pole locations. On the other hand, once the pole and zero locations are specified, much can be said—qualitatively as well as quantitatively—about the behavior of the system, including in particular the general trends of its frequency response $H(j\omega)$. These are the topics of this chapter. With these topics we shall essentially complete our study of circuits as interconnections of the most primitive electrical elements. In the next chapter we shall start to explore the behavior of systems whose building blocks are themselves systems or circuits of some complexity.

4.1 Pole-Zero Diagrams

Often the most convenient and evocative way to summarize the characterization of a system function is to plot its pole and zero locations graphically in the complex s-plane; this is called a *pole-zero diagram*. An example is shown in Figure 4.1–1.

The type of elements used as well as the network structure constrain the regions of the s-plane in which the poles and zeros may lie. First, it is important

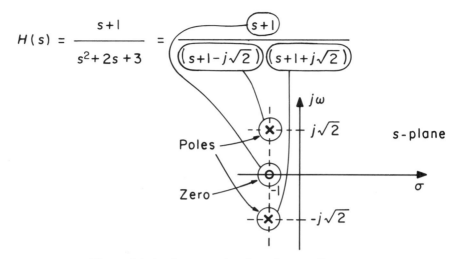

Figure 4.1–1. An example of a pole-zero diagram.

to note (as illustrated in Chapter 2) that the fact that the element values are real numbers forces the pole and zero locations either to lie along the real σ-axis (e.g., the zero in Figure 4.1–1) or to occur in conjugate complex pairs at mirror-image points with respect to the σ-axis (e.g., the poles in Figure 4.1–1). Moreover, if the network is composed of positive R's, L's, and C's only (coupled coils and transformers are allowed, but no controlled sources), then the poles of the system function must lie in the left half-s-plane or on the $j\omega$-axis (i.e., $\sigma \le 0$); see Problem 4.4 for a proof. Networks composed entirely of positive R's, L's, and C's are said to be *passive*. System functions whose poles do not lie in the right half-s-plane are called *stable*; a passive LTI network is stable.*

For special classes of passive circuits the locations of the poles are even more restricted (see Problem 4.4 for a further discussion):

1. For LTI networks composed of positive R's and L's only, or positive R's and C's only (no controlled sources), the poles must all lie on the negative real axis. RC or RL networks are sometimes called *relaxation networks* because their ZIR is a weighted sum of monotonically decaying exponentials.

2. For LTI networks constructed from positive L's and C's only (no controlled sources), the poles must all lie on the $j\omega$ axis. Idealized LC networks are called *lossless networks* and are marginally stable; their ZIR is a weighted sum of undamped sinusoids that neither grow nor decay.

---

*The converse is not true; a stable network need not be passive, as we have already seen in various op-amp examples and as we shall discuss in detail in Chapter 6. More generally, passivity implies that the energy one can extract from an element or circuit cannot exceed a finite bound determined by the initial state—see, e.g., J. L. Wyatt, Jr., et al., *IEEE Trans. Cir. & Sys.*, *CAS-28* (1981): 48–61. Stability implies that the effects of small perturbations remain small; an LTI system is clearly unstable if its ZIR contains growing exponentials—if the poles of the system function lie in the right half-s-plane—because then any disturbance, no matter how small, will ultimately yield a large effect. For a further discussion of stability, including more careful definitions, see Chapter 6.

The zeros of a passive system function—unlike the poles—can in general lie anywhere in the complex plane, although again special system types impose restrictions:

1. If $H(s)$ is a driving-point impedance, that is, if

$$V(s) = H(s)I(s)$$

where $V(s)$ is the $\mathcal{L}$-transform of the ZSR voltage across the impedance $H(s)$ induced by a current source $I(s)$, then (as pointed out in Section 3.2) $1/H(s)$ is also a system function, namely a driving-point admittance satisfying the equation

$$I(s) = \frac{1}{H(s)}V(s)$$

where $I(s)$ is the $\mathcal{L}$-transform of the ZSR current flowing through the impedance $H(s)$ in response to a voltage source $V(s)$. In this case the zeros of $H(s)$, which are the poles of $1/H(s)$, must satisfy the same types of constraints as the poles: In general they must lie in the left half-plane for passive RLC networks, along the negative σ-axis for passive RC and RL networks, or along the $j\omega$-axis for passive LC networks.*

2. A *ladder network* is a circuit that looks topologically like a "ladder" with alternating series and shunt branches. It can be shown that the zeros of any transfer function of a ladder network must lie in the left half-plane. For an RC or RL ladder network the zeros must lie on the negative σ-axis; for an LC ladder network they must lie on the $j\omega$-axis.

Example 4.1–1

The *Twin-T network* shown in Figure 4.1–2 is an important and useful RC circuit.

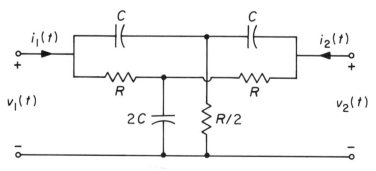

Figure 4.1–2. A Twin-T network.

---

*These conditions on the zeros are necessary but not sufficient for $H(s)$ to be a driving-point impedance. Thus for RC, RL, or LC driving-point impedances, the poles and zeros must in fact alternate, as can be shown by extension of the methods of Problems 4.5 and 4.6. For the general case of an RLC driving-point impedance, restrictions on the zeros can be derived from the fact that the impedance must be a "positive-real function" as discussed in Problem 4.5.

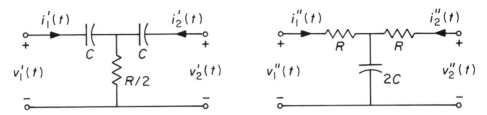

Figure 4.1–3. Component T-networks from which the Twin-T of Figure 4.1–2 can be constructed.

The short-circuit driving-point and transfer admittances of the 2-port shown in Figure 4.1–2 (see the appendix to Chapter 3) can easily be found as the sums of the short-circuit admittances of the two simple ladder networks (called *T-networks* because of their topology) that in parallel comprise it, as shown in Figure 4.1–3:

$$Y'_{11}(s) = Y'_{22}(s) \qquad\qquad Y''_{11}(s) = Y''_{22}(s)$$

$$= \frac{Cs(s+2/RC)}{2(s+1/RC)} \qquad\qquad = \frac{s+1/2RC}{R(s+1/RC)}$$

$$Y'_{12}(s) = Y'_{21}(s) \qquad\qquad Y''_{12}(s) = Y''_{21}(s)$$

$$= \frac{-Cs^2}{2(s+1/RC)} \qquad\qquad = \frac{-1}{2R^2C(s+1/RC)}.$$

Hence

$$Y_{11}(s) = Y_{22}(s) = \frac{Cs(s+2/RC)}{2(s+1/RC)} + \frac{s+1/2RC}{R(s+1/RC)} = \frac{C\left(s^2 + 4s/RC + 1/R^2C^2\right)}{2(s+1/RC)}$$

and

$$Y_{12}(s) = Y_{21}(s) = \frac{-Cs^2}{2(s+1/RC)} - \frac{1}{2R^2C(s+1/RC)} = \frac{-C\left(s^2 + 1/R^2C^2\right)}{2(s+1/RC)}.$$

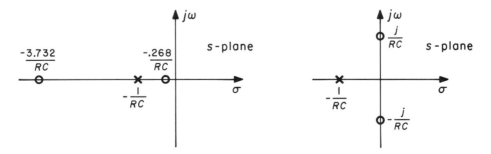

Figure 4.1–4. Pole-zero diagrams for $Y_{11}(s)$ (left) and $Y_{12}(s)$ (right).

We note that $Y_{11}(s)$ is the driving-point admittance of an RC circuit—both poles and zeros (as shown in Figure 4.1–4) are on the negative real axis (and alternate as required by the footnote on page 107). The same properties also characterize $Y'_{11}(s)$ and $Y''_{11}(s)$. On the other hand, $Y_{12}(s)$ is the transfer admittance of an RC circuit—the pole is on the negative real axis as required, but the zeros are on the $j\omega$-axis. (The Twin-T is not a ladder network; thus the zeros of the transfer admittance need not lie on the negative real axis. Note, however, that $Y'_{12}(s)$ and $Y''_{12}(s)$, which are the transfer admittances of RC ladder networks, have their zeros at $s = 0$ or $s = \infty$, which do lie on the negative real axis, at least in a limiting sense.) Because it has a zeros on the $j\omega$-axis, the Twin-T network can be used to block or "trap" input sinusoids with frequencies near $\omega = 1/RC$, preventing such input components from appearing at the output. (For an application see Problem 4.13.)

▶ ▶ ▶

4.2 Vectorial Interpretation of $H(j\omega)$

Many of the important features of $H(j\omega)$ as a function of $j\omega$ can be deduced directly from the pole-zero diagram of $H(s)$. A basis for this use of the pole-zero diagram is the observation that the complex number $s - s_0$, represented as a vector in the complex plane, is the vector joining the tips of the vectors s and s_0. The length of this vector is $|s - s_0|$; its angle is $\angle(s - s_0)$. These ideas are illustrated in Figure 4.2–1.

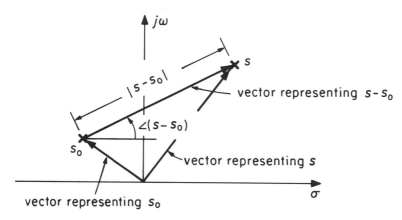

Figure 4.2–1. Vectorial interpretation of the complex numbers s, s_0, and $s - s_0$.

Suppose, now, that we write

$$H(s) = K \frac{(s - s_{z1})(s - s_{z2}) \cdots (s - s_{zN})}{(s - s_{p1})(s - s_{p2}) \cdots (s - s_{pM})}. \qquad (4.2\text{–}1)$$

Then, expressing each term in magnitude-angle form and setting $s = j\omega$, we have

$$|H(j\omega)| = K\frac{\prod\limits_{i=1}^{N}|j\omega - s_{zi}|}{\prod\limits_{i=1}^{M}|j\omega - s_{pi}|} \tag{4.2-2}$$

$$\angle H(j\omega) = \sum_{i=1}^{N}\angle(j\omega - s_{zi}) - \sum_{i=1}^{M}\angle(j\omega - s_{pi}). \tag{4.2-3}$$

That is, $|H(j\omega)|$ is K times the ratio of the product of the lengths of the vectors from each zero to the point $j\omega$ on the $j\omega$-axis divided by the product of the lengths of the vectors from each pole to the point $j\omega$. Similarly, the angle of $H(j\omega)$ is the difference of the sums of the individual term angles. The usefulness of this vectorial interpretation is best illustrated by an example.

Example 4.2–1

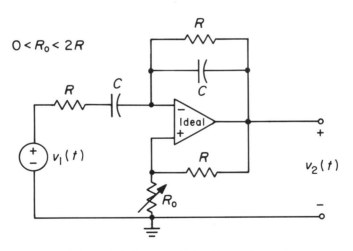

Figure 4.2–2. Circuit for Example 4.2–1.

Let us find the system function of the op-amp circuit shown in Figure 4.2–2. Employing the usual ideal op-amp approximations and using superposition and voltage-divider rules, we can readily find a relation between $V_1(s)$ and $V_2(s)$ by equating the voltages at the inverting and noninverting terminals:

$$\frac{V_2(s)\left(R+\dfrac{1}{Cs}\right)}{R+\dfrac{1}{Cs}+\dfrac{R(1/Cs)}{R+1/Cs}} + \frac{V_1(s)\dfrac{R(1/Cs)}{R+1/Cs}}{R+\dfrac{1}{Cs}+\dfrac{R(1/Cs)}{R+1/Cs}} = V_2(s)\frac{R_0}{R_0+R}. \tag{4.2-4}$$

Rearranging (4.2–4) gives

$$\frac{V_2(s)}{V_1(s)} = H(s) = K\frac{\frac{1}{Q}\left[\frac{s}{\omega_0}\right]}{\left[\frac{s}{\omega_0}\right]^2 + \frac{1}{Q}\left[\frac{s}{\omega_0}\right] + 1}$$

(4.2–5)

where

$$\omega_0 = \frac{1}{RC}$$

(4.2–6)

$$Q = \frac{R}{2R - R_0} \quad \left(\text{so that } \frac{1}{2} < Q < \infty \text{ if } 0 < R_0 < 2R\right)$$

(4.2–7)

$$K = -\frac{R_0 + R}{2R - R_0} = H(j\omega_0).$$

(4.2–8)

The pole-zero diagram for $H(s)$ is shown in Figure 4.2–3; as R_0 is changed, the poles move along a circle of radius ω_0. When $Q = 1/2$ ($R_0 = 0$), the poles are coincident on the negative σ-axis at $\sigma = -\omega_0 = -1/RC$. When $Q \to \infty$ ($R_0 \to 2R$), the poles approach the points $\pm j\omega_0$ (and indeed if $R_0 > 2R$, the poles move into the right half-plane, which implies an unstable system).

The pole-zero diagram for this circuit as R_0 is adjusted behaves in precisely the same way as the pole-zero diagram for the impedance $Z(s)$ of the parallel resonant circuit shown in Figure 4.2–4 as the shunt resistance R is varied:

$$Z(s) = \frac{1}{\frac{1}{R} + \frac{1}{Ls} + Cs} = R\frac{\frac{1}{Q}\left[\frac{s}{\omega_0}\right]}{\left[\frac{s}{\omega_0}\right]^2 + \frac{1}{Q}\left[\frac{s}{\omega_0}\right] + 1}$$

where

$$K = R, \quad \omega_0 = \frac{1}{\sqrt{LC}}, \quad Q = R\sqrt{\frac{C}{L}} = \omega_0 RC = \frac{R}{\omega_0 L}.$$

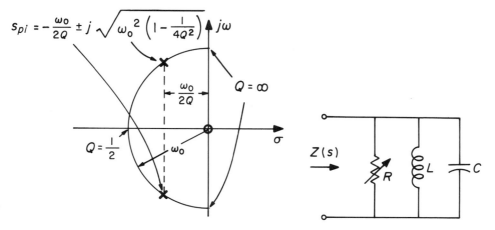

Figure 4.2–3. Poles and zeros for (4.2–5).   **Figure 4.2–4.** Parallel resonant circuit.

4.3 Potential Analogies

The vectorial interpretation of the pole-zero factors of $H(s)$ helps to make it clear how $|H(s)|$ becomes large for s near a pole and small for s near a zero. Sometimes it is also helpful to think in terms of a physical model. Imagine a rubber sheet with σ and $j\omega$ axes drawn on it. Arrange "tent poles" to push the sheet up at pole locations and "stakes" or "thumbtacks" to tack it down to the ground at zeros. The resulting surface roughly approximates $|H(s)|$. Such an imaginary experiment is often useful for visualizing the effect on $|H(j\omega)|$ of a proposed arrangement of poles and zeros.

An analogous experiment can actually be set up to yield accurate quantitative results. Consider a two-dimensional conducting surface such as a specially coated paper or a shallow tank filled with an electrolyte. Draw a pair of perpendicular axes, labelled σ and ω, on this surface to represent the complex plane. It is an elementary exercise in field theory to show that the potential at a point (σ, ω) on this surface resulting from a current probe touching the surface at the location (σ_0, ω_0) is proportional to $\ln |s - s_0|$, where $s = \sigma + j\omega$ and $s_0 = \sigma_0 + j\omega_0$. If several inward-directed current probes are simultaneously applied to points $(\sigma_{zi}, \omega_{zi})$, several outward-directed current probes are simultaneously applied to points $(\sigma_{pi}, \omega_{pi})$, and the currents are all equal in magnitude, then the potential at (σ, ω) is

$$\phi(\sigma, \omega) \sim \ln \left| \frac{(s - s_{z1})(s - s_{z2}) \cdots (s - s_{zM})}{(s - s_{p1})(s - s_{p2}) \cdots (s - s_{pN})} \right|.$$

That is, the potential is proportional to the log magnitude of the system function $H(s)$ having the corresponding zero and pole locations.

Today the digital computer has largely replaced such analog techniques for determining the quantitative relation between pole-zero locations and the magnitude and phase of $H(s)$. Yet the qualitative value of potential analogies remains. Sometimes known potential fields can be used without experiment to suggest network designs. Suppose, for example, we want to design a circuit with a frequency response that is approximately constant in magnitude over the band $|\omega| < 2\pi W$ and decays rapidly to zero outside, so that it approximates an ideal lowpass filter such as we shall study extensively in later chapters. The potential field produced by a conducting ring with the collector at infinity, as in Figure 4.3–1a, is an appropriate starting point; from potential theory, the potential will be uniform inside the ring and will fall to a low value outside. We can approximate such a potential with uniformly spaced discrete probes as in Figure 4.3–1b; the corresponding poles would not be a very satisfactory design because half of them are in the right half-plane. However, by symmetry (or by cascading the network with poles as in Figure 4.3–1b with the all-pass structure of Figure 4.3–1c), we conclude that uniformly spaced poles as in Figure 4.3–1d should provide a stable design whose magnitude along the $j\omega$-axis is the square root of the magnitude obtained in Figure 4.3–1b. (See Problem 4.2 for a discussion of all-pass networks.) More densely spaced poles will, of course, provide a better approximation.

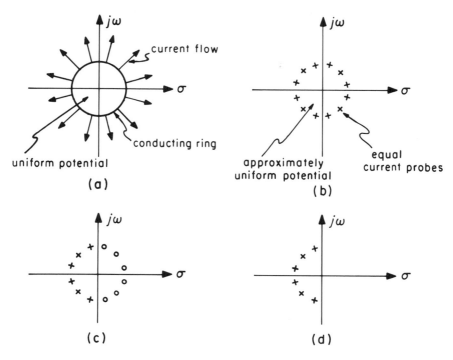

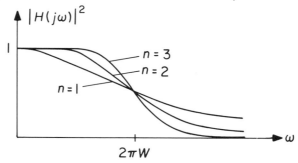

Figure 4.3–1. Pole locations for a lowpass filter.

The resulting filter is one of the class of Butterworth lowpass filters discussed earlier. It can easily be shown to have the frequency response

$$|H(j\omega)|^2 = \frac{1}{1 + \omega^{2n}/(2\pi W)^{2n}}$$

where n is the number of poles. $|H(j\omega)|^2$ is plotted in Figure 4.3–2 for several values of n. (See Problem 4.12 for further discussion.)

Figure 4.3–2. $|H(j\omega)|^2$ for several Butterworth filters.

Example 4.4–1

To illustrate Bode's method for sketching $|H(j\omega)|$ and $\angle H(j\omega)$, suppose that

$$H(s) = 10\frac{s}{(1+s)\left(1+\dfrac{s}{10}\right)}.$$

Then

$$20\log_{10}|H(j\omega)| = 20\log_{10}(10) + 20\log_{10}|\omega| - 20\log_{10}|1 + j\omega| - 20\log_{10}\left|1 + \frac{j\omega}{10}\right|.$$

The asymptotes for each term are shown in the upper part of Figure 4.4–2. The dotted line in the lower part of Figure 4.4–2 shows the superimposed asymptotes; the solid line is a close approximation to the actual response curve for this bandpass filter. The corresponding phase asymptotes and approximate characteristic are shown in Figure 4.4–3.

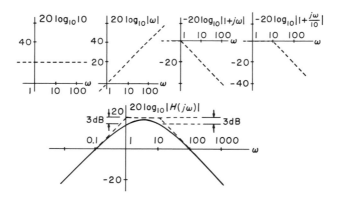

Figure 4.4–2. Bode magnitude plot for Example 4.4–1.

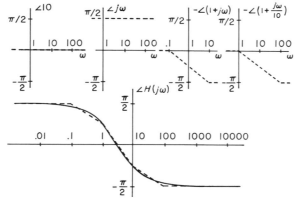

Figure 4.4–3. Bode phase plot for Example 4.4–1.

▶ ▶ ▶

Note that every rational function is proportional to some integer power of frequency, $(j\omega)^n$, as ω approaches either 0 or ∞. Hence Bode magnitude plots for large or small frequency are asymptotically straight lines with slopes $6n$ dB/octave (or $20n$ dB/decade). Simultaneously, the angle plots approach $90n$ degrees (modulo 360°). Conversely, if the dB-vs.-$\log_{10}\omega$ plot of an experimentally determined $|H(j\omega)|$ falls at, say, 18 dB/octave as $\omega \to \infty$, then $|H(j\omega)| \sim 1/\omega^3$ at high frequencies. These asymptotic properties apply, of course, to Bode plots for any rational system function, $H(s)$, although the simple techniques we have been describing for constructing such plots are restricted to $H(s)$ having all poles and zeros on the σ-axis.

Example 4.4–2

As a second example, consider

$$H(s) = \frac{\mu\left(1 + \dfrac{s}{30}\right)^2}{(1+s)^3}.$$

The Bode diagram for $(1+\tau s)^n$ is simply that for $(1+\tau s)$ multiplied by n. The overall result is shown in Figure 4.4–4.

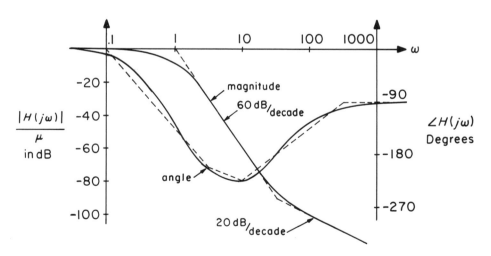

Figure 4.4–4. Bode plots for Example 4.4–2.

▶ ▶ ▶

c) Input-output frequency responses, $|H(j2\pi f)| = \dfrac{V_1(j2\pi f)}{V_0(j2\pi f)}$:

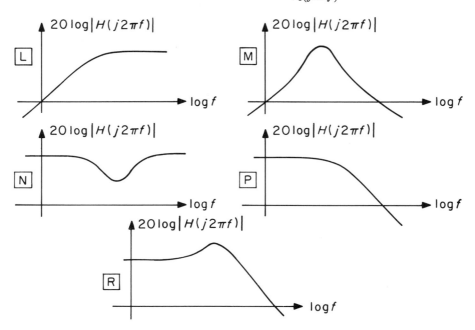

d) Responses to the unit step, $v_0(t) = u(t)$:

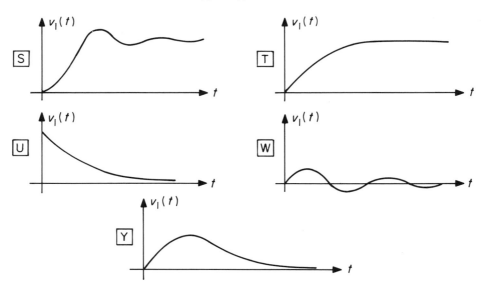

The correct answers are: (1) $\boxed{A}\ \boxed{J}\ \boxed{P}\ \boxed{T}$; (2) $\boxed{B}\ \boxed{F}\ \boxed{M}\ \boxed{W}$; (3) $\boxed{E}\ \boxed{G}\ \boxed{L}\ \boxed{U}$.

Exercise 4.2

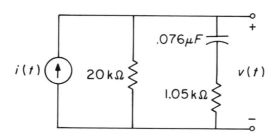

a) Show that the system function of the circuit above is approximately

$$H(s) = \frac{V(s)}{I(s)} = 10^3 \frac{s + 2\pi \times 2000}{s + 2\pi \times 100} \qquad \text{(ZSR)}.$$

b) Locate the poles and zeros of $H(s)$ on a sketch of the s-plane.

c) Use Bode's method to show that the magnitude and phase of the frequency response $H(j2\pi f)$ of the above circuit vs. $\log_{10} f$ (where $f = \omega/2\pi$) are as sketched below.

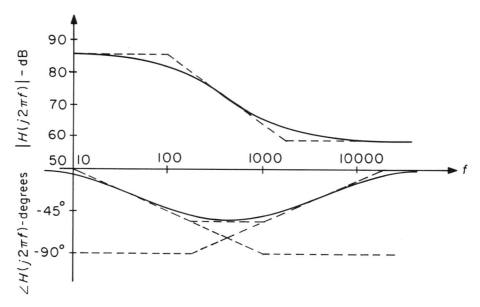

PROBLEMS FOR CHAPTER 4

Problem 4.1

The system function $H(s) = V_3(s)/V_0(s)$ for the circuit of Example 3.3–1 and Problem 3.1 was found to be

$$H(s) = \frac{\dfrac{-\alpha s}{(1+\alpha)R_1 C_1}}{s^2 + \dfrac{s}{1+\alpha}\left(\dfrac{1}{R_1 C_1} + \dfrac{1}{R_2 C_2} + \dfrac{1}{R_2 C_1}\right) + \dfrac{1}{(1+\alpha)R_1 C_1 R_2 C_2}}.$$

a) Show that non-negative values of R_1, R_2, C_1, C_2, and α can always be chosen to position the poles of $H(s)$ at any desired locations on the negative real axis or as a conjugate pair anywhere in the left half-s-plane.

b) In particular show that poles at $s = -10^3$ and -2×10^3 sec^{-1} can be realized with $R_1 = 1000 \ \Omega$, $R_2 = 10,000 \ \Omega$, $\alpha = 1$, and appropriate values of C_1 and C_2.

c) Use Bode's method to sketch $|H(j\omega)|$ for the pole locations achieved in (b).

d) Assuming R_1 and R_2 as in (b), find the minimum value of α and appropriate values of C_1 and C_2 to position the poles at $s = -10^2 \pm j10^3$ sec^{-1}.

e) Sketch $|H(j\omega)|$ for the pole locations achieved in (d).

Problem 4.2

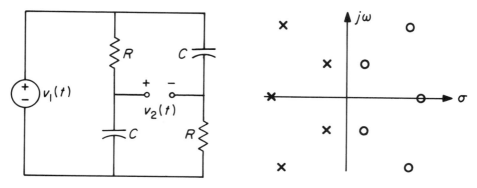

a) The circuit above is an example of an *all-pass network*. Compute and sketch $|H(j\omega)|$. Why is this called an "all-pass" network?

b) Using the vectorial interpretation of $H(s)$ developed in Section 4.2, argue that any system function having poles and zeros at mirror image points as shown has the following properties:
   i) $|H(j\omega)| = $ constant
   ii) $\angle H(j\omega)$ is a non-increasing function of ω.
Such a pole-zero diagram is a general characterization of an all-pass network.

Problem 4.3

A second-order Butterworth lowpass filter with a cutoff frequency of 6000 rad/sec will have an $H(s)$ described by the pole-zero plot shown below.

a) Determine analytically and sketch $|H(j2\pi f)|/H(0)$ vs. f. (HINT: It is easier to start by finding $|H(j\omega)|^2$.)

b) Find values of L and C to realize

$$\frac{V_2(s)}{V_1(s)} \sim H(s) \quad \text{(ZSR)}$$

with the form of circuit shown below.

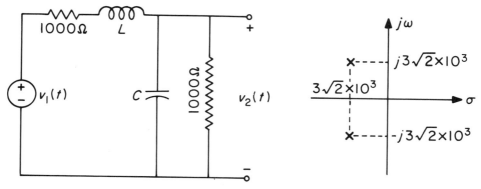

Problem 4.4

Nonzero branch currents and voltages of the form

$$i_\ell = I_\ell e^{s_p t}, \qquad v_\ell = V_\ell e^{s_p t}$$

where s_p is a natural frequency can exist in a circuit under conditions in which all external sources are zero—this, indeed, is the characteristic significance of a natural frequency. This problem derives limitations on the complex numbers s_p that can be natural frequencies for linear circuits composed entirely of positive R's, L's, and C's.

a) Use Tellegen's Theorem* to show that if all the branch voltages and currents have the form above and if all sources and mutual inductances† are zero, then

$$s_p \sum_{\substack{all \\ inductors}} L_i |I_i|^2 \;+\; s_p^* \sum_{\substack{all \\ capacitors}} C_j |V_j|^2 \;+\; \sum_{\substack{all \\ resistors}} R_k |I_k|^2 \;=\; 0.$$

---

*Tellegen's Theorem states that if $\{v_m\}$ is a complete set of branch voltages in a network that satisfy KVL and $\{i_m^*\}$ is a complete set of branch currents in the same network that satisfy KCL ($\{v_m\}$ and $\{i_m^*\}$ do not have to be sets that would simultaneously satisfy the dynamic equations for the network), then $\sum v_m i_m^* = 0$, where the sum is taken over all branches of the network. (See, e.g., C. A. Desoer and E. S. Kuh, Basic Circuit Theory (New York, NY: McGraw-Hill, 1969) p. 393ff.)

†The results to be derived remain true if mutual inductances are not zero, but the proof is slightly more complex.

Problem 4.7

Prior to being fed to the recording head that cuts the master for a conventional phonograph record, the audio signal is passed through a network that attenuates the lower frequencies (to avoid overdriving) and boosts the higher frequencies (to help combat scratch noise). A compensating network must then be inserted into the reproducing equipment. The standard RIAA equalizer characteristic is shown in the figure below.

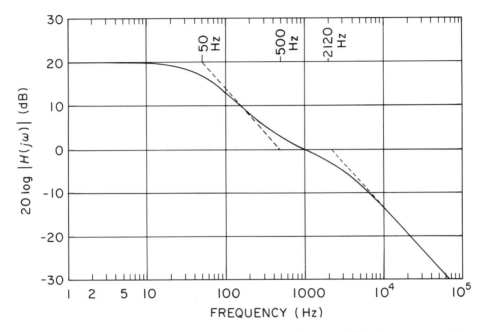

a) Find a simple pole-zero diagram and a rational function $H(s)$ that corresponds to the magnitude characteristic shown. Assume that $H(s)$ contains one zero and two poles, all on the negative real axis.

b) Show that the impedance of the circuit below can be made proportional to $H(s)$ above. Choose the proportionality factor so that the largest resistor is 1.1 MΩ and find all four parameter values.

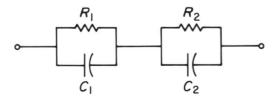

c) A common phonograph-cartridge preamplifier circuit is shown in the figure below. The 739 is a dual low-noise op-amp designed for audio applications. The 47 kΩ resistor is the standard load resistance for magnetic cartridges. C_3 and C_4 are blocking capacitors that are necessary because the 739 is usually powered from a single-sided power supply; assume (for the moment) that they are so large that their effects can be ignored at most frequencies of interest. Show that this circuit approximates the RIAA characteristic. (Note that, except for rounding off to standard 5% values, the impedance of the dashed box is that derived in (b).) What is the gain of this circuit at 1 kHz?

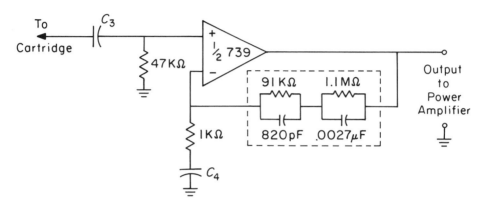

d) In 1978 the RIAA characteristic was modified to specify equalization 3 dB down from the previous standard at 20 Hz, rolling off at 6 dB per octave below 20 Hz. Show that this can be approximately achieved with the above circuit by setting $C_4 = 8.2 \ \mu F$. (HINT: Treat the 820 pF and 0.0027 μF capacitors as open circuits at the lower frequencies where C_4 is important, whereas C_4 is approximately a short circuit at the higher frequencies where the other capacitors are important.)

Problem 4.8

The intermediate-frequency (IF) amplifier of a superheterodyne radio receiver has two important functions—to provide the bulk of the amplification needed for the desired station or frequency channel, and to attenuate all other frequency channels. To accomplish these goals, several stages of tuned amplification are usually employed. A simple model for such an amplifier might appear as shown in the figure below.

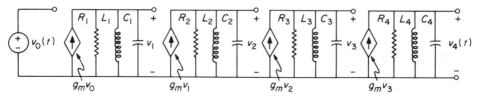

response of this circuit for the three positions of the potentiometer shown in (1), (2), and (3). Make reasonable approximations.

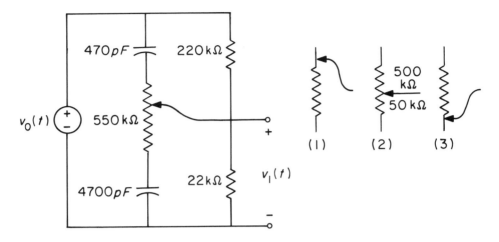

b) Discuss the "taper" of the potentiometer—the curve of resistance between the tap and one end as a function of geometric tap position or shaft angle—that would probably be appropriate for this circuit. (With the addition of several more elements this circuit can provide both treble and bass control as shown below.)

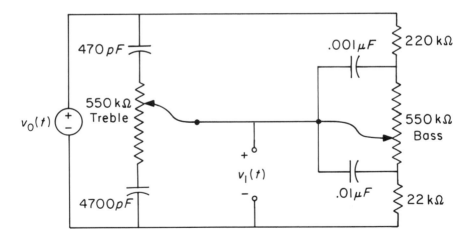

Problem 4.11

The circuit shown on the next page is frequently employed as a *crossover network* to apply the appropriate ranges of frequencies to the low- and high-frequency speakers (represented by the resistances R) of a hi-fi system.

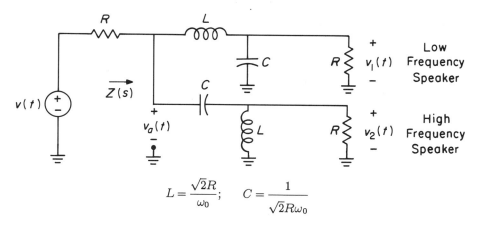

$$L = \frac{\sqrt{2}R}{\omega_0}; \quad C = \frac{1}{\sqrt{2}R\omega_0}$$

a) Show that the impedance $Z(s)$ looking into the crossover network is in fact a constant resistance R, independent of frequency.

b) Find the system functions relating each of the outputs to the input. Locate their poles and zeros on a sketch of the s-plane. (HINT: Note that as a result of (a), the voltage $v_a(t)$ is simply a fixed fraction of $v_0(t)$, independent of frequency. You should find that the poles of both system functions are located at $s = \omega_0 e^{\pm j3\pi/4}$.)

c) Sketch the magnitudes of the frequency transfer characteristics $|V_i(j\omega)/V(j\omega)|$ of the two speakers. What is the significance of the parameter ω_0?

Problem 4.12

If $H(s)$ is a rational function (i.e., a ratio of polynomials in s with real coefficients), then it is easy to show that $|H(j\omega)|^2$ will be a rational function in ω^2 with real coefficients. To go backwards from $|H(j\omega)|^2$ to $H(s)$ one can use the following procedure:

i) Observe that

$$|H(j\omega)|^2 = H(j\omega)H(-j\omega) = [H(s)H(-s)]_{s=j\omega}.$$

Hence, if we substitute $\omega^2 = -s^2$ in $|H(j\omega)|^2$, we can identify the result as $H(s)H(-s)$.

ii) Locate the poles and zeros of $H(s)H(-s)$ on a sketch of the s-plane. In general they will occur in quadruplets, except that poles or zeros on the real axis need occur only in pairs, as shown on the left in the figure on the next page.

iii) Identify the poles in the left half-plane with $H(s)$. Their images in the right half-plane then correspond to $H(-s)$. The zeros of $H(s)$ may be taken (in conjugate pairs) from either half-plane; if they are all taken in the left half-plane, the result is called the minimum-phase $H(s)$.

a) Find the poles and zeros of a system function $H(s)$ having approximately the frequency response shown on the preceding page.

b) Find a differential equation relating the input $x(t)$ and the output $y(t)$ of the LTI system described in (a).

Problem 4.15

The following circuit appears in an op-amp manufacturer's data manual with the suggestion that it may be useful as a scratch filter in a phonograph amplifier.

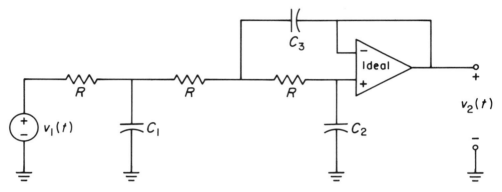

a) Show that the system function $H(s) = V_2(s)/V_1(s)$ (ZSR) is given by

$$H(s) = \frac{1}{(RC_1 RC_2 RC_3)s^3 + 2RC_2(RC_1 + RC_3)s^2 + (RC_1 + 3RC_2)s + 1}.$$

b) Locate the poles of $H(s)$ on a plot of the s-plane for $C_1 = 0.0022$ μF, $C_2 = 330$ pF, $C_3 = 0.0056$ μF, $R = 10$ kΩ. (HINT: One of the poles should be near $s = -2\pi \times 10^4$ sec^{-1}.)

c) Sketch $|H(j\omega)|$. (HINT: Consider $H(s)$ as approximating one of the class of Butterworth lowpass filters described in Section 4.3.)

d) The op-amp manufacturer suggests that the half-power frequency of this lowpass filter (the frequency at which $|H(j\omega)|^2 = 0.5$) can be adjusted by using a 3-ganged switch or potentiometer to change the common value, R, of the three resistors. Explain how this works in terms of the algebraic structure of $H(s)$. Describe the pole locations and determine the value of the half-power frequency if R is doubled (to 20 kΩ) with no change in the values of the capacitors. Repeat for R halved (to 5 kΩ). (It should be possible to answer these questions without recomputing or refactoring the denominator polynomial of $H(s)$.)

5

INTERCONNECTED SYSTEMS
AND FEEDBACK

5.0 Introduction

Thus far we have used the words "circuit," "network," and "system" more or less interchangeably. But the connotations of these words are rather different. "Circuit" has perhaps the most clearly electrical overtones of the three—charged carriers circulating around closed loops. "Network" emphasizes the reticulated topological characteristics of the structure with little reflection of the dynamic implications of the constituent elements. "System," on the other hand, suggests that it will be productive to consider the composite *hierarchically*—as a systematic interconnection of subsystems and, at least potentially, as an element in a supersystem. The power and limitations of the hierarchical "systems approach" to complex structures is an important subtheme of this book; this is a good point to begin explaining why.

Physical science has been most effective in dealing with those phenomena that can be successfully *analyzed*—that is, resolved into constituents, taken apart, reduced to or understood as nothing but the interactions of their components. Indeed, this success (together with the possibility of summarizing the effects of past experiences by the present state of energy distribution throughout the system) essentially defines what we mean by a "physical system." And the inverse of this analysis process—*synthesizing* a complex structure through an appropriate interconnection of elements to realize some overall purpose—is the essence of technological design. In contrast, science and engineering have been much less successful in either understanding or manipulating social, economic, political, or biological systems, where the whole has characteristically a tendency to be greater than—or at least to appear different from—the sum of the parts (and where the past often seems to be more significantly reflected in the current structure of the system than in the distribution of energy or its equivalent within a time-invariant structure).

Another characteristic of physical systems is that the analytic and synthetic processes are usually most effective if carried out in stages rather than all at once. A television receiver, for example, is best understood at the highest level as a combination of amplifiers, mixers, oscillators, filters, gates, detectors, etc., each of which is composed of integrated circuits, transistors, resistors, capacitors, etc., which are in turn composed of various basic materials of appropriate sizes,

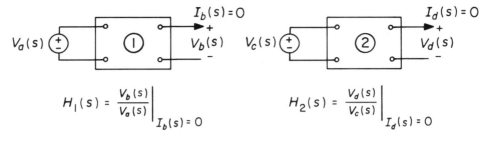

Figure 5.1-2. LTI 2-ports described by open-circuit voltage transfer ratios.

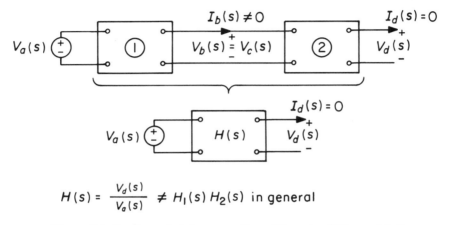

$$H(s) = \frac{V_d(s)}{V_a(s)} \neq H_1(s)\,H_2(s) \text{ in general}$$

Figure 5.1-3. Cascade interconnection of 2-ports of Figure 5.1-2.

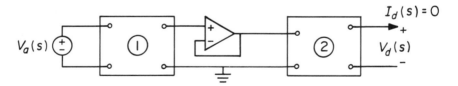

Figure 5.1-4. Use of isolating amplifier in a cascade connection of 2-ports.

between the two 2-ports, as shown in Figure 5.1–4. If this voltage follower is designed into or considered part of the second 2-port, then we are ensuring that the output current of the first 2-port is zero even when it is connected to the second 2-port. If the voltage follower is designed into or considered part of the first 2-port, then we are ensuring that the output voltage of the first 2-port is independent of the current drawn. Either way the result is $H(s) = H_1(s)H_2(s)$. And if we always insert such an isolating amplifier between 2-ports, or if both 2-ports have such voltage followers as input or output stages, then the result of cascading will be independent of the order. An example of the use of isolation in this way is provided by the cascading of Sallen-Key circuits to synthesize

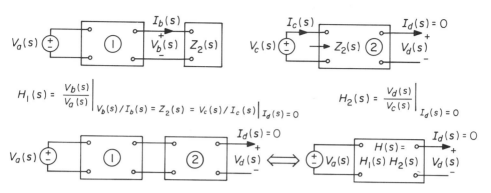

Figure 5.1–5. Redefinition of $H(s)$ to include the effect of loading.

pole-only systems, as discussed in Problem 4.9.*

Another way to make the system function of the cascade equal to the product of the system functions is to measure or calculate the system function $H_1(s)$ of the first 2-port under the condition that it be loaded by an impedance equal to the driving-point impedance of the second 2-port, rather than loaded by an open circuit. Then the system function of the cascade, as shown in Figure 5.1–5, will be $H(s) = H_1(s)H_2(s)$ as desired (although interchanging the order of cascading will not yield the same result unless $H_2(s)$ is similarly defined and appropriate load impedances are provided). Often, if a series of 2-ports are designed to be connected together in various combinations—examples are audio components such as amplifiers, mixers, attenuators, and microphones, and microwave waveguide or coaxial-cable components—it is convenient to design each 2-port so that it achieves its desired characteristics when driven by a Thévenin source and loaded by a resistance of a certain standard value such as 50 Ω or 300 Ω. If this is done, and if the output stage is always loaded by its specified impedance, then the effect of cascading is independent of the order of the components for LTI subsystems.

Cascade and parallel connections of subsystems to make larger systems are extremely important in science and engineering. Nevertheless the class of composite systems is much larger than simply those that can be built up out of successive applications of just these two operations. The simplest example of a more complex system is the *feedback loop* explored in the next section.

---

*In practice the order in which systems are cascaded is often extremely important—even if isolating amplifiers are employed—for reasons having to do with the extent to which our idealizations are valid. Thus if we cascade a large amplifier and a large attenuator, putting the amplifier first may overload the input stages of the attenuator, whereas putting the attenuator first may lead to such a low signal level at the input to the amplifier that incidental "pick-up," power-supply ripple, and thermal noise may become important. To escape this Scylla-Charybdis situation in, for example, long-distance telephone circuits (see Section 5.2), amplifiers are distributed every few kilometers, so that the signal level is never allowed to drop too low or rise too high. Of course, even in principle the effect of cascading is independent of order only if both systems are linear and time-invariant; the design of modulating and detecting systems is critically sensitive to this fact.

5.2 Simple Feedback Loops

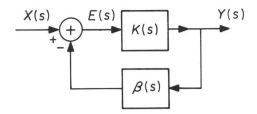

Figure 5.2–1. Simple feedback loop.

An LTI system composed of two LTI subsystems interconnected as shown in Figure 5.2–1 is called a simple *feedback loop*. To find the overall input-output system function $H(s) = Y(s)/X(s)$, write two equations

$$E(s) = X(s) - \beta(s)Y(s), \qquad Y(s) = K(s)E(s)$$

and eliminate the intermediate variable $E(s)$ to obtain

$$H(s) = \frac{K(s)}{1 + \beta(s)K(s)}. \qquad (5.2-1)$$

This is a sufficiently important formula to warrant engraving on your memory. Note that the plus sign in the denominator is a result of choosing the minus sign on the lower (feedback) input to the adder in the block diagram of Figure 5.2–1; if the feedback sign in the diagram had been plus, the denominator sign would have been minus.

There is little in the deceptively simple formula (5.2–1) to suggest the design magic that is hidden there. One additional key step is necessary. Suppose we choose the subsystems $K(s)$ and $\beta(s)$ so that the *loop gain* $K(s)\beta(s)$ has a magnitude much greater than 1. Then*

$$H(s) \approx \frac{K(s)}{\beta(s)K(s)} = \frac{1}{\beta(s)} \quad \text{if } |\beta(s)K(s)| \gg 1. \qquad (5.2-2)$$

---

*Except in certain idealized situations, it is generally impossible to make $|\beta(s)K(s)| \gg 1$ for all complex s. Nevertheless the approximation $H(s) \approx 1/\beta(s)$ will still be valid and useful if the range of s for which $|\beta(s)K(s)| \gg 1$ includes all those s for which the L-transform of the input $X(s)$ has a significant magnitude.

Two aspects of this approximate formula have significance for design:

a) If the loop gain is large, the overall system function is not dependent significantly on the properties of the feedforward path, $K(s)$;

b) If the loop gain is large, the overall system function is approximately equal to the reciprocal of the feedback-path system function, $\beta(s)$.

Historically, the usefulness of the second of these features was not widely recognized until the late 1930's when it became the basis for what is now called classical control theory, which we shall explore in the next chapter. However, the importance of the first feature—that feedback reduces the effect of fluctuations and distortions in the feedforward path—has been at least intuitively understood for a long time. Norbert Wiener, for example, identified the first conscious appreciation of the value of a closed-loop feedback system as appearing in a treatise on the fly-ball steam-engine speed governor published by James Clerk Maxwell in 1868.[*] But much the same idea, applied to the regulation of water-clocks, was described by Archimedes in the third century B.C.; no doubt the first "conscious appreciation" of the value of feedback is far older. Indeed the "inventor" of feedback, whoever he or she was, probably did not consider the "invention" as anything subtle—just a straightforward application of common sense.

The modern theory of feedback systems essentially begins with the work of H. S. Black and his associates (notably H. Nyquist and H. W. Bode) at the Bell Telephone Laboratories in the late 1920's. Black was working on the design of amplifiers for long-distance telephone lines.[†] The first transcontinental system (1914) used #8 (3 mm diameter) copper wire weighing about half a ton per mile. Even so, accumulated losses due to the resistance of 3000 miles of wire amounted to about 60 dB; three to six vacuum tube amplifiers were used to boost the signal amplitude. It was appreciated that if more amplifiers could be used, then the attentuation resulting from smaller wire would be acceptable, leading to potentially significant cost reductions. But the amplifiers of that day had limited bandwidth and introduced substantial nonlinear distortion; the compound effect of cascading more than a very small number of these amplifiers was intolerable. Black's challenge was to invent a better amplifier; his invention (1927)[‡] of the *negative feedback amplifier* was so successful that by 1941 the first coaxial-cable system could use 600 cascaded amplifiers, each with a gain of 50 dB (that is, the cascaded cable losses amounted to a fantastic 30,000 dB!) and a bandwidth so much greater than the 1914 amplifiers that 480 telephone channels were available instead of just one.

In his study, Black distinguished two kinds of feedback—*degenerative* (or *negative*) feedback, in which the feedback signal actually reduces the input to the $K(s)$ block (in our notation, this means $\beta > 0$), and *regenerative* (or *positive*) feedback, in which the feedback signal increases the input ($\beta < 0$ or the sign on

---

[*]J. C. Maxwell, *Proc. Roy. Soc. (London)* (March 5, 1868).

[†]H. W. Bode, *Proc. Symp. Active Networks and Feedback Systems* (Polytechnic Institute of Brooklyn, NY: Polytechnic Press, 1960).

[‡]The first open publication was in *Electrical Engineering, 53* (Jan. 1934): 114–120.

the adder changed to +).* The advantageous effects of regenerative feedback—increasing the gain and (for $|\beta K| > 1$) producing useful oscillations—had been recognized long before Black, but degenerative feedback was generally thought to be deleterious, since obviously it reduced the overall gain.† Black, however, pointed out that if one sought a highly reliable overall system with behavior insensitive to distortions or changes in the always imperfect active elements in the K branch, then it was definitely desirable to design initially an amplifier with more gain than ultimately needed and to reduce the gain by negative (passive) feedback to the desired amount. His argument was essentially that already given: If $|\beta(s)K(s)| \gg 1$, then the overall system function is $H(s) \approx 1/\beta(s)$, independent of $K(s)$ and hence independent of many of the corruptions and limitations of $K(s)$. The best way to appreciate how this works in detail is to consider a number of examples.

5.3 Examples of the Effects of Negative Feedback

Example 5.3–1

Suppose it is desired to build an amplifier having a gain of 10 and capable of supplying some tens of watts to a load such as a loudspeaker. Such an amplifier could be built using a single-stage amplifier employing a power transistor and no feedback, or it could be built using a multistage amplifier employing the same power transistor in the output stage and using feedback to reduce the overall gain to 10. The two possibilities are described by the block diagrams in Figure 5.3–1, where to be specific we have assumed that the part of the multistage amplifier preceding the power stage has a gain of 100.

Now suppose that the gain of the power amplifier is reduced to half its initial value (as a result perhaps of aging of the active elements, or changes in loading, or temperature, or power supply voltages). The overall gain of the non-feedback amplifier is then also reduced by 50%, whereas that of the feedback amplifier has only been changed by 1%.

---

*The distinction between positive and negative feedback is clear enough if β and K are real constant multipliers (gains or attenuations) but is less clear if $\beta(s)$ and $K(s)$ are complex functions of s (as we shall usually assume).

†Black's patent application was delayed for more than nine years in part because the concept was so contrary to established beliefs that the Patent Office initially did not believe it would work. They treated the application "in the same manner as one for a perpetual motion machine." See H. S. Black, "Inventing the negative feedback amplifier," *IEEE Spectrum* (Dec. 1977): 54–60.

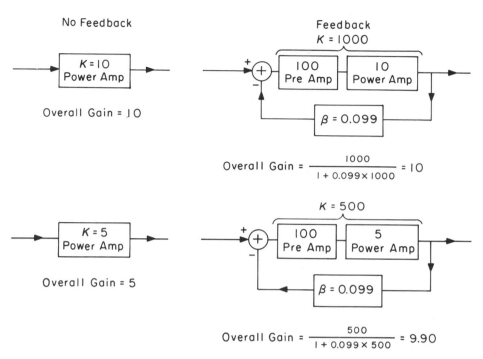

Figure 5.3–1. Gain change effects in a power amplifier with and without feedback.
▶ ▶ ▶

Bode* formalized the effect illustrated by Example 5.3–1 by defining a quantity called the *sensitivity*, S, of the amplifier as

$$S = \frac{\text{fractional change in gain of overall system}}{\text{fractional change in gain of active element}} = \frac{\dfrac{\Delta H}{H}}{\dfrac{\Delta K}{K}}. \qquad (5.3\text{–}1)$$

For an amplifier without feedback, $S = 1$. On the other hand, for small changes the sensitivity of a feedback amplifier is

$$S \approx \frac{K}{H}\frac{\partial H}{\partial K} = \frac{\partial \ln H}{\partial \ln K} = \frac{1}{1 + \beta K} \qquad (5.3\text{–}2)$$

---

*H. W. Bode, *Network Analysis and Feedback Amplifier Design* (New York, NY: Van Nostrand, 1945) p. 52. Our definition is actually the reciprocal of Bode's.

which shows that the effect of negative feedback is to make the system less sensitive (than an unfedback system) by $1/(1 + \beta K)$, or for large loop gain by approximately one over the loop gain. Correspondingly, the effect of regeneration is to increase the sensitivity, often leading for large regeneration to instability and oscillation.

Less formally, as we previously argued, for large loop gain ($|\beta K| \gg 1$) the overall gain is

$$H \approx \frac{1}{\beta} \tag{5.3-3}$$

independent of K and dependent only on β, which is usually determined by passive, linear, cheap, and reliable elements. In this same spirit, we can consider the adder in the diagram as a *comparator*—comparing the input x with βy (the inverse of the desired operation, operating on the output); any *error signal* is so heavily amplified that it must be very small. This, of course, is the approach we have been taking all along toward the analysis of ideal op-amp circuits. Our goal in this chapter is to abstract the general principles of feedback that previously we have illustrated only for specific cases.

Example 5.3–2

Real circuits rarely fit the simple feedback-loop model exactly; because of loading effects, the identification of β and K is usually neither easy nor unique. Fortunately, the difficulties become less when the loop gain βK is large—which, of course, is precisely the range of values for which feedback has a significant effect.

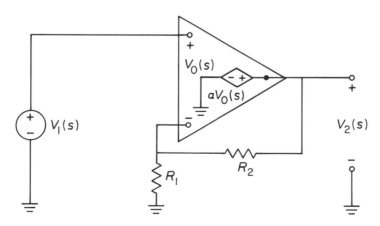

Figure 5.3–2. Non-inverting op-amp amplifier circuit.

a) The non-inverting op-amp amplifier shown in Figure 5.3–2 is a rare exception to this rule. If we model the op-amp as shown with infinite input impedance, zero output impedance, and finite gain α, it is apparent that

$$V_0(s) = V_1(s) - \frac{R_1}{R_1 + R_2} V_2(s), \quad V_2(s) = \alpha V_0(s).$$

These equations correspond exactly with the equations for the simple feedback loop if we identify $K = \alpha$ and $\beta = R_1/(R_1 + R_2)$, so that

$$\frac{V_2(s)}{V_1(s)} = \frac{K}{1 + \beta K} = \frac{\alpha}{1 + \dfrac{\alpha R_1}{R_1 + R_2}} \approx \frac{R_1 + R_2}{R_1} = \frac{1}{\beta}$$

independent of α if the loop gain $\beta K = \dfrac{\alpha R_1}{R_1 + R_2} \gg 1$.

b) On the other hand the more common inverting op-amp amplifier circuit shown in Figure 5.3–3 illustrates the more usual situation. Using the same op-amp model as before, we may write

$$V_0(s) = \frac{-R_2}{R_1 + R_2} V_1(s) - \frac{R_1}{R_1 + R_2} V_2(s), \quad V_2(s) = \alpha V_0(s)$$

which are not precisely in the form of the simple feedback-loop equations. If we choose $K = \alpha$ and $\beta = R_1/(R_1 + R_2)$ as before, we are in effect describing the circuit by the block diagram in Figure 5.3–4.

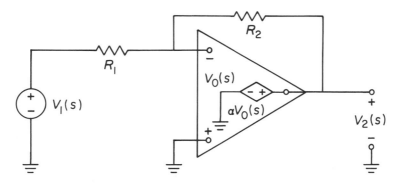

Figure 5.3–3. Inverting op-amp amplifier circuit.

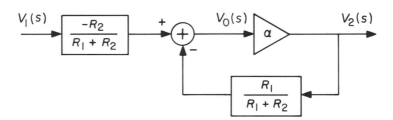

Figure 5.3–4. Block diagram for an inverting op-amp amplifier.

The overall gain is

$$\frac{V_2(s)}{V_1(s)} = -\frac{R_2}{R_1 + R_2} \frac{\alpha}{1 + \alpha\dfrac{R_1}{R_1 + R_2}} \approx -\frac{R_2}{R_1}$$

independent of α if the loop gain $\alpha R_1/(R_1 + R_2) \gg 1$. We note that this large-α gain is not $1/\beta$. Alternatively, we might choose $\beta = R_1/R_2$ and $K = \alpha R_2/(R_1 + R_2)$, which is equivalent to describing the circuit in terms of the block diagram in Figure 5.3–5. Except for a sign, this block diagram is identical to a simple feedback-loop block diagram and does reduce for large K to $-V_2(s)/V_1(s) = 1/\beta$. Obviously, a large number of other choices of β and K corresponding to still other block diagrams are possible; there is no uniquely "correct" or "best" choice. We note, however, that—independent of the choice—the loop gain βK is well-defined and can be computed by "opening" the loop at any convenient spot, applying a unit source, and calculating the signal that returns to the other side of the opening. (Care must be taken that loading effects, if any, are properly accounted for.)

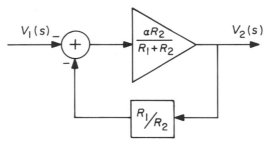

Figure 5.3–5. Another block diagram for an inverting op-amp amplifier.

▶ ▶ ▶

Perhaps the most important conclusion to be drawn from the ambiguity in identifying β and K illustrated in Example 5.3–2 is that feedback is clearly and undeniably a useful concept in the design or synthesis of a new system to achieve some desired performance. Whether it is useful in the analysis and understanding of some existing system depends on the particular case. If the system being analyzed was in fact consciously designed as a feedback system, then it will almost certainly be effective to analyze it in these terms. The feedback paths will usually then turn out to be structurally distinct (or nearly so), and simplifying assumptions will suggest themselves that will both reduce the analytical effort and enhance our understanding. But if the system being studied is merely "complicated" (for example, a biological system in which typically "everything influences everything else"), then the utility of feedback as a guide to analysis is more doubtful. For it is always possible to formulate any system analysis problem—even the voltage divider shown in Figure 5.3–6—in feedback terms, but there is certainly no guarantee that such a view will prove helpful. It is worth keeping in mind that, whatever the block diagram, the useful features of feedback to be described in this chapter usually seem to depend on the fact that the K block is an active element, capable of power gain.

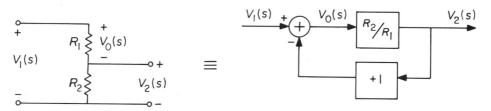

Figure 5.3–6. A voltage divider as a "feedback" system.

Example 5.3–3

To explore the way in which feedback reduces the distorting effects of non-linearities, consider an amplifier whose output $y(t)$ is a non-linear memory-less function of its input $z(t)$:

$$y(t) = f[z(t)].$$

Specifically, suppose $f[\cdot]$ has the shape shown by the solid line in Figure 5.3–7; this graph exhibits *saturation* of the output for large values of the input as well as *dead zone* or *crossover distortion* in that the output is zero until the magnitude of the input exceeds a threshold value. Such characteristics are common in op-amp output stages.

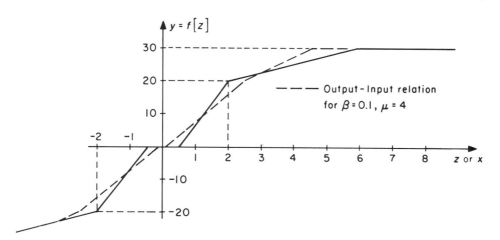

Figure 5.3–7. Non-linear input-output characteristic.

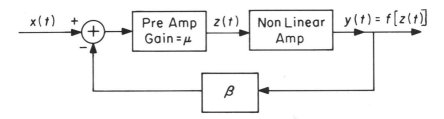

Figure 5.3–8. Feedback amplifier containing a non-linear output stage.

Suppose this amplifier is cascaded with a pre-amplifier whose input contains a fedback fraction of the non-linear amplifier output, as shown in Figure 5.3–8.   We readily compute that

$$z(t) = \mu(x(t) - \beta y(t)) = f^{-1}[y(t)]$$

or

$$x(t) = \beta y(t) + \frac{1}{\mu} f^{-1}[y(t)]$$

where $f^{-1}[\ ]$ is the inverse amplifier function shown in Figure 5.3–9. For $y < 30$ and large enough μ, we get approximately

$$y(t) \approx \frac{1}{\beta} x(t)$$

which is a linear relationship. To explore the nature of the approximation, let $\beta = 0.1$ and $\mu = 4$; the output-input relationship is shown dashed in Figure 5.3–7 and is clearly—within the absolute limits imposed by output saturation—a substantial improvement in linearity over the unfedback amplifier.

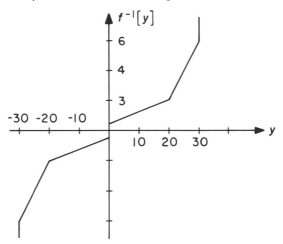

Figure 5.3–9. Inverse amplifier characteristic.

▶ ▶ ▶

Example 5.3–4

Consider a power amplifier that is nearly linear but produces a little distortion at full power output. By this we mean that we can describe such an amplifier approximately by the block diagram in Figure 5.3–10, in which $n(t)$ is the difference between the actual distorted output and the output of a linear amplifier with gain K.

Figure 5.3–10. Block diagram modelling a nearly linear amplifier.

If $n(t)$ is a separate signal, independent of $x(t)$, then this block diagram describes a linear system, and we shall analyze it as if this were the case. But, of course, since the distortion in a non-linear system actually depends upon the input $x(t)$, it must be emphasized that the analysis scheme to be discussed is an acceptable approximation only if the output level is held fixed and if the distortion is small (so that, for example, if we feed back $n(t)$, the "distortion of the distortion" can be ignored).

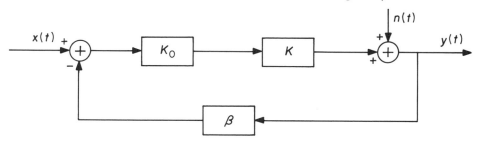

Figure 5.3–11. Feedback added to amplifier model of Figure 5.3–10.

Suppose now we feed back a fraction β of $y(t)$ and add additional gain to K to compensate, as shown in Figure 5.3–11. By superposition (the approximate system is linear), we readily find

$$y(t) = \frac{K_0 K}{1 + \beta K_0 K} x(t) + \frac{1}{1 + \beta K_0 K} n(t).$$

If we adjust K_0 and β so that $K_0/(1 + \beta K_0 K) = 1$, the system has the same gain as before; but for the same amplitude of the desired component of the output, the distortion has been reduced by the factor $1/(1 + \beta K_0 K)$, which can be considerable.
▶ ▶ ▶

Of course, the analysis in Example 5.3–4 applies without the approximation to any situation in which $n(t)$ really is an independent added disturbance or noise. It should be observed, however, that feedback can markedly reduce the effect of noise added to the output, but has no effect on noise added to the input and only an intermediate effect on noise added at an intermediate point (such as between K_0 and K in Figure 5.3–11). One typical application of this effect of feedback is in multistage amplifiers in which the d-c supply voltages are obtained by rectifying a-c. In such an amplifier, the voltages for the final stages do not need to be particularly well filtered if substantial feedback is employed.

Example 5.3–5

One of the most important uses of feedback is to reduce the effect on the K circuit of changes in some impedance. The Watt speed governor for steam engines analyzed by Maxwell can be considered a design of this sort, intended to reduce the effects on speed of changes in the mechanical load. A similar electrical example is the design of a voltage regulator (see Problem 5.1). Another example is the design of an amplifier to drive a loudspeaker in a high-fidelity sound reproduction system. Such an amplifier must cope with the fact that a loudspeaker has an input impedance that is a rather wild function of frequency. To achieve an overall flat frequency response, it is usually considered desirable to keep the voltage across the speaker terminals constant as a function of frequency (for constant input voltage to the amplifier)—for a permanent-magnet type speaker, voltage by Faraday's Law controls voice-coil velocity. Consequently, a high-fidelity amplifier should behave as an ideal voltage source.

The effective output impedance of an amplifier can be reduced by feeding back a signal proportional to the amplifier load voltage and comparing it with the amplifier input signal. Consider, for example, the non-inverting op-amp circuit of Example 5.3–2—reproduced in Figure 5.3–12—in which a fraction β of the output voltage is fed back. Assume $R \gg R_0$ for simplicity. The effective output impedance looking back from the terminals of R_L is the value of R_1 in the Thévenin equivalent to the feedback amplifier shown in Figure 5.3–13. In circuits containing controlled sources, the most effective way to compute the Thévenin resistance is usually

$$R_1 = \frac{V_2(s) \text{ when } R_L = \infty}{I_2(s) \text{ when } R_L = 0} = \frac{\text{open-circuit voltage}}{\text{short-circuit current}}.$$

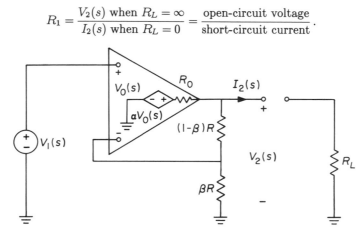

Figure 5.3–12. Non-inverting op-amp amplifier.

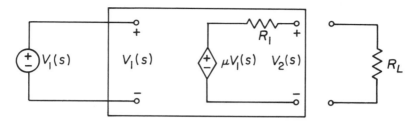

Figure 5.3–13. Thévenin equivalent to non-inverting op-amp amplifier.

The short-circuit current in the circuit of Figure 5.3–12 is easy to find since there is no feedback when the output terminals are shorted. Thus

$$I_2(s) \text{ (short-circuit)} = \frac{\alpha V_1(s)}{R_0}.$$

Since $R \gg R_0$, on open circuit $V_2(s) \approx \alpha V_0(s)$ and $V_0(s) = V_1(s) - \beta V_2(s)$, so that $V_2(s)/\alpha \approx V_1(s) - \beta V_2(s)$. Solving, we find that

$$V_2(s) \text{ (open-circuit)} \approx \frac{\alpha V_1(s)}{1 + \beta \alpha}$$

and consequently

$$\mu = \frac{\alpha}{1 + \beta \alpha} \approx \frac{1}{\beta}.$$

Hence, finally,

$$R_1 = \frac{\alpha V_1(s)}{1 + \beta \alpha} \frac{R_0}{\alpha V_1(s)} = \frac{R_0}{1 + \beta \alpha}$$

which, if $\beta \alpha \gg 1$ (as it usually is), represents a significant reduction from the unfedback amplifier. For example, a 741 op-amp in a voltage-follower circuit ($\beta = 1$) has $R_0 \approx 75$ Ω and $\alpha \approx 2 \times 10^5$. Then

$$R_1 = \frac{75}{1 + 2 \times 10^5} \approx 3.75 \times 10^{-4} \Omega.$$

▶ ▶ ▶

It is perhaps worth observing that the overall gain and the output impedance are identical in theory for the feedback amplifier in Example 5.3–5 and for the unfedback amplifier with the output simply shunted by an appropriate resistor. However, most active elements used as the last stage in a power amplifier must work into an appropriate load impedance to achieve the desired level of power output without saturation or other distortion effects. Thus, in practice, shunting usually cannot be employed to reduce the effective output impedance of the amplifier, but feedback can.

Example 5.3–6

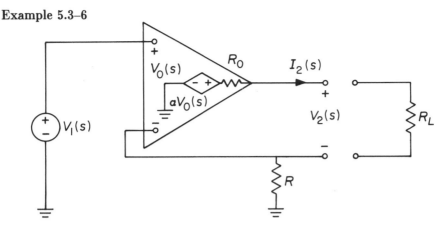

Figure 5.3–14. Amplifier with feedback proportional to output current.

A scheme similar to that of the preceding example can be used to increase the output resistance, that is, to make the amplifier behave more nearly like an ideal current source by feeding back a voltage proportional to the output current. Consider the circuit shown in Figure 5.3–14. On open circuit, there is no feedback. Thus

$$V_2(s) \text{ (open-circuit)} = \alpha V_1(s).$$

On short circuit, $I_2(s) = \alpha V_0(s)/(R_0 + R)$ and $V_0(s) = V_1(s) - RI_2(s)$. Eliminating $V_0(s)$ and solving, we obtain

$$I_2(s) \text{ (short-circuit)} = \frac{\alpha V_1(s)}{R_0 + (1 + \alpha)R}.$$

Hence R_1 in the Thévenin equivalent circuit is

$$R_1 = \frac{V_2(s) \text{ (open-circuit)}}{I_2(s) \text{ (short-circuit)}} = \alpha V_1(s) \frac{R_0 + (1 + \alpha)R}{\alpha V_1(s)} = R_0 + (1 + \alpha)R$$

which can be quite large. This circuit is occasionally used to provide an approximation to an ideal current source, although it has the disadvantage of no common ground between input and output.

▶ ▶ ▶

Example 5.3–7

Feedback can also have a marked effect on the input impedance to various circuits. It is sometimes useful to consider the standard op-amp integrator circuit in Figure 5.3–15 from this point of view.

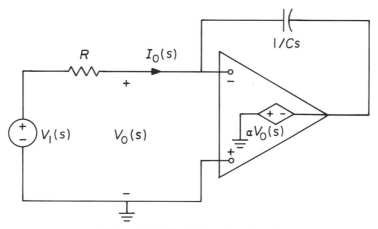

Figure 5.3–15. Integrator circuit.

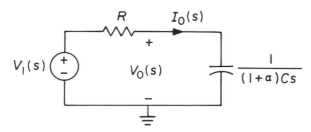

Figure 5.3–16. Equivalent input circuit for the integrator of Figure 5.3–15.

We readily compute that the voltage across the capacitor is $V_0(s)(1+\alpha)$, so that $I_0(s) = CsV_0(s)(1+\alpha)$. Thus from the standpoint of the driving source the op-amp-capacitor combination is equivalent to a capacitor of value $(1+\alpha)C$, as shown in the equivalent input circuit of Figure 5.3–16. The value of $(1+\alpha)C$ can be very large. This, of course, is what makes the circuit function as an approximation to an ideal integrator. But the same idea has other applications. The capacitor C might be simply a stray capacity coupling the output and input of an amplifier stage; the multiplying effect of the gain of the amplifier then creates a large effective input shunt capacitance that can significantly reduce the high-frequency gain of the preceding stage. This is called the *Miller effect* and historically was important in limiting the radio-frequency gain of early vacuum-tube amplifiers, until the invention of the pentode substantially reduced the effective value of C. (For a discussion of the analogous problem in transistors see Problem 5.5.) Another application depends on the fact that the gain of an electronic amplifier can often be changed electrically by adjusting the operating point. If a feedback capacitor is connected around such an amplifier as in Figure 5.3–15, then the combination yields a capacitance whose value can be changed electrically. Such an element is widely useful in electrically-tuned filters, FM modulators, etc.

▶ ▶ ▶

5.4 Summary

The analysis of systems composed of interconnected elements leads naturally to the next hierarchical level—the interconnection of systems to yield supersystems. The arrangement of two subsystems in a simple feedback loop turns out to be particularly interesting as a design tool. Indeed, we agree with J. K. Roberge that "a detailed understanding of feedback is the single most important prerequisite to successful electronic circuit and system design."* In this chapter we have studied the way in which feedback reduces the effects of variations and distortions in the feedforward path because the overall system function for large loop gain is not highly dependent on the characteristics of this path. In the next chapter we shall illustrate how we can use the fact that for large loop gain the overall system function is approximately equal to the reciprocal of the system function describing the feedback path.

---

*J. K. Roberge, *Operational Amplifiers: Theory and Practice* (New York, NY: John Wiley, 1975). This is an excellent text for further study of the topic of this chapter.

EXERCISES FOR CHAPTER 5

Exercise 5.1

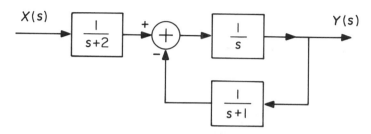

Show that the system function of the block diagram above is

$$H(s) = \frac{Y(s)}{X(s)} = \frac{s+1}{s^3 + 3s^2 + 3s + 2}.$$

Exercise 5.2

In the non-inverting op-amp amplifier circuit of Example 5.3–2, suppose that R_1 and R_2 are chosen such that the ideal $(\alpha = \infty)$ gain is 10. Show that α must be greater than 9.99×10^3 if the actual gain is to be within 0.1% of the ideal value.

PROBLEMS FOR CHAPTER 5

Problem 5.1

If the input to the non-inverting op-amp amplifier of Example 5.3–5 is a constant (e.g., a battery or the voltage across a Zener diode), then the circuit shown below becomes a series *voltage regulator* designed to maintain a fixed output voltage $V_0 \approx V_R/\beta$ independent of load resistance (provided that the output current I_0 is less than some maximum determined by the amplifier). Integrated-circuit voltage regulators such as the LM317 often work on this principle and combine all the relevant elements on a single chip. Sketch the regulation characteristic V_0 vs. I_0 as R_L is varied, and show that the desired properties are achieved as $\alpha \to \infty$. Assume for simplicity that $R \gg R_0$.

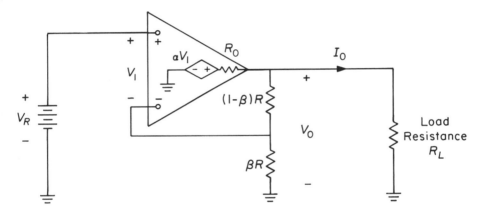

Problem 5.2

Find the Thévenin equivalent circuit at the output of the following feedback amplifier. Assume $i_0(t) = 0$.

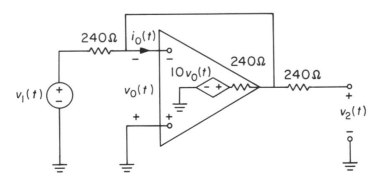

Problem 5.3

Find the input resistance $v_1(t)/i_1(t)$ of the following voltage-follower circuit. Model the op-amp as suggested in the figure.

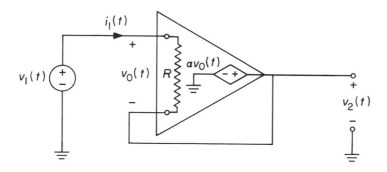

Problem 5.4

The design of satisfactory analog multipliers is a perennial problem. One useful scheme, shown below, employs two identical voltage-controlled amplifiers or attenuators whose gain is (ideally) an instantaneous monotonic (but not necessarily linear) function of the control voltage $v(t)$.

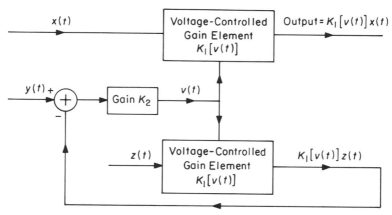

a) Argue that for a sufficiently large value of K_2 the output is approximately equal to $x(t)y(t)/z(t)$ independent of the shape of $K_1[\ ]$.

b) Let $z(t) = 1$, $K_1[v(t)] = 1 + v(t)$, and assume that $x(t)$ and $y(t)$ are restricted to the range $0 \leq x(t) \leq 10$, $0 \leq y(t) \leq 10$. Find the minimum value of K_2 such that the output error is never greater than 1% of the maximum value of the product $x(t)y(t)$.

Problem 5.5

This problem explores the effect of collector-to-base capacitance on the frequency response of the common-emitter amplifier stage shown in the figure. Assuming that the npn transistor can be represented near its operating point by a hybrid-π model, the overall equivalent circuit at middle and higher frequencies might appear approximately as shown on the right.

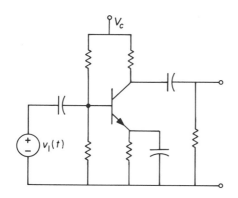
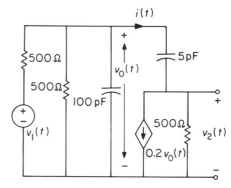

Circuit Diagram	**Equivalent Circuit**
Capacitors shown are low-frequency blocking and bypass capacitors	Blocking and bypass capacitors have been replaced by short circuits

   The two capacitors describe various charge-storage effects in the base region. In particular the 5 pF capacitor represents primarily the capacitance of the back-biased collector-base junction; although its value is small, its effect is amplified because of its location in the feedback path. As mentioned in Example 5.3–7, this is called the *Miller effect*.

a) Calculate the input impedance looking to the right at the point where the current is labelled $i(t)$ and thus show that the impedance presented to the source $v_1(t)$ can be represented by the following equivalent circuit.

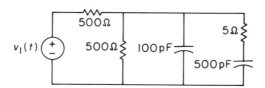

b) Determine approximately the half-power frequency of this amplifier stage and compare with the half-power frequency if the collector-base junction capacitance had been zero. The effective multiplication of the value of this capacitance by the gain of the stage can thus be a serious problem.

6

THE DYNAMICS OF FEEDBACK SYSTEMS

6.0 Introduction

Adding feedback around an existing system, $K(s)$, yields a new system that not only may be less sensitive to variations and distortions of $K(s)$ (if, as we saw in Chapter 5, the loop gain is large), but also in general has different pole-zero locations than $K(s)$ and thus a different dynamic behavior. The effect of feedback on system dynamics was extensively studied in the 1940's and 1950's as an approach to the design of *control systems*. Indeed, Norbert Wiener* and his followers made feedback control under the name "cybernetics" one of the cornerstones of a virtually complete philosophical system embracing everything from automation, to physiological and psychological processes, the behavior of economic and social systems, and even the discovery of principles of ethics. Time and the development of alternative perspectives have somewhat diminished this glorious vision, but feedback control remains both an important concept and a useful design technique.

6.1 Inverse Systems

The dynamic effect of feedback depends on the magnitude of the loop gain, $\beta(s)K(s)$, over the interesting range of frequencies. If the amount of feedback is "small" $(|\beta(s)K(s)| \ll 1)$, then

$$H(s) = \frac{K(s)}{1 + \beta(s)K(s)} \approx K(s) \qquad (6.1-1)$$

and the dynamic effect of feedback is also small; the poles and zeros of $H(s)$ will usually be in nearly the same locations as the poles and zeros of $K(s)$. On the other hand, if the amount of feedback is "large" $(|\beta(s)K(s)| \gg 1)$, then

$$H(s) = \frac{K(s)}{1 + \beta(s)K(s)} \approx \frac{1}{\beta(s)}. \qquad (6.1-2)$$

---

*N. Wiener, *Cybernetics, or Control and Communication in the Animal and the Machine* (New York, NY: John Wiley, 1948).

The dynamic effect of feedback in this case can be enormous, since the natural frequencies of $H(s)$ are approximately the poles of $1/\beta(s)$ rather than the poles of $K(s)$.

In the case of a large loop gain, the system function of the feedback system is approximately the inverse of the system function of the feedback path. Employing a feedback system with a large loop gain to realize a system function inverse to that of a given system is often a useful design technique. An inverse system is typically desired to help overcome or compensate for the dynamic deficiencies of some communication, measurement, or control device. For example, suppose a sensor responds sluggishly to the changes in some quantity $x(t)$. In such cases, the transform of the sensor output can often be described as the product, $X(s)H_0(s)$, rather than simply the transform, $X(s)$, of the desired signal $x(t)$. The system function $H_0(s)$ thus describes the corrupting effect of the sluggish sensor. Suppose, however, we could arrange to connect the sensor in cascade with a system whose system function is $1/H_0(s)$; then the transform of the overall response would be $X(s)H_0(s)(1/H_0(s)) = X(s)$, that is, the inverse system $1/H_0(s)$ would recover the uncorrupted signal $x(t)$ from the distorted sluggish output waveform of the sensor.* Feedback is a particularly effective way to realize an inverse system if the details of $H_0(s)$ are unknown or if they change with time or environmental conditions; a replica of the original system in the feedback path should presumably behave similarly under similar conditions. Feedback may also be useful for synthesizing an inverse system if a system with system function $H_0(s)$ is easier to build for some reason than a system with system function $1/H_0(s)$. The following examples illustrate several applications of this idea.

Example 6.1–1

It is easy to show by impedance manipulations that the open-circuit voltage transfer ratio of the *bridged-T network* in Figure 6.1–1 is as given with frequency response and pole-zero locations as shown (for large values of a). The inverse of $H_0(s)$ can be realized approximately by placing the bridged-T network in the feedback path around an amplifier, as shown in Figure 6.1–2. This circuit will have the system function $H_2(s) \approx 1/H_0(s)$, with pole-zero locations and frequency response as shown. Since the poles of $H_2(s)$ are located at complex frequencies, a passive circuit realizing $H_2(s)$ would have to contain inductors. Several other circuits using feedback to achieve complex poles with RC elements have been described previously (see Examples 1.3–2, 1.4–2, 4.2–1, and Problems 1.6, 3.11, 4.9, 4.15).

---

*The process of operating on the output of an LTI system to recover its input is also called *deconvolution*, for reasons to be described in Chapter 10. Note that, for reasons discussed in Chapter 3, an inverse system cannot usually be constructed by simply driving the output terminals of the original system and observing the signal at the input terminals.

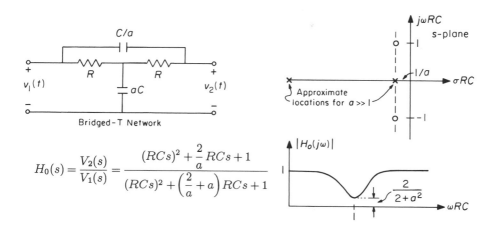

Figure 6.1–1. Bridged-T network.

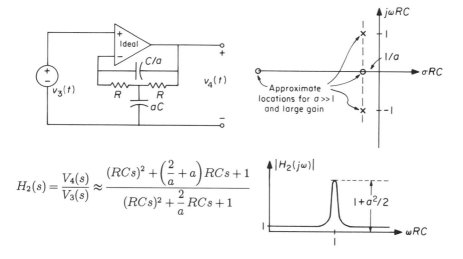

Figure 6.1–2. Approximate inverse of bridged-T transfer function.

▶ ▶ ▶

Note that, since the zeros of a passive transfer function may be in the right half-plane, attempts to realize the inverse of such a system may lead to an unstable structure with right half-plane poles. Instabilities in feedback systems can arise in many ways and set limits on the benefits that can be achieved with feedback, as we shall discuss in Section 6.3.

Example 6.1–2

The use of feedback to construct inverse systems is not limited to linear systems. Thus many types of transistors show exponential dependence of collector current on base-emitter voltage near zero collector-base voltage. The exponential law holds for many decades. (A similar relationship holds for diodes, but over a smaller dynamic range.) The circuit shown in Figure 6.1–3 then yields an accurate logarithmic characteristic that can be exploited in analog multipliers and many other devices.

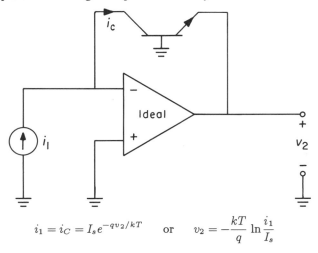

$$i_1 = i_C = I_s e^{-qv_2/kT} \quad \text{or} \quad v_2 = -\frac{kT}{q} \ln \frac{i_1}{I_s}$$

Figure 6.1–3. A logarithmic amplifier.

▶ ▶ ▶

6.2 Feedback Effects on Bandwidth and Response Time

If the magnitude of the loop gain $|\beta(s)K(s)|$ is not much greater than 1, then the overall feedback system response is not $1/\beta(s)$, independent of $K(s)$. Nevertheless, the overall response may still be dramatically different from the response of the feedforward path alone, as the following examples illustrate.

Example 6.2–1

Figure 6.2–1 shows the open-loop gain of a typical 741 op-amp. This is a Bode plot with low-frequency gain $\approx 2 \times 10^5$, 6 dB/octave high-frequency slope, and a half-power bandwidth of about 6 Hz ≈ 40 rad/sec. We may thus represent the op-amp up to fairly high frequencies by the equivalent circuit shown in Figure 6.2–2 with $K(s) = \dfrac{8 \times 10^6}{s + 40}$.

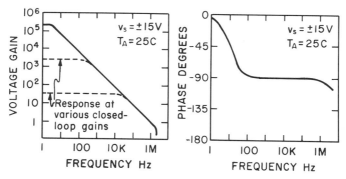

Figure 6.2–1. Open-loop voltage gain and phase response of a 741 op-amp as functions of frequency.

If the op-amp is now connected in the non-inverting amplifier circuit shown in Figure 6.2–2 with

$$\beta = \frac{R_1}{R_1 + R_2}$$

then the closed-loop system function is

$$H(s) = \frac{V_2(s)}{V_0(s)} = \frac{K(s)}{1 + \beta K(s)} = \frac{8 \times 10^6}{s + (40 + 8 \times 10^6 \beta)}.$$

The half-power bandwidth has thus been raised to $(40 + 8 \times 10^6 \beta)$ rad/sec, which for typical values of β is very much larger than the open-loop bandwidth of 40 rad/sec.

It is interesting and useful to observe that the product of the low-frequency gain,

$$H(0) = \frac{8 \times 10^6}{40 + 8 \times 10^6 \beta}$$

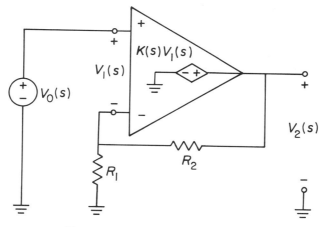

Figure 6.2–2. Non-inverting amplifier.

and the half-power bandwidth is a constant,

$$\text{gain} \times \text{bandwidth (Hz)} = \frac{8 \times 10^6}{2\pi} \approx 1.2 \times 10^6 \text{ Hz}$$

independent of β. (The value of this constant is often given in op-amp specification listings as the *unit-gain bandwidth*.) The Bode plots for the closed-loop frequency response thus have the form shown by the dotted lines in the preceding figure—the smaller the required gain, the greater the resulting bandwidth. For example, a 741 op-amp in a feedback circuit yielding a closed-loop gain of 100 will have a half-power bandwidth equal to $1.2 \times 10^6/100 = 12$ kHz. For further discussion of the frequency response effects of op-amps, see Problem 6.9.

▶ ▶ ▶

Example 6.2–2

As a second example we shall study the effect of feedback on the sluggish response of a motor speed controller. Mathematically, this system is almost identical to the feedback amplifier examined in the last example, but the context is different and we shall emphasize the step response rather than the frequency response.

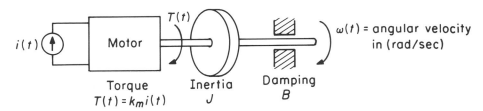

Figure 6.2–3. Description of an electric motor and its load.

The basic system to be controlled is shown in Figure 6.2–3. For simplicity, the electric motor will be idealized as producing a torque $T(t)$ proportional to armature current. Mechanically, the motor and load together will be represented by an inertia J and a damping term B. Let us first study the way in which the shaft speed $\omega(t)$ responds to changes in the current $i(t)$. Balancing drive and reaction torques yields

$$T(t) = k_m i(t) = J\frac{d\omega(t)}{dt} + B\omega(t).$$

The system function relating shaft speed and current is then

$$H_m(s) = \frac{\Omega(s)}{I(s)} = \frac{k_m}{Js + B}.$$

If a step of current $i(t) = Iu(t)$ is applied from rest, then $I(s) = I/s$ and the ZSR is

$$\Omega(s) = H_m(s)I(s) = \frac{k_m I/J}{s(s + B/J)} = \frac{k_m I/B}{s} - \frac{k_m I/B}{s + B/J}$$

or

$$\omega(t) = \omega_\infty(1 - e^{-t/t_0})u(t)$$

where the final velocity is $\omega_\infty = k_m I/B$ and the time constant is $t_0 = J/B$. The step drive and motor response are shown in Figure 6.2–4. If the inertia J of the system is large, the step response will be sluggish. Moreover, both the response time constant and the final speed depend directly on the mechanical damping B, which is frequently an unreliable parameter, sensitive to aging, temperature, etc.

Figure 6.2–4. Motor speed response to a step current input.

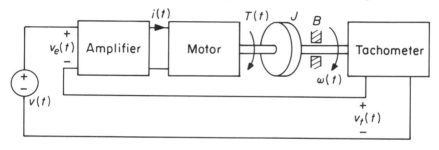

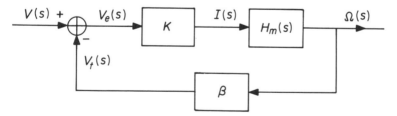

Figure 6.2–5. Motor controller with velocity feedback.

A more rapid and reliable response can be obtained by adding feedback to the motor as shown in the upper part of Figure 6.2–5. A tachometer is a measuring device (basically an electrical generator) whose output voltage is proportional to input shaft velocity:

$$v_t(t) = \beta\omega(t).$$

The amplifier is described by the equation

$$i(t) = Kv_e(t).$$

Hence the system can be represented by the block diagram in the lower part of Figure 6.2–5. The form of the overall system function $H(s)$ is the same as for the unfedback system $H_m(s)$, but the pole is at $-(B+\beta K k_m)/J$ instead of $-B/J$:

$$H(s) = \frac{\Omega(s)}{V(s)} = \frac{K H_m(s)}{1+\beta K H_m(s)} = \frac{K k_m/J}{s+(B+\beta K k_m)/J}.$$

Consequently, for $\beta K > 0$ the bandwidth is larger and the step response

$$w(t) = \hat{\omega}_\infty\left(1 - e^{-t/\hat{t}_0}\right)u(t) \qquad \text{where} \qquad \hat{\omega}_\infty = \frac{K k_m V}{B+\beta K k_m}, \quad \hat{t}_0 = \frac{J}{B+\beta K k_m}$$

is faster than before, as shown in Figure 6.2–6. Note also that both the time constant of the response and the final motor speed in the feedback system are largely independent of the damping B if βK is large enough.

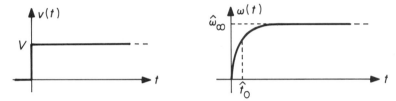

Figure 6.2–6. Response of the feedback system to a step input.

As $\beta K \to \infty$, the response time of the feedback system to an input step change in desired speed theoretically goes to zero, and the final speed depends only on the tachometer calibration β. What limits the improvements that can be achieved in practice? One problem is that as βK grows, so does the peak input current to the motor. To see this, treat $I(s)$ as the output of the feedback system and compute

$$\begin{aligned}
I(s) &= \frac{KV(s)}{1+\beta K H_m(s)} = \frac{K(s+B/J)V}{s(s+B/J+\beta K k_m/J)} \\
&= \frac{\hat{\omega}_\infty B/k_m}{s} + \frac{\hat{\omega}_\infty \beta K}{s+B/J+\beta K k_m/J}.
\end{aligned}$$

The corresponding time function $i(t) = \hat{\omega}_\infty[B/k_m + \beta K e^{-t/\hat{t}_0}]u(t)$ is sketched in Figure 6.2–7.

Physically, a large initial current is necessary to provide sufficient torque to quickly overcome the inertia of the motor and load. However, burnout and saturation effects

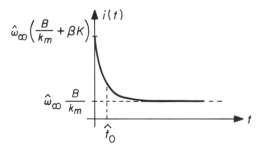

Figure 6.2–7. Armature current for a step input to the feedback controller.

limit the applicable range of our simplified representation of the motor to input currents less in magnitude than some maximum I_{max}. Assuming that the worst transient results from a sudden reversal from maximum velocity ω_{max} in one direction to the same speed in the opposite direction, we leave it as a simple exercise to show that the loop gain must not exceed

$$\beta K < \frac{I_{max}}{2\omega_{max}} - \frac{B}{k_m}.$$

▶ ▶ ▶

Example 6.2–3

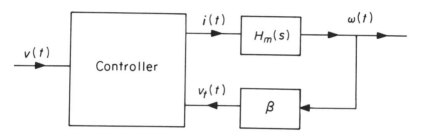

Figure 6.2–8. A non-linear motor controller.

In control systems of the kind described in Example 6.2–2, where certain internal variables are constrained, improved performance can often be obtained by replacing the linear feedback system with one containing a non-linear *controller*, as illustrated in Figure 6.2–8. The properties of the controller are described by the following equation:

$$i(t) = \begin{cases} +I_{max}, & v(t) > v_t(t) \\ \dfrac{Bv(t)}{\beta k_m}, & v(t) = v_t(t) \\ -I_{max}, & v(t) < v_t(t). \end{cases}$$

Whenever the speed differs from that desired, maximum current (torque) is applied to correct the error as soon as possible. As soon as the speed reaches the desired value, the current is set equal to the value required to maintain that speed in the steady state.* The performance of this system in response to a step change from rest is readily calculated as in Example 6.2–2 and is shown in Figure 6.2–9.

---

*In practice, such an extreme example of a "bang-bang" controller would probably not be very satisfactory—drift in the value of $B/\beta k_m$ or small rapid fluctuations (noise) in either the input or the output in the steady state would yield large rapid "hunting" swings in $i(t)$. A more practical scheme would include a region where the error, $v(t) - v_t(t)$, is small but not zero and in which $i(t)$ is proportional to the error, as in the linear feedback system previously proposed.

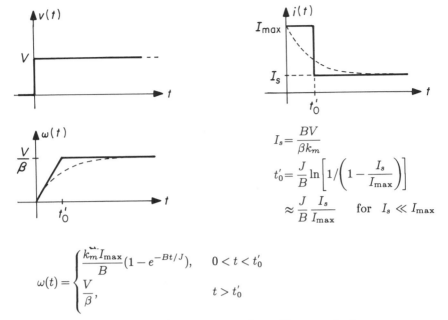

$$I_s = \frac{BV}{\beta k_m}$$

$$t_0' = \frac{J}{B} \ln\left[1 \Big/ \left(1 - \frac{I_s}{I_{max}}\right)\right]$$

$$\approx \frac{J}{B} \frac{I_s}{I_{max}} \quad \text{for} \quad I_s \ll I_{max}$$

$$\omega(t) = \begin{cases} \dfrac{k_m I_{max}}{B}(1 - e^{-Bt/J}), & 0 < t < t_0' \\[2mm] \dfrac{V}{\beta}, & t > t_0' \end{cases}$$

Figure 6.2–9. Step response of non-linear controller.

For comparison, the response of the linear feedback system is shown dotted in Figure 6.2–9, assuming β and K are adjusted to give the same final velocity and the same maximum current. Under these circumstances the time constant of the linear response is $\hat{t}_0 = JI_s/BI_{max}$. Thus, provided that $I_s \ll I_{max}$, the time t_0' that the non-linear controller takes to reach the desired final speed is equal to the time $\hat{t}_0$ that the linear feedback controller takes to get within $1/e$ of the desired final value. The initial rate of change of $\omega(t)$ is the same for both systems. However, the non-linear-controller system maintains (almost) this same initial rate for some time, and thus reaches the desired speed in substantially less time than is required for the linear feedback system to approach this speed. And the comparison would be even more favorable to the non-linear controller if the gain of the linear system had not been optimally chosen for precisely this speed change. It is because of improvements of this kind that "controller" approaches to control-system design have today largely replaced the "feedback" approach popular a few decades ago.

▶ ▶ ▶

6.3 Stability

The available dynamic range of certain internal system variables often limits the practical improvements in performance that can be achieved with linear feedback—as Example 6.2–2 showed. But another difficulty is perhaps even more common. As the loop gain is increased, poles of the closed-loop system may be moved into the right half-s-plane, yielding an unstable system.

In general, a physical system is said to be *stable* if any small perturbation in the conditions under which it is being operated (in the drives or initial state) produces only a small perturbation in the behavior (response) of the system. For linear systems, an equivalent statement is that *B*ounded *I*nputs in the zero state yield *B*ounded *O*utputs (BIBO).* Evidently, a necessary condition for BIBO stability (as mentioned in Chapter 4) is that the natural frequencies (the poles of the system function) must lie in the left half-plane (l.h.p.); otherwise a step function (a bounded input) would initiate an unbounded response. For circuits such as we have been considering—composed of a finite number of elements and having system functions that are well-behaved at $s = \infty$—the l.h.p. condition is also sufficient. (We shall discuss a generalization in Chapter 10.) Circuits composed of positive R, L, and C elements (passive circuits) are always stable, as noted in Chapter 4. (Pure lossless circuits—containing L's and C's but no R's— are perhaps marginally stable, since such circuits have poles on the $j\omega$ axis and will give unbounded responses to bounded sinusoidal excitation at their natural frequencies; see Example 10.3–1.) Circuits containing controlled sources, however, may be either stable or unstable; typically they will be stable if the controlled-source gains are small enough (since for zero gains the circuit is composed of positive R's, L's, and C's only) but may become unstable as the gain(s) are increased beyond some critical value(s). Thus it is not surprising that stability problems often limit the advantages of feedback, since such improvements depend, as we have seen, on large loop gains.

For a simple feedback system with system function

$$H(s) = \frac{K(s)}{1 + \beta(s)K(s)}$$

the natural frequencies are the zeros of $1 + \beta(s)K(s)$.† An important part of the feedback system design process is to explore the zero locations of $1 + \beta(s)K(s)$ as a function of various design parameters (such as the magnitude of the forward gain or the fraction of the output fed back) and to discover (if possible) ways to modify the design if the desired performance is not realized with a stable system.

Example 6.3–1

Let $\beta(s)K(s) = \mu/(\tau s + 1)^3$, which might describe the moderate-frequency behavior of the loop gain of a 3-stage feedback amplifier. The quantity μ is the loop gain at low frequencies; to reduce output distortion we wish μ to be as large as possible. Stability requires that the zeros of $1 + \beta(s)K(s) = [(\tau s + 1)^3 + \mu]/(\tau s + 1)^3$ have negative real

---

*More fundamentally, we should demonstrate for a stable system that all internal variables as well as the designated outputs remain bounded. It is enough to show that the state variables remain bounded; this is called *stability in the sense of Liapunov* as contrasted with BIBO stability. A non-linear system may be stable for some inputs and initial states but not for others. We shall give an example of the importance of Liapunov stability in the next section.

†Since $K(s)$ appears in both numerator and denominator, poles of $K(s)$ will not be poles of $H(s)$ unless they are also zeros of $\beta(s)$. It is also possible for a zero of $1 + \beta(s)K(s)$ not to be a natural frequency of $H(s)$ if $K(s)$ and $\beta(s)$ have a zero and a pole, respectively, at the zero of $1 + \beta(s)K(s)$.

parts. There are general formulas for finding the roots of a cubic polynomial, but they are clumsy. In the present case, it is clear that the numerator is zero when

$$\tau s + 1 = (-\mu)^{1/3}.$$

That is, the roots of $(\tau s + 1)^3 + \mu = 0$ are

$$s = \frac{1}{\tau}\left(-1 - \mu^{1/3}\right), \; \frac{1}{\tau}\left(-1 + \mu^{1/3}e^{j\pi/3}\right), \; \frac{1}{\tau}\left(-1 + \mu^{1/3}e^{-j\pi/3}\right).$$

The loci of the roots, as functions of μ, are shown in Figure 6.3–1. For $-1 < \mu < 8$ the roots all lie in the l.h.p.; for $\mu < -1$ there is one real zero in the r.h.p., and for $\mu > 8$ there are two conjugate zeros in the r.h.p. If the system is to be stable, then, the low-frequency loop gain cannot exceed 8. In practice, a value substantially less than 8 would probably be chosen—to give a stability *margin*, and to yield a transient or step response without the substantial *ringing* or *overshoot* that often results from a system pole pair close to the $j\omega$-axis.

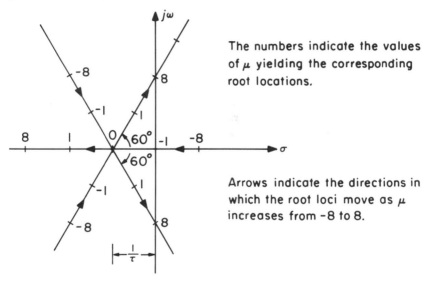

The numbers indicate the values of μ yielding the corresponding root locations.

Arrows indicate the directions in which the root loci move as μ increases from –8 to 8.

Figure 6.3–1. Root loci of $(\tau s + 1)^3 + \mu$.

▶ ▶ ▶

Sketching the root loci of $1 + \beta(s)K(s) = 0$ as above can be an important aid to feedback system design. Fortunately, simple techniques exist for finding the approximate pattern of the root loci without actually finding the roots analytically—see, for example, J. K. Roberge, *Operational Amplifiers: Theory and Practice* (New York, NY: Wiley, 1975) Chapter IV, or almost any book on linear control theory.

A test, called the *Routh test*, is also available to determine whether all the roots of a polynomial lie in the l.h.p. The Routh test proceeds directly from the coefficients of the various terms in the polynomial and does not require finding the roots analytically. Unfortunately, the Routh test gives no information about the location of the roots except the half-plane in which they lie; it thus provides no insight into relative measures of stability that depend on the proximity of the poles to the $j\omega$-axis, or the balancing of time and frequency responses. Nevertheless, it is frequently useful in conjunction with other procedures.

Certain special aspects of the Routh test, however, have broad applicability and can be summarized as follows:

1. Necessary conditions for a polynomial to have all its roots in the l.h.p. are:
   i) All of the terms must have the same sign;
   ii) All of the powers between the highest and the lowest must have nonzero coefficients, unless all even-power or all odd-power terms are missing.
2. For quadratic polynomials these conditions are also sufficient.
3. For a cubic polynomial, $s^3 + \alpha s^2 + \beta s + \gamma$, necessary and sufficient conditions (NASC) are $\alpha, \beta, \gamma > 0$ and $\beta > \gamma/\alpha$.

The following example illustrates the application of these rules.

Example 6.3–2

1. $s^2 - 3s + 2$ does not have all its roots in the l.h.p. because two of the terms have + signs and one has a − sign. Indeed, $s^2 - 3s + 2 = (s - 2)(s - 1)$.
2. $s^5 + s^3 + 10s^2$ does not have all its roots in the l.h.p. because the s^4 term is missing. Indeed, $s^5 + s^3 + 10s^2 = s^2(s+2)(s-1+j2)(s-1-j2)$. Also we note that $s^5 + s^3 + 10s^2 = s^2(s^3 + s + 10)$ and the cubic factor does not satisfy the NASC for cubics in the third rule above.
3. $s^3 + s^2 + 4s + 30$ satisfies the necessary conditions under the first rule above, but in fact $s^3 + s^2 + 4s + 30 = (s+3)(s-1+j3)(s-1-j3)$ so these conditions are not sufficient. Of course, this polynomial does not satisfy the NASC in the third rule.
4. $s^4 + 13s^2 + 36$ satisfies all the necessary conditions and in fact has all of its roots on the $j\omega$ axis: $s^4 + 13s^2 + 36 = (s+j2)(s-j2)(s+j3)(s-j3)$. Note that the roots of the polynomial $z^2 + 13z + 36$ are negative real, as is clearly required if the s-roots are to be pure imaginary.
5. $s^4 + 6s^2 + 25$ satisfies all the necessary conditions but has its roots in both half-planes at mirror-image points: $s^4 + 6s^2 + 25 = (s-1+j2)(s-1-j2)(s+1+j2)(s+1-j2)$. Note that the polynomial $z^2 + 6z + 25$ has complex roots.

▶ ▶ ▶

Example 6.3–3

The characteristic polynomial in Example 6.3–1 was

$$(\tau s + 1)^3 + \mu = \tau^3 \left(s^3 + \frac{3}{\tau}s^2 + \frac{3}{\tau^2}s + \frac{1+\mu}{\tau^3} \right).$$

The NASC from the third rule for cubics above gives

$$\frac{3}{\tau^2} > \frac{1+\mu}{\tau^3}\frac{\tau}{3}$$

or
$$\mu < 8$$

when μ is positive. When μ is negative, we must satisfy $1 + \mu > 0$ or the last term will have a negative coefficient. Hence the stable region is

$$-1 < \mu < 8$$

as previously determined by plotting the locus of the roots.

▶ ▶ ▶

Example 6.3–4

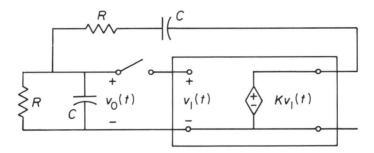

Figure 6.3–2. A simple RC audio oscillator.

Figure 6.3–2 is a much simplified diagram of one of the first successful products of the Hewlett-Packard Company, an RC audio oscillator. To determine the condition for oscillation, open the switch and compute the loop gain:

$$\frac{V_0(s)}{V_1(s)} = \frac{K\dfrac{R\left(\dfrac{1}{Cs}\right)}{R + \left(\dfrac{1}{Cs}\right)}}{\dfrac{R\left(\dfrac{1}{Cs}\right)}{R + \left(\dfrac{1}{Cs}\right)} + R + \left(\dfrac{1}{Cs}\right)} = \frac{K(RCs)}{(RCs)^2 + 3(RCs) + 1}.$$

With the switch closed, the circuit will be unstable if the values of s for which the open-loop gain $V_0(s)/V_1(s) = 1$ lie in the r.h.p. Equivalently these are the values of s for which

$$(RCs)^2 + (3 - K)(RCs) + 1 = 0.$$

If $K = 3$, the natural frequencies will lie precisely on the $j\omega$-axis, so that—once started—sinusoidal oscillations will be produced. In practice, K is designed to be slightly greater than 3 so that the poles lie inside the right half-plane and the oscillator is self-starting. As the amplitude of the oscillations builds up, a resistor (not shown) in the amplifier changes value slightly, reducing the gain; the amplitude stabilizes at the value that yields an effective K of 3. The frequency of the oscillation is then the root of

$$(RCs)^2 + 1 = 0$$

or $\omega = 1/RC$. The frequency can be changed over a very wide range by making the two resistors variable and ganging them together.

▶ ▶ ▶

    The root-locus and Routh stability tests for feedback systems require that $\beta(s)$ and $K(s)$ be known rational functions of s. But often information about $\beta(s)$ and $K(s)$ may be given in other forms, such as non-rational functions or experimentally determined magnitude and angle plots of open-loop frequency response. In such cases a graphical stability test due to Nyquist is most useful, as discussed in the appendix to this chapter.

    If we examine the behavior of a closed-loop system in the sinusoidal steady state, it is easy to explain physically (rather than just mathematically) why large values of loop gain may lead to instabilities. Thus suppose there is a frequency $s = j\omega$ for which the loop gain $\beta(j\omega)K(j\omega)$ has an angle of $180°$. Then (including the minus sign in the comparator) the signal returned and added to the input will—at this frequency—be precisely in phase with the input. If, furthermore, the magnitude of the loop gain at this frequency is equal to unity, then the amplitude of this returned signal will be equal to the input amplitude. Once such a returned signal is established, the input could be set equal to zero and the signal in the loop and at the output would maintain itself! For larger values of loop gain, any incidental disturbance will start an oscillation at this frequency that will grow; that is, the system will be unstable. To avoid instabilities, then, we must ensure that the magnitude of the loop gain is less than unity at any frequency at which the angle of the loop gain is $180°$. (There are some interesting exceptions to this rule—see the appendix.)

    The value of the loop gain at lower frequencies can be increased—and the effectiveness of feedback at these frequencies enhanced—if a *compensating network* is included in the loop to reduce the loop gain at higher frequencies where the phase angle of the loop gain approaches $180°$. This scheme is explored further in the appendix and in Problem 6.6.

6.4 Feedback Stabilization of Unstable Systems

Feedback around a stable system can produce instabilities by in effect moving poles from the left to the right half-plane. On the other hand, feedback can also be used to stabilize an inherently unstable system, as the following example illustrates.

Example 6.4–1

Consider an inverted pendulum attached by a hinge to a car free to move along a horizontal track. Our objective is to design a driving system for the car that will use signals derived from the car and pendulum motions to stabilize the pendulum in its inverted position. The whole scheme may be considered a one-dimensional simulation of the common parlor trick of balancing a ruler or broom stick on end on your palm by moving your hand appropriately. The details of what we have in mind are shown in Figure 6.4–1.

It is straightforward to derive from the elementary mechanics of rigid bodies that

$$mg\ell \sin\theta - m\ell\ddot{x}\cos\theta = I\ddot{\theta}$$

where the dots imply differentiation and the variables and parameters have the meanings defined in Figure 6.4–1. For small angles this equation may be linearized to

$$I\ddot{\theta} - mg\ell\theta = -m\ell\ddot{x}$$

which implies that $\theta(t)$ and $x(t)$ are related by the system function

$$H(s) = \frac{\Theta(s)}{X(s)} = \frac{-m\ell s^2}{Is^2 - mg\ell}$$

which of course is unstable.

As a first attempt at stabilization, connect a rotary potentiometer as shown in Figure 6.4–1 to the pivot shaft of the inverted pendulum from which a voltage proportional to $\theta(t)$ can be derived. Apply the difference between $\theta(t)$ and the desired value $\theta_0(=0)$ through an amplifier to drive the motor in such a direction as to increase $x(t)$ if $\theta(t) - \theta_0$ is positive. Assume that the motor and pulley system can be described by the system function

$$M(s) = \frac{X(s)}{V(s)} = \frac{k_m}{s(s+\alpha)}.$$

That is, the motor produces in the steady state a velocity $\dot{x}(t)$ proportional to the applied voltage $v(t)$ (independent of load) and the velocity responds to changes in $v(t)$ as a first-order system with time constant $1/\alpha$.

The block diagram for this closed-loop system is shown in Figure 6.4–2. If we take θ_0 as the input, the system function is

$$\hat{H}(s) = \frac{\Theta(s)}{\Theta_0(s)} = \frac{-KM(s)H(s)}{1 - KM(s)H(s)} = \frac{Kk_m\dfrac{m\ell}{I}s}{s^3 + \alpha s^2 + \dfrac{m\ell}{I}(Kk_m - g)s - \dfrac{m\ell}{I}g\alpha}.$$

Because of the minus sign in the constant term of the denominator, the system of Figure 6.4–2 remains unstable for any value of K. Physically, the trouble is that a steady angular error in the system of Figure 6.4–2 leads only to a steady velocity of the car. But movement of the car at constant velocity has no effect on the angular motion of the pendulum; to produce a restoring torque, the car must be accelerated.

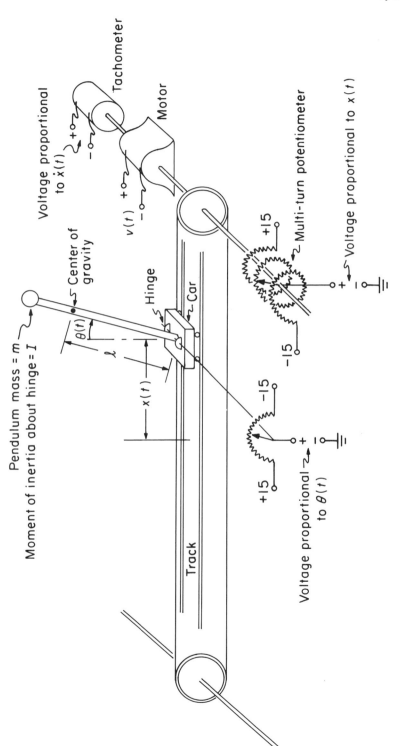

Figure 6.4-1. Inverted pendulum simulation.

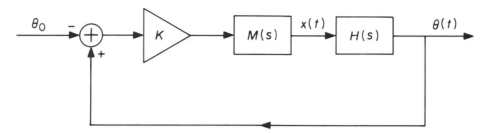

Figure 6.4–2. Block diagram for a first attempt at stabilization.

Such an effect can be obtained by adding an additional feedback path proportional to the integral of $\theta(t)$. The resulting block diagram has the form shown in Figure 6.4–3 and is characterized by the system function

$$\hat{H}(s) = \frac{\Theta(s)}{\Theta_0(s)} = \frac{-KM(s)H(s)}{1 - \left(1 + \dfrac{a}{s}\right)KM(s)H(s)}$$

$$= \frac{Kk_m\dfrac{m\ell}{I}s}{s^3 + \alpha s^2 + \dfrac{m\ell}{I}(Kk_m - g)s + \dfrac{m\ell}{I}(Kk_m a - g\alpha)}.$$

From the special case of the Routh test given earlier for the cubic, it follows that this system will be stable if $a < \alpha$ and $Kk_m a > g\alpha$, conditions that obviously can be met by appropriate choices of a and K.

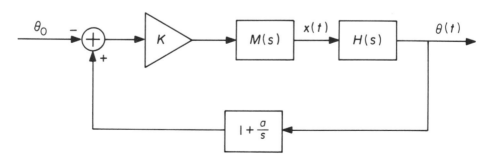

Figure 6.4–3. An improved stabilization system.

However, despite the fact that the feedback system of Figure 6.4–3 has theoretically "stabilized" the inverted pendulum, the scheme would not work very well in
practice. There are two difficulties. First, inserting practical numbers for the various
parameters will show that, although the system poles can be placed in the left half-
s-plane, it will probably be difficult to move them very far from the $j\omega$-axis. As a
result, transient disturbances of the pendulum would be only slowly corrected, and in
an oscillatory manner—a series of modest but rapid disturbances could lead to failure
by driving the system beyond its linear range. The factor primarily responsible for
this effect is the probably small value of α, describing the frequency response of the
motor. But this difficulty is not fundamental since, as in Example 6.2–2, the response of
the motor can be speeded up by connecting a tachometer to provide velocity feedback
around the motor, as shown in Figure 6.4–4. It is easy to show that the system function
of this block diagram is $\hat{H}(s)$ of Figure 6.4–3 with α replaced by $\alpha' = \alpha + K k_m b$. The
"settling time" of the new system can consequently be significantly reduced.

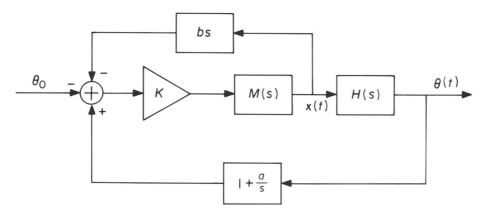

Figure 6.4–4. Further improvements.

A second difficulty with the system of Figure 6.4–3 (or Figure 6.4–1) is more subtle.
Consider the system function describing the displacement $x(t)$ of the car, which is

$$\frac{X(s)}{\Theta_0(s)} = \frac{\hat{H}(s)}{H(s)} = \frac{-K k_m \left(s^2 - g\dfrac{m\ell}{I} \right)}{s\left[s^3 + \alpha' s^2 + \dfrac{m\ell}{I}(K k_m - g)s + \dfrac{m\ell}{I}(K k_m a - g\alpha') \right]}.$$

This system has a pole at $s = 0$ and is thus only marginally stable—a succession of
random disturbances will induce a "random walk" in the car's position that will sooner
or later cause it to go off one end or the other of the track. This can be avoided by
adding still another feedback path, as in Figure 6.4–5.

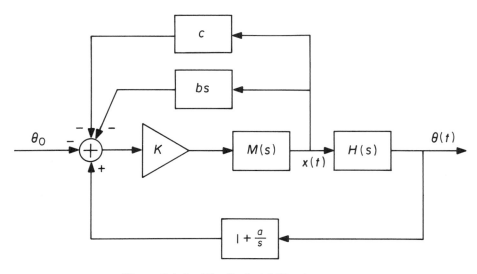

Figure 6.4–5. The final stabilization system.

The input-output system function now becomes

$$H(s) = \frac{\Theta(s)}{\Theta_0(s)}$$

$$= \frac{K k_m \dfrac{m\ell}{I} s^2}{s^4 + \alpha' s^3 + \left[\dfrac{m\ell}{I}(K k_m - g) + K k_m c\right] s^2 + \dfrac{m\ell}{I}(K k_m a - g\alpha')s - cg K k_m \dfrac{m\ell}{I}}.$$

It is apparent from the sign of the last term in the denominator that stability has been upset unless c is negative—which implies that deviations of $x(t)$ from its zero position will induce motor inputs in a direction that makes the error worse. But a little reflection on how you move your hand balancing a pointer will make it clear that this counterintuitive result is indeed correct. To achieve an ultimate motion of your hand to the right, you must first move it sharply to the left, displacing the pendulum angle to the right so that you can then steadily move your hand to the right under the pendulum. A full study of the Routh conditions for this 4th-order polynomial shows that stability now requires

$$-\frac{m\ell}{Ic}\left(1 - \frac{a}{\alpha'}\right)\left(1 - \frac{g\alpha'}{K k_m a}\right) > 1$$

in addition to the conditions $a < \alpha'$ and $K k_m a > g\alpha'$ already given for the earlier cubic. The coefficient c, in addition to being negative, must thus not be too big. Examination of the system function for $X(s)/\Theta_0(s)$ shows that the pole at $s = 0$ has in fact now been removed, so that with appropriate parameter choices our final design (Figure 6.4–5) should give satisfactory behavior for both angle $\theta(t)$ and car position $x(t)$.

This example illustrates the importance of looking at the stability of internal variables as well as input-output variables—which, as mentioned earlier, is called stability in the sense of Liapunov.

▶ ▶ ▶

6.5 Summary

Feedback can both reduce the sensitivity of a system to various parameter changes and loading effects, and also dramatically change the dynamic response of the system by altering the locations of the system poles. The latter effect can be useful, for example, in speeding up the response (or increasing the bandwidth) of a control system or amplifier, or in stabilizing an inherently unstable system such as the inverted pendulum. But feedback can also move system poles into the right half-s-plane and thereby change a stable system into an unstable one. Often stability considerations set limits on the sensitivity reductions that are realizable with feedback.

APPENDIX TO CHAPTER 6

The Nyquist Stability Criterion

The *Nyquist stability criterion* is based on an examination of the locus or polar plot (called the *Nyquist plot*) of the complex number $\beta(s)K(s)$ as s ranges along a contour following the $j\omega$-axis and around the right half-plane. The Nyquist plot, as shown in Figure 6.A–1, is thus a mapping into the βK-plane of the closed contour C shown in the s-plane.

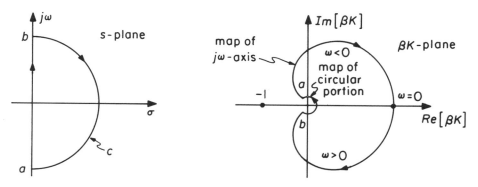

Figure 6.A–1.   Contours in the s-plane and βK-plane.

The foundation for the Nyquist criterion is a theorem from function theory called Cauchy's Principle of the Argument. This theorem in general compares the contour generated by some function $X(s)$ in the X-plane as s traces clockwise around any simple contour in the s-plane. It states that the net number of clockwise encirclements of the point $X(s) = 0$ by the closed contour in the X-plane is equal to $Z - P$, where Z is the number of zeros of $X(s)$ and P the number of poles of $X(s)$ enclosed by the contour in the s-plane. Cauchy's Principle of the Argument is easily demonstrated for a rational function

$$X(s) = A\frac{(s - z_1)(s - z_2)\cdots}{(s - p_1)(s - p_2)\cdots}$$

by recalling from Chapter 4 that the angle (or argument) of $X(s)$ (assuming A is real) is the sum of the angles of the vectors $(s - z_i)$ minus the sum of the angles of the vectors $(s - p_i)$, as illustrated in Figure 6.A–2. To be specific, suppose the contour in the s-plane is the contour C shown to the left in Figure 6.A–1. Consider a zero (or pole)

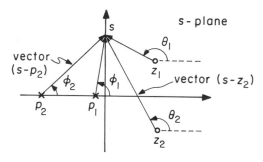

Figure 6.A–2. $\angle X(s) = \theta_1 + \theta_2 - \phi_1 - \phi_2$.

outside the contour C, as shown to the left in Figure 6.A–3. As the point s traverses the contour, the angle of $(s - z_1)$ (or $(s - p_1)$) varies, but it returns to its initial value when the contour is completely traced. If, however, z_1 (or p_1) lies inside the contour, as shown to the right in Figure 6.A–3, there is a net decrease in angle of 2π radians as the contour is traversed. Since the contributions of the individual terms add (subtract), the desired result follows immediately.*

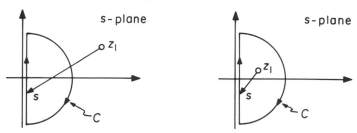

Figure 6.A–3. Zeros outside and inside a contour.

To derive Nyquist's criterion, set $X(s) = 1 + \beta(s)K(s)$ and require that the radius of the semicircular part of the contour C be very large. Then the number of clockwise encirclements of the origin by the $X(s)$ contour, or equivalently the number of clockwise encirclements of the point $\beta(s)K(s) = -1$ by the $\beta(s)K(s)$ contour, is equal to the difference between the number of zeros and the number of poles of $X(s)$ in the right half-plane. Since zeros of $X(s)$ are poles of the closed-loop feedback system and poles of $X(s)$ are poles of the loop gain $\beta(s)K(s)$, the closed-loop system will be unstable if the number of clockwise encirclements of $\beta(s)K(s) = -1$ by the $\beta(s)K(s)$ contour is greater than the number of poles of $\beta(s)K(s)$. This is Nyquist's criterion.

To apply this test, it is often helpful to carry out the following imaginary experiment. Pretend that there is a nail driven into the βK-plane at $\beta(s)K(s) = -1$ and that a string is tied to the nail so tightly that it cannot slip. The other end of the string is tied to a pencil. Starting anywhere, trace with the pencil entirely around the contour in the direction implied by a clockwise encirclement of the right half-s-plane. The net number of times that the string ends up wrapped around the nail in the clockwise direction is equal to the number of -1's of $\beta(s)K(s)$ in the right half-plane (minus the number of poles of $\beta(s)K(s)$ in the right half-plane, if any). In the sketches in

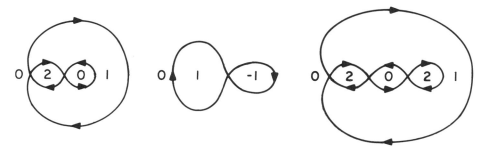

Figure 6.A–4. The numeral is the number of clockwise encirclements.

---

*The theorem actually applies to a wider function class than rational functions. For a more complete discussion see H. W. Bode, *Network Analysis and Feedback Amplifier Design* (New York, NY: Van Nostrand, 1945) p. 147ff.

Figure 6.A–4, the numeral indicates the number of clockwise encirclements if the point $\beta(s)K(s) = -1$ lies in the corresponding region.

Example 6.A–1

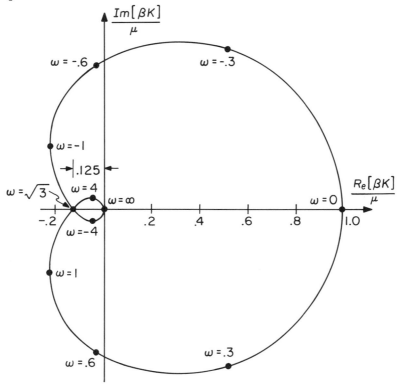

Figure 6.A–5. Polar plot of $\mu/(s+1)^3$.

Continuing Example 6.3–1, Figure 6.A–5 shows a polar plot of

$$\beta(s)K(s) = \frac{\mu}{(s+1)^3}.$$

For $\mu = 1$, the -1 point is well to the left outside the contour. For $\mu = 8$, the scales of ordinate and abscissa are multiplied by 8, and the -1 point coincides with the point on the contour labelled $\omega = \sqrt{3}$; for large μ the point -1 lies inside the contour and in fact is encircled twice, implying two zeros in the right half-plane. For negative μ, the contour is flipped over right for left; the point -1 will lie inside the contour, encircled once (implying one zero in the r.h.p.) for $\mu < -1$. Thus the closed-loop system will be stable for $-1 < \mu < 8$, as was previously demonstrated directly. Note that although we used an analytic formula for $\beta(s)K(s)$ to calculate the values of the loop gain for various frequencies, measured sinusoidal steady-state values would have worked as well. This point is illustrated in the next example.

▶ ▶ ▶

Example 6.A–2

Figure 6.A–6 is a sketch of $|\beta K|/\mu$ and $\angle \beta K/\mu$ vs. ω as might have been measured as the loop gain of some feedback system. (Actually these curves are the Bode plot for

$$\frac{\beta(s)K(s)}{\mu} = \frac{\left(1 + \dfrac{s}{30}\right)^2}{(1+s)^3} .$$

as constructed in Example 4.4–2.)

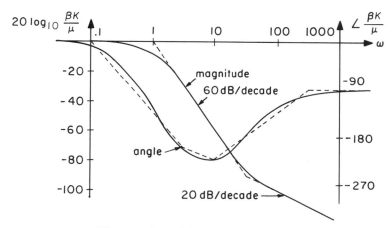

Figure 6.A–6. Measured loop gain.

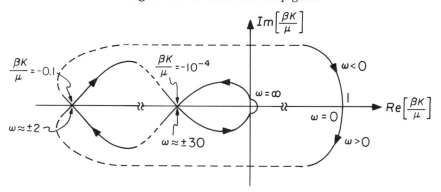

Figure 6.A–7. The shape of the Nyquist plot for $\beta K/\mu$ of Figure 6.A–6.

For stability analysis, interest centers on the Nyquist plot of $\beta K/\mu$ near the real axis, as shown schematically in Figure 6.A–7. Clearly for $\mu < 10$ (approximately) the system is stable, whereas for $10 < \mu < 10^4$ (again approximately) it is unstable. But, remarkablly, if $\mu > 10^4$, the system is again stable! This type of *conditional* (or *Nyquist*) *stability* is occasionally observed. Prior to Nyquist no one had understood how it was possible for an amplifier to be stable in spite of the fact that there was a frequency at which the returned sinusoidal signal was in phase with and larger than the sinusoidal input. Conditional stability may help explain the observation that overdriving an amplifier will sometimes cause it to oscillate—despite the fact that (because the dominant non-linearity is usually a saturation) the effect of overdriving is to a first approximation often an effective reduction in gain.

▶ ▶ ▶

Example 6.A–3

Continuing Example 6.3–1, it should be clear from the Nyquist plot of Example 6.A–1 (reproduced in Figure 6.A–8 below) that performance would be improved if a compensating network could be inserted in cascade with either $\beta(s)$ or $K(s)$ that would add a positive phase shift to the loop gain for frequencies around $\omega = 1$ without substantially changing the magnitude for $\omega < 1$—for example, as shown by the dashed contour in Figure 6.A–8.

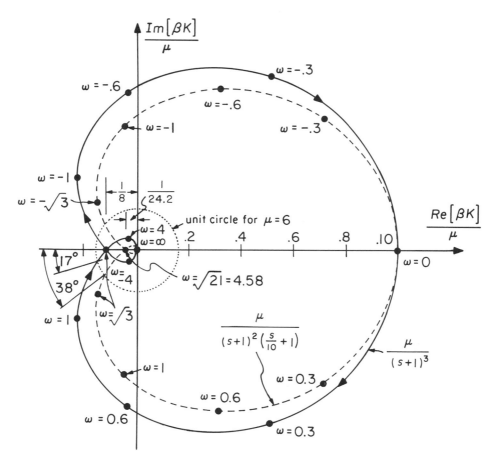

Figure 6.A–8. Nyquist plot of Example 6.A–1 with compensation (shown dashed).

The improvement in performance that could be provided by compensation can be measured in several ways:

a) The maximum value of μ yielding stable performance would be increased from just less than $1/0.125 = 8$ to just less than $1/0.041 = 24.2$.

b) For a gain less than the maximum allowed stable value, the *gain margin* is defined as the ratio of the maximum allowed stable value to the actual gain. (That is, for any selected value of μ, the gain margin is the reciprocal of the actual gain at a phase angle of $-180°$.) Thus if the chosen value of gain were 6, the gain margin would go from $8/6 = 1.33$ without compensation to $24.2/6 = 4.03$ with compensation. As a rule of thumb for electronic amplifiers, gain margins of 3 or more are considered desirable.

c) Again for a gain less than the maximum allowed stable value, the *phase margin* is defined as the difference between $-180°$ and the angle of the loop gain $\beta(s)K(s)$ at the frequency ω where the magnitude of the loop gain is 1. For a chosen gain of 6, the phase margin goes, as shown, from about $17°$ without compensation to about $38°$ with compensation. Again as a rule of thumb, phase margins of $30°$ to $60°$ are considered desirable.

As we shall see in Chapter 16, the magnitude and angle of the sinusoidal frequency response of a causal network are not independently selectable. Hence arbitrary adjustments of Nyquist plots are not possible. The compensation shown in the dashed line of Figure 6.A–8 was obtained by cascading $\beta(s)K(s)$ with a *lead network* whose system transfer function is

$$G(s) = \frac{s+1}{\dfrac{s}{10}+1}.$$

A Bode plot of $G(j\omega)$ is shown in Figure 6.A–9; the name "lead network" comes from the fact that the positive phase angle of $G(j\omega)$ implies that a sinusoidal output waveform "leads" (reaches its peak value before) the input. The key to the success of the lead network is that it produces substantial phase shift ($\sim 40°$ at $\omega = 1$) before it produces a significant increase in gain ($|G(j1)| \approx 1.4$).

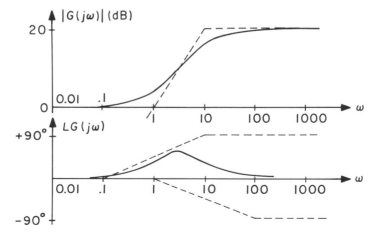

Figure 6.A–9. Magnitude and angle of the lead network response $G(j\omega)$.

With the lead network inserted in the loop, the loop gain becomes

$$\beta(s)K(s)G(s) = \frac{\mu}{(s+1)^2\left(\dfrac{s}{10}+1\right)}$$

and

$$1+\beta(s)K(s)G(s) = \frac{s^3 + 12s^2 + 21s + 10(1+\mu)}{(s+1)^2(s+10)}.$$

Stable positive values of μ (from the Routh test) thus satisfy

$$21 > \frac{10(1+\mu)}{12} \quad \text{or} \quad \mu < 24.2.$$

For the maximum value of μ,

$$s^3 + 12s^2 + 21s + 10(1+\mu) = (s+12)(s^2+21).$$

These calculations determine the magnitude and frequency of the Nyquist plot at the point where it crosses the negative real axis. For further consideration of this compensation scheme see Problem 6.5.

▶ ▶ ▶

Example 6.A-4

Suppose that

$$\beta(s)K(s) = \frac{\mu(s+2)}{(s+1)(s-3)}.$$

The loop-gain system function is unstable because of the pole at $s = +3$. However, the Nyquist diagram appears as shown in Figure 6.A–10. The values in the table in Figure 6.A–10 take into account the fact that βK has a pole in the right half-s-plane. Hence, for $1+\beta K$ not to have zeros in the right half-plane, the Nyquist contour must have one counterclockwise encirclement of -1. This occurs for $\mu > 2$. For these values of gain the overall amplifier with the loop closed is stable, in spite of the fact that if the loop were opened the amplifier would oscillate. This is another example of how feedback can be used to stabilize an unstable system.

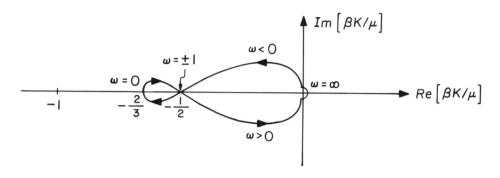

Range of μ	Number of CW encirclements of $\beta K = -1$	Number of r.h.p. closed-loop poles
$0 < \mu < 3/2$	0	1
$3/2 < \mu < 2$	1	2
$\mu > 2$	-1	0

Figure 6.A–10. Nyquist plot for Example 6.A–4.

▶ ▶ ▶

EXERCISES FOR CHAPTER 6

Exercise 6.1

For the circuit below, find the values of μ and α such that the response, $v_2(t)$, to a unit step, $v_1(t) = u(t)$, has the same final value but rises ten times faster than the step response that would be obtained with $\alpha = 0$ and $\mu = 1$. Answers: $\alpha = 0.9$, $\mu = 10$.

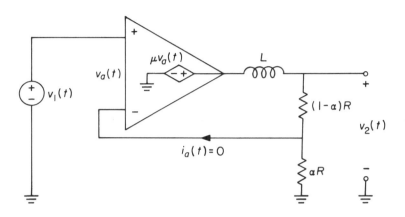

Exercise 6.2

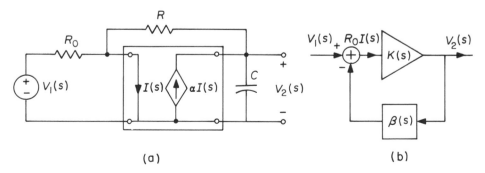

(a)   (b)

a) Show that the feedback amplifier of (a) can be represented by the block diagram of (b) with $K(s) = (\alpha/R_0C)/(S + 1/RC)$ and $\beta(s) = -R_0/R$. (Note that the output of the summation box in (b) is specified to be the quantity $R_0 I(s)$.)

b) Show that the amplifier is stable if $-\infty < \alpha < 1$.

Exercise 6.3

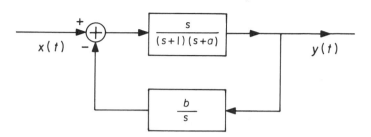

a) Determine a and b for the system above so that the overall system function becomes

$$H(s) = \frac{s}{(s+2)(s+3)}.$$

b) If $a = 2$, what is the range of values of b for which the system is input-output stable?

c) Determine the unit step response of this system if the system function is as in (a).

Answers: (a) $a = 4$, $b = 2$;   (b) $b > -2$;   (c) $y(t) = \left(e^{-2t} - e^{-3t}\right)u(t)$.

Exercise 6.4

a) An (unstable) LTI system has the response $y(t)$ to a unit step input $x(t)$ as shown below. Show that the system function for this system is

$$H(s) = \frac{1}{s - \alpha}.$$

b) To stabilize this system, apply feedback of the form

$$\beta(s) = (1 + \alpha) + s.$$

Show that the overall system now has the unit step response shown on the next page with $\tau = 2$ sec.

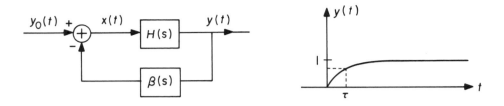

Exercise 6.5

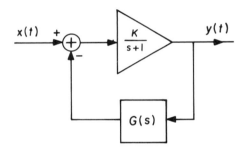

For what values of $K, -\infty < K < \infty$, will the overall system be stable if $G(s)$ has the following forms:

a) $G(s) = \dfrac{s^2}{s+1}$.

b) $G(s) = \dfrac{s^3}{(s+1)^2}$.

c) $G(s) = \dfrac{s+2}{s}$.

Answers: (a) $K > -1$;   (b) $8 > K > -1$;   (c) $K > 0$.

PROBLEMS FOR CHAPTER 6

Problem 6.1

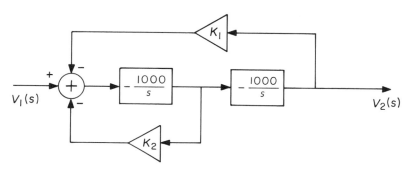

a) Find the zero-frequency gain of this system, assuming that K_1 and K_2 are picked so that it is stable.

b) For what range of (real) values of K_1 and K_2 will the system be stable?

c) Find and sketch the unit step response for $K_2 = -3$, $K_1 = 2$.

Problem 6.2

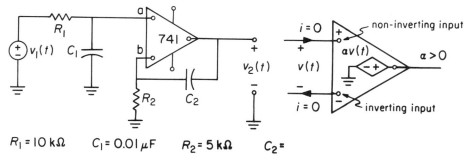

$R_1 = 10\ k\Omega$ $C_1 = 0.01\ \mu F$ $R_2 = 5\ k\Omega$ $C_2 =$

The diagram to the left above is intended to function as an integrator. Unfortunately the draftsman forgot to label the inverting and non-inverting terminals at the input to the op-amp, and he neglected to give one of the capacitor values.

a) Modelling the op-amp by the equivalent circuit shown to the right above with wire "a" connected to the non-inverting terminal of the op-amp, find the system function of the integrator circuit.

b) Should wire "a" be connected to the inverting or to the non-inverting terminal to ensure stable performance? Explain your answer. Does your answer depend on the magnitude of α?

c) Assuming that $\alpha \to \infty$ in the stable connection of the op-amp, determine the value of the unlabelled capacitor to yield ideal integrator behavior.

Problem 6.3

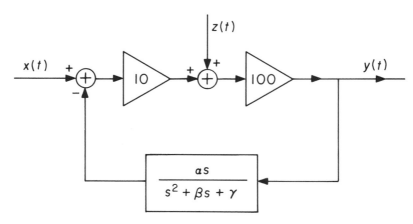

a) Find the ZSR $Y(s) = L[y(t)]$ in terms of $X(s) = L[x(t)]$ and $Z(s) = L[z(t)]$ for the system above.

b) What conditions on α, β, and γ will ensure stability of this system?

Problem 6.4

Let $K(s) = \mu/(s+1)$ and $\beta(s) = s/(s+\alpha)$ in the conventional Black block diagram of a negative feedback system.

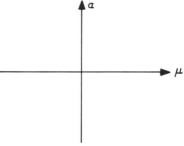

a) On a sketch of the α-μ-plane (as at the right), cross-hatch the region within which the feedback system will be stable.

b) For $\alpha = 4$, sketch in the complex s-plane the locus of the zeros of $1 + \beta(s)K(s)$ as a function of μ, $-\infty < \mu < \infty$.

Problem 6.5

Consider the inverting op-amp circuit shown below. Assume that the op-amp is ideal to the extent that its input impedance is infinite ($i_0(t) = 0$) and its output voltage is independent of load (output impedance $= 0$), but that it is non-ideal in that its gain $K(s)$ is not infinite and not a constant.

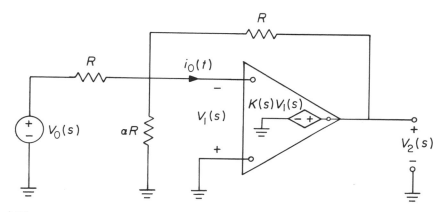

a) The asymptotes of the Bode plot for $|K(j\omega)|$ are shown below. Assuming that the op-amp is stable and has a rational system function with $K(0) > 0$, find $K(s)$.

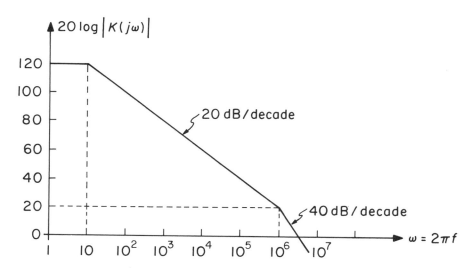

b) Find the overall system function, $H(s) = V_2(s)/V_0(s)$, for this circuit.

c) Find the value of α for which $H(s)$ has a second-order pole on the negative real axis. Make reasonable approximations.

Problem 6.6

The circuit below shows one way of adding *lead compensation* to a non-inverting amplifier. $K(s)$ represents the system function of the op-amp itself, reflecting primarily the high-frequency shunting effect of various internal capacitances.

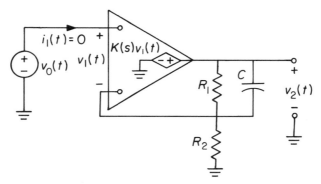

a) Show that before compensation, with $C = 0$, the system can be described by the block diagram below to the left.

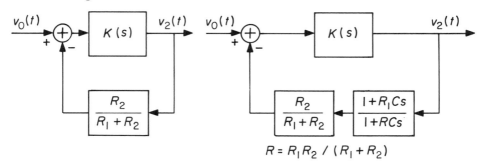

$$R = R_1 R_2 / (R_1 + R_2)$$

b) Show that with compensation the system can be described by the block diagram above to the right.

c) Let $K(s) = 60/(s+1)^3$, $R_1 C = 1$ sec, $R_1 = 9R_2$. Show that the loop gains, compensated and uncompensated, are given by precisely the formulas assumed in Example 6.A–3 ($K(s)$ as given here is not, of course, a very good representation of actual op-amp behavior).

d) Show that the overall system functions for the feedback amplifier using the values in (c) are

$$\frac{V_2(s)}{V_0(s)} = H(s) = \begin{cases} \dfrac{60}{(s+2.82)(s^2+0.18s+2.49)} & \text{(uncompensated)} \\[3mm] \dfrac{60(s+10)}{(s+1)(s+10.65)(s^2+1.35s+6.62)} & \text{(compensated).} \end{cases}$$

e) Discuss the character of the unit step response (final values, rise times, overshoot, etc.) for each case in (d). (Feel free to actually find the step responses if you wish!)

Problem 6.7

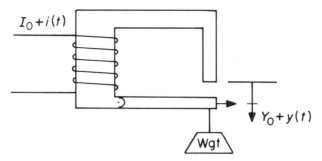

Systems to suspend a weight in a magnetic field are often inherently unstable because the magnetic force on an armature grows rapidly as the gap is reduced. For the system shown above, let Y_0 be the position at which the magnetic force due to the current I_0 just balances the gravitational force due to the weight. For small changes from this (unstable) equilibrium, assume that the incremental current and position satisfy the differential equation

$$\frac{d^2y(t)}{dt^2} - 4y(t) = -10i(t).$$

To stabilize this system, suppose we measure movement from the equilibrium point with a pickup coil whose output voltage is proportional to velocity,

$$v(t) = 2\frac{dy(t)}{dt}.$$

We then propose to compare this signal with a desired velocity signal $x(t)$ and drive the coil with the amplified difference as shown below.

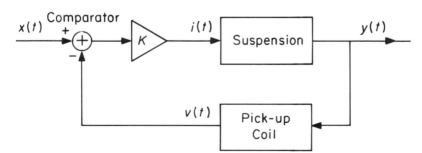

a) Show that this system is unstable for any value of K.

b) The circuit on the next page is suggested as a compensator for the system, to be inserted in the feedback path between the pickup coil and the comparator. Determine the transfer function $V_c(s)/V(s)$ of the circuit, using assumptions appropriate for an ideal op-amp.

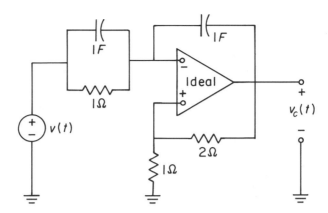

c) Compute the system function for the suspension system with the compensator in the feedback loop.

d) Find the range of values of K for which the compensated system is stable.

Problem 6.8

Stonewell International has hired you as a consultant to study schemes for regulating the altitude $y(t)$ of their Space Buggy by controlling the vertical thrust $x(t)$ (which can be either positive or negative). Assume that $x(t)$ and $y(t)$ can be related by the LTI model shown below to the left; m is the mass of the craft.

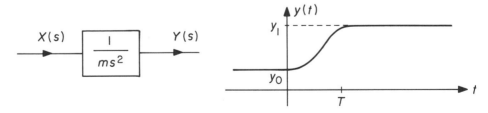

a) Sketch the thrust waveform that would be necessary to change the craft altitude smoothly but rapidly from one value to another (as shown above to the right) by direct control of vertical thrust.

b) One proposed control scheme involves adding a feedback loop and controller signal $v(t)$ as shown below. The mass m of the spacecraft changes as the fuel is used. Describe the locus of the closed-loop poles of this system assuming that β and K are positive constants and m varies from 0 to ∞.

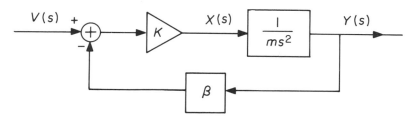

c) Is the system BIBO stable under these conditions?

d) To improve performance, derivative feedback is added; the β-block is replaced by the dashed block shown below, where d/dt indicates an ideal differentiator. Describe the stability of this improved system as a function of the constant $K, -\infty < K < \infty$.

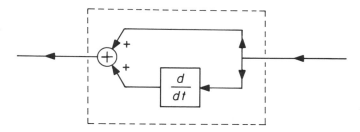

e) Using the improved design with very large $K > 0$, determine the controller signal $v(t)$ necessary to produce the altitude change described in (a). Do you think the "improved" design would in fact be easier or harder to control manually than direct control of the vertical thrust?

Problem 6.9

As suggested in Example 6.2–1, the frequency response effects of op-amps can be approximately described by an equivalent circuit consisting of a controlled source whose gain is a function of frequency, $K(s)$. For two popular op-amps, $K(s)$ is given approximately by

$$K(s) = \begin{cases} \dfrac{10^5}{\left(\dfrac{s}{20\pi}+1\right)\left(\dfrac{s}{2\times 10^6\pi}+1\right)^2} & \text{(Type 741)} \\[4ex] \dfrac{10^5}{\left(\dfrac{s}{200\pi}+1\right)\left(\dfrac{s}{2\times 10^6\pi}+1\right)^2} & \text{(Type 748).} \end{cases}$$

a) Sketch the Bode plot for each of these op-amps. Which would you say was the better op-amp? What criterion are you using for "better"?

b) Discuss the stability of each of these op-amps if used in the voltage-follower circuit shown below to the left.

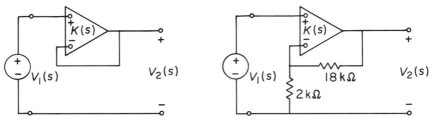

c) Discuss the stability of each of these op-amps if used in the non-inverting amplifier circuit with low-frequency gain ≈ 10 shown above to the right.

d) Now which op-amp do you think is better? What criterion are you now using for "better"? Why is the voltage-follower circuit a more severe test of stability than the non-inverting amplifier circuit?

The 741 and 748 are identical devices except that one of the time constants of the 741 is intentionally made larger; that is, the 741 is *internally compensated* to avoid the difficulty explored in this problem.

Problem 6.10

The simple linear feedback system shown below is intended to perform as a tracking system. Ideally the error signal, $\epsilon(t)$, should be identically zero, or equivalently $y(t)$ should be equal to $x(t)$.

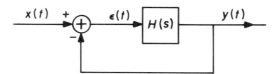

a) Suppose $H(s) = 1/s$. Determine the steady-state form of $\epsilon(t)$ if the input is a unit ramp function

$$x(t) = \begin{cases} t, & t \geq 0 \\ 0, & t < 0. \end{cases}$$

Will the tracking error be zero in the steady state?

b) Repeat (a) for $H(s) = 1/s^2$. Describe the nature of the tracking error in the steady state.

c) Suppose $H(s) = A(s)/s^2$. Choose $A(s)$ so that the problems uncovered in (a) and (b) are resolved and $\epsilon(t) \to 0$ for a unit ramp input. Sketch $\epsilon(t)$, $x(t)$, and $y(t)$ for your choice on the same coordinates.

Problem 6.11

It can be shown that any two-terminal admittance, $Y(s)$, composed of positive R's, L's, and C's must satisfy two conditions:

1. The poles of $Y(s)$ cannot lie in the right half-s-plane.
2. $\Re[Y(s)] \geq 0$ for all $\Re[s] \geq 0$.

(For a discussion of these conditions, see Problem 4.5.)

a) Show that the following circuit is stable for any $Y(s)$ satisfying these conditions if $g_m > 0$.

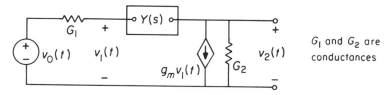

G_1 and G_2 are conductances

b) What is the smallest magnitude of g_m for which there might be instability for some $Y(s)$ satisfying these conditions if g_m is negative?

Problem 6.12

In the circuit shown below, the large box represents a two-stage amplifier idealized to have infinite input impedance and a current-source output as shown. The smaller boxes are one-port circuits characterized by their impedances, $Z(s)$ and $2Z(s)$.

a) With the switch open and a voltage source $v_2(t)$ applied, compute the loop gain, $V_1(s)/V_2(s)$, and show that it equals $g_m Z(s)/4$.

b) For $Z(j\omega)$ as shown in the figure, plot the Nyquist diagram (or polar plot of the loop gain).

c) If $Z(j\omega)$ is as shown, what is the maximum (positive) value that g_m can have if the circuit is to be stable when the switch is closed?

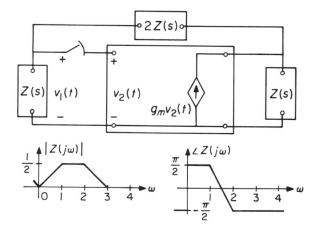

Problem 6.13

The measured sinusoidal steady-state frequency response, $K(j2\pi f)$, of a certain band-pass amplifier is shown in the polar plot on the page opposite. This amplifier is cascaded with an ideal amplifier with gain $\mu > 0$, and a fraction β of the output is fed back as shown in the insert on the opposite page.

a) What is the largest value of $\mu\beta$ for which the overall system will be stable?

b) Let $\mu\beta$ equal half the maximum value found in (a). Find μ and β such that the overall gain at $f = 1$ is 50. What are the gain and phase margin under these conditions?

c) To improve the performance, a *lag network* is inserted at the position of the dotted box with frequency response

$$G(j2\pi f) = \frac{1}{(1+jf/10)^2}.$$

Draw directly on a copy of the polar diagram a sketch of $G(j2\pi f)K(j2\pi f)$—approximately to scale. The simplest procedure is to compute the magnitude and angle of $G(j2\pi f)$ for $f = 1$, $10/\sqrt{3}$, 10, $10\sqrt{3}$, etc. Pay particular attention to the asymptotic behavior as $f \to 0$ and $f \to \infty$.

d) Determine approximately the largest value of $\mu\beta$ for which the new system will be stable.

e) Repeat part (b) for this compensated system, letting $\mu\beta$ be the same value as before (half the value found in (a)).

Problem 6.14

The system shown below is usually called a *phase-shift oscillator*. The transistor current amplifier A has negligibly small input impedance, high output impedance, and a current amplification $i_2/i_1 = -A$, where A is a positive constant. In practice, $C_1 < C_2 < C_3$ and $R_1 > R_2 > R_3$, but for purposes of this problem let $C_1 = C_2 = C_3 = C$ and $R_1 = R_2 = R_3 = R$.

a) Plot the Nyquist diagram of the loop transmission.

b) Find the critical value of A for which the circuit just oscillates.

c) Calculate the frequency of oscillation.

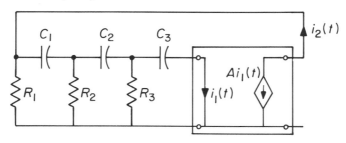

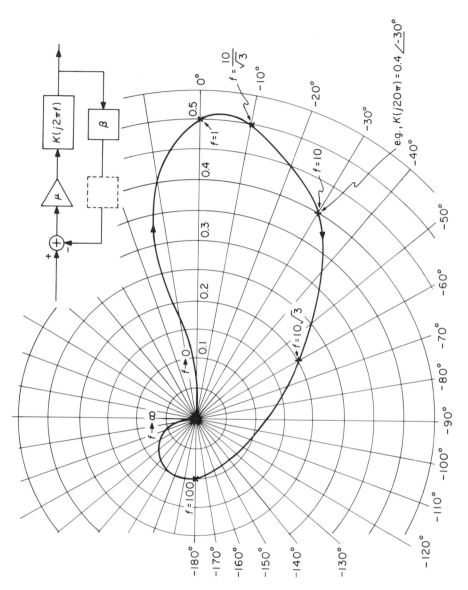

Polar Plot of $K(j2\pi f)$ for Problem 6.13.

Problem 6.15

If there were no acoustic feedback from the loudspeaker to the microphone, an out-door public address system would have the overall frequency response $G(j\omega)$ shown below, where $G(j\omega)$ is the ratio of the sound pressure at the loudspeaker to the sound pressure at the microphone. But for a fixed angle of orientation between speaker and microphone, it is found that the microphone picks up a fraction $1/r^2$ (where r is the distance between speaker and microphone in feet) of the sound from the loudspeaker delayed by the time it takes sound to travel r feet (assume the speed of sound to be 1000 ft/sec). It is also found that the system is not stable for $r_1 \le r \le r_2$. Use Nyquist's criterion to calculate r_1 and r_2.

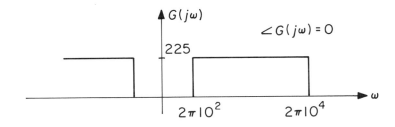

7

DISCRETE-TIME SIGNALS AND
LINEAR DIFFERENCE EQUATIONS

7.0 Introduction

Thus far our study of signals and systems has emphasized voltages and currents as signals, and electric circuits as systems. To be sure, we have pointed out various analogies to simple mechanical systems, and we have explored in at least an introductory way systems composed of larger blocks than elementary R's, L's, and C's. But, mathematically, all of our signals have been specified as functions of a continuous variable, t, and almost all of our systems have been described by sets of linear, finite-order, total differential equations with constant coefficients (or equivalently by system functions that are rational functions of the complex frequency s).

With this chapter we shall begin the process of extending our mathematical models to larger classes of both signals and systems. Such extensions are interesting in part because they will permit us to analyze and design a wider range of practical systems and devices. But an equally important goal is to learn how to brush aside certain less fundamental characteristics of our system models so that we may concentrate on the deep, transcendent significance of their linearity and time-invariance.

Specifically, we shall consider in the next few chapters systems in which the signals are indexed *sequences* rather than functions of continuous time. We shall identify such sequences by $x[n]$, $y[n]$, etc., where the square brackets indicate that the enclosed index variable, called *discrete time*, takes on only integral values: $\ldots, -2, -1, 0, 1, 2, \ldots$. Discrete-time (DT) signals, like continuous-time (CT) signals, may be defined in many ways—by bar diagrams as in Figure 7.0–1, by formulas such as $x[n] = 2^n$, by tables of values, or by combinations of these.

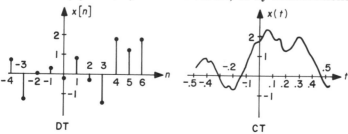

Figure 7.0–1. Comparison of DT and CT signals.

Many systems inherently operate in discrete time. Some examples are certain banking situations ("regular monthly payments"), medical therapy regimes ("two pills every four hours"), econometric models utilizing periodically compiled indices, and evolutionary models characterizing population changes from generation to generation. Moreover, a variety of regular structures in space, rather than time—such as cascaded networks, tapped delay lines, diffraction gratings, surface-acoustic-wave (SAW) filters, and phased-array antennas—lead to similar mathematical descriptions. In other cases, DT signals are constructed by periodic sampling of a CT signal. If the sampling is done sufficiently rapidly and the signal is sufficiently smooth, the loss in information can be small, as we shall demonstrate in Chapter 14. Motion picture and television images are examples of this kind—the image in two space dimensions and continuous time is sampled every 30–40 msec to yield a sequence of frames.

Sometimes the reason for the transformation from continuous to discrete time is to permit time-sharing of an expensive communications or data-processing facility among a number of users. Examples range from a ward nurse measuring patient temperatures sequentially to telemetry systems transmitting interleaved samples of a variety of data from scattered oil wells or weather stations or interplanetary space probes. But the most common reason today for replacing a CT signal by an indexed sequence of numbers is to make it possible for the signal processing to be carried out by digital computers or similar special-purpose logical devices. Examples include such disparate areas as speech analysis and synthesis; radio, radar, infrared, and x-ray astronomy; the study of sonar and geophysical signals; the analysis of crystalline and molecular structures; the interpretation of medical signals such as electrocardiograms, CAT scans, and magnetic resonance imaging; and the image-enhancement or pattern-recognition processing of pictures such as satellite or space probe photos, x-ray images, blood smears, or printed materials. Processing by computers requires not only that the signal be discrete-time, but that the numbers representing each sample be rounded off or quantized—a potential additional source of error. In exchange, however, one gains great flexibility and power. For example, once the signals in a complex radar receiver have been sampled and quantized, all further processing involves only logical operations, which are inherently free of the parameter drift, sensitivity, noise, distortion, and alignment problems that often limit the effectiveness of analog devices. Thus digital filters can process extremely low-frequency signals that in analog filters would be hopelessly corrupted by the effects of aging and drift. Moreover, digital computers can accomplish certain tasks, such as the approximate solution of large sets of non-linear differential equations, that are virtually impossible to do in any other way. And a change in the task to be carried out requires only a change in instructions, not a rebuilding of the apparatus. Thus computer simulation is increasingly replacing the "breadboard" stage in complex system design because it is faster, cheaper, and permits more flexible variation of parameters to optimize performance. Indeed, sometimes (as in the design of integrated circuits) a simulated "breadboard" may be more accurate than one constructed with "real" elements such as lumped transistors that are not the ones that will ultimately be used in the actual device.

A natural vehicle for describing a system intended to process or modify discrete-time signals—a *discrete-time system*—is frequently a set of *difference equations*. Difference equations play for DT systems much the same role that differential equations play for CT systems. Indeed, as we shall see, the analysis of linear difference equations reflects in virtually every detail the analysis of linear differential equations. The next few chapters will thus also serve as a review of much of our development to this point. In addition, DT systems are in certain mathematical respects simpler than CT systems. The extension from difference/differential equation systems to general LTI systems is thus easiest if we first carry it out for DT systems, as we shall in Chapter 9.

7.1 Linear Difference Equations

A linear N^{th} order constant-coefficient difference equation relating a DT input $x[n]$ and output $y[n]$ has the form*

$$\sum_{k=0}^{N} a_k y[n+k] = \sum_{\ell=0}^{N} b_\ell x[n+\ell]. \qquad (7.1-1)$$

Some of the ways in which such equations can arise are illustrated in the following examples.

Example 7.1–1

A $50,000$ mortgage is to be retired in 30 years by equal monthly payments of p dollars. Interest is charged at 15%/year on the unpaid balance. Let $P[n]$ be the unpaid principal in the mortgage account just after the nth monthly repayment has been made. Then

$$P[n+1] = (1+r)P[n] - p, \qquad n \geq 0 \qquad (7.1-2)$$

where $r = 0.15/12 = 0.0125$ is the monthly interest rate. Initially $P[0] = 50,000$, and we seek the value of p such that $P[360] = 0$.

We shall return to this problem in Example 7.3–1, but for the moment notice that (7.1–2) has the form of (7.1–1) with $N = 1$ (first order) and

$$y[n] = P[n] \qquad\qquad\qquad x[n] = p, \quad 0 < n \leq 360$$
$$a_0 = -(1+r) \qquad\qquad\quad b_0 = -1$$
$$a_1 = 1 \qquad\qquad\qquad\qquad b_\ell = 0 \text{ otherwise}$$
$$a_k = 0 \text{ otherwise.}$$

▶ ▶ ▶

---

*There is no loss of generality in assuming the same number $(N+1)$ of terms on each side in (7.1–1), since we may always set certain of the coefficients to zero.

Example 7.1–2

The numerical integration of differential equations typically involves difference equations as an intermediate step resulting from replacing derivatives by formulas involving differences, such as

$$\dot{x}(t) = \frac{dx(t)}{dt} \approx \frac{x(t + \Delta t) - x(t)}{\Delta t}$$

where Δt is the *step size* , a small time increment that we assume fixed. Thus consider Figure 7.1–1, which is the circuit of Example 1.3–3 with two of the sources set to zero.

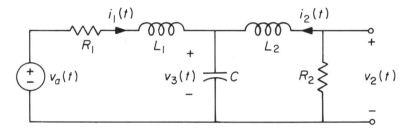

Figure 7.1–1. Circuit for Example 7.1–2.

From Example 1.3–3, the dynamic equations in state form are

$$\frac{di_1(t)}{dt} = -\frac{R_1}{L_1}i_1(t) - \frac{1}{L_1}v_3(t) + \frac{1}{L_1}v_a(t)$$

$$\frac{di_2(t)}{dt} = -\frac{R_2}{L_2}i_2(t) - \frac{1}{L_2}v_3(t)$$

$$\frac{dv_3(t)}{dt} = \frac{1}{C}i_1(t) + \frac{1}{C}i_2(t)$$

with the output equation

$$v_2(t) = -R_2 i_2(t).$$

Replacing derivatives as above yields

$$i_1(t + \Delta t) \approx \left(1 - \frac{R_1 \Delta t}{L_1}\right)i_1(t) - \frac{\Delta t}{L_1}v_3(t) + \frac{\Delta t}{L_1}v_a(t)$$

$$i_2(t + \Delta t) \approx \left(1 - \frac{R_2 \Delta t}{L_2}\right)i_2(t) - \frac{\Delta t}{L_2}v_3(t)$$

$$v_3(t + \Delta t) \approx v_3(t) + \frac{\Delta t}{C}i_1(t) + \frac{\Delta t}{C}i_2(t)$$

$$v_2(t) = -R_2 i_2(t).$$

Substituting $t = n\Delta t$, $i_1(t) = i_1(n\Delta t) = i_1[n]$, etc., and using the numerical values $R_1 = R_2 = 1\,\text{k}\Omega$, $L_1 = L_2 = 0.1\,\text{H}$, $C = 0.2\,\mu\text{F}$, $\Delta t = 10\,\mu\text{sec}$, yields a set of three simultaneous difference equations

$$i_1[n+1] = 0.9i_1[n] - 10^{-4}v_3[n] + 10^{-4}v_a[n]$$
$$i_2[n+1] = 0.9i_2[n] - 10^{-4}v_3[n]$$
$$v_3[n+1] = v_3[n] + 50i_1[n] + 50i_2[n]$$
$$v_2[n] = -10^3 i_2[n].$$

We shall return to this example in Example 7.3–2.
▶ ▶ ▶

There are many different ways to approximate a set of differential equations by a set of difference equations as above; these correspond to different integration algorithms. The choice in Example 7.1–2 (called the *forward Euler algorithm*) is perhaps the most straightforward, but it is not usually the best in terms of minimizing the approximation error for a given step size. Indeed, for many systems (including this one—see Example 8.3–2) the forward Euler algorithm is *numerically unstable* in that the total accumulated error in the approximate solution grows with time unless the step size is sufficiently small. This difficulty can be overcome with various *implicit* integration algorithms such as the *backward Euler algorithm*, which substitutes $\dot{x}(t+\Delta t) \approx \dfrac{x(t+\Delta t) - x(t)}{\Delta t}$ and yields in the present case the equations

$$1.1\, i_1[n+1] + 10^{-4}\, v_3[n+1] = i_1[n] + 10^{-4} v_a[n+1]$$
$$1.1\, i_2[n+1] + 10^{-4}\, v_3[n+1] = i_2[n]$$
$$-50i_1[n+1] - 50i_2[n+1] + v_3[n+1] = v_3[n]$$

which must be solved at each iteration because new values of $i_1[n]$, $i_2[n]$, and $v_3[n]$ at $n+1$ depend on one another (this is why the backward Euler algorithm is called "implicit"). Other examples of simple implicit algorithms are given in Problem 7.1. For a comprehensive discussion of the issues involved in selecting among such algorithms, see, for example, L. O. Chua and P.-M. Lin, *Computer-Aided Analysis of Electronic Circuits* (Englewood Cliffs, NJ: Prentice-Hall, 1975).

Example 7.1–3

As an example of a case in which the index variable n corresponds to space rather than time, consider the uniform cascaded ladder network, a section of which is shown in Figure 7.1–2.

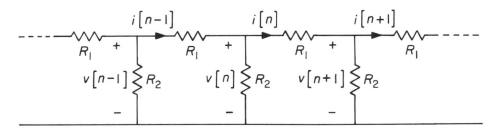

Figure 7.1–2. Uniform ladder network for Example 7.1–3.

The voltages and currents along the ladder satisfy the equations

$$v[n+1] = v[n] - i[n]R_1$$
$$i[n+1] = i[n] - \frac{v[n+1]}{R_2}.$$

The first equation can be solved for $i[n]$ and substituted into the second for $i[n]$ and $i[n+1]$ to yield a single second-order homogeneous difference equation for $v[n]$:

$$R_2 v[n+2] - (2R_2 + R_1)v[n+1] + R_2 v[n] = 0. \qquad (7.1\text{–}3)$$

This example is continued in Example 7.4–1.
▶ ▶ ▶

7.2 Block Diagrams and State Formulations for DT Systems

Circuit and block diagrams provide useful alternatives to differential equations as descriptions of CT systems. Similar diagrams can be devised for the description of DT systems. One such diagram is directly analogous to the integrator-adder-gain class of CT block diagrams introduced in Section 1.4; it employs a DT system called an *accumulator* in place of the integrator. Another kind of DT block diagram—on the whole, a more useful one—uses *delay* elements rather than accumulators as the memory elements. Both types of block diagrams provide significant insights into DT system behavior—particularly the notion of state—and their study yields opportunities for further practice with difference equations.

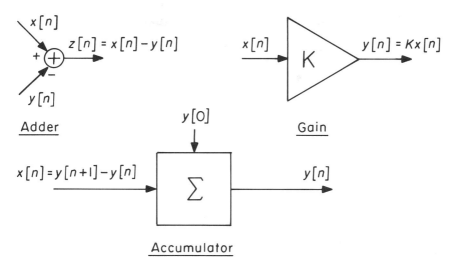

Figure 7.2–1. Some basic discrete-time block-diagram elements.

The elements described in Figure 7.2–1 are those used in what are called accumulator-adder-gain DT block diagrams. The difference equation in this figure defines an accumulator* in the same implicit way one might define an integrator by stating that its input is the derivative of its output. In Example 9.1–1 we shall show that this version of an accumulator can be defined explicitly by

$$y[n] = y[0] + \sum_{m=0}^{n-1} x[m]$$

which clarifies both the name "accumulator" and its relation to a CT integrator. Note that specification of the initial output value $y[0]$ is necessary to yield a unique solution to the accumulator difference equation, which is why it is explicitly shown as a separate input in Figure 7.2–1.

Indeed, the accumulator output $y[n_0]$ at time $n = n_0$ defines the state of the accumulator at that time. That is, knowledge of $y[n_0]$ together with the input $x[n]$ for $n \geq n_0$ determines the output $y[n]$ for $n > n_0$. Hence, if a DT system is described by a block diagram of interconnected accumulators, adders, and gain elements, the current state of the system is defined by the current values of the outputs of the accumulators. To derive the dynamic difference equations for the system in state form, it is sufficient to choose the accumulator outputs as variables and to determine from the block diagram how these variables are combined to yield the input to each accumulator. The procedure parallels exactly the corresponding procedure for CT systems composed of integrators, adders, and gain elements, and is most easily explained through an example.

---

*As implied in Section 7.1, there are many possible DT analogs of a CT integrator. The selection made here implements the forward Euler algorithm as before. See Problem 8.7.

Example 7.2–1

The CT circuit of Example 7.1–2 led (for the parameter values given in that example) to the differential state equations

$$\frac{di_1(t)}{dt} = -10^4 i_1(t) - 10v_3(t) + 10v_a(t)$$

$$\frac{di_2(t)}{dt} = -10^4 i_2(t) - 10v_3(t)$$

$$\frac{dv_3(t)}{dt} = 5\times10^6 i_1(t) + 5\times10^6 i_2(t)$$

with the output equation

$$v_2(t) = -10^3 i_2(t).$$

This circuit is thus equivalent to the block diagram of Figure 7.2–2. Each of the state equations corresponds directly to the equation describing the adder at the input to the corresponding integrator.

In Example 7.1–2, DT difference equations approximating this system were derived by replacing the derivatives by differences,

$$i_1[n+1] = 0.9i_1[n] - 10^{-4}v_3[n] + 10^{-4}v_a[n]$$

$$i_2[n+1] = 0.9i_2[n] - 10^{-4}v_3[n]$$

$$v_3[n+1] = v_3[n] + 50i_1[n] + 50i_2[n]$$

and

$$v_2[n] = -10^3 i_2[n].$$

These equations also describe the accumulator-adder-gain block diagram shown in Figure 7.2–3. Again each difference equation corresponds directly to the equation describing the output of each adder multiplied by $\Delta t = 10^{-5}$. Not surprisingly, the block diagrams in Figures 7.2–2 and 7.2–3 are identical except that each integrator is replaced by a gain Δt in cascade with an accumulator. The difference equations above are said to be in *state form* because they provide an explicit way to calculate the next values of the *state variables*, $i_1[n+1]$, $i_2[n+1]$, and $v_3[n+1]$, from the present values, $i_1[n]$, $i_2[n]$, and $v_3[n]$, and the system input $v_a[n]$.

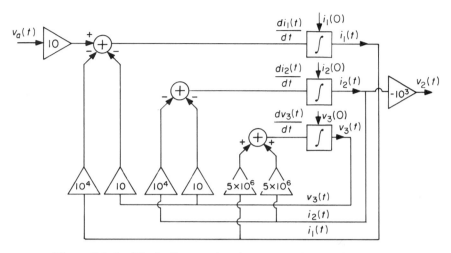

Figure 7.2–2. Block diagram for the circuit of Example 7.1–2.

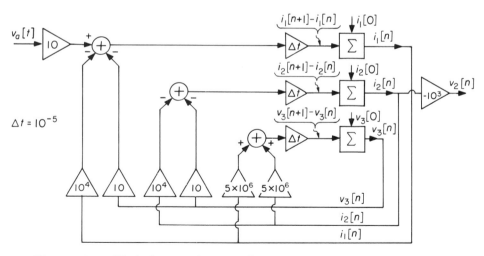

Figure 7.2–3. Block diagram for the difference equations of Example 7.1–2.

▶ ▶ ▶

As Example 7.2–1 shows, it is easy to go from an accumulator-adder-gain block diagram to a set of difference equations in state form, choosing as state variables the outputs of the accumulators. The inverse statement is also true: It is generally possible to go from a set of difference equations or a higher-order input-output difference equation to a block diagram in accumulator-adder-gain form.

However, as implied above, an alternative DT block-diagram representation in which delay elements replace accumulators as the memory elements is also general and often much more convenient. The *unit delay element* is represented schematically as in Figure 7.2–4 and is characterized by the property that its present output is equal to its input one unit of time earlier. In a block diagram made up of delay-adder-gain elements the state of the system is characterized by the present values of the delay-element outputs. Although delay elements can be combined with adders and gains in many configurations, the most useful arrangements seem to result if the delay elements are cascaded—the output of one becoming the input to the next—to form what is sometimes called a DT *tapped delay line.* The following example illustrates some of the things one can do with tapped delay lines.

Figure 7.2–4. Unit delay element.

Example 7.2–2

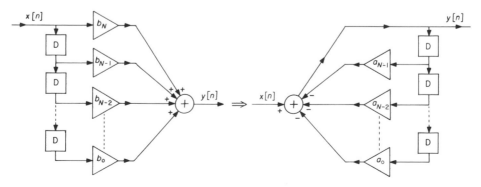

Figure 7.2–5. Feedforward (a) and feedback (b) DT delay-line systems.

The simple feedforward or *transversal* structure of Figure 7.2–5(a) and the feedback or *recursive* arrangement of Figure 7.2–5(b) realize two special cases of the general difference equation (7.1–1). Thus, combining the signals of the adders, it is evident that Figure 7.2–5(a) is characterized by the difference equation

$$y[n] = b_o x[n - N] + \cdots + b_{N-2} x[n - 2] + b_{N-1} x[n - 1] + b_N x[n] \qquad (7.2\text{--}1)$$

or, equivalently,

$$y[n + N] = b_0 x[n] + \cdots + b_{N-2} x[n + (N - 2)] + b_{N-1} x[n + (N - 1)] + b_N x[n + N] \quad (7.2\text{--}2)$$

which corresponds to (7.1–1) with $a_N = 1$ and $a_i = 0$, $0 \le i < N$. Similarly, Figure 7.2–5(b) is characterized by

$$y[n] = -a_{N-1} y[n - 1] - a_{N-2} y[n - 2] - \cdots - a_0 y[n - N] + x[n] \qquad (7.2\text{--}3)$$

or, equivalently,

$$y[n + N] + a_{N-1} y[n + (N - 1)] + a_{N-2} y[n + (N - 2)] + \cdots + a_0 y[n] = x[n + N] \quad (7.2\text{--}4)$$

which corresponds to (7.1–1) with $a_N = b_N = 1$ and $b_i = 0$, $0 \le i < N$. Notice, however, that if the output of the feedforward system of Figure 7.2–5(a) is made the input to the system of Figure 7.2–5(b), that is, if the two systems are cascaded as implied by the double arrow in Figure 7.2–5, then the overall system is characterized by

$$y[n+N] + a_{N-1} y[n + (N-1)] + \cdots + a_0 y[n] = b_0 x[n] + b_1 x[n+1] + \cdots + b_N x[n + N] \quad (7.2\text{--}5)$$

which is precisely (7.1–1) except for the (non-constraining) choice $a_N = 1$. Thus the input-output behavior of *any* DT system can be simulated in this way.

An even more interesting block diagram results if the order in which the systems are cascaded in Figure 7.2–5 is reversed, as in Figure 7.2–6. Such a reversal does not change the overall input-output ZSR behavior. (We are not quite in a position to prove this statement easily at this point—see Section 8.3. The analogous result for CT systems was discussed in Section 5.1.) But observe that if we label the signal at the midpoint $w[n]$, then the signals at successive points in each of the delay lines are the same—$w[n - 1]$, $w[n - 2]$, ..., $w[n - N]$. Hence only one of the two lines is really necessary; an equivalent arrangement is shown in Figure 7.2–7 and is called the *canonical* form because it employs the minimum possible number of delay elements. The representation of the general input-output difference equation (7.1–1) in the form of Figure 7.2–7 is particularly convenient because the coefficients in the difference equation are directly identifiable as corresponding gains in the block diagram. A similar block diagram for CT systems was derived in Problem 1.6.

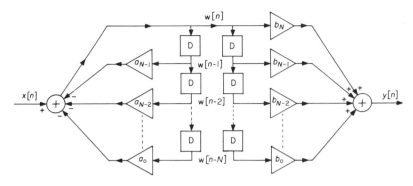

Figure 7.2–6. Cascade of systems of Figure 7.2–5 in reversed order.

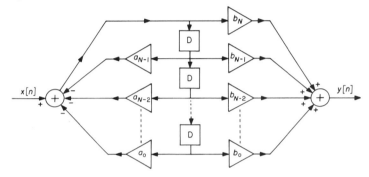

Figure 7.2–7. The canonical form of delay-line realization of the input-output behavior of a general DT LTI system.

Difference equations in state form are readily derived for the system of Figure 7.2–7 by taking the outputs of the delay elements as state variables:

$$\lambda_0[n+1] = \lambda_1[n]$$
$$\lambda_1[n+1] = \lambda_2[n]$$
$$\vdots$$
$$\lambda_{N-2}[n+1] = \lambda_{N-1}[n]$$
$$\lambda_{N-1}[n+1] = -a_{N-1}\lambda_{N-1}[n] - a_{N-2}\lambda_{N-2}[n] - \cdots$$
$$\cdots - a_0\lambda_0[n] + x[n].$$

The output equation is then

$$y[n] = b_N\lambda_{N-1}[n+1] + b_{N-1}\lambda_{N-1}[n] + b_{N-2}\lambda_{N-2}[n] + \cdots + b_0\lambda_0[n]$$
$$= (b_{N-1} - b_N a_{N-1})\lambda_{N-1}[n] + (b_{N-2} - b_N a_{N-2})\lambda_{N-2}[n] + \cdots$$
$$\cdots + (b_0 - b_N a_0)\lambda_0[n] + b_N x[n].$$

▶ ▶ ▶

The canonical form of Figure 7.2–7 is well-adapted to exploit the capabilities of a class of cheap monolithic MOS devices called *charge-transfer devices* (CTD). Several categories of CTD's are available—such as bucket-brigade devices (BBD) or charge-coupled devices (CCD)—that are broadly similar but differ in details of structure and performance characteristics. These devices are driven by a periodic chain of clock pulses whose rate typically may range from perhaps 100 pulses per second (pps) to 10^7 pps or more. On each clock pulse, the analog voltage on the input capacitor of each of a succession of cascaded MOS stages is transferred to become the voltage on the input capacitor of the next stage. Charge transfer requires part of the interpulse period; during the rest of the period, the voltage remains nearly constant. The voltage at each stage can be sampled (through buffers), weighted, and added to the weighted outputs of other stages as in the canonical system structure (fixed weights can be built into the chip; adjustable weighting requires external potentiometers), or even fed back to add to the voltage of earlier stages (general recursive devices). Hundreds or even thousands of stages can be fabricated on a single chip. Some of the uses of CTD's as filters, delay lines, memory devices, correlators, frequency changers, Fourier transformers, etc., will be discussed in later chapters and problems.

7.3 Direct Solution of Linear Difference Equations

As with linear differential equations, linear difference equations describe the input-output behavior of a DT system only implicitly—they must be "solved" to find the response to a specific input. And again as in the CT case, the difference equation and the input in the semi-infinite interval $0 \leq n < \infty$ yield only a partial description of the output for $n \geq 0$. It is necessary to provide N additional pieces of information for an Nth-order system, corresponding to the initial conditions or the initial state in the continuous case. However, unlike the situation with differential equations, the solution to the difference-equation problem is trivial if the information given is the first N values of the output, since it is always possible to rewrite (7.1–1) in the form*

$$y[n + N] = \sum_{\ell=0}^{N} b_\ell x[n + \ell] - \sum_{k=0}^{N-1} a_k y[n + k]. \qquad (7.3–1)$$

We may thus immediately find $y[N]$ from the known input and the known values of $y[0]$, $y[1]$, ..., $y[N-1]$. Iteratively, we find $y[N+1]$ from the input and $y[1]$, $y[2]$, ..., $y[N]$, and so on. This technique is illustrated in the following examples.

---

*Without loss of generality, we have again set $a_N = 1$.

Example 7.3–1

Continuing Example 7.1–1, observe that (7.1–2) already has the form of (7.3–1), that is,

$$P[n+1] = (1+r)P[n] - p \qquad (7.3\text{--}2)$$

with $P[0] = \$50,000$. That is all the information we need, since the equation is first-order. We readily calculate

$$P[1] = (1+r)P[0] - p$$
$$P[2] = (1+r)P[1] - p$$
$$= (1+r)^2 P[0] - (1 + (1+r))p$$
$$P[3] = (1+r)^3 P[0] - (1 + (1+r) + (1+r)^2)p$$

from which we deduce the general term

$$P[n] = (1+r)^n P[0] - \frac{(1+r)^n - 1}{r}p, \quad n \geq 0. \qquad (7.3\text{--}3)$$

To write $P[n]$ in closed form we have used the important formula for the partial sum of a geometric series,

$$1 + a + a^2 + \cdots + a^m = \frac{1 - a^{m+1}}{1-a}. \qquad (7.3\text{--}4)$$

For a further discussion of (7.3–4), see Example 8.1–4.

From (7.3–3) we can directly compute the value of p necessary to pay off this mortgage in 30 years. Setting

$$P[360] = 0$$

gives

$$p = \frac{r(1+r)^{360}}{(1+r)^{360} - 1}P[0].$$

For $r = 0.0125$ (15%/year) and $P[0] = \$50,000$ this becomes

$$p = \$632.22.$$

The overall return to the bank on the $\$50,000$ mortgage is $360p = \$227,599.20$, which illustrates rather clearly why banks make loans.

▶ ▶ ▶

Example 7.3–2

The simultaneous difference equations replacing the differential state equations for the circuit of Example 7.1–2 are already in the form of (7.3–1) and can readily be solved by iteration if $i_1[0]$, $i_2[0]$, and $v_3[0]$ (characterizing the initial state) are known. The results for $i_1[0] = i_2[0] = v_3[0] = 0$ and $v_a[n] = 1$, $n \geq 0$, are shown in Figure 7.3–1 for several values of Δt. Further features of this example are considered in Example 8.3–2.

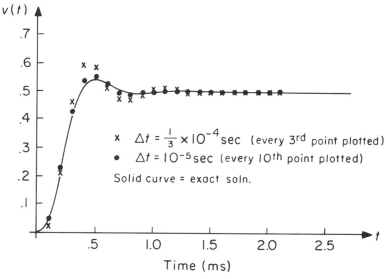

Figure 7.3–1. Solution of the equations of Example 7.1–2 for several values of Δt.

In using difference equations to obtain approximate solutions to differential equations, the choice of sampling interval or step size, Δt, is critical. If Δt is too large, the accuracy of the discrete approximation is poor; if Δt is too small, the number of iterations required to describe the interesting range of the output becomes large. In general, Δt must be significantly smaller than the interval over which the CT state changes substantially. Sometimes it is useful to change Δt during the evolution of the solution to better match the speed at which the state is changing.

▶ ▶ ▶

Unfortunately, the boundary conditions are not always given in a form that is convenient for the direct solution method discussed in this section. And even when they are, identifying a closed-form expression for $y[n]$ is not always easy (although this is not necessarily a serious problem if the equations are being solved by a computer). There are, however, alternative solution schemes that parallel the ZSR and ZIR procedures for CT linear differential equations and that provide much insight as well as workable solutions for many problems. We shall begin to explore these in the next section.

7.4 Zero Input Response

The ladder network of Example 7.1–3 is a typical situation in which the direct method of Section 7.3 fails because the boundary conditions describe a global rather than a local characteristic of the solution.

Example 7.4–1

Continuing Example 7.1–3, suppose we seek the input resistance to, and voltage distribution along, the semi-infinite ladder structure of 1Ω resistors shown in Figure 7.4–1.

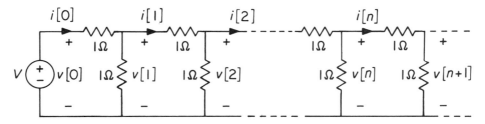

Figure 7.4–1. Resistance ladder network.

In Example 7.1–3 we concluded that $v[n]$ satisfies the homogeneous difference equation

$$v[n+2] - 3v[n+1] + v[n] = 0 \tag{7.4-1}$$

with boundary conditions $v[0] = V$ and $v[n] \to 0$ as $n \to \infty$. Since the difference equation is 2nd-order, whereas only one boundary condition is given for n near zero, the direct-solution method is inapplicable.

Constant-coefficient linear homogeneous difference equations, such as (7.4–1), have solutions of the form

$$v[n] = Az^n \tag{7.4-2}$$

where z is an appropriate (generally complex) number; that is, the homogeneous solutions generally are *discrete-time exponentials*. To show this in the case at hand, substitute (7.4–2) into (7.4–1) to obtain

$$Az^{n+2} - 3Az^{n+1} + Az^n = Az^n(z^2 - 3z + 1) = 0.$$

If $v[n] = Az^n \neq 0$, then z must be a root of

$$\left(z^2 - 3z + 1\right) = \left(z - \frac{3+\sqrt{5}}{2}\right)\left(z - \frac{3-\sqrt{5}}{2}\right) = 0$$

or

$$z = \frac{3+\sqrt{5}}{2}, \frac{3-\sqrt{5}}{2}.$$

Since (7.4–1) is linear, the most general solution is the sum of two terms such as (7.4–2):

$$v[n] = A\left(\frac{3-\sqrt{5}}{2}\right)^n + B\left(\frac{3+\sqrt{5}}{2}\right)^n.$$

Since $(3 - \sqrt{5})/2 = 0.38 < 1$, whereas $(3 + \sqrt{5})/2 = 2.62 > 1$, the second term grows with n and the first dies away. For a semi-infinite resistive ladder driven only at $n = 0$, then, we must have $B = 0$ and $A = V$; the desired solution to our problem is thus

$$v[n] = V\left(\frac{3 - \sqrt{5}}{2}\right)^n, \quad n \geq 0.$$

To find the current $i[0]$, and hence the input resistance $V/i[0]$, we observe that

$$V = i[0] \cdot 1 + v[1] = i[0] + V\left(\frac{3 - \sqrt{5}}{2}\right)$$

so that*

$$\frac{V}{i[0]} = \frac{2}{\sqrt{5} - 1} = \frac{1 + \sqrt{5}}{2} = 1.62\,\Omega.$$

▶ ▶ ▶

The lesson of Example 7.4–1 can be readily generalized and summarized as follows. The zero input response (ZIR) of the system characterized by the linear constant-coefficient difference equation

$$\sum_{k=0}^{N} a_k y[n + k] = \sum_{\ell=0}^{N} b_\ell x[n + \ell] \tag{7.4-3}$$

is the solution to the homogeneous equation

$$\sum_{k=0}^{N} a_k y[n + k] = 0. \tag{7.4-4}$$

This solution has the form

$$y[n] = \sum_{k=1}^{N} A_k z_k^n = 0 \tag{7.4-5}$$

---

*The input resistance (if that is all we seek) can be obtained more directly as follows. If R is the input resistance to the infinite ladder, then adding one more section should not change its value, so that R must be the resistance looking into one section terminated in R.

Thus, $R = 1 + \dfrac{1 \times R}{1 + R}$, or $R^2 - R - 1 = 0$. The roots of this equation are $R = \dfrac{1 \pm \sqrt{5}}{2}$; the negative root is extraneous in the physical situation being analyzed. It is interesting that the value of R is the same as the "golden section" that divides a line in such a way that the ratio of the whole to the longer segment is the same as the ratio of the longer segment to the shorter.

$$\frac{a}{b} = \frac{b}{a - b} = \frac{1 + \sqrt{5}}{2}$$

This same ratio appears in the study of Fibonacci numbers (see Problem 7.2).

where $\{z_k\}$ is the set of N roots of the characteristic equation*

$$\sum_{k=0}^{N} a_k z^k = 0 \qquad (7.4-6)$$

and $\{A_k\}$ is a set of N constants chosen to match the initial conditions. Further examples will be given in succeeding chapters.

7.5 Summary

Linear difference equations bear much the same relationship to DT signals and systems as linear differential equations bear to CT signals and systems. In particular, they can be described by block diagrams in which accumulators become the DT equivalents of integrators whose outputs describe the state of the system. A more convenient block diagram for representing the input-output behavior of any LTI DT system uses delay elements instead of accumulators. Difference equations can be solved in basically the same ways as differential equations. Thus the solutions of homogeneous difference equations (zero input responses) are sums of DT exponentials,

$$y[n] = \sum_{k=1}^{N} A_k z_k^n$$

with $\{z_k\}$ equal to the set of roots of the characteristic equation and the set $\{A_k\}$ chosen to match boundary conditions. The forced response to exponential drives can also be found as in the CT case—substitute an assumed solution having the same form as the drive and match coefficients. But, also as in the CT case, it is more convenient to solve driven problems by transform methods. The DT analog of the $\mathcal{L}$- transform is the Z-transform, which will be the topic of the next chapter.

---

*If the characteristic equation (7.4–6) has multiple-order roots, then terms of the form $n^P z_k^n$ must be added to (7.4–5) in precise analogy to the continuous-time case.

Exercise 7.1

A savings bank advertises "deposit $100 each month for 12 years and we'll pay you $100 a month forever!" Treat the bank as a DT system where the input $x[n]$ is your monthly payment (positive) or withdrawal (negative) and the output $y[n]$ is the principal in your account just after the nth monthly payment has been received or made. Assume that no withdrawals are made during the first 12 years and that only $100/month is withdrawn thereafter.

a) Sketch $x[n]$ in accordance with the terms of the advertisement.

b) Describe the general shape of $y[n]$ if $100/month is precisely the largest payment that can be guaranteed forever given a monthly interest rate r.

c) Find a difference equation relating $y[n]$ and $x[n]$.

d) Determine directly the first few terms of $y[n]$. Infer the general formula

$$y[n] = 100\frac{(1+r)^{n+1} - 1}{r}$$

during the first 12 years, $0 \leq n < 144$, and show that it satisfies the difference equation.

e) Show that the yearly interest rate implied by this scheme is approximately 5.8%.

PROBLEMS FOR CHAPTER 7

Problem 7.1

Two other common simple integration algorithms are:

TRAPEZOIDAL RULE:

$$\frac{\lambda(t+\Delta t)-\lambda(t)}{\Delta t} \approx \frac{1}{2}\left[\dot{\lambda}(t)+\dot{\lambda}(t+\Delta t)\right]$$

SIMPSON'S RULE:

$$\frac{\lambda(t+2\Delta t)-\lambda(t)}{2\Delta t} \approx \frac{1}{6}\left[\dot{\lambda}(t)+4\dot{\lambda}(t+\Delta t)+\dot{\lambda}(t+2\Delta t)\right]$$

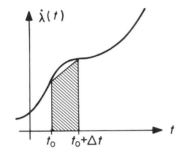

Trapezoidal
Approximation

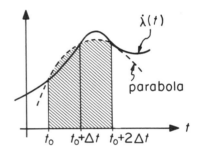

Simpson's
Approximation

a) Show that the trapezoidal rule is equivalent to approximating

$$\int_{t_0}^{t_0+\Delta t} \dot{\lambda}(t)\,dt = \lambda(t_0 + \Delta t) - \lambda(t_0)$$

as the area of the trapezoid shown in the figure above to the left.

b) Show that Simpson's rule is equivalent to approximating

$$\int_{t_0}^{t_0+2\Delta t} \dot{\lambda}(t)\,dt = \lambda(t_0 + 2\Delta t) - \lambda(t_0)$$

as the area under the parabola passing through the points $\dot{\lambda}(t_0)$, $\dot{\lambda}(t_0 + \Delta t)$, and $\dot{\lambda}(t_0 + 2\Delta t)$, as shown in the figure above to the right.

c) Use the trapezoidal rule to convert the differential equations of Example 7.1–2 into a set of difference equations. (Simpson's rule is an excellent algorithm for computing the area under a curve, but it leads to an unstable set of difference equations if applied to the approximate solution of differential equations—see Problem 8.7.)

Problem 7.2

a) The first few terms of a sequence of numbers called the *Fibonacci numbers* are

$$\{y[n]\} = \{0, 1, 1, 2, 3, 5, 8, \cdots\}, \quad n \geq 0.$$

This sequence is generated by the difference equation

$$y[n + 2] = y[n + 1] + y[n]$$

with initial conditions $y[0] = 0$ and $y[1] = 1$. Treating the sequence as the ZIR of an LTI discrete system, find an explicit formula for $y[n]$.

b) Suppose that newborn rabbits become fertile at the age of one month. Assume that the gestation period for rabbits is also one month and that each litter consists of precisely two rabbits—one male and one female. Suppose further that once a pair becomes fertile it will continue to produce a new litter each month indefinitely. Starting with one newborn pair of rabbits, how many rabbits will there be one year later (assuming that all survive)?

Problem 7.3

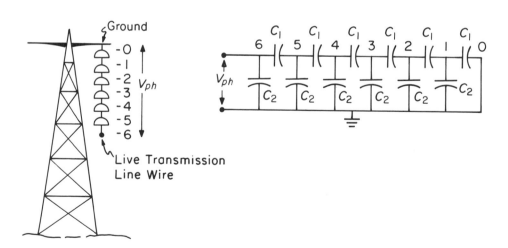

Assume that a string of six insulators supporting a transmission-line wire from a tower can be represented by the equivalent circuit of capacitors shown above for purposes of calculating the voltage distribution along the string. For simplicity, assume that $C_1 = C_2$. Find an expression for the voltage across the kth insulator. If $V_{ph} = 76$ kilovolts, show explicitly how to evaluate the arbitrary constants.

Problem 7.4

a)

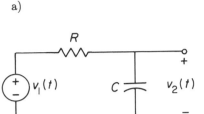
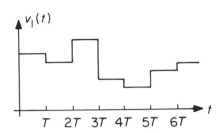

If the input to the CT circuit above is the *staircase function* $v_1(t)$, which is piecewise-constant over intervals of length T,

$$v_1(t) = v_1(nT) = v_1[n], \quad nT \le t < (n+1)T,$$

show that it is possible to describe $v_2(t)$ at times $t = nT$ by the difference equation

$$v_2[n+1] = \alpha v_2[n] + \beta v_1[n],$$

where

$$v_2[n] = v_2(nT).$$

Find the constants α and β.

b)

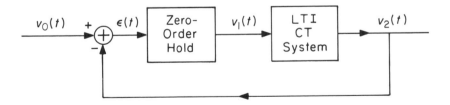

The CT circuit above is connected in a feedback arrangement as shown. The *zero-order hold* block has the property that

$$v_1(t) = v_1(nT) = v_1[n] = \epsilon(nT), \quad nT \le t < (n+1)T.$$

Show that for any continuous waveform $v_0(t)$

$$v_2[n+1] = \gamma v_2[n] + \delta v_0[n]$$

where $v_0[n] = v_0(nT)$. Find γ and δ in terms of α and β.

Problem 7.5

This problem deals with a simple model for the price adjustment process in a single commodity that is traded at discrete times. Suppose the *supply* $s[k]$ at time k is determined by the *price* $p[k-1]$ at time $k-1$ (reflecting a delay due to production time) according to the supply law

$$s[k] = s_0 + bp[k-1]$$

and let the *demand* $d[k]$ at time k be determined by the demand law

$$d[k] = d_0 - ap[k]$$

(where a and b are constants that measure the sensitivities of consumers and producers, respectively, to price changes).

a) At each trading time, market forces cause the price $p[k]$ to adjust so that demand equals supply. Use this fact to determine a first-order difference equation that governs price evolution from time $k-1$ to k.

b) Show that a particular solution of this difference equation is

$$p[k] = \frac{d_0 - s_0}{a+b} = \text{constant.}$$

c) Suppose that the initial price has some arbitrary value $p[0]$. Derive an expression for $p[k]$, $k \geq 0$, if $d_0 = 4$, $s_0 = 1$, $a = 2$, $b = 1$. Sketch the result.

d) Repeat (c) for $d_0 = 4$, $s_0 = 1$, $a = 1$, $b = 2$.

e) Observe that in one case the price evolves toward an equilibrium value, whereas in the other case it does not. Determine the condition on a and b required for convergence to an equilibrium value.

Problem 7.6

The following diagrams describe a *commutating* or *switched-capacitor filter*. In practice, the switches are MOS gates driven by a clock with period T, so that they are alternately open and closed as shown below. Assume that the circuit and switch resistances are so small that the capacitor charging times are essentially zero compared with the clock period. Hence, during the time that S_1 is closed, $v_1(t) = v_0(t)$; and during the time that S_2 is closed, $v_2(t) = v_1(t) = $ constant. Moreover, when S_2 is open, C_2 holds its charge.

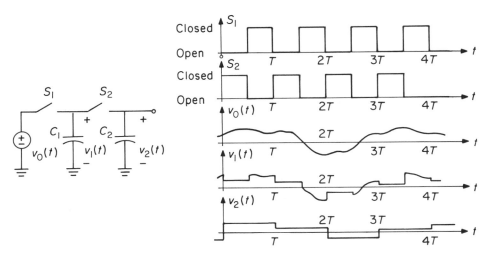

a) Define $v_0[n] = v_0(nT)$, $v_2[n] = v_2((n+0.5)T)$. Show that $v_2[n]$ and $v_0[n]$ satisfy a difference equation of the form

$$v_2[n] = a\, v_0[n] + b\, v_2[n-1]$$

and find the values of the constants a and b in terms of C_1 and C_2.

b) Suppose that $C_2 = 4C_1$ and that both capacitors are uncharged at $t = 0$. Let $v_0(t) = V_0 u(t)$ (assume the step rises just after S_1 opens). Show that

$$v_2[n] = V_0\left[1 - \left(\frac{4}{5}\right)^n\right]u[n].$$

c) Let $T = 1$ msec. Sketch $v_2(t)$. On the same scale, sketch $v_2(t)$ if T were 0.25 msec.

d) Let $v_2'(t)$ be the ZSR to some input $v_0'(t)$. Similarly, let $v_2''(t)$ be the ZSR to some other input $v_0''(t)$. Is it true that $v_2(t) = v_2'(t) + v_2''(t)$ will be the ZSR to the input $v_0(t) = v_0'(t) + v_0''(t)$? That is, does this device satisfy the superposition principle?

8

THE UNILATERAL Z-TRANSFORM
AND ITS APPLICATIONS

8.0 Introduction

L-transforms and system functions provided the most convenient method for finding the zero state response of continuous-time systems. A similar technique is applicable to discrete-time systems. The discrete-time exponential function, z^n, plays the role of the kernel, e^{st}, and the Z-transform replaces the L-transform. Our development of these topics will be both brief and restricted in scope, but it will provide an orderly method for solving linear time-invariant difference equations, as well as introducing such useful notions as the discrete-time *system function* and the characterization of discrete-time systems in terms of pole-zero locations in the complex z-plane.

8.1 The Z-Transform

The (unilateral) *Z-transform* of a sequence $x[n]$ is defined by the formula

$$\tilde{X}(z) = \sum_{n=0}^{\infty} x[n]z^{-n} . \tag{8.1-1}$$

If $|x[n]|$ grows no faster than exponentially, this series will converge for all z outside some circle in the complex z-plane whose radius r_0 is called the *radius of convergence* (see Figure 8.1–1). As in the case of the L-transform, the usefulness of the Z-transform depends on the fact that the relationship between $\tilde{X}(z)$ and the sequence $x[n]$ is *biunique*—to each $x[n]$ defined for $n \geq 0$ there corresponds one and only one $\tilde{X}(z)$ defined for $|z| > r_0$, and vice versa. For Z-transforms,

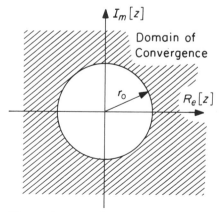

Figure 8.1–1. Typical domain of convergence.

this uniqueness theorem is basically a reinterpretation of the central theorem concerning the uniqueness and convergence of power-series expansions of analytic functions of a complex variable.*

Example 8.1–1

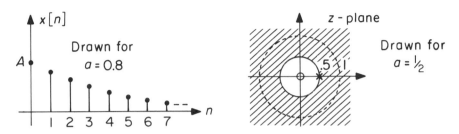

Figure 8.1–2. A DT exponential function and its domain of convergence.

Suppose that $x[n]$ is a DT exponential function

$$x[n] = Aa^n, \quad n \geq 0 \qquad (8.1\text{--}2)$$

as shown in Figure 8.1-2. Then

$$\tilde{X}(z) = A \sum_{n=0}^{\infty} a^n z^{-n}$$

which converges to

$$\tilde{X}(z) = \frac{A}{1 - az^{-1}} \qquad (8.1\text{--}3)$$

if $|az^{-1}| < 1$ or $|z| > |a|$. These formulas remain valid if a is complex. We note that $\tilde{X}(z)$ has a zero at $z = 0$ and a pole at $z = a$ on the circle bounding the region of convergence.

One important special case results if $a = 1$ so that $x[n] = A =$ constant, $n \geq 0$. The pole of $\tilde{X}(z)$ is now located at $z = 1$. This situation is illustrated in Figure 8.1–3.

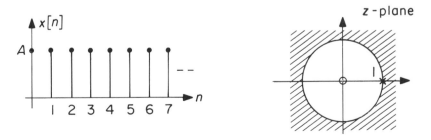

Figure 8.1–3. Z-transform of $x[n] = A$, $n \geq 0$.

---

*See, e.g., E. B. Saff and A. D. Snider, *Fundamentals of Complex Analysis for Mathematics, Science, and Engineering* (New York, NY: Prentice-Hall, 1976).

Another interesting situation arises if a is negative, since then the signs of successive values of $x[n]$ alternate. In this case the pole is on the negative real axis. This situation is illustrated in Figure 8.1–4.

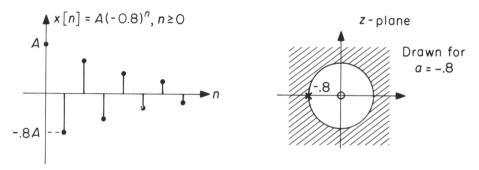

Figure 8.1–4. Z-transform of $x[n] = A(-0.8)^n$, $n \geq 0$.

▶ ▶ ▶

Example 8.1–2

Consider, for $|z| > 1/2$,

$$\tilde{X}(z) = \frac{30z^2}{6z^2 - z - 1} = \frac{5}{1 - \frac{1}{6}z^{-1} - \frac{1}{6}z^{-2}}.$$

By the uniqueness theorem this should correspond to a unique $x[n]$, $n \geq 0$. How can we carry out the inverse Z-transformation? One way to find $x[n]$, $n \geq 0$, is to expand $\tilde{X}(z)$ in a power series in z^{-1}. This can be accomplished, for example, by long division:

$$
\require{enclose}
\begin{array}{r}
5 + \dfrac{5}{6}z^{-1} + \dfrac{35}{36}z^{-2} + \cdots \\[4pt]
1 - \dfrac{1}{6}z^{-1} - \dfrac{1}{6}z^{-2} \enclose{longdiv}{5}
\end{array}
$$

$$5 - \frac{5}{6}z^{-1} - \frac{5}{6}z^{-2}$$

$$\frac{5}{6}z^{-1} + \frac{5}{6}z^{-2}$$

$$\frac{5}{6}z^{-1} - \frac{5}{36}z^{-2} - \frac{5}{36}z^{-3}$$

$$\frac{35}{36}z^{-2} + \frac{5}{36}z^{-3}$$

$$\frac{35}{36}z^{-2} - \frac{35}{216}z^{-3} - \frac{35}{216}z^{-4}$$

$$\cdots$$

(Note that both the numerator and the denominator of $\tilde{X}(z)$ are written as series of descending powers of z.) Thus for sufficiently large $|z|$ (in fact, for $|z| > 0.5$) we can write

$$\tilde{X}(z) = 5 + \frac{5}{6}z^{-1} + \frac{35}{36}z^{-2} + \cdots .$$

Since in general $\tilde{X}(z) = \sum_{n=0}^{\infty} x[n]z^{-n}$, we conclude that

$$x[0] = 5,$$

$$x[1] = \frac{5}{6},$$

$$x[2] = \frac{35}{36},$$

etc.

Although some variant of this procedure will always work to recover to $x[n]$ from $\tilde{X}(z)$, it is obviously clumsy. A more powerful technique parallels the partial-fraction method for $\mathcal{L}$-transforms. Thus we may write

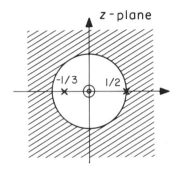

Figure 8.1–5. Pole-zero plot.

$$\tilde{X}(z) = \frac{30z^2}{6z^2 - z - 1} = \frac{5}{1 - \frac{1}{6}z^{-1} - \frac{1}{6}z^{-2}}$$

$$= \frac{5}{\left(1 - \frac{1}{2}z^{-1}\right)\left(1 + \frac{1}{3}z^{-1}\right)}$$

$$= \frac{3}{1 - \frac{1}{2}z^{-1}} + \frac{2}{1 + \frac{1}{3}z^{-1}}$$

where the coefficients of each fraction are obtained as before, that is,

$$\left.\frac{5\left(1 - \frac{1}{2}z^{-1}\right)}{\left(1 - \frac{1}{2}z^{-1}\right)\left(1 + \frac{1}{3}z^{-1}\right)}\right|_{z^{-1}=2} = 3; \qquad \left.\frac{5\left(1 + \frac{1}{3}z^{-1}\right)}{\left(1 - \frac{1}{2}z^{-1}\right)\left(1 + \frac{1}{3}z^{-1}\right)}\right|_{z^{-1}=-3} = 2.$$

Then, since the Z-transform is a linear operation, we conclude from uniqueness and Example 8.1–1 that

$$x[n] = 3\left(\frac{1}{2}\right)^n + 2\left(-\frac{1}{3}\right)^n, \quad n \geq 0.$$

This formula checks our preceding results for $n = 0$, 1, and 2 but is clearly much more effective than the power-series method if we are interested in values of $x[n]$ for n much greater than 2. Notice the special way in which $\tilde{X}(z)$ is written in terms of negative powers of z with the constant term in the denominator equal to 1. Note also the particular form of the partial-fraction expansion, which is chosen so that terms of the form $1/(1 - az^{-1})$ can be recognized as corresponding to the sequence a^n, $n \geq 0$.

▶ ▶ ▶

Example 8.1–3

The partial-fraction expansion procedure of the preceding example apparently fails if the numerator of $\tilde{X}(z)$ is of equal or higher degree in z^{-1} than the denominator. Consider, for example,

$$\tilde{X}_1(z) = \frac{11 - z^{-1} - z^{-2}}{1 - \frac{1}{6}z^{-1} - \frac{1}{6}z^{-2}} = \frac{11 - z^{-1} - z^{-2}}{\left(1 - \frac{1}{2}z^{-1}\right)\left(1 + \frac{1}{3}z^{-1}\right)}.$$

If we attempt as above to write (incorrectly, as we shall see)

$$\tilde{X}_1(z) = \frac{k_1}{1 - \frac{1}{2}z^{-1}} + \frac{k_2}{1 + \frac{1}{3}z^{-1}}$$

with

$$k_1 = \left. \frac{\left(11 - z^{-1} - z^{-2}\right)\left(1 - \frac{1}{2}z^{-1}\right)}{\left(1 - \frac{1}{2}z^{-1}\right)\left(1 + \frac{1}{3}z^{-1}\right)} \right|_{z^{-1} = +2} = 3$$

$$k_2 = \left. \frac{\left(11 - z^{-1} - z^{-2}\right)\left(1 + \frac{1}{3}z^{-1}\right)}{\left(1 - \frac{1}{2}z^{-1}\right)\left(1 + \frac{1}{3}z^{-1}\right)} \right|_{z^{-1} = -3} = 2$$

we obtain the same coefficients as in Example 8.1–2, which corresponds to the partial-fraction expansion of $\dfrac{5}{1 - \frac{1}{6}z^{-1} - \frac{1}{6}z^{-2}}$ rather than the expression we sought to describe.*

One clue to the difficulty is that each of the terms in the attempted expansion vanishes as $z^{-1} \to \infty$, whereas the given $\tilde{X}_1(z) \to 6$ as $z^{-1} \to \infty$. Indeed,

$$\tilde{X}_1(z) = \frac{11 - z^{-1} - z^{-2}}{1 - \frac{1}{6}z^{-1} - \frac{1}{6}z^{-2}} = \frac{5}{1 - \frac{1}{6}z^{-1} - \frac{1}{6}z^{-2}} + 6$$

so we could write (correctly)

$$\tilde{X}_1(z) = \frac{3}{1 - \frac{1}{2}z^{-1}} + \frac{2}{1 + \frac{1}{3}z^{-1}} + 6.$$

In general, we can obtain an expansion of this kind by first dividing the denominator into the numerator, reducing the degree in z^{-1} of the remainder until it is less than the degree of the denominator. Thus if we seek an expansion of

$$\tilde{X}_2(z) = \frac{1 + z^{-1} + z^{-2}}{1 - z^{-1}}$$

---

*This same difficulty arises, of course, if we attempt to use partial fractions to take the inverse transform of an improper $\mathcal{L}$-transform. The correct partial-fraction expansion is readily obtained in that case as described here. The interpretation of the results, however, requires special techniques, as we shall explain in Chapter 11.

we divide

$$
-z^{-1}+1 \enclose{longdiv}{}
$$

$$
\begin{array}{r}
-z^{-1}-2 \\
-z^{-1}+1 \overline{)\; z^{-2}+\; z^{-1}+1} \\
\underline{z^{-2}-\; z^{-1}} \\
2\,z^{-1}+1 \\
\underline{2\,z^{-1}-2} \\
3
\end{array}
$$

to obtain the expansion

$$
\tilde{X}_2(z) = -z^{-1} - 2 + \frac{3}{1 - z^{-1}} .
$$

The general result is thus a polynomial in z^{-1} plus a proper fraction in z^{-1} that can be expanded in partial fractions in the ordinary way.

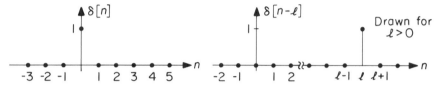

Figure 8.1–6. Unit sample function.   **Figure 8.1–7.** Delayed sample function.

It should be evident from the basic definition of the Z-transform (8.1–1) that the inverse transform of $k_\ell z^{-\ell}$ is a DT function $f[n]$ that is zero for all n except $n = \ell$, at which point $f[\ell] = k_\ell$. It is more convenient, however, to introduce a special function, the *unit sample function* $\delta[n]$, defined* by

$$
\delta[n] = \begin{cases} 1, & n = 0 \\ 0, & n \neq 0 . \end{cases} \tag{8.1–4}
$$

The Z-transform of $\delta[n]$ is obviously

$$
\delta[n] \Longleftrightarrow 1 . \tag{8.1–5}
$$

The *delayed* unit sample function,[†] $\delta[n - \ell]$, is defined by

$$
\delta[n - \ell] = \begin{cases} 1, & n = \ell \\ 0, & n \neq \ell \end{cases} \tag{8.1–6}
$$

and has the Z-transform

$$
\delta[n - \ell] \Longleftrightarrow z^{-\ell}. \tag{8.1–7}
$$

We can write the inverse transforms of $\tilde{X}_1(z)$ and $\tilde{X}_2(z)$ in terms of unit sample functions since

$$
\tilde{X}_1(z) = \frac{11 - z^{-1} - z^{-2}}{1 - \dfrac{1}{6}z^{-1} - \dfrac{1}{6}z^{-2}} = \frac{3}{1 - \dfrac{1}{2}z^{-1}} + \frac{2}{1 + \dfrac{1}{3}z^{-1}} + 6
$$

---

*Note that $\delta[n]$ is defined for *all* n, $-\infty < n < \infty$, not just for $n \geq 0$.

[†]The delayed unit sample function is often called *Kronecker's delta* and indicated by the notation $\delta_{n\ell}$.

so that for $n \geq 0$,

$$x_1[n] = 3\left(\frac{1}{2}\right)^n + 2\left(-\frac{1}{3}\right)^n + 6\delta[n]$$

$$= \begin{cases} 11, & n = 0 \\ 3\left(\frac{1}{2}\right)^n + 2\left(-\frac{1}{3}\right)^n, & n > 0 \end{cases}$$

and since

$$\tilde{X}_2(z) = \frac{1 + z^{-1} + z^{-2}}{1 - z^{-1}} = -z^{-1} - 2 + \frac{3}{1 - z^{-1}}$$

so that for $n \geq 0$,

$$x_2[n] = -\delta[n-1] - 2\delta[n] + 3$$

$$= \begin{cases} 1, & n = 0 \\ 2, & n = 1 \\ 3, & n > 1. \end{cases}$$

▶ ▶ ▶

As with $\mathcal{L}$-transforms, most applications of Z-transforms involve manipulation of a few basic transform pairs using a small number of properties and theorems. Some important theorems are:

SUPERPOSITION (LINEARITY):

$$ax[n] + by[n] \iff a\tilde{X}(z) + b\tilde{Y}(z). \tag{8.1-8}$$

MULTIPLICATION BY AN EXPONENTIAL:

$$a^n x[n] \iff \tilde{X}(a^{-1}z). \tag{8.1-9}$$

MULTIPLICATION BY n:

$$nx[n] \iff -z\frac{d\tilde{X}(z)}{dz}. \tag{8.1-10}$$

DELAY BY $N \geq 0$:

$$x[n-N]u[n-N] \iff z^{-N}\tilde{X}(z). \tag{8.1-11}$$

In the Delay Theorem, $u[n]$ is the discrete-time *unit step function*

$$u[n] = \begin{cases} 1, & n \geq 0 \\ 0, & n < 0. \end{cases} \tag{8.1-12}$$

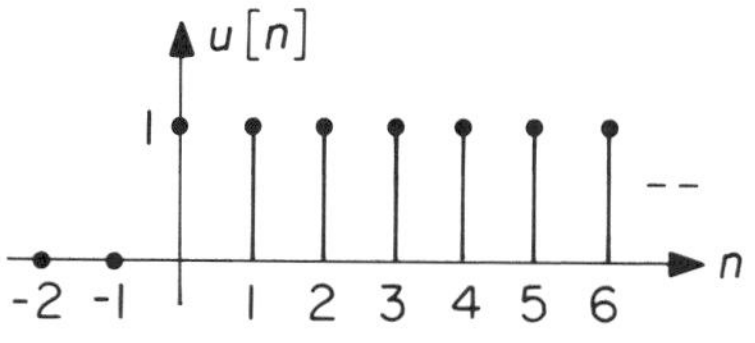

Figure 8.1-8. Unit step function.

For each theorem the proof follows almost immediately from the basic definition (8.1-1). A table of simple Z-transforms and basic Z-transform theorems is given in the appendix to this chapter.

Example 8.1–4

From Example 8.1–1 and the Multiplication-by-n Theorem, the Z-transform of

$$x[n] = na^n, \quad n \geq 0$$

is

$$X(z) = -z\frac{d}{dz}\left(\frac{1}{1 - az^{-1}}\right)$$

$$= \frac{az^{-1}}{(1 - az^{-1})^2}.$$

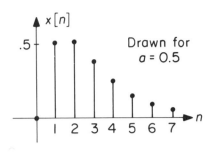

Figure 8.1–9. $x[n] = na^n$, $n \geq 0$.

▶ ▶ ▶

Example 8.1–5

From Example 8.1–1 and the Delay Theorem, we have the Z-transform pair

$$a^{n-N}u[n - N] \Longleftrightarrow \frac{z^{-N}}{1 - az^{-1}}.$$

In particular, the transform of the delayed unit step function is

$$u[n - N] \Longleftrightarrow \frac{z^{-N}}{1 - z^{-1}}.$$

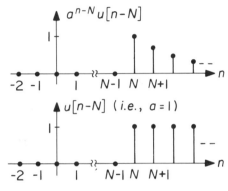

Figure 8.1–10. $a^{n-N}u[n - N]$.

As an application of this result, note that we may write the discrete-time pulse function

$$p_N[n] = \begin{cases} 1, & 0 \leq n < N \\ 0, & n \geq N \end{cases}$$

in the form

$$p_N[n] = u[n] - u[n - N].$$

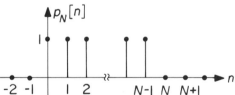

Figure 8.1–11. The function $p_N[n]$.

We may then apply the Linearity Theorem and the Delay Theorem to obtain

$$p_N[n] \Longleftrightarrow \frac{1 - z^{-N}}{1 - z^{-1}} = \tilde{P}_N(z)$$

which we recognize as precisely formula (7.3–4) for the partial sum of the geometric series,

$$\tilde{P}_N(z) \stackrel{\Delta}{=} \sum_{n=0}^{\infty} p_N[n]z^{-n} = \sum_{n=0}^{N-1} z^{-n}$$

which, of course, is also the result obtained if $\tilde{P}_N(z)$ were evaluated directly by (8.1–1).

▶ ▶ ▶

8.2 The Z-Transform Applied to LTI Discrete-Time Systems

To apply the Z-transform to the analysis of discrete-time systems, we need another theorem which plays much the same role for Z-transforms that the Differentiation Theorem does for $\mathcal{L}$-transforms:

> **FORWARD-SHIFT THEOREM:**
> If $\tilde{X}(z)$ is the (unilateral) Z-transform of $x[n]$,
> then $z(\tilde{X}(z) - x[0])$ is the Z-transform of $x[n+1]$.

The proof is immediate from the basic definition of $\tilde{X}(z)$ and the pictures in Figure 8.2–1. The following examples illustrate the broad usefulness of this theorem.

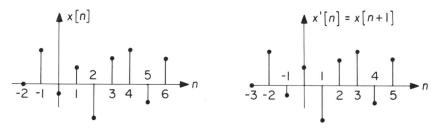

$$\tilde{X}(z) = \sum_0^\infty x[n]z^{-n}$$
$$= x[0] + x[1]z^{-1} + x[2]z^{-2} + \cdots$$

$$\tilde{X}'(z) = \sum_0^\infty x'[n]z^{-n} = \sum_0^\infty x[n+1]z^{-n}$$
$$= x[1] + x[2]z^{-1} + x[3]z^{-2} + \cdots$$
$$= z(\tilde{X}(z) - x[0])$$

Figure 8.2–1. Illustration of the Forward-Shift Theorem.

Example 8.2–1

The mortgage problem of Example 7.1–1 led to the difference equation

$$P[n+1] = (1+r)P[n] - p, \quad n \geq 0.$$

Taking the Z-transforms of both sides, using the Forward-Shift Theorem, yields

$$z(\tilde{P}(z) - P[0]) = (1+r)\tilde{P}(z) - \frac{p}{1 - z^{-1}}.$$

Solving for $\tilde{P}(z)$ and expanding in partial fractions yields

$$\tilde{P}(z) = \frac{p/r}{1 - z^{-1}} + \frac{P[0] - p/r}{1 - (1+r)z^{-1}}.$$

Inverse transforming yields

$$P[n] = p/r + (P[0] - p/r)(1+r)^n, \quad n \geq 0$$

which is the result derived by induction in Example 7.2–1.

▶ ▶ ▶

Example 8.2–2

If $\tilde{X}(z)$ is the Z-transform of $x[n]$, then $z^2\tilde{X}(z) - z^2x[0] - zx[1]$ is the Z-transform of $x[n+2]$. This is easy to show, either directly or by considering $x[n+2]$ as the forward shift of $x[n+1]$ and applying the Forward-Shift Theorem twice to obtain $z\big(z[\tilde{X}(z) - x[0]] - x[1]\big)$. Extensions to still larger forward shifts proceed similarly. Such extensions permit Z-transforms to be applied directly to the solution of input-output LTI difference equations of arbitrary order. Thus, suppose we have a system described by

$$y[n+2] - \frac{1}{6}y[n+1] - \frac{1}{6}y[n] = 2x[n].$$

We seek the response to $x[n] = 1$, $n \geq 0$, with initial conditions $y[0] = 0$, $y[1] = 1$. Taking the Z-transforms of both sides, using the Forward-Shift Theorem and its extension, we find

$$\left(z^2\tilde{Y}(z) - z^2y[0] - zy[1]\right) - \frac{1}{6}\left(z\tilde{Y}(z) - zy[0]\right) - \frac{1}{6}\tilde{Y}(z) = 2\tilde{X}(z).$$

Substituting

$$\tilde{X}(z) = \frac{1}{1 - z^{-1}}$$

and inserting the given values for $y[0]$ and $y[1]$ yields

$$\tilde{Y}(z) = \frac{z^{-1} + z^{-2}}{\left(1 - z^{-1}\right)\left(1 - \dfrac{z^{-1}}{6} - \dfrac{z^{-2}}{6}\right)}$$

$$= \frac{3}{1 - z^{-1}} - \frac{3.6}{1 - \dfrac{1}{2}z^{-1}} + \frac{0.6}{1 + \dfrac{1}{3}z^{-1}}.$$

Inverse transforming gives

$$y[n] = 3 - 3.6\left(\frac{1}{2}\right)^n + 0.6\left(-\frac{1}{3}\right)^n, \quad n \geq 0.$$

It is easy to check directly that this satisfies the difference equation and has the required values at $n = 0$ and $n = 1$.

▶ ▶ ▶

Transforming the input-output difference equation directly (as above) may lead to some difficulties of interpretation if the equation contains terms proportional to $x[n+1]$, $x[n+2]$, ..., corresponding to shifted input sequences. In this case, knowledge of $y[0]$, $y[1]$, ..., $y[N-1]$, together with $x[n]$, $n \geq 0$, determines a unique response, but the values of $y[0]$, $y[1]$, ..., $y[N-1]$ do not define the state of the system at $n = N-1$. (It is necessary to know $x[0]$, $x[1]$, ..., $x[N-1]$ as well.) The situation is precisely analogous to that discussed in Problem 3.3 for continuous-time systems.

8.3 Frequency-Domain Representations of Discrete-Time Systems

The description or analysis of CT LTI circuits or block diagrams in preceding chapters was much simplified by transforming from the time domain, in which elements are described by differential equations, to the frequency domain, in which elements are described by impedances or system functions. Structural constraints among subsystems (Kirchhoff's Laws, cascade connections, feedback, etc.) then lead to algebraic equations that can readily be solved to give system-function descriptions of the overall input-output ZSR behavior. And various attributes of system functions—particularly pole-zero locations—permit one to say a great deal about the general characteristics of system behavior even without explicitly solving for the response.

System functions and frequency-domain methods have similar advantages for discrete-time systems, as we can illustrate by developing a frequency-domain form of the delay-adder-gain diagrams of Section 7.2. To begin, recall that the unit delay element was defined in Section 7.2 by the difference equation

$$y[n+1] = x[n] \,. \tag{8.3-1}$$

Z-transforming, using the Forward-Shift Theorem, leads to an equivalent description:

$$z\left(\tilde{Y}(z) - y[0]\right) = \tilde{X}(z)$$

or

$$\tilde{Y}(z) = \frac{\tilde{X}(z)}{z} + y[0] \,. \tag{8.3-2}$$

We may thus replace the delay block in block diagrams by the transform representation shown in Figure 8.3–1. Adder and gain elements transform into the frequency domain without alteration.

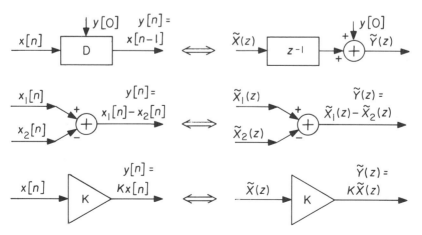

Figure 8.3–1. Frequency-domain representations of delay, adder, and gain elements.

Replacing each block as above, any delay-adder-gain block diagram becomes a frequency-domain representation of the DT system having precisely the same properties as frequency-domain representations of CT systems. Some of the principal consequences are illustrated in the following example.

Example 8.3–1

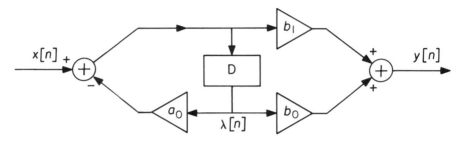

Figure 8.3–2. 1st-order system for Example 8.3–1.

The block diagram of Figure 8.3–2 is the form taken by the general canonical block diagram of Figure 7.2–7 in the 1st-order case. It is equivalent to the difference equation

$$y[n+1] + a_0 y[n] = b_1 x[n+1] + b_0 x[n] .$$

Replacing the blocks and variables by their frequency-domain equivalents yields the block diagram of Figure 8.3–3.

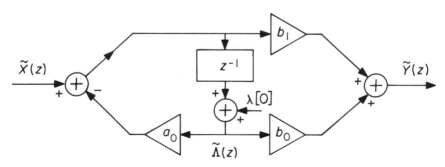

Figure 8.3–3. Frequency-domain equivalent of Figure 8.3–2.

Using superposition and Black's formula for the feedback loop, we obtain, almost by inspection,

$$\tilde{\Lambda}(z) = \frac{z^{-1}\tilde{X}(z)}{1 + a_0 z^{-1}} + \frac{\lambda[0]}{1 + a_0 z^{-1}}$$

and

$$\tilde{Y}(z) = b_0 \tilde{\Lambda}(z) + b_1 (\tilde{X}(z) - a_0 \tilde{\Lambda}(z)) .$$

Eliminating $\tilde{\Lambda}(z)$ yields the input-output relation

$$\tilde{Y}(z) = \frac{b_0 + b_1 z}{a_0 + z}\tilde{X}(z) + \frac{(b_0 - b_1 a_0)z}{a_0 + z}\lambda[0]$$

which can, of course, be checked by applying the Forward-Shift Theorem directly to the difference equation, using the fact (derivable from Figure 8.3–2) that

$$y[0] - b_1 x[0] = (b_0 - b_1 a_0)\lambda[0].$$

To obtain a complete solution for $y[n]$, we need only substitute an appropriate expression for $\tilde{X}(z)$ and inverse transform. This is most conveniently done for a numerical example. Thus suppose

$$a_0 = 0.5, \quad b_0 = 2, \quad b_1 = 2, \quad \lambda[0] = 1$$

and

$$x[n] = \left(\frac{1}{3}\right)^n, \quad n \geq 0.$$

Then

$$\tilde{X}(z) = \frac{1}{1 - \frac{1}{3}z^{-1}}$$

and

$$\tilde{Y}(z) = \frac{2(1 + z^{-1})}{\left(1 + \frac{1}{2}z^{-1}\right)\left(1 - \frac{1}{3}z^{-1}\right)} + \frac{1}{1 + \frac{1}{2}z^{-1}}$$

$$= \frac{-0.2}{1 + \frac{1}{2}z^{-1}} + \frac{3.2}{1 - \frac{1}{3}z^{-1}}$$

so that

$$y[n] = -0.2\left(-\frac{1}{2}\right)^n + 3.2\left(\frac{1}{3}\right)^n, \quad n \geq 0.$$

▶ ▶ ▶

It should be reasonably obvious from this example that the following comments apply to DT linear time-invariant systems in general. In each case the close parallel between DT LTI system behavior and CT LTI system behavior should be carefully noted.

1. The total response time can be considered as the sum of the *zero state response* (ZSR)—the first term in the equation for $\tilde{Y}(z)$ in the example above—and the *zero input response* (ZIR)—the remaining term. The ZSR depends only on the input $x[n]$, $n \geq 0$; the ZIR depends only on the initial state, for example, the outputs of the delay elements at $n = 0$.

2. If there is more than one input, the ZSR is a superposition of terms describing the separate effects of each input. Each such term is a product of the Z-transform of that input and a *system function* $\tilde{H}_i(z)$ relating the ith input to the output. Thus, in Example 8.3–1 the system function relating the output to the external input $x[n]$ is

$$\tilde{H}(z) = \frac{b_0 + b_1 z}{a_0 + z} .$$

The inverse transform of the product $\tilde{H}(z)\tilde{X}(z)$ is the ZSR response to the input $x[n]$.

3. The system function and the input-output difference equation imply one another through the replacement

$$z \Longleftrightarrow \text{forward shift.}$$

In general the finite-order DT LTI system described by the difference equation

$$\sum_{k=0}^{N} a_k y[n+k] = \sum_{\ell=0}^{N} b_\ell x[n+\ell] \tag{8.3–3}$$

is also described by the system function

$$\tilde{H}(z) = \frac{\displaystyle\sum_{\ell=0}^{N} b_\ell z^\ell}{\displaystyle\sum_{k=0}^{N} a_k z^k} . \tag{8.3–4}$$

Since $\tilde{H}(z)$ is a rational function of z, it is characterized (except for a multiplicative constant) by the locations of its *poles* and *zeros*.

4. The poles of $\tilde{H}(z)$ are also the roots of the characteristic equation describing the solutions of the homogeneous difference equation

$$\sum_{k=0}^{N} a_k z^k = 0 . \tag{8.3–5}$$

If the N roots of this equation, that is, the N poles of $\tilde{H}(z)$, are labelled z_1, z_2, ..., z_N, then the ZIR has the form (assuming no multiple-order roots)

$$y[n] = \sum_{k=1}^{N} A_k z_k^n \quad \text{(ZIR)} \tag{8.3–6}$$

where the values of the N constants A_k depend on the initial state. If $|z_k| < 1$ for all poles, then the ZIR dies away and the system described by $\tilde{H}(z)$ is *stable* (in the input-output sense).

5. The domain of convergence of $\tilde{H}(z)$ is the region *outside* the smallest circle enclosing all the poles of $\tilde{H}(z)$. The system is stable if the domain of convergence includes the unit circle, $|z| = 1$.
6. DT LTI systems can be combined in cascade, in parallel, in feedback arrangements, etc., just as CT LTI systems can be. The rules for determining the combined system functions are the same for systems described by Z-transforms as for systems described by L-transforms. For example, the system function of the cascade connection of two DT systems is the product of their individual system functions and is independent of the order of the cascade as shown in Figure 8.3–4. And Black's feedback formula applies to the arrangement shown in Figure 8.3–5.

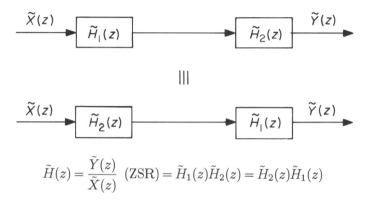

$$\tilde{H}(z) = \frac{\tilde{Y}(z)}{\tilde{X}(z)} \ (\text{ZSR}) = \tilde{H}_1(z)\tilde{H}_2(z) = \tilde{H}_2(z)\tilde{H}_1(z)$$

Figure 8.3–4. Cascade connection of DT systems.

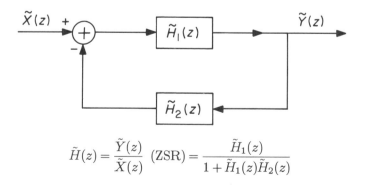

$$\tilde{H}(z) = \frac{\tilde{Y}(z)}{\tilde{X}(z)} \ (\text{ZSR}) = \frac{\tilde{H}_1(z)}{1 + \tilde{H}_1(z)\tilde{H}_2(z)}$$

Figure 8.3–5. Feedback connection of DT systems.

Example 8.3–2

To illustrate further some of these features of the frequency-domain description of

DT system behavior, consider once again the numerical integration of the equations describing the circuit of Examples 1.3–3 and 7.1–2, redrawn in Figure 8.3–6.

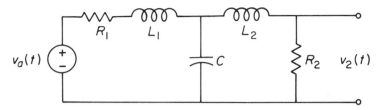

Figure 8.3–6. Circuit of Example 1.3–3.

In Example 7.2–1 we derived block diagrams both for the CT state equations of this circuit and for a set of DT state equations that were approximately equivalent. The diagrams differed only in that each integrator in the CT diagram was replaced by the cascade of a gain element, Δt, and an accumulator in the DT version. In the frequency domain, the corresponding representations are shown in Figure 8.3–7; the representation of the accumulator follows directly from the Forward-Shift Theorem applied to the defining difference equation. Hence the frequency-domain diagrams for the DT and CT state equations are identical under ZSR conditions, except for the replacement of $1/s$ by $\Delta t/(z-1)$ in each integrator-accumulator block. Thus, any system function relating two transforms in the DT diagram will be identical with the corresponding system function in the CT diagram provided that s is replaced by $(z-1)/\Delta t$ wherever it appears. In particular we readily conclude from impedance methods

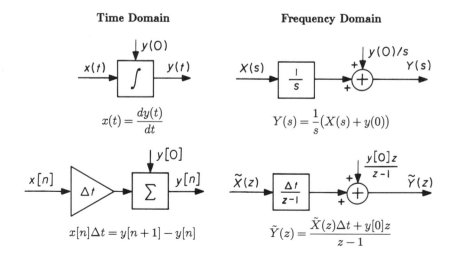

Figure 8.3–7. Integrators and accumulators in the time and frequency domains.

applied to the circuit diagram above that the input-output CT system function is

$$H(s) = \frac{V_2(s)}{V_a(s)} \quad \text{(ZSR)}$$

$$= \frac{\dfrac{R_2}{R_2 + L_2 s} \dfrac{(R_2 + L_2 s)\dfrac{1}{Cs}}{R_2 + L_2 s + \dfrac{1}{Cs}}}{\dfrac{(R_2 + L_2 s)\dfrac{1}{Cs}}{R_2 + L_2 s + \dfrac{1}{Cs}} + R_1 + L_1 s}$$

$$= \frac{0.5}{\left(\dfrac{s}{10^4} + 1\right)\left(\left(\dfrac{s}{10^4}\right)^2 + \left(\dfrac{s}{10^4}\right) + 1\right)} \qquad (8.3\text{--}7)$$

where we have substituted the element values given in Example 7.1–2. The pole locations are thus as shown in Figure 8.3–8. Obviously, the CT circuit is stable; its slowest normal modes have a decay time constant of 2×10^{-4} sec. Following the scheme suggested above, we need only replace s wherever it appears in (8.3–7) by $(z-1)/\Delta t$ to obtain the input-output DT system function

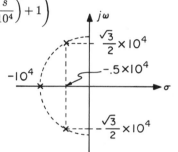

Figure 8.3–8. Pole locations for $H(s)$.

$$\tilde{H}(z) = \frac{\tilde{V}_2(z)}{\tilde{V}_a(z)} \quad \text{(ZSR)} = H\left(\frac{z-1}{\Delta t}\right)$$

$$= \frac{0.5}{\left(\dfrac{z-1}{10^4 \Delta t} + 1\right)\left(\left(\dfrac{z-1}{10^4 \Delta t}\right)^2 + \left(\dfrac{z-1}{10^4 \Delta t}\right) + 1\right)}. \qquad (8.3\text{--}8)$$

The poles of $H(s)$ are located at

$$s = -10^4, \quad -0.5 \times 10^4 \pm j\frac{\sqrt{3}}{2} \times 10^4.$$

Consequently the poles of $\tilde{H}(z)$ are located at

$$\frac{z-1}{\Delta t} = -10^4, \quad -0.5 \times 10^4 \pm j\frac{\sqrt{3}}{2} \times 10^4$$

or

$$z = 1 - 10^4 \Delta t, \quad 1 - 0.5 \times 10^4 \Delta t \pm j\frac{\sqrt{3}}{2} \times 10^4 \Delta t.$$

The pole locations thus depend on Δt, as shown in Figure 8.3–9.

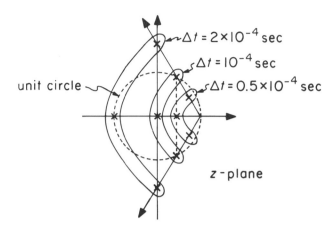

Figure 8.3–9. Pole locations for $\tilde{H}(z)$.

We saw in Example 7.3–2 that the DT system accurately describes the behavior of the CT system for $\Delta t \ll 10^{-4}$ sec. For very small Δt, the geometrical relationship of the poles of $\tilde{H}(z)$ to the point $z = 1$ and the circle $|z| = 1$ is similar to the relationship of the poles of $H(s)$ to the point $s = 0$ and the line $s = j\omega$. Such a similarity is in fact necessary and sufficient for the step responses of the DT and CT systems to be similar, as is discussed more fully in Problem 8.9. For larger Δt, the similarity deteriorates, and for $\Delta t > 10^{-4}$ sec the poles of $\tilde{H}(z)$ have magnitude greater than 1 so that the DT "approximation" actually becomes unstable. (The step response of the DT system with $\Delta t = 1.11 \times 10^{-4}$ sec is shown in Figure 8.3–10. Compare this with the step response for small Δt derived in Example 7.3–2.) As stated in Example 7.1–2, this numerical instability is a result of the choice of the simple forward Euler algorithm to characterize the discrete approximation to the integrator; the effect of other choices is discussed in Problem 8.7.

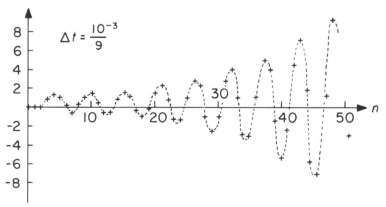

Figure 8.3–10. Step response of $\tilde{H}(z)$ for $\Delta t = 1.11 \times 10^{-4}$ sec.

▶ ▶ ▶

8.4 Summary

The unilateral Z-transform

$$\tilde{X}(z) = \sum_{n=0}^{\infty} x[n]z^{-n}$$

plays the same role for DT systems that the unilateral $\mathcal{L}$-transform plays for CT systems:

a) Because the relationship between $\tilde{X}(z)$ and $x[n]$, $n \geq 0$, is biunique, power-series or partial-fraction expansions may be used to rewrite $\tilde{X}(z)$ in a form from which $x[n]$ is evident.

b) A variety of theorems simplify the manipulation and derivation of transforms and their inverses.

c) The Forward-Shift Theorem reduces the analysis of DT systems characterized by difference equations or block diagrams to an algebraic process.

d) The total response of LTI DT systems can be considered as the sum of a ZSR and a ZIR. The Z-transform of the ZSR has the form

$$\tilde{Y}(z) = \tilde{H}(z)\tilde{X}(z)$$

where the system function, $\tilde{H}(z)$, for LTI systems of the type described in (c), is a rational function of z characterized by its poles and zeros. The poles also determine the form of the ZIR. The system is input-output stable if the poles lie inside the circle $|z| = 1$.

e) The input-output difference equation and the system function are closely related. An accumulator-adder-gain or delay-adder-gain block diagram is readily synthesized to correspond to any set of LTI difference equations or system functions.

The DT systems analyzed in the last two chapters are linear and time-invariant, but they are not the most general LTI discrete-time systems. The most general class requires extension to system functions that are not rational functions of z, and to a *convolution* rather than difference-equation characterization in the time domain. Such an extension is our goal in the next chapter.

APPENDIX TO CHAPTER 8

Table VIII.1—Short Table of Unilateral Z-Transforms

$$\tilde{X}(z) = \sum_{n=0}^{\infty} x[n]z^{-n}$$

$x[n],\ n \geq 0$		$\tilde{X}(z)$
$\delta[n] = \begin{cases} 1, & n = 0 \\ 0, & n \neq 0 \end{cases}$	$\Longleftrightarrow$	1
$u[n-N] = \begin{cases} 1, & n \geq N \geq 0 \\ 0, & n < N \end{cases}$	$\Longleftrightarrow$	$\dfrac{z^{-N}}{1 - z^{-1}}$
a^n	$\Longleftrightarrow$	$\dfrac{1}{1 - az^{-1}}$
n	$\Longleftrightarrow$	$\dfrac{z^{-1}}{(1 - z^{-1})^2}$
na^n	$\Longleftrightarrow$	$\dfrac{az^{-1}}{(1 - az^{-1})^2}$
$a^n \cos n\theta$	$\Longleftrightarrow$	$\dfrac{1 - a\cos\theta z^{-1}}{1 - 2a\cos\theta z^{-1} + a^2 z^{-2}}$
$a^n \sin n\theta$	$\Longleftrightarrow$	$\dfrac{a\sin\theta z^{-1}}{1 - 2a\cos\theta z^{-1} + a^2 z^{-2}}$

Note: $x[n]$ is defined by $\tilde{X}(z)$ for $n \geq 0$ only.

Table VIII.2—Important Unilateral Z-Transform Theorems

Linearity	$ax[n] + by[n]$	$\Longleftrightarrow$	$a\tilde{X}(z) + b\tilde{Y}(z)$
Forward Shift	$x[n+1]$	$\Longleftrightarrow$	$z(\tilde{X}(z) - x[0])$
Delay	$x[n-N]u[n-N]$	$\Longleftrightarrow$	$z^{-N}\tilde{X}(z),\ N \geq 0$
Multiplication by a^n	$a^n x[n]$	$\Longleftrightarrow$	$\tilde{X}(a^{-1}z)$
Multiplication by n	$nx[n]$	$\Longleftrightarrow$	$-z\left(\dfrac{d\tilde{X}(z)}{dz}\right)$

Convolution*

$$x[n]u[n] * h[n]u[n] = \sum_{m=0}^{n} x[m]h[n-m] \quad \Longleftrightarrow \quad \tilde{X}(z)\tilde{H}(z)$$

---

*See Chapter 9.

EXERCISES FOR CHAPTER 8

Exercise 8.1

Derive the following Z-transform pairs, $\tilde{X}(z) = \sum_{n=0}^{\infty} x[n]z^{-n}$:

$x[n]$, $n \geq 0$		$\tilde{X}(z)$
a) $2\delta[n-1] = \begin{cases} 2, & n=1 \\ 0, & \text{otherwise} \end{cases}$	$\Longleftrightarrow$	$2z^{-1}$
b) $1 + \left(\dfrac{1}{2}\right)^n$	$\Longleftrightarrow$	$\dfrac{z(4z-3)}{2z^2 - 3z + 1}$
c) $\left(\dfrac{1}{2}\right)^n \cos\dfrac{n\pi}{3}$	$\Longleftrightarrow$	$\dfrac{4 - z^{-1}}{4 - 2z^{-1} + z^{-2}}$
d) n^2	$\Longleftrightarrow$	$\dfrac{z(z+1)}{(z-1)^3}$

Exercise 8.2

Complete the following table of unilateral Z-transforms by finding formulas for $x[n]$, $n \geq 0$, or $\tilde{X}(z) = \sum_{n=0}^{\infty} x[n]z^{-n}$ as required. For each of the pairs, sketch both $x[n]$, $n \geq 0$, and the pole-zero plot corresponding to $\tilde{X}(z)$.

$x[n]$, $n \geq 0$		$\tilde{X}(z)$
a) 2, all n	$\Longleftrightarrow$	?
b) ?	$\Longleftrightarrow$	$6z^{-1} - z^{-2}$
c) $\cosh 2n$	$\Longleftrightarrow$	?
d) $(n+1)3^{-n}$	$\Longleftrightarrow$	?
e) ?	$\Longleftrightarrow$	$\dfrac{1}{4z^2 - 1}$
f) $x[n] = \begin{cases} 1, & n \leq 2 \\ (1/2)^{n-2}, & n > 2 \end{cases}$	$\Longleftrightarrow$	?

Answers: (a) $\dfrac{2z}{z-1}$ (b) $0, 6, -1, 0, 0, \ldots$ (c) $\dfrac{1 - (\cosh 2)z^{-1}}{1 - 2(\cosh 2)z^{-1} + z^{-2}}$

(d) $\dfrac{z^2}{(z - (1/3))^2}$ (e) $\begin{cases} 0, & n = 0,\, n \text{ odd} \\ (1/2)^n, & \text{otherwise} \end{cases}$ (f) $\dfrac{z^2 + (1/2)z + (1/2)}{z(z - (1/2))}$

PROBLEMS FOR CHAPTER 8

Problem 8.1

The delay-adder-gain synthesis schemes for discrete-time systems discussed in Section 7.2 are only a few of many ways to realize equivalent structures. Several other approaches are explored in this problem.

a) Determine a delay-adder-gain block diagram of the canonic type described in Section 7.2 for realizing the system function

$$\tilde{H}(z) = \frac{3 - 3z^{-1}}{1 + 0.5z^{-1} - 0.5z^{-2}} \,.$$

b) Develop an equivalent block diagram by first expanding $\tilde{H}(z)$ in partial fractions, then realizing each term separately as in Section 7.2, and finally connecting the separate parts in parallel (with adders) to realize $\tilde{H}(z)$.

c) Develop another equivalent block diagram by first writing $\tilde{H}(z)$ as a product of factors, that is,

$$\tilde{H}(z) = \left[\frac{K}{1 - \alpha z^{-1}} \right] \left[\frac{1 - \beta z^{-1}}{1 - \gamma z^{-1}} \right],$$

then realizing each factor separately as in Section 7.2, and finally connecting the separate parts in cascade to realize $\tilde{H}(z)$.

d) The example above had real poles and hence the gains required were all real. Suggest a modification of the procedures in (b) and (c) that will allow a realization of $\tilde{H}(z)$ having more than two, possibly complex, poles while still utilizing only amplifiers with real gains.

Problem 8.2

Each female in a certain rare species of insect lays eggs twice in her lifetime, one week apart, and then immediately dies. Careful experimental studies have uncovered the curious facts that all the females of this species lay their eggs on the same day of the week, Monday, and that each female lays precisely 80 eggs the first time and 500 eggs the second time. It has also been shown that 50% of the eggs hatch in a day or so (the remainder are eaten by turtles), and that half of these are females who reach maturity in time to lay eggs for the first time on Monday of the following week.

a) How many weeks does it take the population of mature female insects (ignore the males, who don't bite anyway) to increase by a factor of more than 10^6? (To be specific, make the count on Monday mornings.)

b) Suppose an insecticide (nonresidual, of course) is applied once a week (on Saturdays) and kills a fraction α of the insects who have just hatched. It has no effect on mature females who have already laid eggs once. How large must α be to hold the total population stationary?

Problem 8.3

A discrete-time system is described by the difference equation

$$y[n+2] = -y[n+1] + 2y[n] + x[n+2] + x[n+1].$$

a) Find the system function $\tilde{H}(z)$ characterizing this system. Show its poles and zeros on a sketch of the z-plane.

b) Find the response of this system to the input

$$x[n] = 3^n u[n].$$

c) Find the response of this system to the input $x[n] = 3^n$, $n \geq 0$, if $y[0] = 0$, $y[1] = 0$. Why is this not the same as the ZSR response to this same input found in (b)?

Problem 8.4

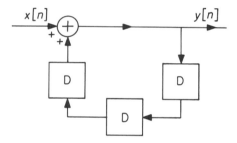

a) Find and sketch the response of the discrete-time system above to the input

$$x[n] = \begin{cases} 1, & n = 0, \\ 0, & n \neq 0. \end{cases}$$

b) For an input $x[n] = (1/2)^n u[n]$, find the output $y[n]$ for $n = 0, 1, 2, 3, 4, 5, 6$. Also find $y[n]$ for $n = 100, 101, 102, 103$. (Answers correct to 1% are acceptable.)

c) The system above is modified by inserting an amplifier with gain 0.9, as shown below. Find and sketch the response to the same input as in (a).

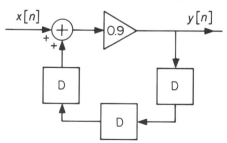

d) Write a difference equation relating $x[n]$ and $y[n]$ for the system of part (c).

e) Find the system function $\tilde{H}(z)$ and the transform of the output $\tilde{Y}(z)$ for the system of part (c) with the input of part (b).

Problem 8.5

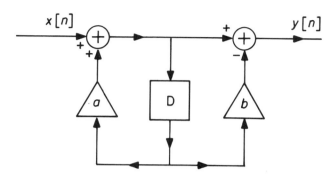

a) What is the system function $\tilde{H}(z)$ of this discrete-time system?

b) Find by any means the response to the unit step, $x[n] = u[n]$.

Problem 8.6

a) Prove the following theorems for the (unilateral) Z-transform

INITIAL- AND FINAL-VALUE THEOREMS:

i) $x[0] = \lim_{z \to \infty} \tilde{X}(z)$.

ii) $x[\infty] = \lim_{z \to 1} (z - 1)\tilde{X}(z)$ (if the limit exists).

(HINT: (i) should give no trouble. (ii) follows by an argument that is essentially the discrete analog of the Final-Value Theorem for the $\mathcal{L}$-transformation. Argue first that the Z-transform of $x[n + 1] - x[n]$ is

$$\lim_{N \to \infty} \sum_{n=0}^{N} [x[n+1] - x[n]]z^{-n} = z\tilde{X}(z) - zx[0] - \tilde{X}(z).$$

Taking the limit as $z \to 1$ on both sides and assuming orders of passing to the limit on the left may be interchanged, obtain

$$\lim_{z \to 1} (z - 1)\tilde{X}(z) - x[0] = \lim_{N \to \infty} \sum_{n=0}^{N} [x[n+1] - x[n]].$$

By writing out a few terms of the sum on the right, convince yourself that the limit (if it exists) is $x[\infty] - x[0]$, which gives the desired result.)

b) Test these theorems by applying them to all of the Z-transform pairs in the table in the appendix to this chapter.

Problem 8.7

a) Section 7.1 and Problem 7.1 discuss several integration algorithms in addition to the forward Euler algorithm. Show that employing one of these other algorithms is equivalent to replacing the integrators in a block-diagram representation of the CT system with a variant of the DT accumulator whose frequency-domain description is one of the following:

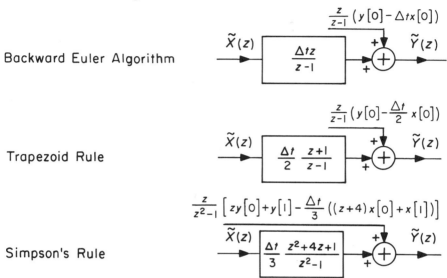

Backward Euler Algorithm

Trapezoid Rule

Simpson's Rule

b) Show that the difference equations derived in this way using the backward Euler or the trapezoid algorithms describe stable DT systems for any Δt and any stable CT system. (HINT: determine the region in the s plane corresponding to the region $|z| \leq 1$ if $s = 1/\tilde{H}_i(z)$ where $\tilde{H}_i(z)$ is the system function of the DT accumulator described by the algorithm. This is the region in which s-plane poles must lie if the DT approximation is to be stable. To determine these regions, exploit the fact that for mappings of this type circles (and straight lines, which are circles of infinite radius) map into circles or straight lines; three points determine a circle.)

c) Show that your results in (b) imply that the trapezoid rule has the important property that, for any value of Δt, the DT system is unstable if and only if the CT system from which it is derived is unstable.

d) If Simpson's rule is used in this way, each s-plane pole corresponds to two z-plane poles. Find formulas for the z poles in terms of the s pole. Show that one of these z poles always has $|z| \geq 1$ so that the DT system is always unstable, independent of the value of Δt or the nature of the CT system. Hence, Simpson's rule is not used for this purpose. (HINT: the left- and right-half s-planes map doubly into the z-plane regions shown to the right.)

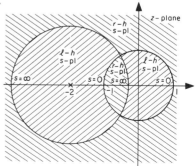

Problem 8.8

A useful technique for the design of DT systems is to choose $\tilde{H}(z)$ so that the response $y[n]$ to a DT step input, $x[n] = u[n]$, is the same as samples of the response $y(t)$ of some CT system $H(s)$ to a CT unit step, $x(t) = u(t)$. That is, we seek

$$y[n] = y(nT)$$

where T is some appropriate sampling interval. In this way, known desirable features of the CT system can be extended to the DT system. The relationship of $\tilde{H}(z)$ and $H(s)$ in this case is said to be a *step-invariant transformation*.

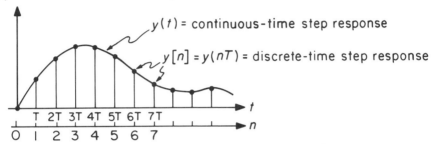

a) Derive the step-invariant $\tilde{H}(z)$ corresponding to $H(s) = \dfrac{1}{s+\alpha}$, $\alpha > 0$. Describe the locations of the poles and zeros of $\tilde{H}(z)$ as functions of α and T. Compare with a pole-zero plot of $H(s)$.

b) Draw a block-diagram realization of $\tilde{H}(z)$ using gain-adder-accumulator blocks as in Sections 7.2 and 8.3. Compare with a corresponding realization of $H(s)$ using gain-adder-integrator blocks.

c) Draw a block-diagram realization of $\tilde{H}(z)$ using gain-adder-delay blocks as in Sections 7.2 and 8.3.

d) Repeat (a) for $T = 0.02$ and a sharply resonant system, $H(s) = \dfrac{s}{s^2 + 2s + 101}$.

Problem 8.9

In the *sampled-data control system* shown in Figure 1 on the next page, the output $y(t)$ of a CT system $H(s)$ is sampled, some processing is done on the resulting sequence of samples, and the processed sequence $r[n]$ is converted back to a CT signal $r(t)$ that is subtracted from the control input $x(t)$ to generate the error signal $e(t)$, which becomes the input to $H(s)$. Usually the discrete-to-continuous converter (D/C) is a zero-order hold (see Figure 2 and Problem 7.4). Because of the effect of the clock on the converters, the system of Figures 1 and 2 is linear but time-varying. If, however, $x(t)$ has the same staircase character as $r(t)$ in Figure 2 (or if we are willing to approximate the actual smooth $x(t)$ by such a waveform), then it is easy to show that the system of Figure 1 is equivalent to the LTI DT system of Figure 3 (where $x[n]$ and $e[n]$ bear the same relationships to $x(t)$ and $e(t)$ that $r[n]$ bears to $r(t)$). Let $\tilde{H}(z)$ be the system function of the equivalent DT system in the dashed box of Figure 3.

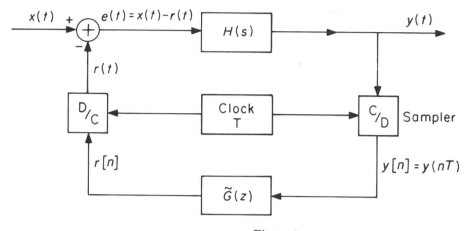

Figure 1.

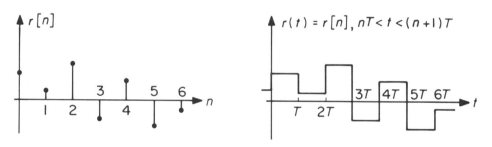

Figure 2.

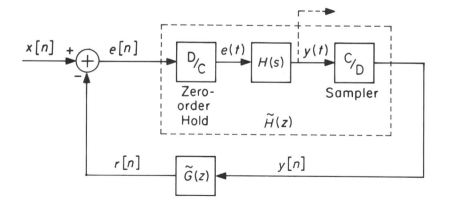

Figure 3.

a) Argue that $\tilde{H}(z)$ is related to $H(s)$ by a step-invariant transformation, that is, if $y_s(t)$ is the response of $H(s)$ to a CT unit step input, $e(t) = u(t)$, and if $y_s[n]$ is the response of $\tilde{H}(z)$ to a DT unit step, $e[n] = u[n]$, then $y_s[n] = y_s(nT)$, all n. See also Problem 8.8.

b) Suppose that $H(s)$ is unstable; specifically, suppose that

$$H(s) = \frac{1}{s-1}, \quad \Re e[s] > 1.$$

Show that

$$\tilde{H}(z) = \frac{(e^T - 1)z^{-1}}{1 - e^T z^{-1}}, \quad |z| > e^T$$

which is also unstable since $T > 0$.

c) Suppose that $H(s)$ is as in (b) and that $\tilde{G}(z) = K$. Find the range of values of K for which the closed-loop DT system of Figure 3 is stable.

Problem 8.10

a) The measured response $h[n]$ to a unit sample input $x[n] = \delta[n]$ of a certain DT system is shown in the figure below. Sketch the response $y[n]$ of this system to a unit step input $x[n] = u[n]$. Evaluate the response for $n \leq 8$.

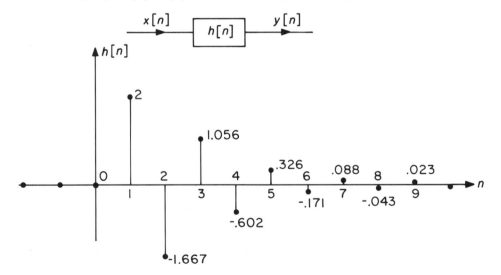

b) A close fit to these observations is provided by the formula

$$h[n] = 12\left[\left(-\frac{1}{3}\right)^n - \left(-\frac{1}{2}\right)^n\right], \qquad n \geq 0.$$

Find a closed-form expression for the system function $\tilde{H}(z)$ corresponding to this formula,

$$\tilde{H}(z) = \sum_{n=0}^{\infty} h[n]z^{-n}.$$

c) The overshoot in the step response of this system is troublesome in many applications. It is proposed to compensate the given system by cascading it (as shown below) with another system described by the difference equation

$$w[n] = ay[n] + by[n-1] + cy[n-2].$$

Find values of the constants a, b, and c such that the overall unit step response of the cascade is simply a delayed unit step, $u[n-1]$.

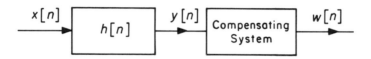

d) Devise a realization of the compensating system described by the difference equation in (c) in terms of delay lines, gain elements, and adders.

Problem 8.11

You are engaged in a game of "matching pennies" with your roommate. Each of you has a stack of pennies. You compare the pennies on the tops of the stacks. If they are both "heads" or both "tails" (i.e., if they "match"), you win and place both pennies on the bottom of your stack. Otherwise your roommate wins and places both pennies on the bottom of his or her stack. You then compare the next pennies in each stack. The game is over when either you or your roommate holds all the pennies.

Suppose at some point you have n pennies. At this point your roommate has $m = N - n$ pennies. Assume that N, the total number of pennies, is fixed. We seek to determine $p[n]$, the probability at this point that you will win the game. Evidently $p[0] = 0$ (you have lost) and $p[N] = 1$ (you have won). In general, $p[n]$ must satisfy the difference equation

$$p[n+1] = \frac{1}{2}p[n+2] + \frac{1}{2}p[n].$$

(Starting from the situation of having $n+1$ pennies, one either wins—with probability 0.5—and thus arrives at a point where one has $n+2$ pennies, or loses—with probability 0.5—and thus arrives at a point where one has n pennies.) Use Z-transform methods to solve this equation subject to the given boundary conditions, finding $p[n]$ for any n, $0 \leq n \leq N$.

9

UNIT SAMPLE RESPONSE AND DISCRETE-TIME CONVOLUTION

9.0 Introduction

The unit sample* function

$$\delta[n] = \begin{cases} 1, & n = 0 \\ 0, & n \neq 0 \end{cases}$$

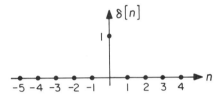

introduced in Example 8.1–3 plays an extremely important role in the theory of LTI discrete-time systems. Indeed,

Figure 9.0–1. Sketch of $\delta[n]$.

the ZSR to $\delta[n]$—called the *unit sample response*† and represented by the symbol $h[n]$—completely characterizes the ZSR to any other input. This result follows at once (for the class of systems we have been studying) from the following:

i) The Z-transform of the unit sample is

$$\sum_{n=0}^{\infty} \delta[n]z^{-n} = 1 \,.$$

ii) The Z-transform of the unit sample response is consequently $\tilde{H}(z) \times 1$. Thus the unit sample response and the system function are a Z-transform pair

$$h[n] \quad \Longleftrightarrow \quad \tilde{H}(z)$$

(which, of course, accounts for the choice of the symbol $h[n]$ to represent the unit sample response).

iii) Given $h[n]$, we can find the ZSR to an arbitrary input $x[n]$ by transforming $h[n]$ and $x[n]$ and then inverse transforming the product $\tilde{H}(z)\tilde{X}(z)$.

---

*In some treatments $\delta[n]$ is called the (discrete-time) unit *impulse* function.

†Since the unit sample is defined to be zero for all $n \neq 0$, including in particular all $n < 0$, we adopt the convention that $h[n]$ is implicitly a ZSR. This is consistent with our earlier conventions concerning the unit step responses of both continuous-time and discrete-time systems. If $\delta[n]$ is called the discrete-time unit impulse function, then $h[n]$ is called the unit impulse response.

But (as we shall show) given $h[n]$, we can also compute the ZSR to an arbitrary input $x[n]$ directly in the time domain—without transforming, multiplying, and inverse transforming—by a process called *convolution*:

$$y[n] = \sum_{m=0}^{n} x[m]h[n-m]\,.$$

Studying how this convolution process works, that is, how it can be equivalent in an input-output sense to the behavior of a set of difference equations or a block diagram, will be our first task in this chapter. Such a study is important for several reasons:

1. Thinking about the ZSR behavior of LTI systems in terms of convolution frequently is simpler or provides insights more powerful than those obtained from difference equations, system functions, or block diagrams.

2. The convolution formula often suggests effective realizations of LTI system operations in computer algorithms or special-purpose hardware.

3. The convolution operation readily generalizes to describe a useful larger class of LTI discrete-time systems than we have yet considered.

In Chapter 10 we shall extend this generalization to the similar but mathematically more subtle situation of general LTI systems in continuous time.

9.1 The Convolution Theorem for Z-Transforms

We shall begin by proving the following theorem:

Z-TRANSFORM CONVOLUTION THEOREM:

   Let

$$y[n] = \sum_{m=0}^{n} x[m]h[n-m]\,. \qquad (9.1\text{--}1)$$

   Then the Z-transform of $y[n]$ is

$$\tilde{Y}(z) = \tilde{X}(z)\tilde{H}(z)$$

   where $\tilde{X}(z)$ and $\tilde{H}(z)$ are the Z-transforms of $x(n)$ and $h[n]$ respectively.

The Convolution Theorem is included in the list of Z-transform properties in the appendix to Chapter 8. To prove this theorem, evaluate $\tilde{Y}(z)$ directly:

$$\tilde{Y}(z) = \sum_{n=0}^{\infty} y[n]z^{-n} = \sum_{n=0}^{\infty}\left[\sum_{m=0}^{n} x[m]h[n-m]\right]z^{-n}. \qquad (9.1\text{--}2)$$

Rewrite (9.1–2) in the form

$$\tilde{Y}(z) = \sum_{n=0}^{\infty} \left[\sum_{m=0}^{\infty} x[m]h[n-m]u[n-m] \right] z^{-m} z^{-(n-m)}. \tag{9.1–3}$$

Taking the summation in (9.1–3) over n first* gives

$$\tilde{Y}(z) = \sum_{m=0}^{\infty} x[m] z^{-m} \left[\sum_{n=0}^{\infty} h[n-m]u[n-m] z^{-(n-m)} \right]. \tag{9.1–4}$$

But because the unit step function vanishes for negative values of its argument, the bracketed sum in (9.1–4) is equal to

$$\sum_{n=0}^{\infty} h[n] z^{-n} = \tilde{H}(z) \tag{9.1–5}$$

independent of m. The remaining sum then yields

$$\tilde{Y}(z) = \tilde{X}(z)\tilde{H}(z) \tag{9.1–6}$$

as we sought to show.

Example 9.1–1

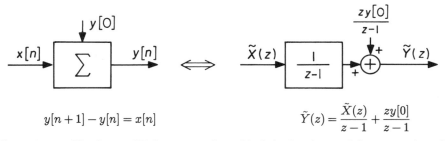

$$y[n+1] - y[n] = x[n] \qquad\qquad \tilde{Y}(z) = \frac{\tilde{X}(z)}{z-1} + \frac{zy[0]}{z-1}$$

Figure 9.1–1. The forward Euler accumulator block in the time and frequency domains.

In Section 8.3 definitions of the forward Euler accumulator block were given in both the time and frequency domains, as shown in Figure 9.1–1. The system function of the accumulator is

$$\tilde{H}(z) = \frac{1}{z-1} = \frac{z^{-1}}{1-z^{-1}}. \tag{9.1–7}$$

---

*Interchanging the order of summing requires that the terms being summed vanish sufficiently rapidly as $n, m \to \infty$. This will always be the case if z is inside a common domain of convergence of $\tilde{H}(z)$ and $\tilde{X}(z)$.

From the Z-transform table in the appendix to Chapter 8, the unit sample response of the accumulator is

$$h[n] = u[n-1]$$

$$= \begin{cases} 1, & n \geq 1 \\ 0, & \text{otherwise.}* \end{cases} \tag{9.1-8}$$

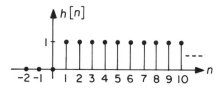

Figure 9.1-2. Unit sample response of the forward Euler accumulator.

From the Convolution Theorem, the ZSR response to an arbitrary input $x[n]$, $n \geq 0$, is then

$$y[n] = \sum_{m=0}^{n} x[m]h[n-m] = \sum_{m=0}^{n} x[m]u[n-m-1]. \tag{9.1-9}$$

As a function of m,

$$u[n-m-1] = \begin{cases} 1, & m < n \\ 0, & m \geq n \end{cases} \tag{9.1-10}$$

which is shown graphically in Figure 9.1-3. Hence

$$y[n] = \begin{cases} \sum_{m=0}^{n-1} x[m], & n \geq 1 \\ 0, & n = 0. \end{cases} \tag{9.1-11}$$

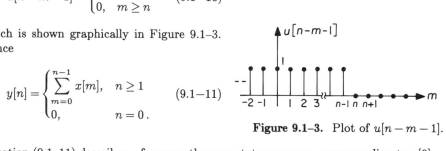

Figure 9.1-3. Plot of $u[n-m-1]$.

Equation (9.1-11) describes, of course, the zero state response, corresponding to $y[0] = 0$. If $y[0] \neq 0$, we must add to the ZSR the inverse transform of

$$\frac{zy[0]}{z-1} = \frac{y[0]}{1-z^{-1}} \tag{9.1-12}$$

which from Table VIII.1 in the appendix to Chapter 8 is the constant $y[0]$ for all $n \geq 0$. The total response is thus

$$y[n] = y[0] + \sum_{m=0}^{n-1} x[m] \tag{9.1-13}$$

which justifies the name "accumulator" and clearly shows its relationship to a continuous-time integrator.

▶ ▶ ▶

---

*Strictly speaking, (9.1-7) determines $h[n]$ only for $n \geq 0$. This is, in fact, all we need in order to evaluate the convolution sum with limits as in (9.1-1). But in the larger context of general LTI DT systems shortly to be introduced, the forward Euler accumulator is a causal system, which implies $h[n] = 0$, $n < 0$.

Example 9.1–2

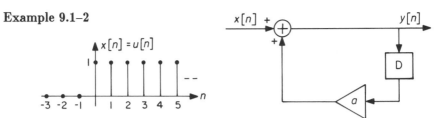

Figure 9.1–4. DT system and input for Example 9.1–2.

We seek the response $y[n]$ of the system shown in Figure 9.1–4 to a unit step input, $x[n] = u[n]$. From the feedback formula, or from the observation that at the output of the adder under ZSR conditions

$$\tilde{Y}(z) = \tilde{X}(z) + az^{-1}\tilde{Y}(z)$$

we conclude that the system function is

$$\tilde{H}(z) = \frac{\tilde{Y}(z)}{\tilde{X}(z)} \text{(ZSR)} = \frac{1}{1 - az^{-1}}.$$

By inverse transforming, we obtain the unit sample response shown in Figure 9.1–5.

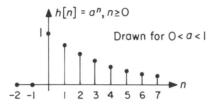

Figure 9.1–5. Unit sample response of the system of Figure 9.1–4.

The same result can be derived directly from the block diagram. Thus in the zero state, $y[0] = x[0] = \delta[0] = 1 = h[0]$. At time $n = 1$, the output of the delay line is $y[0] = 1$ and the signal fed back to the adder is $ay[0] = a$; since $x[1] = \delta[1] = 0$, we conclude that $y[1] = ay[0] = a = h[1]$. Similarly $h[2] = \delta[2] + ah[1] = a^2$, etc.

To compute the response using the convolution formula,

$$y[n] = \sum_{m=0}^{n} x[m]h[n-m]$$

it is helpful to plot $x[m]$ and $h[n-m]$ versus m for a typical value of n, as shown in Figure 9.1–6 for $n = 3$.

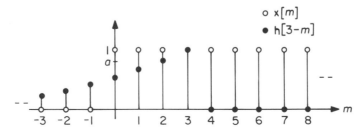

Figure 9.1–6. Plots of $x[m]$ and $h[3-m]$.

Notice how $h[3-m]$ is *folded* over, plotted *backwards* from $m=3$. This accounts in part for the name "convolution," which means literally "fold together." From the plot we can evaluate

$$y[3] = \sum_{m=0}^{3} x[m]h[3-m] = x[0]h[3] + x[1]h[2] + x[2]h[1] + x[3]h[0]$$

$$= 1 \cdot a^3 + 1 \cdot a^2 + 1 \cdot a + 1 \cdot 1 = \frac{1-a^4}{1-a}$$

(where we have used the formula (7.3–4) for the partial sum of a geometric series). In general,

$$y[n] = \frac{1-a^{n+1}}{1-a}, \quad n \geq 0. \tag{9.1–14}$$

As a check, using Z-transforms, we find that

$$\tilde{Y}(z) = \tilde{X}(z)\tilde{H}(z) = \frac{1}{1-z^{-1}} \frac{1}{1-az^{-1}} = \frac{\frac{1}{1-a}}{1-z^{-1}} + \frac{\frac{a}{a-1}}{1-az^{-1}}$$

so

$$y[n] = \frac{1}{1-a} + \frac{a}{a-1}a^n = \frac{1-a^{n+1}}{1-a}, \quad n \geq 0$$

which agrees with (9.1–14). $y[n]$ is plotted in Figure 9.1–7.

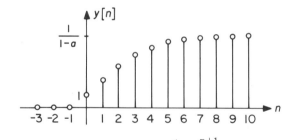

Figure 9.1–7. $y[n] = \dfrac{1-a^{n+1}}{1-a}u[n]$.

▶ ▶ ▶

Example 9.1–3

The process of discrete-time convolution is often aided by a simple device. Write the sequences of numbers $x[n]$ and $h[-n]$ on the edges of separate sheets of paper as shown in Figure 9.1–8. Mark $n=0$ with a small arrow on both sheets. Note that $h[-n]$ is $h[n]$ *reversed in sequence*, that is, plotted *backwards* from $n=0$. Slide one paper past the other. Computing the sum of the products of adjacent numbers at each shift gives the sequence $y[n]$. The illustration shows the placing of the strips for $n=4$; the complete result for all n is shown in Figure 9.1–9.

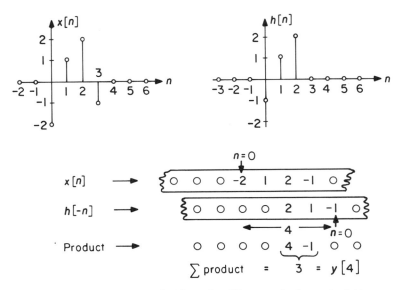

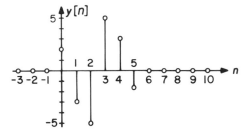

Figure 9.1–8. Mechanical aid to simplify convolution calculations.

Figure 9.1–9. Result of the convolution illustrated in Figure 9.1–8.

▶ ▶ ▶

The process illustrated in Example 9.1–3 is particularly effective if $h[n]$ is nonzero for only a finite number of values of n; such a system is often indicated by the abbreviation FIR (for Finite Impulse* Response). In contrast, $h[n] = a^n$, $n \geq 0$, has nonzero values for arbitrarily large values of n and is classified as IIR (for Infinite Impulse* Response). FIR systems can be simulated by a non-recursive or transversal filter such as that introduced in Figure 7.2–5 and shown in Figure 9.1–10 (for the $h[n]$ above). Recursive (feedback) systems of the type described in Figure 7.2–6 lead to IIR type behavior, as illustrated in Figure 9.1–5 of Example 9.1–2.

---

*These abbreviations reflect the alternate name for $h[n]$—discrete-time unit impulse response—mentioned in the footnote in Section 9.0.

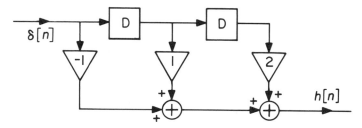

Figure 9.1–10. Transversal filter realization of the $h[n]$ of Example 9.1–3.

In Example 9.1–3 both $h[n]$ and $x[n]$ are of finite duration. Their Z-transforms are thus simply polynomials in z^{-1}, and this provides an easy way of checking the result given above. Thus

$$\tilde{X}(z) = -2 + z^{-1} + 2z^{-2} - z^{-3} \quad \text{and} \quad \tilde{H}(z) = -1 + z^{-1} + 2z^{-2}$$

so

$$\begin{aligned}
\tilde{Y}(z) &= \tilde{H}(z)\tilde{X}(z) \\
&= \left(-2 + z^{-1} + 2z^{-2} - z^{-3}\right)\left(-1 + z^{-1} + 2z^{-2}\right) \\
&= 2 - 3z^{-1} - 5z^{-2} + 5z^{-3} + 3z^{-4} - 2z^{-5}
\end{aligned}$$

from which the $y[n]$ in Figure 9.1–9 follows at once.

9.2 Convolution and General Linear Time-Invariant Systems

The formula

$$y[n] = \sum_{m=0}^{n} x[m]h[n-m] \tag{9.2–1}$$

describes a restricted case of the more general convolution formula

$$y[n] = \sum_{m=-\infty}^{\infty} x[m]h[n-m]. \tag{9.2–2}$$

The general formula describes an operation on $x[n]$ that is both linear and time-invariant, satisfying the two conditions:

LINEARITY (SUPERPOSITION):

If $x_1[n]$ and $x_2[n]$ are any two arbitrary inputs to some operation yielding well-defined outputs $y_1[n]$ and $y_2[n]$ respectively, that is, if

$$x_1[n] \;\rightarrow\; y_1[n]$$
$$x_2[n] \;\rightarrow\; y_2[n]$$

then the operation is said to be *linear* if

$$x[n] = ax_1[n] + bx_2[n] \;\rightarrow\; y[n] = ay_1[n] + by_2[n]$$

for all constants *a* and *b*.

TIME-INVARIANCE:

If $x[n]$ is any arbitrary input to some operation yielding the well-defined output $y[n]$, that is, if

$$x[n] \;\rightarrow\; y[n]$$

then the operation is said to be *time-invariant* if

$$x[n - N] \;\rightarrow\; y[n - N]$$

for all $-\infty < N < \infty$.

The proofs in both cases are almost immediate if "well-defined" is interpreted to mean that the convergence of the infinite summation in (9.2–2) is independent of the order in which the terms are taken.

The linearity and time-invariance conditions on *operations* defined in the preceding paragraph are the same (except for the change from continuous to discrete time) as the conditions specified in Chapter 1 for LTI *elements*. Note, however, that a system composed of LTI elements is described by an LTI operation only under zero-state conditions.

The operation described by (9.2–2) is not only linear and time-invariant, but also the *most general* discrete-time operation satisfying both the linearity and time-invariance conditions. The most direct way to demonstrate this is as follows:

1. The system input $\delta[n]$ yields some output; call it $h[n]$.
2. If the system is time-invariant, then the input $\delta[n-m]$ for any fixed integer m must yield the output $h[n-m]$, $-\infty < m < \infty$.
3. An arbitrary input $x[n]$, $-\infty < n < \infty$, can be written in the form

$$
\begin{aligned}
x[n] &= \cdots + x[-1]\delta[n+1] + x[0]\delta[n] + x[1]\delta[n-1] + \cdots \\
&= \sum_{m=-\infty}^{\infty} x[m]\delta[n-m].
\end{aligned}
\tag{9.2–3}
$$

4. Hence, by linearity, the output for $x[n]$ must be

$$y[n] = \quad \cdots + x[-1]h[n+1] + x[0]h[n] + x[1]h[n-1] + \cdots$$

$$= \sum_{m=-\infty}^{\infty} x[m]h[n-m]. \qquad (9.2\text{--}2)$$

Thus we have shown that any DT LTI operator can be written in the convolution form (9.2–2). One input-output pair, $\delta[n] \rightarrow h[n]$, characterizes the response of a general DT LTI operation to any other input.* The steps in the argument above are illustrated in Figure 9.2–1.

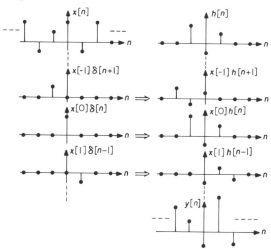

Figure 9.2–1. $y[n] = \quad \cdots x[-1]h[n+1] + x[0]h[n] + x[1]h[n-1] + \cdots$

The general convolution formula (9.2–2) is a summation over an infinite number of terms. The result will be finite only if most of these terms are vanishingly small, and hence at least one of the two sequences, $x[n]$ and $h[-n]$, must vanish sufficiently rapidly as $n \rightarrow \pm\infty$. There are several ways to ensure this. The simplest is to apply the pair conditions leading to the special case (9.2–1) that we derived earlier from Z-transforms:

1. Restrict consideration to systems that are *causal*,[†] that is, those for which

$$h[n] \equiv 0, \quad n < 0. \qquad (9.2\text{--}4)$$

2. Accept only inputs that vanish for $n < 0$, that is,

$$x[n] \equiv 0, \quad n < 0. \qquad (9.2\text{--}5)$$

---

*Note that the response to one input $\delta[n]$ does not in general characterize the behavior of a non-linear system, and the unit sample response describes the behavior of a linear time-varying system only if it is known for samples applied at every time instant.

[†] We shall discuss the choice and implications of this word more fully in the next chapter.

The summand $x[m]h[n-m]$ in (9.2–2) then vanishes for $m < 0$ (because $x[m]$ is zero for $m < 0$), and also for $m > n$ (because $h[n-m]$ is zero for $n-m < 0$). So the summation for finite n is reduced to a finite number of terms, and hence is finite if $x[n]$ and $h[n]$ are sequences of terms of finite amplitudes. Causality is a natural assumption for the class of systems we have thus far been studying; we would not expect them to respond before we stimulate them. Constraining inputs to be zero for $n < 0$ ensures the zero state at $n = 0$, and it is only under such conditions that the behavior of causal LTI systems satisfies the conditions of an LTI operation.

But there are other conditions on $h[n]$ and $x[n]$ that can guarantee the existence of $y[n]$ in (9.2–2) and thus operationally describe LTI systems sharing many of the analytical advantages of the systems we have been studying, as well as possessing additional useful features to be discussed in the following chapters. Perhaps the simplest alternative set of conditions is to require that:

1. $h[n]$ be absolutely summable, that is,

$$\sum_{n=-\infty}^{\infty} |h[n]| = H_0 < \infty ; \tag{9.2–6}$$

2. $x[n]$ be bounded in magnitude, that is,

$$|x[n]| \leq X_{\max} < \infty . \tag{9.2–7}$$

It is easy to see that these conditions are *sufficient* to ensure that $y[n]$ of (9.2–2) be bounded since

$$|y[n]| = \left| \sum_{m=-\infty}^{\infty} x[m]h[n-m] \right|$$

$$\leq \sum_{m=-\infty}^{\infty} |x[m]| \, |h[n-m]| \tag{9.2–8}$$

$$\leq X_{\max} \sum_{m=-\infty}^{\infty} |h[n-m]| = X_{\max}H_0 < \infty .$$

In other words, the condition (9.2–6) guarantees that a bounded input yields a bounded output; it is thus a (BIBO) *stability condition*. Moreover, (9.2–6) is a *necessary* as well as sufficient condition for BIBO stability since if $\sum_{n=-\infty}^{\infty} |h[n]|$ is *not* bounded then there is at least one bounded input, specifically

$$x[n] = \mathrm{sgn}\{h[-n]\} = \begin{cases} 1, & h[-n] > 0 \\ -1, & h[-n] < 0 \end{cases} \tag{9.2–9}$$

that yields

$$y[0] = \sum_{m=-\infty}^{\infty} x[m]h[-m] = \sum_{m=-\infty}^{\infty} |h[-m]| \tag{9.2–10}$$

which is unbounded by hypothesis. Hence, the behavior of BIBO stable LTI systems driven by bounded inputs is also described by the convolution formula (9.2–2). Note that such systems need not be causal and the inputs need not vanish for $n < 0$.

As we have seen, the study of the dynamic ZSR response of LTI systems satisfying the causality condition

$$h[n] \equiv 0, \quad n < 0$$

is closely related to difference equations, system functions, Z-transforms, and characterization of the past in terms of a present state. Such an approach is particularly suited to what might be called DT *control* problems. To ignore causal constraints on a controller is to neglect what is often the most basic problem in control system design—the fact that a non-causal controller could accurately predict its own input and thus completely circumvent the inevitable lag in any real system between the initiation of an action and its actual accomplishment. Furthermore, in practice a control situation can often be broken down into a sequence of responses to discrete commands—"land on runway 21, brake, taxi to loading ramp N2, etc." At the initiation of each action, it is natural to summarize the situation in terms of a present state rather than as the superposition of the present and future consequences of all previous inputs. (Once the plane has lined up on the landing path for runway 21, the details of its entire previous flight history are largely irrelevant.) Typically, a control problem has a natural beginning and a short duration, comparable to the time constants or memory of the system. It is thus to be expected that the formulation of control problems will place heavy emphasis on the concepts of causality and state, and that solution techniques for such problems will extensively exploit unilateral transforms.

In many of the remaining chapters in this book, however, we shall be concerned with a different class of situations that we might describe as *communication* or *data-processing* problems. Here, the causality issue is of lesser significance. The division of the time axis into "before" and "after" some initial moment (with the past summarized by an initial state) frequently seems highly artificial. In communication systems we are usually interested in long strings of responses to long strings of inputs. The signal durations are typically much longer than the time constants of the system. It is often natural to idealize such inputs as bounded waveforms extending indefinitely in time in both directions, $-\infty < t < \infty$. An important class of *operations* on such waveforms is LTI, described (in discrete time—a similar formula applies in continuous time, as we shall see) by the general convolution formula (9.2–2) with a unit sample response $h[n]$ that is stable but not necessarily causal.

In the chapters to come, we shall use the words "LTI system" to describe a device that carries out an LTI operation rather than a device composed of LTI elements. When we discuss a general LTI system in this sense—a system that may be non-causal and that may have inputs ranging over $-\infty < n\,(\text{or } t) < \infty$—such ideas as state, ZSR, and ZIR are not really applicable. In effect, we are taking responsibility for the *total* input—not just the input from now on—and we are usually interested in some output, rather than what is going on inside the "black box." Stability for such systems loosely implies that things that happened "at" $-\infty$, or that won't happen "until" $+\infty$, have negligible influence at finite times. Causality is a constraint we may be prepared to give up, if it simplifies the argument to do so. This is partly because in many cases n does not represent time (but rather, perhaps, space) or because the operations described will be carried out "off line" (that is, not in real time) by computers. But it is mostly because the primary difference between the performance of a causal and a non-causal but stable system is frequently only a small delay, which is usually unimportant for communication as contrasted with control problems. We shall have more to say about causality and its implications in Chapter 15. In the frequency domain, the appropriate tool for handling stable but not necessarily causal systems is the *Fourier transform*, which we shall start to develop in Chapter 12.

If a system is neither causal nor stable, it still may have finite well-defined responses to appropriately constrained inputs, but generally ad hoc techniques are needed to provide anything equivalent to frequency-domain methods.* Fortunately, such systems seem to have limited utility.

9.3 Algebraic Properties of the General Convolution Operation

The general convolution operation

$$y[n] = \sum_{m=-\infty}^{\infty} x[m]h[n-m] \qquad (9.3-1)$$

generates a new sequence $y[n]$ from two given sequences, $x[n]$ and $h[n]$. It is often convenient to represent this operation by a short-hand symbol such as an asterisk,

$$y[n] = x[n] * h[n] \qquad (9.3-2)$$

which is to be read "$y[n]$ equals $x[n]$ convolved with $h[n]$." As an algebraic operation, convolution has a number of useful properties:

COMMUTATIVE LAW:

$$
\begin{aligned}
x[n] * h[n] &= \sum_{m=-\infty}^{\infty} x[m]h[n-m] \\
&= \sum_{m=-\infty}^{\infty} h[m]x[n-m] = h[n] * x[n]\,.
\end{aligned}
\qquad (9.3-3)
$$

---

*The *bilateral* Laplace transform is sometimes useful in such situations, as we shall briefly discuss in Chapter 13.

ASSOCIATIVE LAW:

$$x[n] * \left(h_1[n] * h_2[n]\right) = \left(x[n] * h_1[n]\right) * h_2[n]. \qquad (9.3\text{--}4)$$

DISTRIBUTIVE LAW:

$$\left(x_1[n] + x_2[n]\right) * h[n] = x_1[n] * h[n] + x_2[n] * h[n]. \qquad (9.3\text{--}5)$$

The distributive law is simply the linearity principle stated earlier. The other two properties follow more or less directly on writing out the summations involved— both sides include the same terms added up in different orders.*

The commutative property might have been inferred—at least for the special case of causal $h[n]$ and $x[n] = 0$, $n < 0$—from the fact that the inverse Z-transform of the product of $\tilde{X}(z)$ and $\tilde{H}(z)$ is independent of order:

$$\tilde{X}(z)\tilde{H}(z) = \tilde{H}(z)\tilde{X}(z).$$

Thus if $w_1[n]$ and $w_2[n]$ are two distinct sequences, it makes no difference which we consider the unit sample response and which the input—the resulting output is the same, as illustrated in Figure 9.3–1.

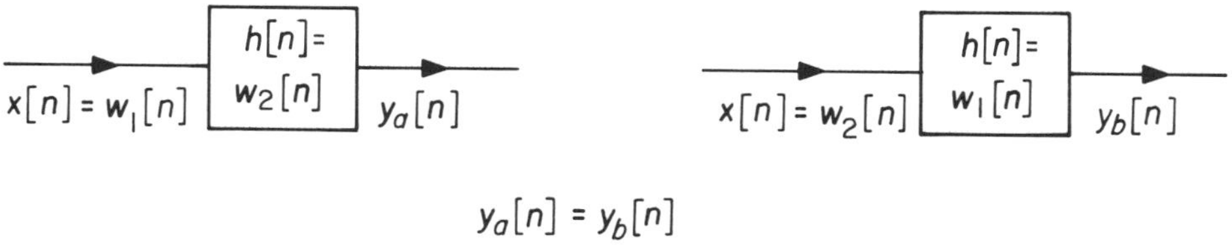

$$y_a[n] = y_b[n]$$

Figure 9.3–1. Illustration of the commutative property of convolution.

In a cascade of two LTI systems, the associative law states that we may compute the overall ZSR output either by first convolving the input with $h_1[n]$ to obtain the input to the second system, which is then convolved with $h_2[n]$ to give the output, or by first convolving $h_1[n]$ and $h_2[n]$ to obtain an equivalent composite unit sample response $h[n] = h_1[n] * h_2[n]$, which is then convolved with the input to give the output. These equivalences are illustrated in Figure 9.3–2. Of course, since the commutative law states that

$$h_1[n] * h_2[n] = h_2[n] * h_1[n] \qquad (9.3\text{--}6)$$

the input-output behavior of a cascade of LTI systems is independent of the order of cascading.

---

*If the summation contains an infinite number of terms, then $x[n]$ and/or $h[n]$ must vanish fast enough so that the sum is independent of the order. By violating this condition we can construct somewhat artificial examples in which convolution is not associative. For a CT example, see Problem 10.8.

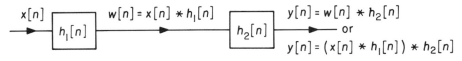

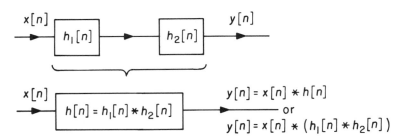

Figure 9.3–2. Illustration of the associative property of convolution.

9.4 An Example of Deconvolution

The following example describes one practical application of the ideas of this chapter.

Example 9.4–1

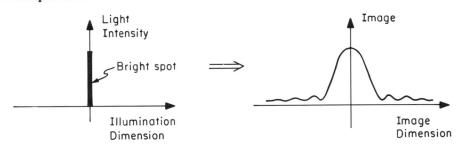

Figure 9.4–1. Transformation of a point spot into a smeared image.

Because of limited aperture, lens imperfections, focusing errors, etc., no camera can produce a perfectly sharp image. The actual situation can, in one dimension rather than two, be described by Figure 9.4–1. We shall assume that an isolated bright point spot in the illumination field is converted by a lens of limited resolving power into a smeared image of nonzero width. If the picture is to be telemetered from some remote location, such as a space probe, it might be desirable to sample it at regularly spaced points before transmission, so that effectively the received smeared image from a bright spot might appear as in the discrete "time" representation shown in Figure 9.4–2. For simplicity we shall assume that the shape of the smearing of a unit sample is $h[n] = a^{|n|}$ as shown in Figure 9.4–2, although an actual measured response would undoubtedly be more complicated.

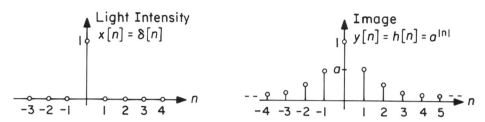

Figure 9.4–2. Sampled DT equivalent of Figure 9.4–1.

The light intensity $x[n]$ describing an interesting scene is a sequence of many spots of various intensities; the image $y[n]$ of the scene is then a superposition of the weighted smeared spot images (assuming the camera output is linear in exposure, which again is a simplification not always justified in practice). In one dimension,

$$y[n] = \sum_{m=-\infty}^{\infty} x[m]h[n-m]. \tag{9.4–1}$$

If the light intensity is a unit step,

$$x[n] = u[n] \tag{9.4–2}$$

representing a sharp black-white edge in the picture, the image will be given by

$$y[n] = \sum_{m=-\infty}^{\infty} u[m]h[n-m] = \sum_{m=0}^{\infty} h[n-m]. \tag{9.4–3}$$

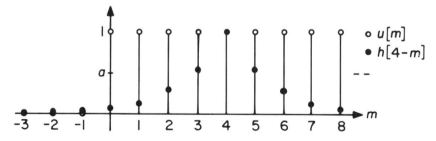

Figure 9.4–3. Plots of $u[m]$ and $h[4-m]$.

The summand terms of (9.4–3) are shown in Figure 9.4–3. Equation (9.4–3) can readily be evaluated, but it is a trifle easier (both analytically and graphically) to work with the commuted form

$$y[n] = \sum_{m=-\infty}^{\infty} h[m]x[n-m] = \sum_{m=-\infty}^{\infty} h[m]u[n-m] = \sum_{m=-\infty}^{n} h[m] \tag{9.4–4}$$

whose terms are shown in Figure 9.4–4. (A good rule to follow in general is to use that form of the convolution formula in which the simpler of the two sequences, $x[n]$ or $h[n]$, is the one that is folded. This will usually lead to the simplest pictures and formulas.)

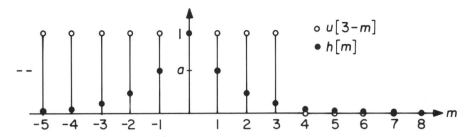

Figure 9.4-4. Plots of $h[m]$ and $u[3-m]$.

For $n \leq 0$ and $h[n] = a^{|n|}$, (9.4-4) yields

$$
\begin{aligned}
y[n] = \sum_{m=-\infty}^{n} h[m] &= \sum_{m=-\infty}^{n} a^{-m} = a^{-n} + a^{-n+1} + a^{-n+2} \cdots \\
&= a^{-n}[1 + a + a^2 + \cdots] \\
&= \frac{a^{-n}}{1-a}, \quad n \leq 0.
\end{aligned}
\tag{9.4-5}
$$

For $n > 0$,

$$
\begin{aligned}
y[n] = \sum_{m=-\infty}^{n} h[m] &= \sum_{m=-\infty}^{0} a^{-m} + \sum_{m=1}^{n} a^{m} \\
&= \frac{1}{1-a} + a + a^2 + \cdots + a^n \\
&= \frac{1}{1-a} - 1 + (1 + a + a^2 + \cdots + a^n) \\
&= \frac{1}{1-a} - 1 + \frac{1 - a^{n+1}}{1-a} \\
&= \frac{1 + a - a^{n+1}}{1-a}, \quad n > 0.
\end{aligned}
\tag{9.4-6}
$$

The result is plotted in Figure 9.4-5. The effect of smearing in producing a fuzzy image is clear.

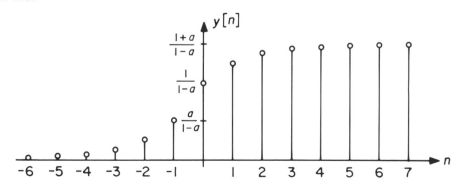

Figure 9.4-5. Image of the step input.

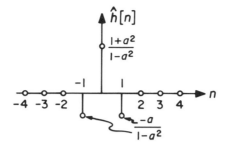

Figure 9.4–6. Sample response $\hat{h}[n]$ of a deconvolving filter.

All is not lost, however. It is, in fact, possible to recover the original sharp image through appropriate processing of the received samples. Specifically, consider the consequences of convolving $y[n]$ with a unit sample response

$$\hat{h}[n] = \frac{a}{1-a^2}\left(-\delta[n+1] + \frac{1+a^2}{a}\delta[n] - \delta[n-1]\right) \tag{9.4–7}$$

as shown in Figure 9.4–6. It is worth noticing as a general rule that convolving with a delayed or advanced unit sample function simply delays or advances the input sequence accordingly:

$$x[n] * \delta[n-N] = \sum_{m=-\infty}^{\infty} x[m]\delta[n-m-N] = x[n-N]. \tag{9.4–8}$$

Hence, for the output (9.4–6) corresponding to a step input,

$$z[n] = y[n] * \hat{h}[n] = -\frac{a}{1-a^2}y[n+1] + \frac{1+a^2}{1-a^2}y[n] - \frac{a}{1-a^2}y[n-1]. \tag{9.4–9}$$

For $n \le -1$, this becomes

$$z[n] = -\frac{a}{1-a^2}\frac{a^{-(n+1)}}{1-a} + \frac{1+a^2}{1-a^2}\frac{a^{-n}}{1-a} - \frac{a}{1-a^2}\frac{a^{-(n-1)}}{1-a} \tag{9.4–10}$$

$$= 0, \quad n \le -1.$$

For $n = 0$,

$$z[0] = -\frac{a}{1-a^2}y[1] + \frac{1+a^2}{1-a^2}y[0] - \frac{a}{1-a^2}y[-1]$$

$$= -\frac{a}{1-a^2}\frac{1+a-a^2}{1-a} + \frac{1+a^2}{1-a^2}\frac{1}{1-a} - \frac{a}{1-a^2}\frac{a}{1-a} \tag{9.4–11}$$

$$= 1.$$

For $n \geq 1$,

$$z[n] = \frac{-a}{1-a^2} \frac{1+a-a^{(n+2)}}{1-a} + \frac{1+a^2}{1-a^2} \frac{1+a-a^{(n+1)}}{1-a} - \frac{a}{1-a^2} \frac{1+a-a^n}{1-a}$$

$$\text{(9.4–12)}$$

$$= 1, \quad n \geq 1.$$

Thus $z[n] = u[n]$. You should have no trouble using the ideas of this chapter to show that in general for an arbitrary $x[n]$, the result of convolving $y[n]$ and $\hat{h}[n]$ will be to yield $z[n] = x[n]$. (HINT: Argue that $h[n] * \hat{h}[n] = \delta[n]$.) Hence, $\hat{h}[n]$ is an inverse system of the type discussed in Section 6.1 (except, of course, $\hat{h}[n]$ and $h[n]$ are non-causal); it undoes or *deconvolves* the operation $x[n] * h[n]$ to recover $x[n]$. Problem 9.5 explains how to devise $\hat{h}[n]$ in general when $h[n]$ is causal. Later, when we study discrete-time Fourier transforms, we shall learn how to design such systems for non-causal $h[n]$.

▶ ▶ ▶

9.5 Summary

A necessary and sufficient condition for an LTI discrete-time operation is that its output $y[n]$ be given by the convolution of its input $x[n]$ and the response $h[n]$ to the unit sample function $\delta[n]$:

$$y[n] = \sum_{m=-\infty}^{\infty} x[m]h[n-m] = x[n] * h[n] . \qquad \text{(9.5–1)}$$

The convolution operation is generally commutative, associative, and distributive.

Although the infinite sum in the convolution formula obviously need not exist for arbitrary sequences $x[n]$ and $h[n]$, two special cases cover most of the important applications:

i) $h[n]$ is *causal*, that is,

$$h[n] \equiv 0, \quad n < 0 \qquad \text{(9.5–2)}$$

and $x[n]$ also vanishes for negative n.

ii) $h[n]$ is *BIBO stable*, that is,

$$\sum_{n=-\infty}^{\infty} |h[n]| < \infty \qquad \text{(9.5–3)}$$

and $|x[n]|$ is bounded.

The first case leads to the special formula

$$y[n] = \sum_{m=0}^{n} x[m]h[n-m] \tag{9.5-4}$$

and includes the ZSR response of the LTI systems dealt with previously, described by difference equations, delay-(or accumulator-)adder-gain block diagrams, and Z-transform system functions. In particular, for causal $h[n]$, convolution in the time domain corresponds to multiplication in the frequency domain,

$$\sum_{m=0}^{n} x[m]h[n-m] \Longleftrightarrow \tilde{X}(z)\tilde{H}(z) \tag{9.5-5}$$

and the unit sample response and system functions are a unilateral Z-transform pair:

$$h[n] \Longleftrightarrow \tilde{H}(z). \tag{9.5-6}$$

Frequency-domain methods for handling the second (stable but non-causal) case will be developed in later chapters.

EXERCISES FOR CHAPTER 9

Exercise 9.1

a) For each of the systems described below, find the Z-transform $\tilde{Y}(z)$ of the ZSR output $y[n]$ in terms of the Z-transform $\tilde{X}(z)$ of the input $x[n]$:

    i) $y[n+1] = x[n+1] - x[n]$.

    ii) $y[n] = \sum_{k=0}^{n} x[k]$.

    iii) $y[n] = n\,x[n]$.

    iv) $y[n+1] = 2\,x[n]$.

b) Which of the above systems are linear? Which are shift-invariant? For those systems that are both linear and shift-invariant, find the unit sample response $h[n]$.

Answers: All four systems are linear, but only (i) and (iv) are also shift-invariant, with $h[n] = \delta[n] - \delta[n-1]$ and $h[n] = 2\delta[n-1]$ respectively.

Exercise 9.2

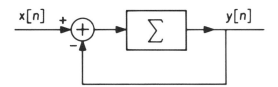

Consider the discrete recursive (feedback) system shown above. The accumulator is of the forward Euler type and thus has unit sample response $u[n-1]$ as discussed in Example 9.1–1.

a) Find the unit sample response of the overall system, $h[n]$, by letting $x[n] = \delta[n]$ and exploring directly (in time) the way in which impulses "circulate around the loop" and superimpose.

b) Find the system function $\tilde{H}(z)$ of the overall system from the feedback formula

$$\tilde{H}(z) = \frac{\tilde{\Sigma}(z)}{1 + \tilde{\Sigma}(z)}$$

(where $\tilde{\Sigma}(z)$ is the system function of the accumulator) and check your answer to (a).

c) Describe an alternate but equivalent realization of this system in terms of adder-gain-delay elements as discussed in Section 7.2.

PROBLEMS FOR CHAPTER 9

Problem 9.1

For each of the following pairs of discrete-time functions, find

$$y[n] = x[n] * h[n] = \sum_{m=-\infty}^{\infty} x[m]h[n-m].$$

Sketch $x[n]$, $h[n]$, and $y[n]$.

a)

$h[n] = (1/2)u[n]$.

b) $x[n] = 2(u[n+1] - u[n-3])$

$$h[n] = \begin{cases} -1, & n = -1 \\ 1, & n = +1 \\ 0, & \text{elsewhere.} \end{cases}$$

c)

$h[n] = (1/2)^n u[n]$.

d) $x[n] = a^n u[n], \quad |a| < 1$

$h[n] = \sin \frac{2\pi n}{8}$.

e) $x[n] = (-b)^n u[-n-1], \quad |b| > 1$

$h[n] = a^n u[n+2], \quad |a| < 1$.

Problem 9.2

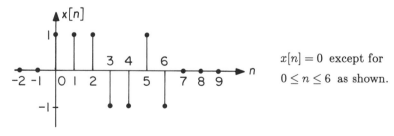

$x[n] = 0$ except for $0 \le n \le 6$ as shown.

a) Let $h[n] = x[6-n]$. Sketch $h[n]$.

b) Find $y[n] = x[n] * h[n] = \sum_{m=0}^{n} x[m]h[n-m]$ directly for the $h[n]$ given in (a).

c) Find $y[n] = x[n] * h[n]$ for the $h[n]$ given in (a) by using the fact that $\tilde{Y}(z) = \tilde{X}(z)\tilde{H}(z)$.

Problem 9.3

a) Using Z-transforms, find the causal unit sample response for the system described by the difference equation

$$6y[n+2] - 5y[n+1] + y[n] = x[n+2] + x[n+1].$$

b) Show by direct substitution that

$$y[n] = -\frac{3}{2}\left(\frac{1}{2}\right)^n u[-n-1] + \frac{4}{3}\left(\frac{1}{3}\right)^n u[-n-1]$$

also satisfies this difference equation when $x[n] = \delta[n]$. This $y[n]$ is sketched below. A difference equation thus does not define a unique LTI system. (However, there is only one causal LTI system corresponding to each difference equation.)

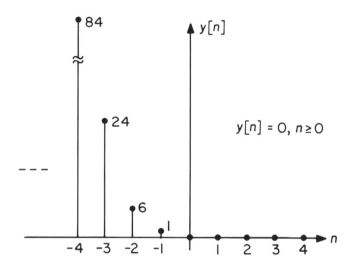

Problem 9.4

Consider the cascade of two discrete-time LTI systems shown below.

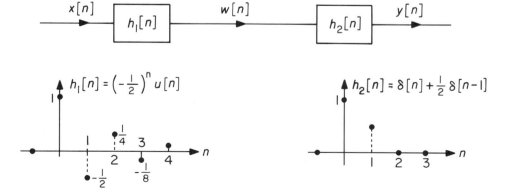

a) Let $x[n] = u[n]$. Compute successively

    i) $w[n] = h_1[n] * x[n]$

    ii) $y[n] = h_2[n] * w[n] = h_2[n] * (h_1[n] * x[n])$.

b) Again for $x[n] = u[n]$, compute successively

    i) $g[n] = h_2[n] * h_1[n]$

    ii) $y[n] = g[n] * x[n] = (h_2[n] * h_1[n]) * x[n]$.

Your results for $y[n]$ should be the same in (a) and (b), illustrating the associative property of discrete convolution:

$$h_2[n] * (h_1[n] * x[n]) = (h_2[n] * h_1[n]) * x[n].$$

The associative property permits the definition of an overall equivalent unit sample response (in this case $g[n]$) for the cascade of two discrete-time LTI systems.

Problem 9.5

In many applications, as illustrated in Example 9.4–1, it may be desirable to recover the *input* sequence given the *output* sequence of a discrete-time system with known unit sample response $h[n]$. One way to do this is to construct an inverse system with unit sample response $\hat{h}[n]$ such that the cascade of the original system and the inverse system yields an overall system with unit sample response equal to the unit sample function $\delta[n]$, that is, such that

$$h[n] * \hat{h}[n] = \delta[n].$$

If both $h[n]$ and $\hat{h}[n]$ are causal, then this formula implies that the corresponding system functions are reciprocals of one another.

a) Suppose that $\tilde{H}(z)$, the system function corresponding to $h[n]$, is

$$\tilde{H}(z) = \frac{z - 1/2}{z - 1}.$$

Find the causal unit sample response $\hat{h}[n]$ for the system inverse to this system and show directly by convolution that $h[n] * \hat{h}[n] = \delta[n]$.

If $h[n]$ is causal and $|\tilde{H}(z)| \to 0$ as $|z| \to \infty$ (as will often be the case), then it will not be possible to find a causal $\hat{h}[n]$ by the reciprocal-system-function procedure since the domain of convergence of $1/\tilde{H}(z)$ cannot be the *outside* of some circle. However, if there exists an integer N such that $z^N \tilde{H}(z)$ is finite and nonzero at ∞, then the reciprocal of $z^N \tilde{H}(z)$ will describe a causal system whose unit sample response can be interpreted as $\hat{h}(n - N)$.

b) Use this procedure to find the unit sample response of a noncausal system inverse to the simple accumulator with system function

$$\tilde{H}(z) = \frac{1}{z - 1}$$

and show directly by convolution that $h[n] * \hat{h}[n] = \delta[n]$.

Problem 9.6

In addition to the effects mentioned in Example 9.4–1, relative motion between the camera and the object being photographed will also produce a smeared image. This problem discusses a simple digital scheme for reconstructing the true image from the smeared image. Let $f[x,y]$ represent the intensity of the sampled 2-dimensional true image, where x and y are integer-valued discrete (space) coordinates describing the location of each picture element or "pixel." For simplicity, assume that the film image intensity is linear in exposure, so that the smeared image can be represented by

$$g[x,y] = \sum_{\xi=0}^{\infty} f[x - \xi, y]h[\xi]$$

where we have assumed that the motion is in the x-direction. To be specific, assume that

$$h[n] = \begin{cases} 1, & n = 0, 1, 2 \\ 0, & \text{elsewhere.} \end{cases}$$

a) Explain with particular examples and appropriate sketches how these formulas describe the effects of motion smearing.

b) Using Z-transforms, determine and sketch a unit sample response $\hat{h}[n]$ such that

$$f[x,y] = \sum_{\xi=0}^{\infty} g[x - \xi, y]\hat{h}[\xi].$$

If the picture is processed by convolving with this $\hat{h}[n]$, the original unsmeared image can be reconstructed.

Problem 9.7

A classical technique for "smoothing" a discrete-time signal $x[n]$ is to fit a polynomial of degree N,

$$p[n] = p_0 + p_1 + p_2 n^2 + \cdots + p_N n^N$$

to a sequence of $M > N$ successive values of $x[n]$ in such a way as to minimize the sum of the squares of the errors $(x[n] - p[n])$ over the M values being fit. The value of the "smoothed" signal $y[n]$ at the time corresponding to the middle of the sequence of length M is set equal to the value of the polynomial at that point. The whole process is then repeated for the next sequence of M values of $x[n]$, of which $M - 1$ values overlap with the preceding sequence. The smoothed signal $y[n]$ presumably preserves "trends" in $x[n]$, while reducing "noise" that causes $x[n]$ to deviate erratically from the trend. The whole scheme is illustrated on the next page for $N = 2$ and $M = 5$. The solid lines in the top three figures are least-square quadratic fits to the five points spanned.

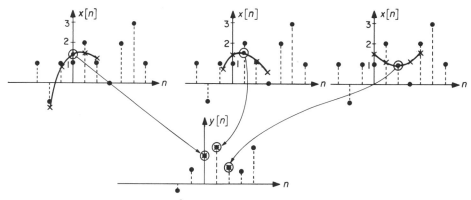

a) Consider a polynomial of 2nd degree,

$$p[n] = p_0 + p_1 n + p_2 n^2$$

fit to the five values $x[-2]$, $x[-1]$, $x[0]$, $x[1]$, $x[2]$. Show that the values of p_0, p_1, and p_2 that minimize the squared error

$$\sum_{n=-2}^{2} (x[n] - p[n])^2$$

are

$$p_0 = \frac{-3x[-2] + 12x[-1] + 17x[0] + 12x[1] - 3x[2]}{35}$$

$$p_1 = \frac{-2x[-2] - x[-1] + x[1] + 2x[2]}{10}$$

$$p_2 = \frac{2x[-2] - x[-1] - 2x[0] - x[1] + 2x[2]}{14}.$$

Sketch $p[n]$ for the values of $x[n]$ in the figure above and show that it agrees with the solid curve in the left part of the figure.

b) The smoothed value $y[0]$ is $p[0] = p_0$. In general, then, it should be evident that

$$y[n] = \frac{1}{35}\Big(-3x[n-2] + 12x[n-1] + 17x[n] + 12x[n+1] - 3x[n+2]\Big).$$

Show that this formula yields the smoothed results in the preceding figure for the $x[n]$ specified there.

c) What is the unit sample response $h[n]$ of the LTI system described by the difference equation in (b)? Is $h[n]$ causal? Is $h[n]$ FIR or IIR? Draw a block diagram using delay elements, gains, and adders that will realize $h[n]$, except possibly for some overall delay.

10

CONVOLUTIONAL REPRESENTATIONS
OF CONTINUOUS-TIME SYSTEMS

10.0 Introduction

The discrete-time convolution operation

$$y[n] = \sum_{m=-\infty}^{\infty} x[m]h[n-m]$$

describes a class of LTI systems that is larger than those specified by linear constant-coefficient difference equations but (as we shall see in later chapters) shares many of the analytical simplicities and structure that make such difference equations so useful. The convolution operation also provides an explicit formula for computing in the time domain the response of LTI systems, complementing the explicit frequency-domain formula

$$\tilde{Y}(z) = \tilde{X}(z)\tilde{H}(z)$$

for the ZSR response of causal systems.

Similar statements hold for continuous-time systems. In this chapter we shall show that the *convolution integral*

$$y(t) = \int_0^t x(\tau)h(t-\tau)\,d\tau$$

is the time-domain equivalent of the $\mathcal{L}$-transform formula

$$Y(s) = X(s)H(s)$$

for LTI circuits or analogous causal systems characterized by linear constant-coefficient differential equations. As in the discrete-time case, $h(t)$ is the inverse $\mathcal{L}$-transform of the system function $H(s)$. We call $h(t)$ the *unit impulse response*, but just what a "unit impulse" may be that $h(t)$ should be its "response" is unfortunately not as straightforward in continuous-time systems as it was in the preceding chapter. We shall accordingly postpone until Chapter 11 consideration of what impulses and their relatives "are." In this chapter we shall concentrate on the most important properties of the convolution integral formula, both in the restricted form above, in which it describes the ZSR response of a lumped causal system, and in the more general form

$$y(t) = \int_{-\infty}^{\infty} x(\tau)h(t-\tau)\,d\tau$$

that (as we shall argue primarily in the following chapter) characterizes the response of any LTI continuous-time system, causal or not.

10.1 The L-Transform Convolution Theorem

We begin by proving a theorem.

$\qquad$ L-TRANSFORM CONVOLUTION THEOREM:
$\qquad\qquad$ Let

$$y(t) = \int_0^t x(\tau)h(t-\tau)\,d\tau . \tag{10.1--1}$$

$\qquad\qquad$ Then the L-transform of $y(t)$ is

$$Y(s) = X(s)H(s)$$

$\qquad\qquad$ where $H(s) = L[h(t)]$ and $X(s) = L[x(t)]$.

The proof parallels that of the corresponding Z-transform theorem in Section 9.1. Thus direct evaluation of $Y(s)$ gives

$$\begin{aligned}
Y(s) &= \int_0^{\infty} y(t)e^{-st}\,dt = \int_0^{\infty}\left[\int_0^t x(\tau)h(t-\tau)\,d\tau\right]e^{-st}\,dt \\
&= \int_0^{\infty}\left[\int_0^{\infty} x(\tau)h(t-\tau)u(t-\tau)\,d\tau\right]e^{-st}\,dt .
\end{aligned} \tag{10.1--2}$$

Interchanging the orders of integration* gives

$$Y(s) = \int_0^{\infty} x(\tau)\left[\int_0^{\infty} h(t-\tau)u(t-\tau)e^{-st}\,dt\right]d\tau . \tag{10.1--3}$$

Since only $\tau > 0$ is involved in the outside integral, $h(t-\tau)u(t-\tau)$ is simply $h(t)u(t)$ delayed, and the inside integral becomes $H(s)e^{-s\tau}$. Carrying out the outside integral then finishes the proof.

$\qquad$ The Convolution Theorem is a relationship that applies to any waveforms and their transforms. If, in particular, $x(t)$, $y(t)$, and $h(t)$ are the input, output, and impulse response of a circuit or lumped system, then the fact that we have previously shown that

$$Y(s) = X(s)H(s)$$

for the ZSR implies that we can also find the ZSR in the time domain from the convolution formula (10.1–1). Several examples will help clarify how this works.

---

*As in the analogous discrete-time argument, it is easy to justify the order interchange inside a common domain of absolute convergence of $H(s)$ and $X(s)$.

Example 10.1–1

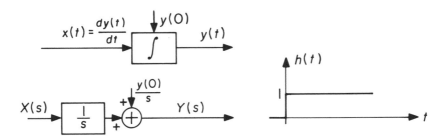

Figure 10.1–1. The integrator block.   **Figure 10.1–2.** Impulse response.

In Problem 2.1, it was suggested that an integrator had the time-domain and frequency-domain block-diagram representations shown in Figure 10.1–1. The system function of the integrator is thus

$$H(s) = \frac{1}{s} \tag{10.1-4}$$

and hence

$$h(t) = u(t) \tag{10.1-5}$$

as shown in Figure 10.1–2.*   The Convolution Theorem then implies that the ZSR component of the response should be

$$y(t) = \int_0^t x(\tau)h(t-\tau)\,d\tau$$
$$= \int_0^t x(\tau)u(t-\tau)\,d\tau = \int_0^t x(\tau)\,d\tau . \tag{10.1-6}$$

The total response thus becomes

$$y(t) = \int_0^t x(\tau)\,d\tau + y(0) \tag{10.1-7}$$

exactly as would be expected for something called an "integrator."

▶ ▶ ▶

---

*Strictly speaking, the fact that $h(t)$ is the inverse $\mathcal{L}$-transform of $H(s) = 1/s$ tells us only that $h(t) = 1$, $t > 0$. This is, in fact, all we need in order to evaluate the convolution integral with integration limits as defined in the $\mathcal{L}$-Transform Convolution Theorem. But in the larger context of general LTI systems shortly to be introduced, an integrator is a causal system, which implies $h(t) = 0$, $t < 0$.

Example 10.1–2

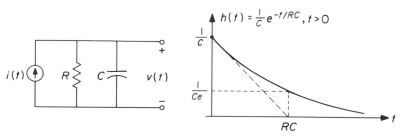

Figure 10.1–3. Circuit and impulse response for Example 10.1–2.

The RC circuit shown in Figure 10.1–3 has the system function

$$H(s) = \frac{V(s)}{I(s)} \, (\text{ZSR}) = \frac{1/C}{s + (1/RC)} \, . \tag{10.1–8}$$

Hence*

$$h(t) = \frac{1}{C} e^{-t/RC}, \quad t > 0 . \tag{10.1–9}$$

Let's now use the convolution formula to find the ZSR response of the circuit of Figure 10.1–3 to the input of Figure 10.1–4.

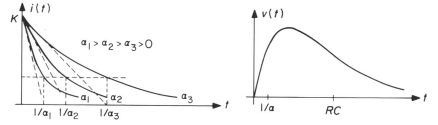

Figure 10.1–4. Input and output of the circuit of Figure 10.1–3.

From (10.1–1) we write

$$v(t) = \int_0^t i(\tau) h(t - \tau) \, d\tau$$

$$= \int_0^t (K e^{-\alpha \tau}) \left(\frac{1}{C} e^{-(t-\tau)/RC} \right) d\tau . \tag{10.1–10}$$

---

*It is worth pausing a moment to consider the units of $h(t)$ in this example, which are evidently $(\text{farads})^{-1}$ = volts/ampere-seconds. If $h(t)$ is to be considered a unit impulse response and if the response is a voltage, then an input "unit impulse" of current must have units of ampere-seconds (rather than amperes), so that

$$\left(\frac{\text{output}}{\text{input}} \right) \times \text{input} = \text{output}$$

or

$$\left(\frac{\text{volts}}{\text{ampere-seconds}} \right) \times (\text{ampere-seconds}) = \text{volts}.$$

This curious conclusion, obviously necessary if the convolution formula is to balance dimensionally, will be clarified in Chapter 11.

Factoring out $(K/C)e^{-t/RC}$ (which is not a function of the integration variable τ) and combining the remaining exponentials in τ gives

$$v(t) = \frac{K}{C} e^{-t/RC} \int_0^t e^{-\left(\alpha - \frac{1}{RC}\right)\tau}\, d\tau$$

$$= \frac{K}{C} e^{-t/RC} \left[\frac{1}{-\left(\alpha - \frac{1}{RC}\right)} e^{-\left(\alpha - \frac{1}{RC}\right)\tau} \right] \Bigg|_0^t$$

$$= \frac{K}{C\left(\alpha - \frac{1}{RC}\right)} \left[e^{-t/RC} - e^{-\alpha t} \right], \quad t > 0. \tag{10.1-11}$$

This answer is sketched in Figure 10.1–4 and is readily checked by transforms. As a special case, we let $\alpha \to 0$ and obtain the step response

$$v(t) = KR\left(1 - e^{-t/RC}\right), \quad t > 0$$

which can be checked by elementary means.

It is illuminating to look at this problem graphically, sketching the integrand of (10.1–10) vs. τ for several values of t and interpreting the integral as the area. (See Figure 10.1–5.) Note that $h(t - \tau)$ plotted against τ is $h(t)$ turned around—plotted *backwards*—with the origin moved to the point where $t - \tau = 0$, or $\tau = t$.

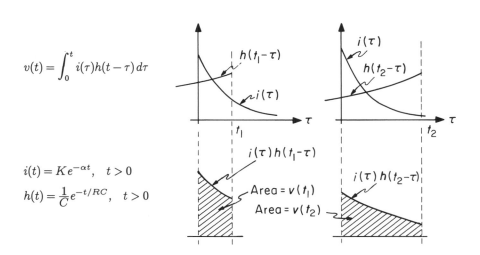

$$v(t) = \int_0^t i(\tau) h(t - \tau)\, d\tau$$

$$i(t) = K e^{-\alpha t}, \quad t > 0$$

$$h(t) = \frac{1}{C} e^{-t/RC}, \quad t > 0$$

Figure 10.1–5. Graphical interpretation of (10.1–10).

▶ ▶ ▶

10.2 Convolution and General LTI Systems

Following precisely the argument in Section 9.2, the formula

$$y(t) = \int_0^t x(\tau)h(t-\tau)\, d\tau \qquad (10.2\text{–}1)$$

may be considered the special case of the general convolution formula

$$y(t) = \int_{-\infty}^{\infty} x(\tau)h(t-\tau)\, d\tau \qquad (10.2\text{–}2)$$

that results when $x(t) \equiv 0$, $t < 0$ (so that the system is in the zero state at $t = 0$) and when $h(t) \equiv 0$, $t < 0$ (so that the system is causal). The general formula (10.2–2) still satisfies the continuous-time versions of the linearity (superposition) and time-invariance conditions of Section 9.2, and indeed is the most general functional system description satisfying these conditions. But, unlike the discrete-time case, showing that (10.2–2) is the most general possible description of an LTI system is far from straightforward; we defer discussion to Chapter 11.

We can extend the asterisk notation, introduced in Section 9.3 to represent discrete-time convolution, to describe general continuous-time convolution as well:

$$y(t) = \int_{-\infty}^{\infty} x(\tau)h(t-\tau)\, d\tau = x(t) * h(t)\,. \qquad (10.2\text{–}3)$$

Assuming that integrals such as (10.2–2) are well enough defined to permit interchanges of integration order, it is easy to show that continuous-time convolution satisfies commutative, associative, and distributive laws.

Example 10.2–1

The commutative law states that

$$x(t) * h(t) = h(t) * x(t) \qquad (10.2\text{–}4)$$

or

$$\int_{-\infty}^{\infty} x(\tau)h(t-\tau)\, d\tau = \int_{-\infty}^{\infty} h(\tau)x(t-\tau)\, d\tau\,. \qquad (10.2\text{–}5)$$

In system terms, this means for example that the output is the same whether we put an input $x(t)$ into a system with impulse response $h(t)$, or put the input $h(t)$ into a system with impulse response $x(t)$.

Commutativity follows for the ZSR responses of causal systems from the corresponding property of the product of L-transforms,

$$X(s)H(s) = H(s)X(s)\,. \qquad (10.2\text{–}6)$$

It can easily be proved in general by a direct change of variable. Thus, if we let $\mu = t-\tau$, $d\mu = -d\tau$, the left-hand side of (10.2–5) becomes

$$\int_{-\infty}^{\infty} x(\tau)h(t-\tau)\, d\tau = -\int_{\infty}^{-\infty} x(t-\mu)h(\mu)\, d\mu = \int_{-\infty}^{\infty} h(\tau)x(t-\tau)\, d\tau \qquad (10.2\text{–}7)$$

where in the second equality we have used the fact that μ as a local variable of integration can be replaced by any other symbol, for example, τ.

Often one order or the other is more convenient, particularly if some further manipulation of the convolution integral is planned. Thus for the circuit of Example 10.1–2 either of the two expressions

$$v(t) = \int_0^t \frac{1}{C} e^{-\tau/RC} i(t-\tau) \, d\tau = \int_0^t i(\tau) \frac{1}{C} e^{-(t-\tau)/RC} \, d\tau \qquad (10.2\text{–}8)$$

describes the ZSR for an arbitrary input $i(t)$, $t > 0$. Thus either should provide a general ZSR to the differential equation describing the circuit:

$$\frac{v(t)}{R} + C\frac{dv(t)}{dt} = i(t). \qquad (10.2\text{–}9)$$

It is somewhat easier to show this for the second expression, though, since in the process of differentiating the first expression derivatives of $i(t)$ would appear that would have to be cleared by integrating by parts. For the second expression, we have simply*

$$\frac{dv(t)}{dt} = i(t)\frac{1}{C} e^{-(t-t)/RC} + \int_0^t i(\tau)\left(-\frac{1}{RC^2}\right) e^{-(t-\tau)/RC} \, d\tau \qquad (10.2\text{–}10)$$

or

$$\frac{dv(t)}{dt} = \frac{i(t)}{C} - \frac{v(t)}{RC} \qquad (10.2\text{–}11)$$

which establishes that the convolution formula is in fact a solution to the differential equation. (In formal discussions of the solution of differential equations, the convolution solution is usually obtained through what is called the *method of integrating factors*.)

▶ ▶ ▶

Example 10.2–2

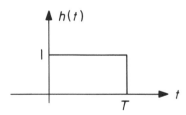

Figure 10.2–1. Impulse response of a finite-time integrator.

Convolution is a particularly effective tool for studying the operation of systems whose impulse responses can be represented, approximately at least, by a sum of delayed pulses of simple geometric shapes. A *finite-time integrator*, for example, is characterized by the impulse response shown in Figure 10.2–1. Such a system cannot be described by an ordinary total differential equation and hence cannot be realized exactly with a finite RLC circuit (although a close approximation could be achieved with a sufficiently large

---

*Recall Leibniz's rule for the derivative of a definite integral with respect to a parameter:

$$\frac{d}{dy}\left\{\int_{a(y)}^{b(y)} f(x,y)\, dx\right\} = f(b,y)\frac{db}{dy} - f(a,y)\frac{da}{dy} + \int_{a(y)}^{b(y)} \frac{\partial f(x,y)}{\partial y}\, dx.$$

number of elements). A simpler realization scheme might employ a distributed element such as a *delay line* (we shall have more to say about delay lines in Chapter 11) in a configuration such as that shown in Figure 10.2–2.

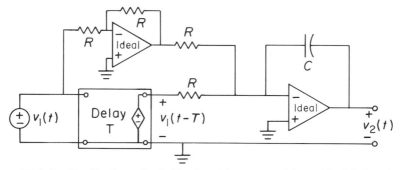

Figure 10.2–2. Realization of a finite-time integrator with an ideal delay element.

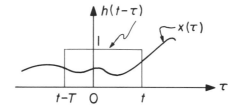

Figure 10.2–3. $h(t - \tau)$ and $x(\tau)$ for a finite-time integrator.

For the purposes of this example we may assume that the impulse response $h(t)$ of Figure 10.2–1 is simply the measured impulse response of some "black box" (presumed to be LTI). For an arbitrary value of t, the plot of $h(t - \tau)$ vs. τ appears as in Figure 10.2–3. Hence the convolution integral for this case can be written in the form

$$y(t) = \int_{-\infty}^{\infty} x(\tau)h(t - \tau)\, d\tau = \int_{t-T}^{t} x(\tau)\, d\tau \,.$$

That is, at any instant t, the output $y(t)$ is the integral of the input over the last T seconds—which explains the name "finite-time integrator."

Suppose now that we seek to evaluate $y(t)$ by convolution when the input is a long pulse,

$$x(t) = \begin{cases} A, & 0 < t < 3T \\ 0, & \text{elsewhere} \end{cases} = A[u(t) - u(t - 3T)] \,.$$

The input $x(\tau)$ and the folded impulse response $h(t - \tau)$ are sketched in Figure 10.2–4 for five values of t corresponding to the five different kinds of overlap that can occur between $x(\tau)$ and $h(t - \tau)$.

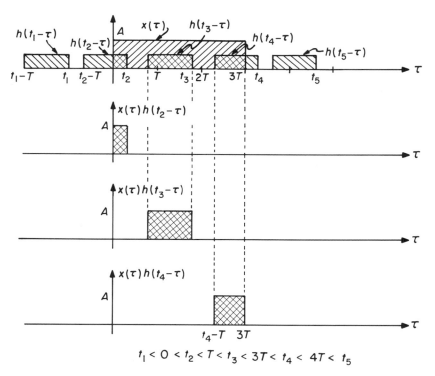

Figure 10.2–4. Graphical evaluation of $x(t) * h(t)$.

We consider each of these cases in turn and evaluate

$$y(t) = \int_{-\infty}^{\infty} x(\tau)h(t-\tau)\,d\tau .$$

a) $t < 0$:

For this range of t (for example, $t = t_1$), there is no overlap between $x(\tau)$ and $h(t-\tau)$, so

$$y(t) = 0, \quad t < 0.$$

b) $0 < t < T$:

For $t = t_2$, the overlapping of $x(\tau)$ and $h(t_2 - \tau)$ produces a pulse of height A in the product, as shown in the second part of Figure 10.2–4; $y(t_2)$, the integral of the product, is simply the area of the pulse, that is, At_2. Thus, for any t in this region,

$$y(t) = At, \quad 0 < t < T.$$

c) $T < t < 3T$:

In this region (for example, $t = t_3$), the time interval in which $h(t-\tau)$ is nonzero lies entirely inside the interval in which $x(\tau)$ is nonzero. The product waveform (third part of Figure 10.2–4) is thus a pulse of height A and duration T independent of t, and $y(t)$, the area of this pulse, is a constant,

$$y(t) = AT, \quad T < t < 3T.$$

d) $3T < t < 4T$:

For $t = t_4$, $h(t_4 - \tau)$ only partially overlaps $x(\tau)$, and the area of the product pulse (fourth part of Figure 10.2–4) is $A[3T - (t_4 - T)] = A(4T - t_4)$, or in general

$$y(t) = A(4T - t), \quad 3T < t < 4T.$$

e) $t > 4T$:

There is no overlap between $h(t_5 - \tau)$ and $x(\tau)$; thus

$$y(t) = 0, \quad t > 4T.$$

Collecting results, we can sketch $y(t)$ as in Figure 10.2–5.

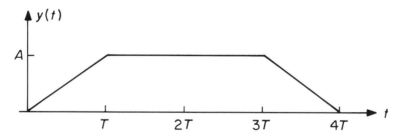

Figure 10.2–5. Result of $x(t) * h(t)$.

▶ ▶ ▶

Note in this example that the convolution of two pulses yields a pulse whose duration is the sum of the durations of the component pulses. This result is generally true even for "pulses" that do not have a simple rectangular shape, and it is at least approximately true even for waveforms that are not strictly pulses, that is, not precisely zero outside some interval, as we shall discuss in Chapter 16. Note also that all of the integrals in the example were evaluated as areas of simple geometrical shapes. This technique is largely restricted to waveforms composed of straight lines; since complicated waveforms can frequently be approximated by sequences of straight lines, however, this can be a useful method for obtaining approximate answers.

Example 10.2–3

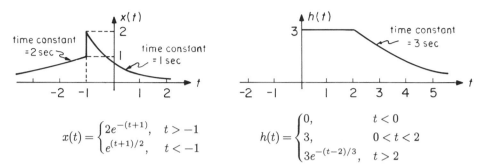

$$x(t) = \begin{cases} 2e^{-(t+1)}, & t > -1 \\ e^{(t+1)/2}, & t < -1 \end{cases} \qquad h(t) = \begin{cases} 0, & t < 0 \\ 3, & 0 < t < 2 \\ 3e^{-(t-2)/3}, & t > 2 \end{cases}$$

Figure 10.2–6. $x(t)$ and $h(t)$ for Example 10.2–3.

The evaluation of convolution integrals requires great care if $x(t)$ and/or $h(t)$ are specified by different formulas over different pieces of their ranges—as, for example, in Figure 10.2–6. Of course, formulas valid for all t can be devised for $x(t)$ and $h(t)$ in Figure 10.2–6 by using unit step functions; for example,

$$x(t) = 2e^{-(t+1)}u(t+1) + e^{(t+1)/2}u(-t-1)$$
$$h(t) = 3[u(t) - u(t-2)] + 3e^{-(t-2)/3}u(t-2).$$

Hence $y(t) = x(t) * h(t)$ can be formally written as

$$\begin{aligned} y(t) &= \int_{-\infty}^{\infty} x(\tau)h(t-\tau)\,d\tau \\ &= \int_{-\infty}^{\infty} \left[2e^{-(\tau+1)}u(\tau+1) + e^{(\tau+1)/2}u(-\tau-1) \right] \\ &\qquad \times \left[3[u(t-\tau) - u(t-\tau-2)] + 3e^{-(t-\tau-2)/3}u(t-\tau-2) \right] d\tau \end{aligned}$$

but this is a clumsy expression to evaluate.

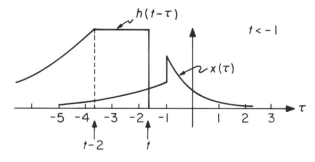

Figure 10.2–7. $x(\tau)$ and $h(t-\tau)$ for $t < -1$.

The simplest approach is usually to start by drawing diagrams similar to those in the preceding example for typical values of t. Thus, for $t < -1$, $x(\tau)$ and $h(t-\tau)$ are as shown in Figure 10.2–7. The convolution integral for $t < -1$ is

$$y(t) = \int_{-\infty}^{t-2} \underset{\substack{\Uparrow \\ \text{formula for} \\ x(\tau),\, \tau < t-2}}{e^{(\tau+1)/2}} \quad \underset{\substack{\Uparrow \\ \text{formula for} \\ h(t-\tau),\, \tau < t-2}}{3e^{-(t-\tau-2)/3}} \quad d\tau$$

$$+ \int_{t-2}^{t} \underset{\substack{\Uparrow \\ \text{formula for} \\ x(\tau),\, t-2 < \tau < t}}{e^{(\tau+1)/2}} \quad \underset{\substack{\Uparrow \\ \text{formula for} \\ h(t-\tau),\, t-2 < \tau < t}}{3} \quad d\tau$$

$$= 3e^{-t/3} e^{7/6} \int_{-\infty}^{t-2} e^{\tau(5/6)} \, d\tau$$

$$+ 3e^{1/2} \int_{t-2}^{t} e^{\tau/2} \, d\tau$$

$$= 3e^{-t/3} e^{7/6} \frac{6}{5} e^{(t-2)(5/6)}$$

$$+ 3e^{1/2} 2\left[e^{t/2} - e^{(t-2)/2}\right]$$

$$= 6e^{(t+1)/2} - 2.4e^{(t-1)/2}, \quad t < -1.$$

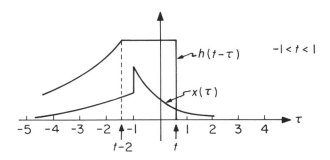

Figure 10.2–8. $x(\tau)$ and $h(t-\tau)$ for $-1 < t < 1$.

Similarly, for $-1 < t < 1$, $x(\tau)$ and $h(t-\tau)$ are as shown in Figure 10.2–8. The convolution integral is

$$y(t) = \int_{-\infty}^{t-2} e^{(\tau+1)/2} \, 3 e^{-(t-\tau-2)/3} \, d\tau + \int_{t-2}^{-1} e^{(\tau+1)/2} \, 3 \, d\tau$$

$$+ \int_{-1}^{t} 2e^{-(\tau+1)} \, 3 \, d\tau$$

$$= 12 - 2.4e^{(t-1)/2} - 6e^{-(t+1)}, \quad -1 < t < 1.$$

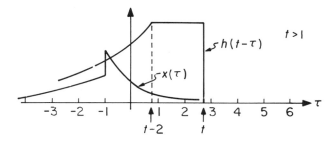

Figure 10.2–9. $x(\tau)$ and $h(t-\tau)$ for $t > 1$.

Finally, for $t > 1$, $x(\tau)$ and $h(t-\tau)$ are as shown in Figure 10.2–9, and

$$y(t) = \int_{-\infty}^{-1} e^{(\tau+1)/2}\, 3\, e^{-(t-\tau-2)/3}\, d\tau + \int_{-1}^{t-2} 2e^{-(\tau+1)}\, 3\, e^{-(t-\tau-2)/3}\, d\tau$$

$$+ \int_{t-2}^{t} 2e^{-(\tau+1)}\, 3\, d\tau$$

$$= 12.6e^{-(t-1)/3} - 3e^{-(t-1)} - 6e^{-(t+1)}, \quad t > 1.$$

Combining, we can sketch $y(t)$ as in Figure 10.2–10.

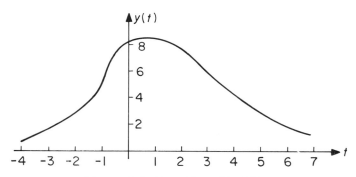

Figure 10.2–10. $y(t) = x(t) * h(t)$.

▶ ▶ ▶

10.3 Causality and Stability

To ensure the existence of the convolution integral (10.2–2) with infinite limits, either $x(t)$ or $h(-t)$ (or both) must vanish sufficiently rapidly as t tends to both $+\infty$ and $-\infty$. As we pointed out in connection with the analogous discrete-time problem, there are several important ways in which this can be guaranteed. Specifically:

1. We can constrain both $h(t) \equiv 0$ and $x(t) \equiv 0$ for $t < 0$.

2. We can constrain $h(t)$ in such a way that $\int_{-\infty}^{\infty} |h(t)|\, dt < \infty$ and consider only inputs that are bounded.

An LTI system for which $h(t) \equiv 0$ for $t < 0$ is said to be *causal*—it does not respond before it is stimulated.* "Causality" is the traditional word used in discussions of system theory to describe this non-anticipatory property, but it is perhaps an unfortunate choice in several respects. On the one hand, since we insist that all our systems have a unique response to each completely specified stimulus situation, all of our systems are "causal" whether or not they are non-anticipatory—the input is the agent that determines the output. On the other hand, the mere fact that one event precedes another does not, of course, imply that the first is the agent that causes the second—this is the well-known "post hoc ergo propter hoc" fallacy in logic. Nevertheless, we shall abide by tradition and use the word "causality" instead of the less euphonious "non-anticipatory."

For causal LTI systems, we can relate the general convolution integral

$$y(t) = \int_{-\infty}^{\infty} x(\tau)h(t-\tau)\,d\tau$$

to our previous convolution expression for the ZSR of a lumped circuit, which in one form was

$$y(t) = \int_{0}^{t} x(\tau)h(t-\tau)\,d\tau\ .$$

If $h(t)$ is causal, then $h(t-\tau) = 0$ for $\tau > t$ and we may replace the upper limit ∞ in the superposition integral by t without any effect. To explain the lower limit, we write

$$y(t) = \int_{-\infty}^{t} x(\tau)h(t-\tau)\,d\tau = \int_{-\infty}^{0} x(\tau)h(t-\tau)\,d\tau + \int_{0}^{t} x(\tau)h(t-\tau)\,d\tau\ .$$

The second integral is the ZSR form previously studied—the response for $t > 0$ to the input $x(t)$ for $t > 0$ if the input (and hence the state) were zero for $t \leq 0$. The first integral is then the ZIR—the response for $t > 0$ to the state established by inputs for $t < 0$. Indeed, if $h(t)$ is a sum of exponentials,

$$h(t) = \sum_{i} a_i e^{s_i t} u(t)$$

(as will be the case for a finite RLC network if the characteristic equation has no repeated roots), then we have immediately

$$\int_{-\infty}^{0} x(\tau)h(t-\tau)\,d\tau = \sum_{i} a_i \left[\int_{-\infty}^{0} x(\tau)e^{-s_i \tau}\,d\tau \right] e^{s_i t}$$

that is, a sum of the natural-frequency terms with amplitudes dependent on the previous input as reflected in the state at $t = 0$, exactly as we saw in the frequency domain in Chapter 3. The extension of frequency-domain methods to

*For a more formal definition of causality, see Exercise 10.3.

non-causal systems requires an approach different from the unilateral Laplace transform of Chapter 2, as we shall begin to discuss in Chapter 12.

In some cases causality can be related to another fundamental system attribute. A system is said to be *passive* if under all possible conditions of excitation the energy absorbed by the system over the entire past is positive.[*] Interestingly, a passive linear circuit must be causal,[†] as we now show. For a two-terminal electrical circuit, the energy absorbed up to time t is $\int_{-\infty}^{t} v(\tau)i(\tau)\,d\tau$, where $v(t)$ and $i(t)$ are the voltage and current at the terminals; if the system is passive, this energy must be positive for any $v(t)$ (taken as the stimulus) and any t. Now let

$$v(t) = v_1(t) + av_2(t)$$

where $v_1(t)$ is an arbitrary voltage, $v_2(t)$ is another voltage that is arbitrary except that $v_2(t) \equiv 0$ for $t < t_0$, and a is an arbitrary constant. If the circuit is linear, we can write

$$i(t) = i_1(t) + ai_2(t)$$

where $i_1(t)$ and $i_2(t)$ are the responses to $v_1(t)$ and $v_2(t)$, respectively. For $t < t_0$, we have $v(t) = v_1(t)$, and the integrated energy input is

$$\int_{-\infty}^{t} v(\tau)i(\tau)\,d\tau = \int_{-\infty}^{t} v_1(\tau)[i_1(\tau) + ai_2(\tau)]\,d\tau$$
$$= \int_{-\infty}^{t} v_1(\tau)i_1(\tau)\,d\tau + a\int_{-\infty}^{t} v_1(\tau)i_2(\tau)\,d\tau$$
$$\geq 0, \quad t < t_0.$$

Since a is arbitrary, this inequality can be satisfied for nonzero $v_1(\tau)$ only if the second integral vanishes for any $v_1(t)$ and all $t < t_0$. But this requires $i_2(t) = 0$, $t < t_0$, which implies causality (see Exercise 10.3.)

Students frequently wonder why, since real physical systems presumably must be causal, we should be interested at all in non-causal systems. One answer (as was suggested in Chapter 9) is that this presumption is incorrect—the independent variable may (as in a lens or an antenna) be space rather than time, in which case points to the left of the origin are just as "real" as those to the right. And, of course, any system that processes stored data (for example, a digital computer) can easily simulate the operation of a non-causal system, although not, to be sure, in "real time." Another answer is that we cannot appreciate the implications and consequences of causality except in a larger context. But there is more to it than that. Causality is a very powerful assumption; it imposes a variety of strong constraints on behavior. This is why it is an important concept. On the other hand, such constraints can also be a complicating nuisance, producing in many cases only a second-order effect on

[*]See Section 4.1.

[†]D. C. Youla, L. J. Castriota, and H. J. Carlin, "Bounded Real Scattering Matrices and the Foundations of Linear Passive Network Theory," *IRE Trans. Circuit Theory*, CT-6 (March 1959): 102–124.

performance but a first-order effect on analytical difficulty by turning a simple and clear problem into a hard and confusing one. It is valuable, therefore, to have the freedom to make "approximations" with respect to causality, just as we do with respect to many other attributes, for the sake of gaining simplicity and insight. We shall see numerous examples of these gains in the chapters that follow.

A BIBO stable system was defined in Chapter 6 as one yielding a bounded zero-state output to every bounded input. We showed that a necessary and sufficient condition for stability in this sense for finite RLC circuits is that all poles of $H(s)$ lie inside the left half-plane. We can extend the BIBO stability condition to systems described by (10.2–2), causal or not:

> BIBO STABILITY:
> A necessary and sufficient condition for BIBO stability of a
> CT LTI system is that the impulse response be absolutely
> integrable, that is,

$$\int_{-\infty}^{\infty} |h(t)|\, dt < \infty. \qquad (10.3-1)$$

The second condition for the existence of the convolution integral with infinite limits (stated at the beginning of this section) is thus a stability condition. The proof of the necessity and sufficiency of the absolute integrability condition is completely analogous to the corresponding proof for the discrete-time case given in Section 9.2.

Example 10.3–1

For an LTI RLC network, $h(t)$ will consist of a finite sum of complex exponential terms (multiplied by powers of t if the characteristic equation contains repeated roots). Obviously $h(t)$ will be absolutely integrable if all the exponents have (nonzero) negative real parts (poles inside the left half-plane), and will not be absolutely integrable if any of the exponents have (nonzero) positive real parts (one or more poles inside the right half-plane). Poles directly on the $j\omega$-axis correspond to networks composed entirely of lossless (LC) elements (or analogous devices such as ideal integrators); the impulse responses of such networks have constant amplitudes for $t > 0$ (or grow as a power of t if there are multiple-order poles on the $j\omega$-axis). Some examples are shown in Figure 10.3–1.

Obviously, $\int_{-\infty}^{\infty} |h(t)|\, dt \to \infty$ for all of these examples. In general, networks with poles on the $j\omega$-axis are unstable in the BIBO sense: There are bounded inputs that yield unbounded outputs in each case. For example, you should have no difficulty demonstrating that the bounded inputs shown in Figure 10.3–2 for each of the three examples (a), (b), and (c) of Figure 10.3–1 yield the unbounded responses shown in Figure 10.3–2.

▶ ▶ ▶

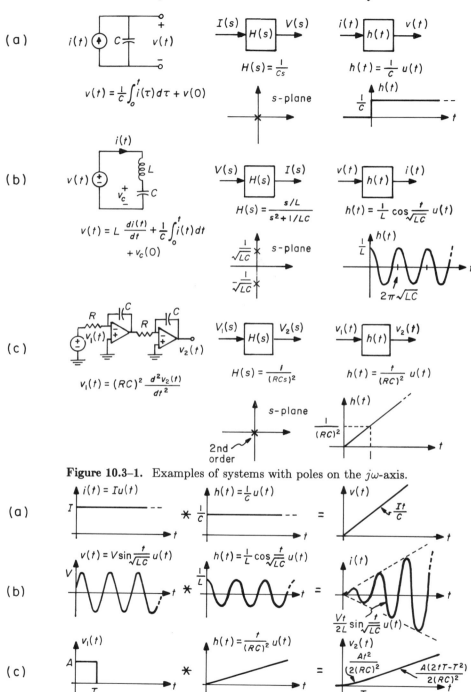

Figure 10.3–1. Examples of systems with poles on the $j\omega$-axis.

Figure 10.3-2. Bounded inputs yielding unbounded outputs for the systems of Figure 10.3–1.

Physical systems approximating ideal lossless networks—in general, LTI systems whose system functions have poles on the $j\omega$-axis and whose impulse responses grow no faster than a power of t—are usually not as badly behaved as systems whose system functions have poles inside the right half-plane and whose impulse responses grow exponentially. It is thus sometimes useful to distinguish the former category as *marginally stable*, even though in a strict BIBO sense both groups are unstable.

10.4 Summary

In a manner almost precisely analogous to the development in the preceding chapter, we have shown that the convolution integral

$$y(t) = \int_{-\infty}^{\infty} x(\tau)h(t-\tau)\,d\tau = x(t) * h(t)$$

describes explicitly a general linear time-invariant operation in continuous time. In the special case in which the system is causal, that is, $h(t) \equiv 0$, $t < 0$, and in which the input $x(t)$ also vanishes for $t < 0$, this formula becomes

$$y(t) = \int_{0}^{t} x(\tau)h(t-\tau)\,d\tau$$

and corresponds directly to multiplication of $\mathcal{L}$-transforms in the frequency domain:

$$Y(s) = X(s)H(s).$$

Another special case in which the general convolution integral is guaranteed to exist results if the magnitude of the input $x[n]$ is bounded and if the system is BIBO stable, so that

$$\int_{-\infty}^{\infty} |h(t)|\,dt < \infty.$$

Since neither $h(t)$ nor $x(t)$ necessarily vanishes for $t < 0$ in this case, unilateral $\mathcal{L}$-transform methods cannot be applied. But the frequency domain is still extremely useful for such LTI systems, as we shall begin to discuss in Chapter 12.

Although we gave $h(t)$ the name "unit impulse response," we have not as yet said anything about what sort of a thing a "unit impulse" might be to produce such a "response." This is the topic of the next chapter.

EXERCISES FOR CHAPTER 10

Exercise 10.1

Given an arbitrary system about which you know nothing, can you devise a sequence of specific stimuli such that the corresponding responses will allow you to claim that the system is (or is not) linear and/or time-invariant? Discuss.

Exercise 10.2

If $y(t)$ is the output of an LTI system, prove that the integral $\int_{-\infty}^{\infty} y(t)\,dt$ (that is, the area under the output waveform) is equal to the product of the areas under the input and the impulse response waveforms.

Exercise 10.3

In general, a (linear or non-linear) system is defined to be causal if, given any two inputs that are equal for $t < t_0$, the corresponding outputs are also equal for $t < t_0$. Show from the general convolution formula (10.2–3) that $h(t) \equiv 0$ for $t < 0$ is a necessary and sufficient condition for an LTI system to be causal in this sense.

Exercise 10.4

Show from the convolution integral that the unit step response of a causal LTI circuit can be written in the form

$$y(t) = \int_0^t h(\tau)\,d\tau\,.$$

From this prove directly in the time domain that the impulse response is the derivative of the step response.

PROBLEMS FOR CHAPTER 10

Problem 10.1

The output $y_0(t)$ shown below is observed when a certain (unknown) input $x_0(t)$ is applied to a certain LTI system with (unknown) impulse response $h_0(t)$.

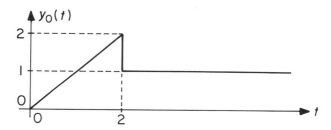

Three other LTI systems with inputs and impulse responses related to $x_0(t)$ and $h_0(t)$ are described below. For each, sketch the corresponding response.

a) $x_1(t) = 2x_0(t)$, $h_1(t) = 0.5h_0(t)$.

b) $x_2(t) = x_0(t) - x_0(t-2)$, $h_2(t) = h_0(t)$.

c) $x_3(t) = x_0(t-2)$, $h_3(t) = h_0(t+1)$.

Problem 10.2

Consider the input and impulse response shown below.

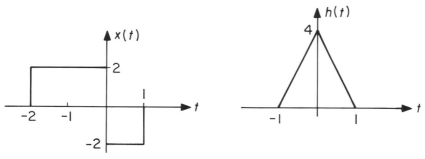

a) What is the complete set of values of t for which $y(t) = h(t) * x(t) = 0$?

b) For what value of t does $y(t)$ have its largest positive value?

c) What is $y(+1)$?

Problem 10.3

The relationship between the convolution of causal time functions and the product of their transforms can be easily extended to

$$L[x_1(t) * x_2(t) * x_3(t)] = L[x_1(t)]L[x_2(t)]L[x_3(t)].$$

a) Use this result to find the inverse transform of

$$X(s) = \frac{1}{s^2(s+a)}.$$

b) Check your answer by evaluating residues.

Problem 10.4

Consider functions $\alpha(t)$, $\beta(t)$, and $\psi(t)$ defined on the infinite line, $-\infty < t < \infty$. Let

$$h(t) = \begin{cases} 0, & t > 1 \\ \psi(t), & 0 < t < 1 \\ 0, & t < 0 \end{cases} \qquad x(t) = \begin{cases} \alpha(t), & t > 0 \\ \beta(-t), & t < 0. \end{cases}$$

In each of several ranges of t write $y(t) = h(t) * x(t)$ as a sum of definite integrals involving $\alpha(t)$, $\beta(t)$, and $\psi(t)$.

Problem 10.5

Discuss whether each of the circuits below appears to be linear and/or time-invariant at its terminals.

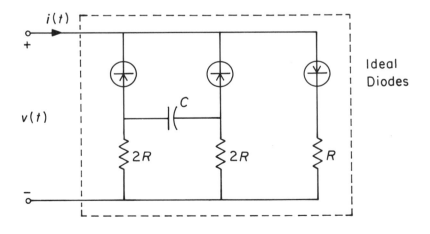

(a)

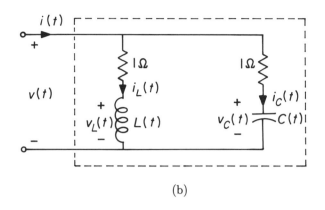

$C(t)$ and $L(t)$ are time-varying elements, so that

$$i_C(t) = \frac{d}{dt}[v_C(t)\,C(t)]$$

$$v_L(t) = \frac{d}{dt}[i_L(t)\,L(t)]$$

and the functions $L(t)$ and $C(t)$ are numerically equal, $L(t) = C(t)$.

(b)

Problem 10.6

Find by convolution for each pair of waveforms the response to the input $x(t)$ of the LTI system with impulse response $h(t)$. Express your results either graphically or analytically as you choose.

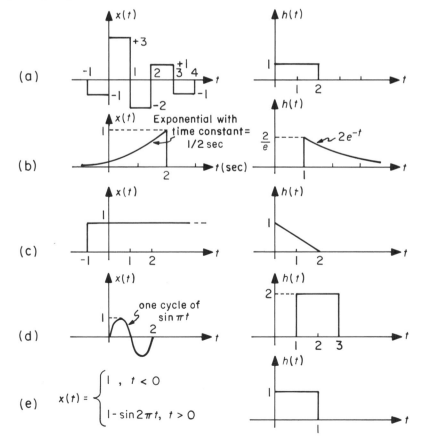

Problem 10.7

Suppose that the temperature response of an electric oven to a change in the current input can be modelled as shown in Figure 1 by a square-law device in cascade with an LTI system. The response of the overall system to a step increment is as shown.

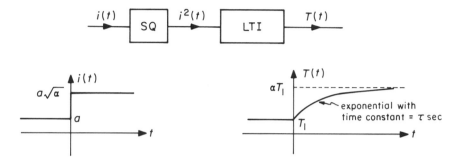

Figure 1.

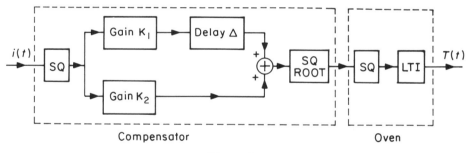

Figure 2.

To speed up the response it is proposed to compensate the system with the device shown in Figure 2.

a) Choose K_1 and K_2 (including their signs) so that

i) The oven reaches its final temperature in response to a step increment after precisely Δ sec.

ii) The steady-state temperature is the same as before.

Sketch the actual input to the oven for a step change in $i(t)$ such as the one shown in Figure 1.

b) In practice, what limits the speed of response that can be obtained in this way?

Problem 10.8

An LTI system with impulse response $h(t)$ as shown in Figure 1 (called a *raised cosine pulse*) is frequently employed as an interpolating or smoothing filter (we shall consider some reasons for this choice in a later chapter). One way to construct such a filter is to cascade a finite-time integrator with impulse response

$$h_1(t) = \begin{cases} 1, & 0 < t < T \\ 0, & \text{elsewhere} \end{cases}$$

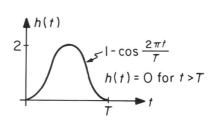

Figure 1.

with a high-Q resonant circuit having an idealized impulse response

$$h_2(t) = \frac{2\pi}{T} \sin \frac{2\pi t}{T} u(t)$$

as shown in Figure 2.

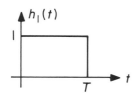

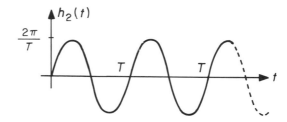

Figure 2.

a) Show that the $h(t)$ of Figure 1 is given by

$$h(t) = h_1(t) * h_2(t).$$

b) Show directly by convolution that the response $y(t)$ of the overall LTI system of Figure 1 to the sinusoid $x(t) = \cos \frac{2\pi t}{T}$, $-\infty < t < \infty$, (the "steady-state" response) is

$$y(t) = -\frac{T}{2} \cos \frac{2\pi t}{T}, \quad -\infty < t < \infty.$$

c) Show that if we consider the system as made up of the cascade of $h_1(t)$ and $h_2(t)$, and if we attempt to calculate the response to the input of (b) in stages, we get into trouble. In particular, show that if the order of the subsystems is $h_1(t)$ followed by $h_2(t)$, the response of $h_1(t)$ alone to $x(t)$ is zero (and hence the overall response is presumably zero), whereas if the order of the subsystems is $h_2(t)$ followed by $h_1(t)$, the response of $h_2(t)$ alone to $x(t)$ is infinite (and hence the overall response is infinite).

Analytically, these difficulties are a result of the fact that the triple convolution $x(t) *$ $h_1(t) * h_2(t)$ is not associative in this particular case. Specifically

$$x(t) * (h_1(t) * h_2(t)) = -\frac{T}{2}\cos\frac{2\pi t}{T}$$

$$(x(t) * h_1(t)) * h_2(t) = 0$$

$$(x(t) * h_2(t)) * h_1(t) = \infty \text{ (or undefined).}$$

The trouble is easy to understand in the frequency domain: $h_1(t)$ has a zero in its $\mathcal{L}$-transform at precisely the frequency at which the ideal lossless tuned circuit $h_2(t)$ has a pole. The difficulties vanish if either the frequency of $x(t)$ or the duration of $h_1(t)$ is slightly altered, or if a little "loss" is added to $h_2(t)$. In practice, if $h(t)$ is to be realized from such a cascade connection and if signals approximately described by $\cos\frac{2\pi t}{T}$ over extended intervals are possible inputs, then better results are likely to be obtained from the cascade $h_1(t)$ followed by $h_2(t)$ than from $h_2(t)$ followed by $h_1(t)$, since the amplitude of the signal at the point between $h_1(t)$ and $h_2(t)$ will be smaller (or have less range) in the former arrangement.

Problem 10.9

The convolution of a signal with itself turned around in time is called the *auto-correlation function* of that signal:

$$r_x(\tau) = \int_{-\infty}^{\infty} x(t)x(t-\tau)dt .$$

a) By comparison with the convolution integral, determine the impulse response of a filter that, given a particular $x(t)$ as its input, will yield $r_x(t)$ as its output. (Such a *matched filter* has interesting noise-reduction properties. See Problem 10.10.)

b) Sketch the impulse response $h(t)$ of the matched filter corresponding to the waveform below.

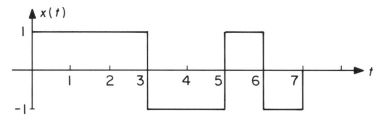

c) Sketch the output of this filter with $x(t)$ as input, that is, the autocorrelation function $r_x(\tau)$ corresponding to $x(t)$.

d) The impulse response obtained in (b) is unrealizable. Show, however, that $h'(t) = h(t-7)$ is the impulse response of a realizable filter, and find the relationship between $r_x'(t)$ (the output of $h'(t)$) and $r_x(t)$ (the output of $h(t)$).

e) Show that $h'(t)$ can be realized with the tapped delay line arrangement shown below, and find the values of the gain factors a_i.

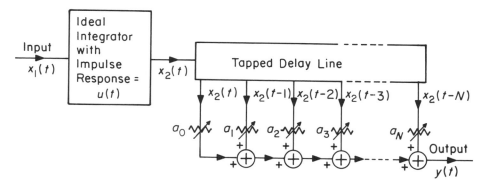

Problem 10.10

For simplicity, consider an alphabet with three letters—a, b, and c. To transmit a message written in this alphabet over a telephone cable at a rate of one letter every T seconds, we propose to associate each letter with a particular electrical waveform of duration T seconds as shown in Figure 1. A sequence of letters or a message can then be represented by a long waveform as shown in Figure 2.

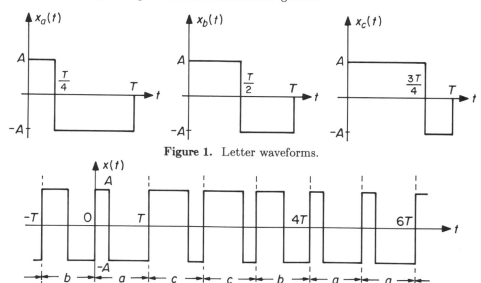

Figure 1. Letter waveforms.

Figure 2. The transmitted waveform corresponding to the sequence *baccbaa*.

The receiver in this system is to operate an automatic typewriter. The signals to activate the keys are obtained as shown in Figure 3. The impulse responses of the three LTI filters are shown in Figure 4; they are called matched filters because their impulse responses are the corresponding signals turned around in time (see Problem 10.9). The waveforms in Figure 3 correspond to the particular $x(t)$ of Figure 2. The threshold circuits are non-linear no-memory devices that pass only signal inputs that exceed $\frac{3AT}{4}$ volts.

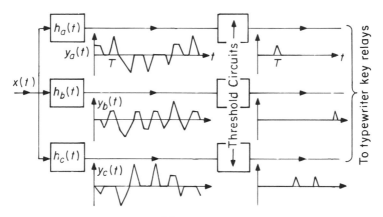

Figure 3.

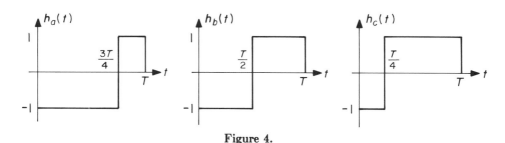

Figure 4.

Clearly the design problem for a system such as this is to choose the waveforms of Figure 1 and the filters of Figure 4 so that the response of each filter to its matched waveform is large and positive while the response of all other filters to that waveform is small or negative. We shall study how well this particular design functions by analyzing the "a" channel in Figure 3.

a) Find the response of $h_a(t)$ separately to each of the waveforms $x_a(t)$, $x_b(t)$, and $x_c(t)$.

b) If you did part (a) correctly, it should be clear that $h_a(t)$ has a larger response to $x_a(t)$ than to either $x_b(t)$ or $x_c(t)$. But this is not enough to guarantee a successful design. Since the response of $h_a(t)$ to each letter waveform lasts twice as long as the letter waveform, the response to successive letters will overlap. Such *intersymbol interference* could cause errors. Find the response to the letter sequence of Figure 2 and show that it is equal to $y_a(t)$ of Figure 3, so that for this sequence at least no errors result from the overlap.

c) Prove that no errors can occur in the "a" channel for any letter sequence.

By a similar analysis of the other channels it can be shown that the largest spurious peak of this sort that can appear in the wrong output channel or at the wrong time is $\frac{AT}{2}$. Hence only the correct peaks (which are always of height AT) will pass the threshold elements and activate the typewriter keys. The requirement that intersymbol interference be eliminated is a rather strong one, as you can show by trying to find equally acceptable waveform designs.

Problem 10.11

The system function of a resonant circuit was shown in Chapter 4 to have the general form

$$\frac{H(s)}{|H(j\omega_0)|} = \frac{\frac{1}{Q}\left(\frac{s}{\omega_0}\right)}{\left(\frac{s}{\omega_0}\right)^2 + \frac{1}{Q}\left(\frac{s}{\omega_0}\right) + 1}\,.$$

a) Show that for large Q

$$h(t) \approx |H(j\omega_0)|\frac{\omega_0}{Q}e^{-\omega_0 t/2Q}\cos\omega_0 t, \quad t > 0\,.$$

b) Sketch $h(t)$ for large Q; label important times and amplitudes.

c) Sketch the response of such a circuit to a unit step drive.

11

IMPULSES AND THE
SUPERPOSITION INTEGRAL

11.0 Introduction

The art of approximation is central to the practice of engineering and science. The problem is that, by and large and to some degree, everything influences everything else in the universe. Modern science has been most successful in those situations where all the effects are small except those resulting from a few primary causes. This is part of the reason why modern science is quantitative—to have a scale on which to measure "small." And it is part of the reason why the methods of modern science have been more successful in explaining physical systems than biological, social, political, or economic phenomena. In these latter fields it is less clear how to make approximations, how to decide what to ignore.

Since the start of this book, we have been trying progressively to back away from our circuits, to let the details get a little out of focus, to lump things, to approximate so that we may come to understand more complex systems. One of the ways in which we can approximate is to talk about idealizations such as linear systems. Another is to consider limiting cases, such as waiting for a long time until the transients die away so that we can explore the steady state. The system function—the limiting steady-state amplitude as a function of s of the response to the special input e^{st}—characterizes the behavior of linear time-invariant systems in the frequency domain. Now we wish to interpret the alternative or dual characterization in the time domain as the limiting response to another sort of input—the *impulse*.

11.1 The Smoothing Effect of Physical Systems

Inevitably, the action of every macroscopic physical system includes some degree of smoothing or averaging of inputs. Indeed, such averaging is implied on fundamental thermodynamic grounds by the word "macroscopic"; on a sufficiently fine (microscopic) scale we could not ignore the fluctuations that arise in all dissipative systems from thermal agitation of discrete charge and mass elements. A system capable of resolving input events arbitrarily close together (in either space or time) could not be analyzed as a deterministic macroscopic system.

Mathematically, smoothing has two principal consequences:

1. Inputs that are zero (or sufficiently small) outside some common, sufficiently short, time interval will usually have essentially the same effect if their areas (time integrals) are the same.* Thus we would expect all the inputs $x_i(t)$ shown in Figure 11.1–1 to give essentially the same output for sufficiently small δ if the shaded areas, $\int_0^\infty x_i(t)\,dt$, are the same.

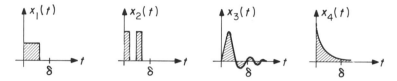

Figure 11.1–1. Examples of "short" time functions.

2. Two inputs that are identical except for finite differences at a small[†] set of discrete times will yield identical outputs. Thus $x_1(t)$ and $x_2(t)$ shown in Figure 11.1–2 should have the same effect on a system for any finite value of K.

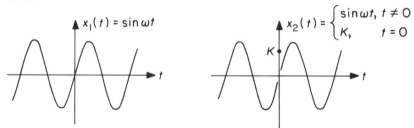

Figure 11.1–2. Waveforms differing only at the point $t = 0$.

For linear systems, these consequences of smoothing are related because the area of the difference waveform $x_1(t) - x_2(t)$ in Figure 11.1–2 is zero.

Example 11.1–1

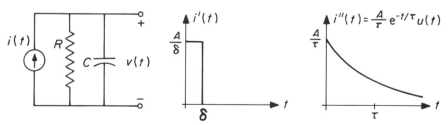

Figure 11.1–3. A simple system with several inputs, $i'(t)$ and $i''(t)$.

---

*We shall explore the possibility of exceptions to this rule in Section 11.5.

[†]Formally, what is required is zero measure.

The system function of the simple circuit shown in Figure 11.1–3 is

$$H(s) = \frac{V(s)}{I(s)} \, (\text{ZSR}) = \frac{1/C}{s + 1/RC}.$$

We seek to compare the responses to the inputs $i'(t)$ and $i''(t)$ shown in the figure. Note that $i'(t)$ and $i''(t)$ have the same area,

$$\int_0^\infty i'(t)\,dt = \int_0^\infty i''(t)\,dt = A$$

independent of the values of τ and δ.

a) Let $V'(s)$ be the response to $I'(s) = L[i'(t)] = \dfrac{A}{\delta s}\left(1 - e^{-s\delta}\right)$. Then

$$V'(s) = H(s)I'(s) = \frac{A}{\delta C}\frac{1 - e^{-s\delta}}{s(s + 1/RC)} = \frac{AR}{\delta}\left(\frac{1}{s} - \frac{1}{s + 1/RC}\right)\left(1 - e^{-s\delta}\right).$$

Hence,

$$v'(t) = L^{-1}[V'(s)] = \frac{AR}{\delta}\left(1 - e^{-t/RC}\right)u(t) - \frac{AR}{\delta}\left(1 - e^{-(t-\delta)/RC}\right)u(t - \delta)$$

$$= \begin{cases} 0, & t < 0 \\[2mm] \dfrac{AR}{\delta}\left(1 - e^{-t/RC}\right), & 0 < t < \delta \\[2mm] \dfrac{AR}{\delta}\left(e^{\delta/RC} - 1\right)e^{-t/RC}, & t > \delta. \end{cases}$$

This is plotted in Figure 11.1–4. Since $e^{\delta/RC} \approx 1 + \delta/RC$ for $\delta \ll RC$, we note that if the pulse duration δ is much smaller than the time constant of the circuit, RC, then $v'(t)$ approaches a limit:

$$\lim_{\delta \to 0} v'(t) = \frac{A}{C}e^{-t/RC}u(t).$$

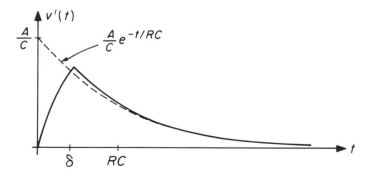

Figure 11.1–4. Response to $i'(t)$.

b) Let $V''(s)$ be the response to $I''(s) = \dfrac{A/\tau}{s + 1/\tau}$. Then

$$V''(s) = H(s)I''(s) = \frac{A/\tau C}{(s + 1/\tau)(s + 1/RC)} = \frac{AR}{RC - \tau}\left(\frac{1}{s + 1/RC} - \frac{1}{s + 1/\tau}\right).$$

Hence,

$$v''(t) = \frac{AR}{RC - \tau}\left(e^{-t/RC} - e^{-t/\tau}\right)u(t).$$

This result is plotted in Figure 11.1–5. We note that

$$\lim_{\tau \to 0} v''(t) = \frac{A}{C}e^{-t/RC}u(t)$$

which is the same as $\lim_{\delta \to 0} v'(t)$. Moreover, observe that (for $A = 1$) the common limiting response is the function

$$h(t) = \frac{1}{C}e^{-t/RC}u(t)$$

that we computed in Example 10.1–2 as the inverse $\mathcal{L}$-transform of the system function for this circuit and called the *unit impulse response*.

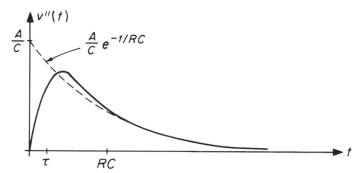

Figure 11.1–5. Response to $i''(t)$.

▶ ▶ ▶

Both $i'(t)$ and $i''(t)$ in Example 11.1–1 thus behave (if their areas are equal to $A = 1$) like unit impulses in the limit as δ, $\tau \to 0$. Nevertheless, it would not be appropriate to define the unit impulse function $\delta(t)$ as

$$\lim_{\delta \to 0} i'(t) = \lim_{\tau \to 0} i''(t) = \delta(t).$$

The trouble is that a function is defined mathematically by assigning it a value for each value of its argument; the limiting procedure above fails to assign a value to $\delta(t)$ in the only really interesting part of its range, the neighborhood of $t = 0$. This is not just a mathematical quibble. Attempts to define the impulse function by statements such as

$$\text{i)} \quad \delta(t) = 0, \quad t \neq 0$$
$$\text{ii)} \quad \delta(0) = \infty \qquad \qquad \qquad (11.1-1)$$
$$\text{iii)} \quad \int_{-\epsilon}^{\epsilon} \delta(t)\, dt = 1$$

lead to ambiguity, inconsistency, and "paradoxical" behavior. The most successful way around these difficulties seems to be to give up trying to assign values to the impulse "function" $\delta(t)$ and instead to concentrate on its effects or properties.*

Functions defined in terms of what they "do" rather than what they "are" are called *distributions* or *generalized functions*. During the last 25 years an extensive mathematical literature dealing with such functions has been published.† This chapter will try to convey a little of the flavor of this approach.

11.2 Impulses and Their Fundamental Properties

If a unit impulse could be defined as an ordinary function $\delta(t)$, then one property it should have—indeed, its most basic and characteristic property—is that the response of an LTI system to the input $\delta(t)$ should be what we have already labelled the unit impulse response, $h(t)$, of that system. Since in general we can describe the response of any LTI system by the convolution integral

$$y(t) = \int_{-\infty}^{\infty} x(\tau)h(t-\tau)\, d\tau = \int_{-\infty}^{\infty} h(\tau)x(t-\tau)\, d\tau$$
$$= x(t) * h(t)$$

we expect that we should be able to set $x(t) = \delta(t)$ and obtain $y(t) = h(t)$, that is,

$$h(t) = \int_{-\infty}^{\infty} \delta(\tau)h(t-\tau)\, d\tau = \int_{-\infty}^{\infty} h(\tau)\delta(t-\tau)\, d\tau \qquad (11.2-1)$$

or, symbolically,

$$\boxed{\delta(t) * h(t) = h(t)\,.} \qquad (11.2-2)$$

---

*S. J. Mason used to tell the story of one student who complained, "You mean everywhere except at the origin an impulse is so small you can't see it, whereas at the origin it's so big you can't see it. In other words, you can't see it at all; at least I can't!" Of course, the student is right: There is no way to define what an impulse "is"—only what it "does."

†Some early, and still useful, references are: L. Schwartz, *Théorie des Distributions*, *Vols. I and II* (Paris: Hermann, 1957–1959); A. H. Zemanian, *Distribution Theory and Transform Analysis* (New York, NY: McGraw-Hill, 1965); M. J. Lighthill, *Fourier Analysis and Generalized Functions* (New York, NY: Cambridge University Press, 1958); I. M. Gel'fand and G. E. Shilov, *Generalized Functions, Vol. I: Properties and Operations* (New York, NY: Academic Press, 1964).

Since we could apply $\delta(t)$ as a test to any LTI system, (11.2–2) must be true for any time function $h(t)$. We take (11.2–2) as the *defining property* of an impulse: *An impulse convolved with any function reproduces that function.* We use the notation $\delta(t)$ to represent a unit impulse as if it were an ordinary function, but we do not assign $\delta(t)$ "values" as we would an ordinary function; whenever we are uncertain about what such notation "means," we can simply convolve it with an arbitrary "test function" and use (11.2–2) to interpret the result—as illustrated in the following examples.*

Example 11.2–1

Of course, it is still often helpful to think of $\delta(t)$ as a tall narrow pulse-like waveform located near $t = 0$. To connect the convolutional definition (11.2–2) with this notion, we can draw pictures of the integrands in (11.2–1) as we did in Chapter 10.

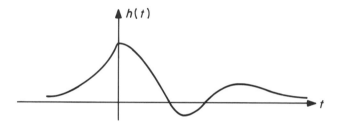

Figure 11.2–1. An "arbitrary" $h(t)$.

If $h(t)$ has the arbitrary form shown in Figure 11.2–1, then the integrands in (11.2–1) appear as in Figure 11.2–2, where we have represented $\delta(t)$ as a narrow rectangular pulse although any narrow tall shape would do as well.

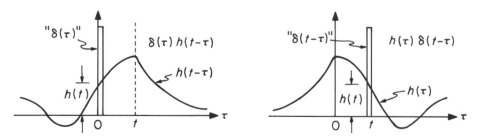

Figure 11.2–2. Integrands of (11.2–1) for $h(t)$ of Figure 11.2–1.

---

*The use of impulse functions in science and engineering was popularized by the English physicist P.A.M. Dirac and by Oliver Heaviside long before impulses became "respectable" mathematically. Indeed we continue to use Dirac's notation, $\delta(t)$, and the unit impulse is often called *Dirac's δ-function*. Both Dirac and Heaviside stressed the idea that $\delta(t)$ was defined in terms of what it "did." Thus Dirac said, "Whenever an improper function [e.g., impulse] appears it will be something which is to be used ultimately in an integrand—the use of improper functions thus does not involve any lack of rigor in the theory, but is merely a convenient notation, enabling us to express in a concise form certain relations which we could, if necessary, rewrite in a form not involving improper functions, but only in a cumbersome way which would tend to obscure the argument."

The two pictures are similar except for a reversal in direction; note that $\delta(\tau)$ is an impulse "at" $\tau = 0$ whereas $\delta(t - \tau)$ is an impulse "at" $\tau = t$, that is, "at" the point where the argument equals zero. The product $\delta(\tau)h(t - \tau)$, consequently, is very small except near $\tau = 0$, where (if the pulse representing $\delta(t)$ is narrow enough and if $h(t-\tau)$ is smooth enough near $\tau = 0$) we have approximately

$$\delta(\tau)h(t - \tau) \approx \delta(\tau)h(t - 0) = \delta(\tau)h(t)$$

so that the response to the impulse is

$$y(t) = \int_{-\infty}^{\infty} \delta(\tau)h(t - \tau)\,d\tau \approx \int_{-\infty}^{\infty} \delta(\tau)h(t)\,d\tau = h(t)\int_{-\infty}^{\infty} \delta(\tau)\,d\tau = h(t)$$

as desired (since $\delta(t)*1 = 1$, so that we may say that $\delta(t)$ has unit area). *Mathematically, the effect of a unit impulse function inside an integral (such as (11.2–1)) is to pick out the value of the remainder of the integrand at the point where the impulse "is."*
▶ ▶ ▶

Example 11.2–2

As another example of the effect of an impulse inside an integral, consider $u(t) * \delta(t)$, which according to (11.2–2) is

$$\int_{-\infty}^{\infty} u(t - \tau)\delta(\tau)\,d\tau = u(t).$$

Since

$$u(t - \tau) = \begin{cases} 1, & \tau < t \\ 0, & \text{otherwise} \end{cases}$$

we have, equivalently,

$$\int_{-\infty}^{\infty} u(t - \tau)\delta(\tau)\,d\tau = \int_{-\infty}^{t} \delta(\tau)\,d\tau = u(t).$$

That is, *the (indefinite) integral of the unit impulse is the unit step.* Conversely, interpreting the indefinite integral as an anti-derivative, we may say operationally that *the unit impulse is the derivative of the unit step*:

$$\delta(t) = \frac{du(t)}{dt}. \tag{11.2–3}$$

To show that such an interpretation is indeed consistent with the definition of the impulse through the property (11.2–2), let $\dot{u}(t)$ stand formally for $du(t)/dt$ and consider

$$g(t) * \dot{u}(t) = \int_{-\infty}^{\infty} \dot{u}(\tau)g(t - \tau)\,d\tau$$

where $g(t)$ is some arbitrary function employed to test what $\dot{u}(t)$ "does" inside an integral. Integrating by parts (with $\dot{g}(t)$ standing formally for $dg(t)/dt$), we have

$$\int_{-\infty}^{\infty} \dot{u}(\tau)g(t-\tau)\,d\tau = \left.u(\tau)g(t-\tau)\right|_{\tau=-\infty}^{\tau=\infty} - \int_{-\infty}^{\infty} u(\tau)(-\dot{g}(t-\tau))\,d\tau$$

$$= [1\cdot g(-\infty) - 0\cdot g(\infty)] + \int_{0}^{\infty} \dot{g}(t-\tau)\,d\tau$$

$$= g(-\infty) - \left.g(t-\tau)\right|_{\tau=0}^{\tau=\infty}$$

$$= g(t).$$

Comparing this result with (11.2–2), we see that $\dot{u}(t)$ *behaves* operationally like $\delta(t)$ as claimed.

Still another way of interpreting (11.2–3) is to think of $u(t)$ as the limit of a sequence of waveforms such as the ramps in Figure 11.2–3. As shown, the sequence of derivatives behaves like an impulse.

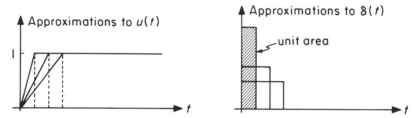

Figure 11.2–3. A sequence of functions approaching a step function and their derivatives.

▶ ▶ ▶

Some comments are suggested by these examples. First, note that any sequence behaving in the limit as $\delta(t)$ must be unbounded in the vicinity of $t = 0$. The symbol shown in Figure 11.2–4—which with minor variations is more or less standard—is thus an appropriate graphical representation for an impulse. Note that the amplitude A is really the *area* of the impulse. If the time scale is changed, then A must be changed correspondingly; that is,

$$\delta(at) = \frac{1}{|a|}\delta(t). \qquad (11.2-4)$$

To show that (11.2–4) is correct, we put $\delta(at)$ inside an integral to see what it "does." Thus we consider

$$\int_{-\infty}^{\infty} \delta(at)g(t)\,dt$$

where $g(t)$ is our arbitrary test function. Changing variable to $\tau = at$, this becomes

$$\int_{-\infty}^{\infty} \delta(\tau)\left[\frac{g(\tau/a)}{|a|}\right]d\tau$$

which by (11.2−1) is equal to $g(0)/|a|$. But this is also the value of the integral

$$\int_{-\infty}^{\infty} \left[\frac{\delta(t)}{|a|} \right] g(t)\, dt \,.$$

That is, $\delta(at)$ behaves like $\delta(t)/|a|$. In particular, with $a = -1$ we conclude from (11.2−4) that $\delta(-t) = \delta(t)$; that is, $\delta(t)$ behaves like an even function even though tall narrow waveforms that approximate an impulse need not be even. Figure 11.2–4 also illustrates how to represent an impulse that is not located at $t = 0$; to find the location of a shifted impulse such as $\delta(t - T)$, determine the value of t that makes the argument zero, that is, $t = T$. The amplitude A of a voltage impulse has units of volt-seconds, not volts. The units of the symbol $\delta(t)$ itself are $\sec^{-1}$ so that $\int_{-\infty}^{\infty} \delta(t)\, dt$ has the dimensionless value unity and $A\delta(t)$ has units of volts.

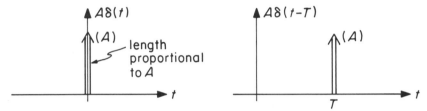

Figure 11.2–4. Symbolic representation of an impulse.

Since any tall narrow waveform behaving in the limit as $\delta(t)$ also behaves as if it were zero for $t \neq 0$, the product of $\delta(t)$ and any sufficiently smooth function $f(t)$ depends only on $f(0)$:

$$\delta(t)f(t) = \delta(t)f(0)\,. \qquad (11.2-5)$$

Again, what is meant by an equality such as (11.2−5) is that both sides behave the same way if placed in an integral with a test function. Equation (11.2−5) is called the *sampling property* of the impulse. (As Heaviside delightfully put it, "The function $[\delta(t)]$ spots a single value of the arbitrary function in virtue of its impulsiveness.") The importance of the words "sufficiently smooth" cannot be overstressed; the key requirement is that the value of $f(t)$ at $t = 0$ and the local average of $f(t)$ near $t = 0$ must be the same. Thus the product of a discontinuous function and an impulse located at the discontinuity, for example, $\delta(t)u(t)$, *cannot be given a consistent meaning and must be avoided*. (For a classic illustration of the difficulties and seeming paradoxes that arise if this warning is ignored, see Problem 11.21.) For essentially the same reason, impulses (and their relatives to be discussed later) cannot be successfully used in non-linear systems; no consistent meaning can be attached, for example, to $\delta^2(t)$. On the other hand, the convolution of an impulse with a discontinuous function (as in (11.2−2)) presents no difficulties, because the precise value of a discontinuous function at the point of discontinuity has no physical importance (recall the smoothing argument in Section 11.1). For example,

$$\delta(t) * u(t) = u(t)\,. \qquad (11.2-6)$$

Example 11.2–3

As another example of the effect of an impulse inside an integral, consider the Laplace transform of a delayed impulse,

$$L[\delta(t - T)] \overset{\Delta}{=} \int_0^\infty \delta(t - T)e^{-st}\, dt$$

$$= \int_{-\infty}^\infty \delta(t - T)[e^{-st}u(t)]\, dt$$

$$= e^{-sT}u(T).$$

That is,

$$\boxed{L[\delta(t - T)] = e^{-sT}, \quad T > 0.} \tag{11.2–7}$$

Conversely, the inverse transform of e^{-sT}, $T > 0$, is an impulse at $t = T$.

The case of an impulse at $t = 0$ presents some difficulties. In effect the lower limit on the defining integral of the unilateral Laplace transform gives us the problem that we want in general to avoid, interpreting a product such as $u(t)\delta(t)$. In the present case, the most straightforward and self-consistent approach is to adopt the convention (for L-tranform applications) that $\delta(t)$ corresponds to a tall narrow waveform that is identically zero for $t < 0$. To indicate that this convention has been adopted, some authors write the defining L-transform integral as

$$\boxed{L[x(t)] = \int_{0-}^\infty x(t)e^{-st}\, dt} \tag{11.2–8}$$

indicating that an impulse at $t = 0$ is to be included inside the integration interval.* In this event we can obtain $L[\delta(t)]$ from the preceding result by letting $T \to 0$; that is,

$$\boxed{L[\delta(t)] = 1.} \tag{11.2–9}$$

This result is, of course, consistent with our previous interpretation of the impulse response $h(t)$ as the inverse L-transform of the system function $H(s)$ for causal systems: If $x(t) = \delta(t)$, then $X(s) = 1$, so $Y(s) = X(s)H(s) = H(s)$, and $y(t) = h(t)$. (This issue is pursued a bit further in Section 11.4 and in Problem 11.17.)

▶ ▶ ▶

11.3 General LTI Systems; The Superposition Integral

In Chapter 10, we introduced the general convolution formula

$$y(t) = \int_{-\infty}^\infty x(\tau)h(t - \tau)\, d\tau \tag{11.3–1}$$

---

*A similar problem arises in discrete time: Is 0 the largest of the negative integers or the smallest of the positive integers? Our definition (8.1–1) of the (unilateral) Z-transform includes 0 with the positive integers.

to describe explicitly a large class of linear time-invariant systems. If $h(t) = 0$, $t < 0$, that is, if the system is causal, this formula can be used to compute the response of the lumped RLC networks (and their mechanical, acoustical, thermal, chemical, economic, and other analogs) that have been our main concern to this point. In such systems, characterized by constant-coefficient finite-order total differential equations or equivalently by rational system functions, the impulse response will be a sum of terms of the form $t^n e^{s_i t}$. But the same formula may also be applied to distributed systems characterized by partial differential equations; in this case $h(t)$ is much less constrained in form. And it also applies to systems such as antennas, optical systems, and x-ray or ultrasound imaging systems in which the independent variable may be space instead of time; such systems need not be causal. In any of these cases it may be that we know enough about the internal structure of the system to compute $h(t)$ from more fundamental measurements of element and parameter values. Or we may simply treat the system as a "black box" accessible only at its terminals; in that case we can at least in principle measure $h(t)$ by the experimental equivalent of Example 11.1–1. In either case, if we know that the system can be described by (11.3–1), then measuring or computing the response to *just one* input—the unit impulse— characterizes the response to *every other* input.

The significance of this last statement depends on how large and interesting a class of systems can be described by the convolution formula (11.3–1). What we propose to argue is that *every* system that satisfies the linearity and time-invariance constraints can be represented in this way, and hence its input-output behavior under any conditions can be characterized by observing it under just one condition—when driven by an impulse.*

Linearity and time-invariance are thus very powerful constraints. In contrast, an attempt to characterize the input-output behavior of a non-LTI system solely from observations of its responses to various stimuli is at best an uncertain undertaking. At one extreme, if the system may capriciously alter its characteristics at any time, then obviously any effort to describe its lawful behavior on the basis of observation is fundamentally hopeless. On the other hand, even if the system is known to be time-invariant (so that we may confidently expect the response to some particular input to be the same today or tomorrow as it was yesterday), we can still in general say nothing about the output corresponding to any input that we have not previously explicitly and specifically studied. It is possible, to be sure, to deduce from observations the characteristics of systems constrained in certain other specific ways than linearity and time-invariance, but no such class of constraints comes even close in power and significance to the LTI class.

The argument justifying (11.3–1) as the most general representation of an LTI continuous-time system parallels the argument in Chapter 9 justifying the

---

*In fact, the linearity and time-invariance constraints are so strong that knowing the response of an LTI system to (almost) *any* one stimulus (not just an impulse) will permit us to calculate the response to (almost) *any other* stimulus; that is, (almost) *any one* input-output pair (essentially) *characterizes* an LTI system! (For an explanation of the"almosts," see Problem 12.1.)

convolution sum (9.2–2) as the most general representation of LTI discrete-time systems. Recall that we first interpreted an arbitrary DT input $x[n]$ as a sum of weighted delayed unit sample functions:

$$x[n] = \sum_{m=-\infty}^{\infty} x[m]\delta[n-m] . \qquad (11.3\text{–}2)$$

Then, exploiting time-invariance, we argued that each delayed unit sample function in (11.3–2) must produce as output the delayed unit sample response:

$$\delta[n-m] \Longrightarrow h[n-m] . \qquad (11.3\text{–}3)$$

Finally, exploiting linearity, we concluded that a weighted sum of inputs must produce the correspondingly weighted sum of responses:

$$x[n] = \sum_{m=-\infty}^{\infty} x[m]\delta[n-m] \quad \Longrightarrow \quad \sum_{m=-\infty}^{\infty} x[m]h[n-m] = y[n] . \qquad (11.3\text{–}4)$$

Hence the input-output behavior of any LTI DT system can be characterized by the convolution formula

$$y[n] = \sum_{m=-\infty}^{\infty} x[m]h[n-m] \qquad (11.3\text{–}5)$$

where $h[n]$ is the observed response to $\delta[n]$.

To extend this argument to CT systems requires one additional step, showing that essentially any CT function $x(t)$ behaves as if it were the limit of a chain of weighted delayed unit impulse functions. Thus consider the smooth function $x(t)$ in Figure 11.3–1 as composed of a sum of delayed pulse functions $x_n(t)$ as shown.

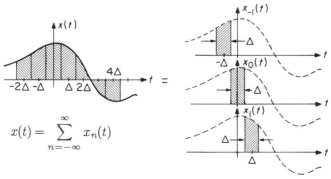

Figure 11.3–1. Decomposition of $x(t)$.

For sufficiently small Δ, $x_n(t)$ behaves like an impulse of area $x(n\Delta)\Delta$ located at $t = n\Delta$:

$$x_n(t) \approx x(n\Delta)\delta(t - n\Delta)\Delta . \tag{11.3-6}$$

Hence for sufficiently small Δ we can say that

$$x(t) \approx \sum_{n=-\infty}^{\infty} x(n\Delta)\delta(t - n\Delta)\Delta \tag{11.3-7}$$

in the sense that both sides of (11.3–7) will have essentially the same effect on a (macroscopic, smoothing) physical system and will look essentially the same to any measuring instrument of limited resolution. Note also that if we pass formally to the limit as $\Delta \to 0$ in (11.3–7), the sum becomes an integral,

$$x(t) = \int_{-\infty}^{\infty} x(\tau)\delta(t - \tau)\,d\tau = x(t) * \delta(t) \tag{11.3-8}$$

which we have already taken as one form of the defining property of $\delta(t)$, but which we are now interpreting as an expansion of $x(t)$ in a chain of weighted delayed impulse functions.

If we accept (11.3–7), then the remainder of the argument is precisely analogous to (11.3–3, 4, 5). Since the system is time-invariant,

$$\delta(t - \tau) \Longrightarrow h(t - \tau) \tag{11.3-9}$$

and since the system is linear (see Figure 11.3–2),

$$\sum_{n=-\infty}^{\infty} x(n\Delta)\delta(t - n\Delta)\Delta \Longrightarrow \sum_{n=-\infty}^{\infty} x(n\Delta)h(t - n\Delta)\Delta \approx y(t) . \tag{11.3-10}$$

Passing to the limit $\Delta \to 0$ yields formally

$$y(t) = \int_{-\infty}^{\infty} x(\tau)h(t - \tau)\,d\tau = x(t) * h(t) \tag{11.3-11}$$

as a general representation for any LTI CT system. Because it interprets $y(t)$ as the weighted superposition of delayed impulse responses, (11.3–11) is often called the *superposition-integral* representation of an LTI system.

The verbal hand-waving in which we have just been indulging provides a nice picture of why we might expect a formula like (11.3–11) to provide a general representation of an LTI CT system, but it is hardly nice mathematics. Nevertheless, L. Schwartz showed in 1957 that our conclusion above is rigorously valid provided $x(t)$, $y(t)$, and $h(t)$ are all defined as generalized functions—in terms of what they "do." Some such interpretation is clearly necessary, since otherwise such elementary LTI systems as ideal amplifiers, delay lines, and differentiators could not be described by (11.3–11)—as we shall shortly illustrate.

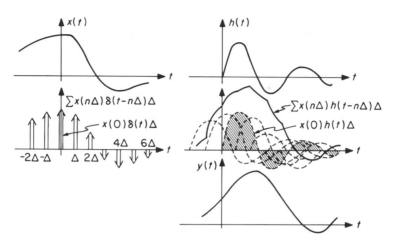

Figure 11.3-2. Construction of $y(t)$.

Arguments similar to the superposition integral derivation occur throughout mathematical physics. Thus the idea of considering the total effect of a continuous density of mass, charge, or force as a superposition of the elementary effects of "lumped" point masses, etc., is a common and powerful one. As a specific example, the electrostatic potential at a point $\mathbf{r}$ due to a point charge q at a point $\mathbf{r}_0$ is simply

$$\psi(\mathbf{r}) = \frac{q}{|\mathbf{r} - \mathbf{r}_0|}.$$

If we seek the potential due to a charge density $\rho(\mathbf{r}_0)$, we may replace the continuous density by lumped charges $\rho(\mathbf{r}_0)dv$ and add the effects to yield

$$\psi(\mathbf{r}) = \oint \frac{\rho(\mathbf{r}_0)\,dv}{|\mathbf{r} - \mathbf{r}_0|}$$

which is the *Poisson integral* solution to Poisson's equation. Obviously, both the form of this result and the nature of the argument leading to it are structurally identical with the superposition integral (11.3–11) and the "derivation" leading to it.

Integral solutions of this type can be found for many of the linear differential equations of mathematical physics (for another example, see Problem 11.3). In general this procedure is called *Green's method* in differential equation theory, and the kernel of the integral analogous to the impulse response is called *Green's function* for the problem at hand.

Linearity is, of course, basic to Green's method and to the superposition integral, but time-invariance (or its spatial equivalent, uniformity or homogeneity) is much less necessary. Thus for a time-varying linear system we may define $h(t, \tau)$ as the response at time t to an impulse applied at time τ and—paralleling the argument leading to (11.3–11)—we may write

$$y(t) = \int_{-\infty}^{\infty} x(\tau) h(t, \tau) \, d\tau \, . \tag{11.3–12}$$

(For a further consideration of time-varying systems, see Problem 11.1.)

Green's method is one of the most powerful tools for advanced studies of differential equations, although it is probably more valuable for discussing the general properties of solutions than as a method for finding specific solutions. A similar comment applies to the superposition integral, as we shall see; in many respects our principal accomplishment in this book will be to provide a rich, poetic language for *talking about* the behavior of systems—as contrasted with providing detailed formal methods of analysis.

Example 11.3–1

An *ideal amplifier* of gain K is characterized by the fact that the output is identical with the input but K times larger:

$$y(t) = K x(t) \, .$$

The impulse response is obviously an impulse of area K,

$$h(t) = K \delta(t)$$

since then

$$y(t) = x(t) * h(t) = x(t) * K \delta(t) = K x(t) \, .$$

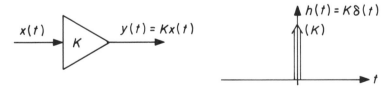

Figure 11.3–3. The ideal amplifier and its impulse response.

▶ ▶ ▶

Example 11.3–2

An *ideal delay line* is represented by the block symbol shown in Figure 11.3–4 and has the impulse response $h(t) = \delta(t - T)$. The output is

$$y(t) = \int_{-\infty}^{\infty} x(\tau) h(t - \tau) \, d\tau = \int_{-\infty}^{\infty} x(\tau) \delta(t - \tau - T) \, d\tau = x(t - T)$$

which is the input delayed T seconds. An ideal delay line is an LTI system but cannot be constructed from a finite number of RLC elements. (Its system function is $H(s) = \mathcal{L}[h(t)] = e^{-sT}$, which is not a rational function.) Networks approximating delay performance as closely as desired can be constructed, however, if sufficient elements are employed. (See, for example, Problem 11.19.) Delays resulting from the finite velocity of sound or electromagnetic radiation may for many purposes be quite accurately described by an ideal delay line.

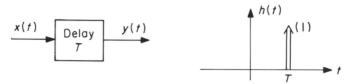

Figure 11.3–4. The ideal delay line and its impulse response.

▶ ▶ ▶

Example 11.3–3

One very useful LTI system is constructed by connecting a number of delay lines in cascade to form a *tapped delay line*, weighting the signals at each tap with individual amplifiers or attenuators, and adding the results to produce a single output as shown in Figure 11.3–5. The result is obviously very similar to the DT transversal filters first discussed in Section 7.2.

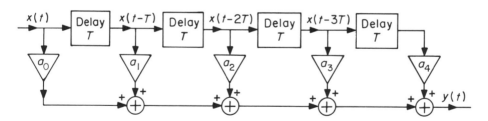

Figure 11.3–5. A tapped delay line system.

The impulse response of this tapped delay line system is obtained by setting $x(t) = \delta(t)$ and adding the outputs of the various taps:

$$h(t) = a_0\delta(t) + a_1\delta(t - T) + \cdots + a_4\delta(t - 4T)$$

as shown schematically in Figure 11.3–6. By appropriate settings of the gains of the tap amplifiers, a tapped delay line can be adjusted to approximate the impulse response of any LTI system provided that:

a) the total length of the delay line is equal to or greater than the duration of the important part of the response one wishes to approximate;

b) the tap spacing is sufficiently small that the inputs of interest are nearly constant over the tap spacing interval.

One way of choosing the tap gain settings is suggested in Figure 11.3–7. The area of each impulse is made equal to the corresponding cross-hatched area of the desired impulse response. (Compare with the argument leading to (11.3–7).)

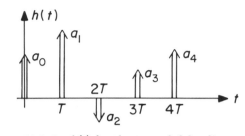

Figure 11.3–6. $h(t)$ for the tapped delay line system.

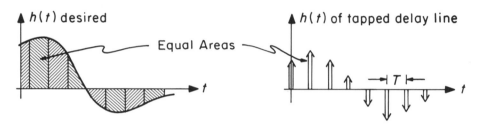

Figure 11.3–7. A method for choosing the tap gains.

To improve the approximation, we can cascade the tapped delay line with an *interpolating* or *smoothing filter*. One common example of such a filter is a *finite-time integrator* or *zero-order hold* with impulse response as shown in Figure 11.3–8 and previously discussed in Example 10.2–2. The result of cascading such a filter with the tapped delay line having the impulse response of Figure 11.3–6 is an LTI system whose impulse response is the "staircase" waveform shown in Figure 11.3–9.

Impulse responses of this same type can also be obtained by cascading a tapped delay line with an ideal integrator whose impulse response is $u(t)$. As an exercise, you might try to find values for the tap gains in Figure 11.3–10 such that the overall impulse response is the same as that shown in Figure 11.3–9.

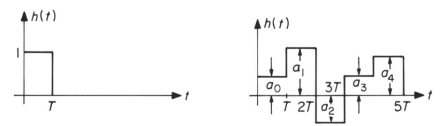

Figure 11.3–8. $h(t)$ for finite-time integrator.

Figure 11.3–9. $h(t)$ for the cascade of the systems of Figures 11.3–6 and 11.3–8.

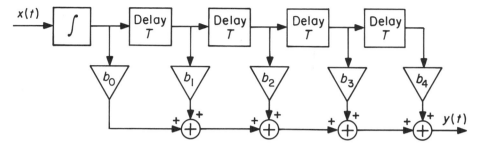

Figure 11.3–10. Tapped delay line system plus integrator.

▶ ▶ ▶

Example 11.3–4

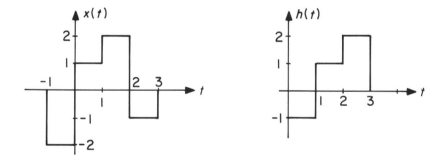

Figure 11.3–11. Staircase waveforms.

The convolution of two "staircase" waveforms such as those shown in Figure 11.3–11 is particularly easy if the construction of Example 11.3–3 is exploited in reverse. Thus suppose we are interested in finding the response of the filter $h(t)$ to the waveform $x(t)$ shown in Figure 11.3–11. We could proceed without too much difficulty as in Example 10.2–2, folding and shifting one of the waveforms, sliding it past the other, and computing the area of the product rectangles for each shift. But it is somewhat simpler to think of each waveform as composed of the *impulse train* waveforms $\hat{x}(t)$ and $\hat{h}(t)$ shown in Figure 11.3–12, each convolved with the pulse waveform $p(t)$. The whole process then can be conceived as finding the impulse response of the multiple-cascade system shown in Figure 11.3–13. It should be evident that $\hat{x}(t) * p(t) = x(t)$, since each impulse in $x(t)$ generates in the second block a pulse $p(t)$ scaled in amplitude by the area of the impulse and delayed or advanced in time in accordance with the timing of the impulse.

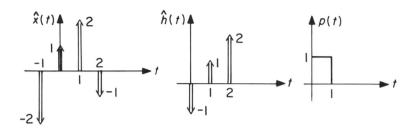

Figure 11.3–12. Impulse trains derived from $x(t)$ and $h(t)$ of Figure 11.3–11.

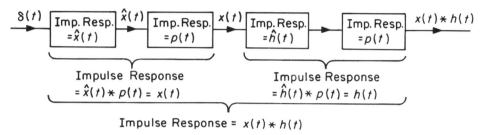

Figure 11.3–13. Analysis of $x(t) * h(t)$.

The trick is to use the associative law of Section 10.2 to rearrange the order in which operations are done. Specifically:

1. convolve $\hat{x}(t) * \hat{h}(t)$;

2. convolve $p(t) * p(t)$;

3. convolve the results of these two operations.

The convolution of two impulse trains is very simple once it is recognized that the fundamental impulse property (11.2–2), $\delta(t) * x(t) = x(t)$, is valid even if $x(t)$ itself contains impulses. This is a general principle that is worth restating.

CONVOLUTION WITH A DELAYED AND SCALED IMPULSE:

> The convolution of any waveform (including a waveform containing impulses and their relatives) with a delayed and scaled impulse simply delays and scales the waveform correspondingly. Formally,

$$x(t) * K\delta(t - T) = K x(t - T).$$

This result is virtually self-evident on observing that $K\delta(t-T)$ is the impulse response of an amplifier of gain K cascaded with a delay line of delay T.

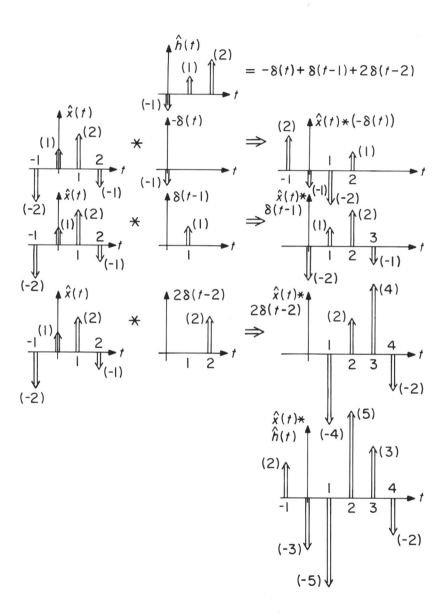

Figure 11.3–14. Convolution of the impulse trains $\hat{x}(t)$ and $\hat{h}(t)$.

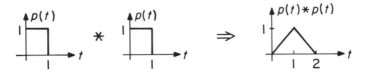

Figure 11.3–15. $p(t) * p(t)$.

Figure 11.3–14 shows graphically the result of decomposing $\hat{h}(t)$ into a sum or weighted delayed impulses, convolving each separately with $\hat{x}(t)$, and adding the results to obtain the impulse train $\hat{x}(t) * \hat{h}(t)$. The convolution of $p(t)$ with itself yields the isosceles triangle shown in Figure 11.3–15 (as in Example 10.2–2). Finally, convolving the impulse train $\hat{x}(t) * \hat{h}(t)$ with the triangle $p(t) * p(t)$ yields the superposition of triangles in Figure 11.3–16, which is the desired convolution $x(t) * h(t)$. Note that the sequence of peaks is simply the sequence of impulse areas in $\hat{x}(t) * \hat{h}(t)$.

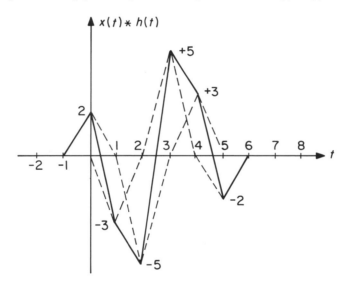

Figure 11.3–16. $y(t) = x(t) * h(t)$.

▶ ▶ ▶

An alternative way of describing the convolution of the two impulse trains, $\hat{x}(t) * \hat{h}(t)$, is often useful. Thus we may write in general for impulse trains

$$\hat{x}(t) = \sum_{n=-\infty}^{\infty} x[n]\delta(t-n), \quad \hat{h}(t) = \sum_{n=-\infty}^{\infty} h[n]\delta(t-n)$$

where the square brackets imply functions of discrete time as in Chapters 7–9. It is then easy to show that

$$\hat{y}(t) = \hat{x}(t) * \hat{h}(t) = \sum_{n=-\infty}^{\infty} y[n]\delta(t-n)$$

where $y[n]$ is the discrete-time convolution of $x[n]$ and $h[n]$, that is,

$$y[n] = x[n] * h[n] = \sum_{m=-\infty}^{\infty} x[m]h[n-m].$$

Then $y[n]$ may be readily evaluated by any of the methods of Chapter 9, including Z-transforms.

11.4 Impulses and Sudden Changes in Initial State

The impulse response of a circuit is not only (by definition) a ZSR but also (since effectively $\delta(t) = 0$ for $t > 0$) it is in a sense a ZIR. Both input and state are zero, but the impulse response is not zero! The resolution of this apparent paradox is that an applied impulse establishes suddenly at $t = 0+$ a particular nonzero state whose decay is the impulse response. The determination of this state is an interesting example of impulse manipulations.

Example 11.4–1

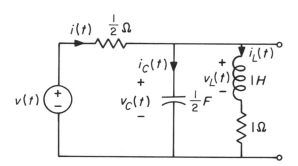

Figure 11.4–1. Circuit for Example 11.4–1.

A version of the circuit in Figure 11.4–1 was first studied in Example 2.5–2. Let $v(t) = \delta(t)$ and $i_L(0-) = v_C(0-) = 0$. At $t = 0$ the "infinite" voltage $v(t)$ must appear almost entirely across the 0.5 Ω resistor; even an impulse of current (implying a finite integral of current, that is, a finite charge) can produce only a finite change in $v_C(t)$ from $t = 0-$ to $t = 0+$, and Kirchhoff's Voltage Law must be satisfied around the left-hand loop. Thus $i(t)$ will contain an impulse at $t = 0$, all of which must flow into the capacitor (since the current in the inductor can at most change its slope by a finite amount) to yield a discontinuous voltage across the inductor equal to $v_C(0+)$. The state at $t = 0+$ is thus $i_L(0+) = i_L(0-) = 0$, and

$$v_C(0+) = \frac{1}{0.5\,\text{F}} \int_{0-}^{0+} i_C(t)\,dt = 2 \int_{0-}^{0+} i(t)\,dt$$

$$= 2 \int_{0-}^{0+} \frac{v(t)}{0.5\,\Omega}\,dt = 4 \text{ volts.}$$

The branch voltages and currents for this circuit with $v(t) = \delta(t)$ are shown in Figure 11.4–2. Notice how Kirchhoff's Laws are satisfied at each instant of time, including in particular the impulses. Notice also how the "derivative" of a discontinuous function, such as $v_C(t)$, contains an impulse of area equal to the height of the discontinuity; this is a generalization of the discussion in Section 11.2 showing that the "derivative" of the unit step function "is" (that is, behaves like) a unit impulse function.

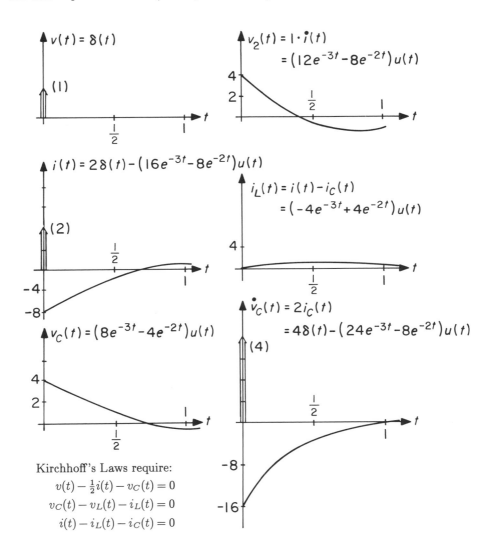

Kirchhoff's Laws require:

$$v(t) - \tfrac{1}{2}i(t) - v_C(t) = 0$$
$$v_C(t) - v_L(t) - i_L(t) = 0$$
$$i(t) - i_L(t) - i_C(t) = 0$$

Figure 11.4–2. Voltage and current waveforms in the circuit of Figure 11.4–1.
▶ ▶ ▶

Example 11.4–1 shows that we could think of the impulse response as the ZIR to the state suddenly established by the impulse. However, we shall choose instead to think of the impulse response as a ZSR. That is, when the drive contains impulses at $t = 0$, we define the "initial state" to be the one existing at $t = 0-$; this is consistent with our redefinition of the Laplace transform integral as having a lower limit $t = 0-$.

In Chapter 2 and Problem 2.3 we argued that the proper relationships between the Laplace transforms of the branch variables resulted if we replaced each capacitor (dual arguments apply to inductors) by either of the impedance representations to the left in Figure 11.4–3. The first of these (a) has the obvious interpretation (as explored in Problem 2.2) that a capacitor with initial voltage $v(0)$ is indistinguishable for $t > 0$ from an uncharged capacitor at $t = 0$ in series with a battery of constant voltage $v(0)$. We are now able to give a similar time-domain interpretation to the alternative circuit (b). A current source whose transform is a constant $Cv(0)$ is an impulse source $Cv(0)\delta(t)$. Such a source suddenly establishes a voltage $v(0)$ across the uncharged capacitor C at $t = 0$. Obviously with such impulse sources connected to each energy storage element in a circuit we can arrange any initial conditions we desire. On the other hand, with a single impulse source at the input to the circuit we can establish only one specific state that decays into one specific response, the impulse response of the circuit.

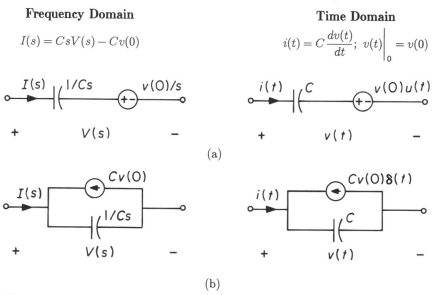

Figure 11.4–3. Alternative representations of the initial state of a capacitor.

11.5 Doublets and Other Generalized Functions; Impulse Matching

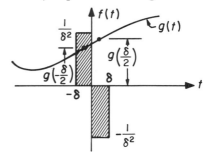

Figure 11.5–1. A function $f(t)$ approximating a doublet.

The class of useful generalized functions—defined by what they "do" rather than what they "are"—contains many more members than just impulses. Consider, for example, the shaded waveform $f(t)$ in Figure 11.5–1. To find out what $f(t)$ "does," we can multiply it by a smooth test function $g(t)$ and integrate:

$$\int_{-\infty}^{\infty} f(t)g(t)\,dt = \int_{-\delta}^{0} \frac{1}{\delta^2} g(t)\,dt - \int_{0}^{\delta} \frac{1}{\delta^2} g(t)\,dt .$$

Approximating each integral for small δ by the area of a trapezoid, we have

$$\int_{-\infty}^{\infty} f(t)g(t)\,dt \approx \frac{g(-\delta/2) - g(\delta/2)}{\delta} \xrightarrow{\delta \to 0} - \frac{dg(t)}{dt}\bigg|_{t=0} .$$

That is, for small δ, $f(t)$ picks out (minus) the value of the derivative of the test function $g(t)$ at $t = 0$.

The function $f(t)$ is said to behave in the limit like a *doublet*, which we shall designate by the symbol $\dot{\delta}(t)$ and define by the property $x(t) * \dot{\delta}(t) = \dot{x}(t)$ or

$$\int_{-\infty}^{\infty} x(\tau)\dot{\delta}(t - \tau)\,d\tau = \dot{x}(t) \tag{11.5–1}$$

or equivalently by the property

$$\int_{-\infty}^{\infty} x(t)\dot{\delta}(t)\,dt = -\dot{x}(0) . \tag{11.5–2}$$

(Note the signs in (11.5–1) and (11.5–2): $\dot{\delta}(t)$ behaves like an odd function; that is, $\dot{\delta}(-t) = -\dot{\delta}(t)$.)

Equation (11.5–1) states that $\dot{\delta}(t)$ is the impulse response of an *ideal differentiator*. The symbol $\dot{\delta}(t)$ is appropriate since operationally the doublet is the derivative of the impulse. This follows formally from (11.5–1) if $x(t) = \delta(t)$. It can also be demonstrated by finding out what $d\delta(t)/dt$ "does" when it is convolved with a test function:

$$g(t) * \frac{d\delta(t)}{dt} = \int_{-\infty}^{\infty} \frac{d\delta(\tau)}{d\tau} g(t - \tau)\,d\tau .$$

Integrating by parts, we obtain

$$g(t) * \frac{d\delta(t)}{dt} = \delta(\tau)g(t-\tau)\big|_{\tau=-\infty}^{\infty} - \int_{-\infty}^{\infty} \delta(\tau)(-\dot{g}(t-\tau))\,d\tau$$
$$= \delta(t) * \dot{g}(t) = \dot{g}(t)$$

which is just what (by definition) a doublet "does." So the two are operationally equivalent and might as well share the same symbol.

The Laplace transform of the doublet follows immediately from the basic definition:

$$\mathcal{L}\left[\dot{\delta}(t)\right] = \int_{0^-}^{\infty} \dot{\delta}(t)e^{-st}\,dt = -\frac{d}{dt}(e^{-st})\bigg|_{t=0} = s \,.$$

As expected from the differentiation relationship, the Laplace transform of the doublet is just s times the Laplace transform of the impulse.

Example 11.5–1

A useful rule for LTI systems is that the ZSR response to the derivative of an input $x(t)$ is the derivative of the response $y(t)$ to $x(t)$. This follows at once from the fact that the effect of the cascade of the systems shown in Figure 11.5–2 is independent of order.

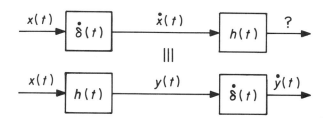

Figure 11.5–2. Demonstration that $h(t) * \dot{x}(t) = \frac{d}{dt}(h(t) * x(t))$.

▶ ▶ ▶

Example 11.5–2

One way to find $h(t)$ that is particularly effective for simple circuits is first to find the unit step response and then to differentiate it. To illustrate, the unit step response of the circuit in Example 11.1–1 is, by inspection,

$$v(t) = R\big(1 - e^{-t/RC}\big)u(t)\,.$$

This is sketched in Figure 11.5–3. The derivative of this response can be found by differentiating the formulas for $v(t)$ separately in the two regions $t < 0$ and $t > 0$.

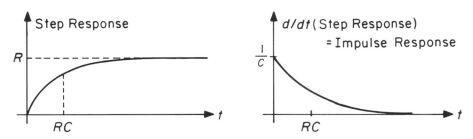

Figure 11.5–3. Step and impulse responses of the circuit of Example 11.1–1.

That is,

$$\frac{dv(t)}{dt} = h(t) = \begin{cases} \dfrac{d}{dt}\big[R\big(1 - e^{-t/RC}\big)\big] = \dfrac{1}{C}e^{-t/RC}, & t > 0 \\[2mm] \dfrac{d}{dt}(0) = 0, & t < 0. \end{cases}$$

The step response has a discontinuity in slope at $t = 0$; hence the impulse response has a discontinuity in value there.

Alternately, $h(t)$ can be found formally by applying the product rule of differentiation to the formula for the step response including $u(t)$ and using the fact that $\dot{u}(t) = \delta(t)$:

$$\begin{aligned} \frac{dv(t)}{dt} = h(t) &= \frac{d}{dt}\big[R\big(1 - e^{-t/RC}\big)u(t)\big] \\[2mm] &= \frac{d}{dt}\big[R\big(1 - e^{-t/RC}\big)\big]u(t) + R\big(1 - e^{-t/RC}\big)\frac{du(t)}{dt} \\[2mm] &= \frac{1}{C}e^{-t/RC}u(t) + \underbrace{R\big(1 - e^{-t/RC}\big)\delta(t)}_{\substack{\| \\ R\big(1 - e^{-0/RC}\big)\delta(t) = 0.}} \end{aligned}$$

Either way the result is the same as before.

▶ ▶ ▶

Other higher-order singularity functions (*triplets, quadruplets,* etc.) can be similarly defined as successive derivatives of $\delta(t)$; their Laplace transforms are higher powers of s. There are no standard symbols for doublets or higher-order singularity functions. The amplitudes of the higher-order singularities are called *moments* and have the units volt-sec^2 for voltage doublets, volt-sec^3 for voltage triplets, etc. The doublet symbol $\dot{\delta}(t)$ itself, of course, has the dimensions sec^{-2}. Spatial equivalents of the doublet and other higher-order singularity functions are the dipole, quadrupole, etc., of electrostatics.

A waveform such as the one in Figure 11.5–4 would for sufficiently small δ behave as an impulse plus a doublet, $\delta(t)+\dot{\delta}(t)$. Here is an example of a waveform that is "zero for $t \neq 0$," "very large near $t = 0$," and that has "unit area" but whose behavior is more complicated than an impulse. Once again we see the difficulty of trying to define what an impulse "is."

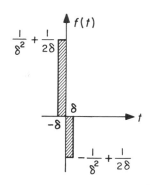

Figure 11.5–4. $f(t) \approx \delta(t) + \dot{\delta}(t)$.

Adding the singularity functions and their transforms to our table of Laplace transform pairs permits us to extend and complete the discussion in Chapter 2 of the Heaviside Expansion Theorem for computing inverse transforms. At that time we restricted the theorem to transforms that are *proper* rational fractions—with a numerator polynomial of lower degree than the denominator. The reason is that the inverse transform of an *improper* rational fraction—with a numerator polynomial of the same or higher degree than that of the denominator—contains impulses or higher-order singularity functions as well as exponentials. These ideas are readily explained through an example. (We have already handled the analogous problem for Z-transforms in Example 8.1–3.)

Example 11.5–3

Consider

$$X(s) = \frac{s^3 + 5s^2 + 9s + 1}{s^2 + 3s + 2}.$$

This transform has poles at $s = -1, -2$. The residues in these poles are

$$\left. \frac{s^3 + 5s^2 + 9s + 1}{(s+1)(s+2)}(s+1) \right|_{-1} = -4$$

and

$$\left. \frac{s^3 + 5s^2 + 9s + 1}{(s+1)(s+2)}(s+2) \right|_{-2} = 5.$$

But clearly

$$X(s) \neq \frac{-4}{s+1} + \frac{5}{s+2} = \frac{s-3}{s^2 + 3s + 2}.$$

These terms describing the finite poles vanish as $s \to \infty$, whereas the given function actually has a pole at ∞. To complete the partial-fraction expansion, we must add terms describing this behavior for large s. The simplest way to find these terms is to divide the denominator into the numerator until the degree of the remainder is less than the degree of the denominator:

$$
\begin{array}{r}
s+2 \\
s^2+3s+2\ \overline{\smash{\big)}\ s^3+5s^2+9s+1} \\
\underline{s^3+3s^2+2s} \\
2s^2+7s+1 \\
\underline{2s^2+6s+4} \\
s-3
\end{array}
$$

Thus

$$X(s) = s + 2 + \frac{s-3}{s^2+3s+2} = s + 2 - \frac{4}{s+1} + \frac{5}{s+2}$$

and

$$x(t) = \dot{\delta}(t) + 2\delta(t) - 4e^{-t}u(t) + 5e^{-2t}u(t).$$

▶ ▶ ▶

   One important feature of singularity functions is that an equation containing ordinary functions and singularity functions of various orders must balance separately for the ordinary functions and for each order of singularity function. There is no way to equate ordinary functions on one side of an equation to singularity functions* on the other side, or impulses on one side to doublets on the other, etc. (except, of course, in some limiting sense). We shall illustrate this point (sometimes called *impulse matching*) with an example.

Example 11.5–4

The differential equation relating the input $v(t)$ and the output $v_C(t)$ in the circuit of Example 11.4–1 is

$$\frac{d^2 v_C(t)}{dt^2} + 5\frac{dv_C(t)}{dt} + 6v_C(t) = 4\frac{dv(t)}{dt} + 4v(t). \qquad (11.5\text{–}3)$$

If $v(t) = \delta(t)$, then the right-hand side of this equation is $4\dot{\delta}(t) + 4\delta(t)$. To match the doublet we conclude that $v_C(t)$ must have a discontinuity of 4 volts at $t = 0$ so that $\dot{v}_C(t)$ can contain an impulse at $t = 0$ of area 4 volt-sec and $\ddot{v}_C(t)$ can contain a doublet of moment 4 volt-sec^2. Thus $v_C(0+)$ must equal 4 volts. But the initial slope $\dot{v}_C(0+)$ must also have an appropriate value if the $\delta(t)$ terms on the two sides are to match. Because of the 4 volt discontinuity in $v_C(t)$, the term $5dv_C(t)/dt$ must contain an impulse of area $4 - 20 = -16$ volt-sec in addition to the term that comes from the doublet. An impulse in the second derivative comes from a discontinuity in *slope*; thus we determine $\dot{v}_C(0+) = -16$ volt-sec^{-1}.

   The roots of the characteristic equation are $s = -2, -3$. Hence the ZIR must have the form

$$v_C(t) = Ae^{-2t} + Be^{-3t}.$$

---

* At least to a finite number of singularity functions.

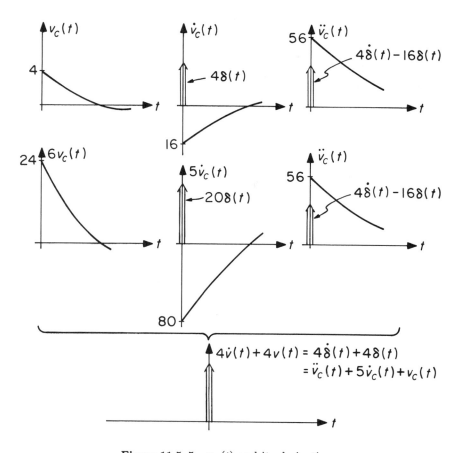

Figure 11.5–5. $v_C(t)$ and its derivatives.

Using the information deduced above from matching impulses, we conclude that $A + B = v_C(0+) = 4$ and $-2A - 3B = \dot{v}_C(0+) = -16$, or $A = -4$, $B = +8$, which is the result previously given for the response of $v_C(t)$ to an impulse. Sketches of the three terms on the left in (11.5–3) are shown in Figure 11.5–5. Note that their sum equals the two terms on the right. Note in particular how derivatives of discontinuities give rise to impulses, and derivatives of impulses to doublets. This is appropriate since we can always consider a discontinuous function to be the sum of a continuous function (whose derivative may be discontinuous but is otherwise generally well-behaved) and a step function (whose derivative is an impulse).

▶ ▶ ▶

11.6 Summary

Pulses of various shapes generally have the same effect on an LTI system if they are sufficiently short in duration and if they have the same area. If the area is unity, they are all equivalent to a unit impulse, $\delta(t)$. An impulse cannot be specified as an ordinary function; instead it is defined as an operator—it is determined by what it "does" rather than what it "is." The fundamental property of an impulse is that the result of convolving any waveform with an impulse reproduces that waveform. Other generalized functions can also be described in terms of what they "do"; thus $\dot{\delta}(t)$, the doublet, is defined by the property that convolving it with any waveform produces the derivative of that waveform. The doublet behaves as if it were the derivative of the impulse, and the impulse behaves as if it were the derivative of the unit step function.

Since any waveform can be considered to be a sum of short-duration pulses with various weights and delays, the response of any LTI system can be thought of as a sum of weighted, delayed, short-duration pulse (impulse) responses. This argument reinterprets the convolution integral as a superposition integral and demonstrates that it is a general representation of the input-output behavior of any LTI system, provided that the impulse response may itself contain impulses, doublets, or other generalized functions.

The study of LTI systems (as hinted in earlier chapters) is extremely important for three interrelated reasons:

1. Linear systems are easy to analyze and to characterize efficiently. Or to say much the same thing in another way, the study of linear systems leads to a rich body of mathematics.

2. Many interesting natural systems are closely approximated by linear systems, at least for sufficiently small stimuli.

3. It is relatively easy to design and construct physical systems based on linear models to perform a variety of non-trivial, important tasks.

In the final analysis the third reason is probably the most important. To paraphrase Voltaire: If there were no linear physical systems, it would be necessary to invent them. Many component manufacturers go to extraordinary lengths to make their products more nearly linear over a wider range in order to extend their usefulness in system design.

The impulse response and the superposition integral are an extremely effective way of describing the behavior of a general LTI system. But our earlier experience with $\mathcal{L}$-transforms should suggest that frequency-domain methods may be effective, too, even when the system is non-causal. The appropriate tool here is the *Fourier transform*—our next topic.

EXERCISES FOR CHAPTER 11

Exercise 11.1

A system has an output $y(t) = F[x(t)]$ that is zero for all times at which the input $x(t) \neq 0$. At each zero crossing of the input, the output contains an impulse of area equal to the derivative of the input at that instant. (Assume that all admissible inputs are continuous with continuous derivatives.)

a) Sketch a typical input and the corresponding output (indicate the area of an impulse conventionally by the height of an arrow).

b) Show that this system is both time-invariant and *homogeneous*, that is, $F[Kx(t)] = KF[x(t)]$.

c) Show that nevertheless the system is *not* linear by constructing a counterexample to superposition.

Exercise 11.2

Show that each of the following circuits or systems has the given impulse response.

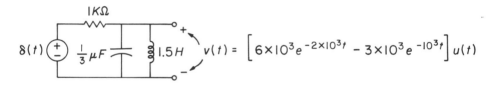

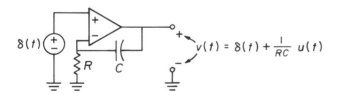

Exercise 11.3

   a) Find by convolution the step response of the system with the impulse response $h(t)$ shown below.

   b) Check your work by using the principle that the impulse response is the derivative of the step response.

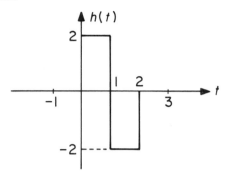

Exercise 11.4

Show in each case below that $y(t) = x(t) * h(t)$.

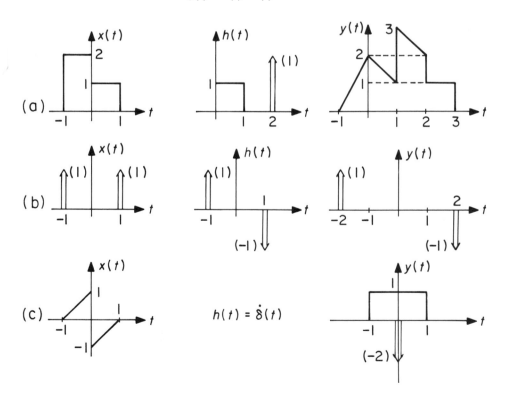

Exercise 11.5

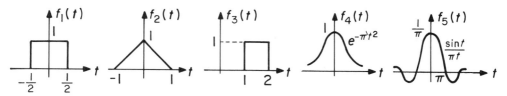

Each of the waveforms above has unit area,

$$\int_{-\infty}^{\infty} f_i(t)\,dt = 1\,.$$

(Actually $f_5(t)$ is not absolutely integrable; the improper integral must be appropriately interpreted to give unit area, for example as $\lim_{T\to\infty}\int_{-T}^{T} f_5(t)\,dt$. We shall have much more to say about this time function in later chapters. See, in particular, Example 13.1–4.)

a) Pick one of the waveforms above and sketch on a common set of axes several members of the waveform sequence $nf_i(nt)$, $n = 1, 2, 3, \ldots$, illustrating how for large n the members of this sequence might behave like a unit impulse.

b) Using an appropriate change of variable, argue that

$$\lim_{n\to\infty}\int_{-\infty}^{\infty} nf(nt)g(t)\,dt = g(0)$$

for any waveform $f(t)$ of unit area and for $g(t)$ sufficiently smooth near $t = 0$.

c) Let $\dot{f}(t) = df(t)/dt$. For several of the waveforms above (say, $f_2(t)$ and $f_4(t)$), sketch on a common set of axes several members of the waveform sequence $n^2\dot{f}_i(nt)$, $n = 1, 2, 3, \ldots$, illustrating how for large n members of this sequence behave like a unit doublet.

d) Using an appropriate change of variable and integration by parts, argue that

$$\lim_{n\to\infty}\int_{-\infty}^{\infty} n^2 \dot{f}(nt)g(t)\,dt = -\dot{g}(0)$$

for any waveform $f(t)$ of unit area and for $g(t)$ sufficiently smooth near $t = 0$.

Higher-order singularity functions, corresponding to higher-order derivatives of $\delta(t)$, can be defined by an extension of this scheme.

Exercise 11.6

Consider the function

$$x(t) = \begin{cases} 1, & t < 0 \\ e^{-3t}, & t > 0. \end{cases}$$

a) Sketch the function $x(t)u(t)$ and its derivative $\dfrac{d}{dt}[x(t)u(t)]$.

b) The general formula for the derivative of a product is

$$\frac{d}{dt}[r(t)s(t)] = r(t)\frac{ds(t)}{dt} + s(t)\frac{dr(t)}{dt}.$$

Apply this formula to $x(t)u(t)$ with $r(t) = x(t)$, $s(t) = u(t)$. Sketch and discuss the various terms on the right-hand side for $x(t)$ as defined above.

c) Certain difficulties arise in applications of this general formula if $x(t)$ has a discontinuity at $t = 0$, for example, if

$$x(t) = \begin{cases} 2, & t < 0 \\ e^{-3t}, & t > 0. \end{cases}$$

Discuss.

Exercise 11.7

Define

$$L = \frac{1}{T}\int_0^T x(t)\sin\frac{2\pi t}{T}\,dt$$

where $x(t)$ must satisfy two conditions:

i) $x(t) \geq 0$,

ii) $\dfrac{1}{T}\displaystyle\int_0^T x(t)dt = 1$.

a) If $x(t)$ is chosen to be an appropriately located impulse of appropriate area, show that it is possible to make $L = 1$.

b) Demonstrate with examples (or prove if you can) that for any other $x(t)$ meeting the required conditions, $L < 1$.

PROBLEMS FOR CHAPTER 11

Problem 11.1

Two notations for a time-varying impulse response are common in the literature:

$h(t, \tau) =$ response at time t to an impulse applied at time τ;

$\hat{h}(t, \tau) =$ response at time t to an impulse applied τ seconds earlier, at time $t - \tau$.

a) Fill in the blanks in the following expressions:

   i) $h(t, \tau) = \hat{h}(\_, \_)$.

   ii) $y(t) = \displaystyle\int_{-\infty}^{\infty} x(\tau) h(\_, \_) \, d\tau = \int_{-\infty}^{\infty} x(\tau) \hat{h}(\_, \_) \, d\tau = \int_{-\infty}^{\infty} x(\_) \hat{h}(t, \tau) \, d\tau.$

b) For each notation, determine the region in the (t, τ) plane where the impulse response must be zero if the time-varying system is to be causal.

c) Using either notation, find a formula for the impulse response of the cascade of two linear time-varying systems and show that the result depends on the order of cascading.

Problem 11.2

In the circuit shown below, the resistance $r(t)$ varies with time in a known fashion, independent of any voltages and currents in the circuit; the voltage drop across $r(t)$ is $v_r(t) = r(t) i(t)$.

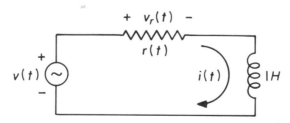

If this circuit is considered as a system with input $v(t)$ and output $i(t)$:

a) Is the system linear?

b) Show that $i(t) = \int_{-\infty}^{t} v(\tau) \exp[-\int_{\tau}^{t} r(\xi) d\xi] \, d\tau$ by inserting this formula in the differential equation relating $i(t)$ and $v(t)$.

c) Find a formula for $h(t, \tau)$, the response at time t to a unit impulse at time τ. Set $r(t) = t$ and sketch on a single graph the response of the system to impulses at a few representative times.

d) Find a formula for $\hat{h}(t, \tau)$, the response at time t to a unit impulse applied τ seconds earlier, at time $t - \tau$.

Problem 11.3

The partial differential equation characterizing the temperature $T(x,t)$ as a function of position, x, and time, t, in a thermally insulated thin rod is

$$\frac{\partial^2 T(x,t)}{\partial x^2} = \frac{1}{k^2}\frac{\partial T(x,t)}{\partial t}$$

where k is a constant dependent on the properties of the material. (This same equation, the *diffusion equation*, also describes many other physical phenomena, such as the voltage on an RC transmission line, the motion of charged carriers in a semiconductor, and the intermixing of two gases.)

a) Show that

$$T(x,t) = \int_{-\infty}^{\infty} T(\xi,0)\frac{1}{2k\sqrt{\pi t}}e^{-(x-\xi)^2/4k^2 t}\,d\xi$$

is a solution to this differential equation in the interval $-\infty < x < \infty$ and for $t > 0$. Here $T(x,0)$ is the temperature distribution vs. x at $t = 0$.

b) Discuss the analogous relationship of this equation to the superposition integral.

Problem 11.4

If $x_1(t)$ below is the input to a certain LTI system, then $y_1(t)$ is the response. Find and plot the response of this system to $x_2(t)$. (HINT: Find a way to represent $x_2(t)$ as a sum of weighted delayed versions of $x_1(t)$.)

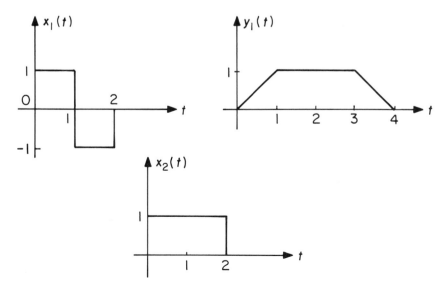

Problem 11.5

An LTI system has the response $q(t)$ to the input $p(t)$ shown in Figure 1.

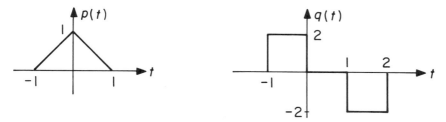

Figure 1.

a) Sketch the input $3p(t-2)$ and the corresponding response.

b) Show that the input function $x(t)$ of Figure 2 can be represented by

$$x(t) = \sum_{n=-\infty}^{\infty} a_n p(t - n)$$

and find the values of a_n.

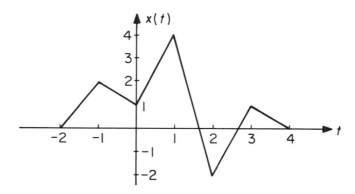

Figure 2.

c) Write an expression for the response $y(t)$ to $x(t)$ of Figure 2 in terms of the responses to elementary inputs, $p(t-n)$. Sketch $y(t)$.

d) Assuming that the system is continuous, find its response to a unit step input. (HINT: Find the response to a unit ramp input and then differentiate.)

e) Create a block-diagram representation of the system in terms of ideal integrators, differentiators, adders, amplifiers, delay lines, etc.

Problem 11.6

Find by convolution for each pair of waveforms the response to the input $x(t)$ of the LTI system with impulse response $h(t)$. Express your results both graphically and analytically.

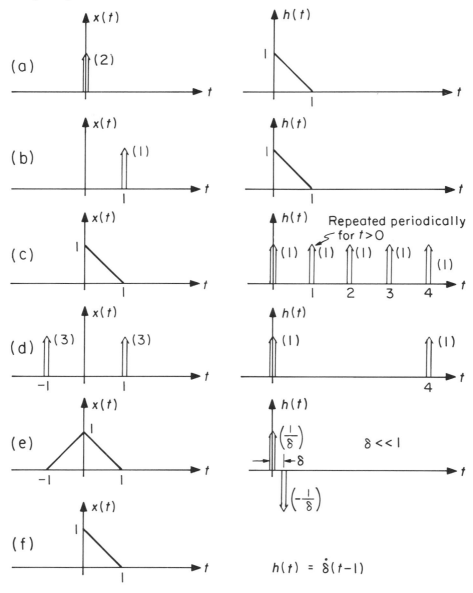

Problem 11.7

A reasonably smooth waveform $x(t)$ can be built up of delayed steps as well as impulses, the height of each step being the increment (or decrement) of $x(t)$ between sample points. Derive an integral expression similar to the superposition integral for the output of a linear time-invariant system as a sum (integral) of weighted, delayed unit step responses $k(t)$, sometimes called the *indicial response* or *indicial admittance* of the system. The resulting integral is known as *Duhammel's integral*.

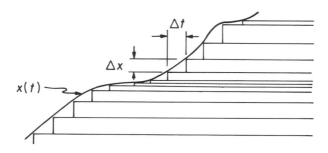

Problem 11.8

Find the inverse Laplace transform of each of the following functions:

a) $\dfrac{(s+1)^2}{s^2+1}$.   b) $\dfrac{s^3+4s^2+4s+2}{(s+1)^2}$.   c) $\dfrac{s^2(1-e^{-s})+2}{s^2+1}$.

Problem 11.9

If $x_1(t)$ as shown below is the input to a causal LTI system, then $y_1(t)$ is the output.

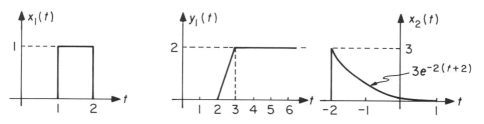

a) Find the impulse response of the system.

b) Find the response $y_2(t)$ to the input $x_2(t)$.

Problem 11.10

The following block diagram can be used to model many simple situations that produce echoes.

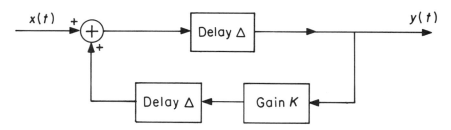

a) Find an expression for the impulse response of this system.

b) For what (positive and negative) values of K is the system stable?

c) For $K = -1$, find the response of this system to the input $x(t) = u(t) - u(t - 2\Delta)$.

Problem 11.11

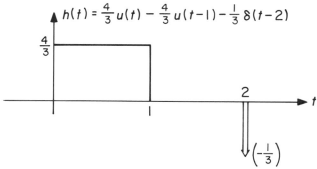

$$h(t) = \frac{4}{3}u(t) - \frac{4}{3}u(t-1) - \frac{1}{3}\delta(t-2)$$

The LTI system with the impulse response $h(t)$ shown above is excited by the input

$$x(t) = a_0 + a_1 t, \qquad -\infty < t < \infty$$

where a_0 and a_1 are constants. (Note that $x(t)$ is defined for all time.)

a) Find the output.

b) Show in general that if we restrict the input to an LTI system to be a polynomial of degree n in t, $-\infty < t < \infty$, and if we want the output $y(t)$ to be identical to the input, it is sufficient to require that

i) $\int_{-\infty}^{\infty} h(t)\, dt = 1$,

ii) $\int_{-\infty}^{\infty} t^k h(t)\, dt = 0, \quad 1 \le k \le n$.

c) Find an $h(t)$ (not simply equal to an impulse) that will pass without distortion any polynomial of the second degree.

Problem 11.12

a) Show directly by convolution that

$$h_1(t) = \begin{cases} 0, & t < 0 \\ e^{-\alpha t}, & t > 0 \end{cases} \quad \text{and} \quad h_2(t) = \dot{\delta}(t) + \alpha\delta(t)$$

are a pair of *inverse systems*, in the sense that $h_1(t) * h_2(t) = \delta(t)$.

b) What is the relationship of the Laplace transforms of $h_1(t)$ and $h_2(t)$ if they are the impulse responses of a pair of causal inverse systems? Illustrate for the particular functions of (a).

A pair of inverse LTI systems may be inserted anywhere in a cascade of LTI systems without altering the overall system response. This is frequently useful for simplifying calculations.

Problem 11.13

The transfer function of a causal LTI system is $H(s) = \dfrac{Y(s)}{X(s)} = \dfrac{s-2}{s+2}$. Find and plot the system output $y_2(t) = h(t) * x(t)$ when the input is

$$x(t) = \begin{cases} -e^t, & t \le 0 \\ 0, & t > 0. \end{cases}$$

Problem 11.14

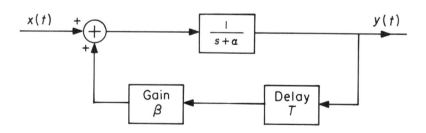

a) Argue that the positive feedback system above is stable if α, β, and T are all positive and $\beta < \alpha$. (HINT: Show that

$$\left| \frac{\beta e^{-sT}}{s+\alpha} \right| < 1 \quad \text{for} \quad \sigma > 0$$

so that the system function can be expanded in a series each term of which is easily inverse transformed.)

b) Find and sketch the impulse response of the overall system under these conditions.

c) Show directly by substitution that the impulse response obtained in (b) is the solution to the differential equation

$$\frac{dy(t)}{dt} + \alpha y(t) - \beta y(t - T) = x(t) = \delta(t)$$

assuming that $y(t) = 0$ for $t < 0$.

Problem 11.15

Suppose that $x(t) = \dot{\delta}(t) = d\delta(t)/dt$ is the input to an LTI system and that $y(t) = e^{-t}u(t)$ is the corresponding output.

a) Find the system function $H(s)$ of a causal system that would yield this response to this input.

b) Find the impulse response $h(t)$ of this system.

c) Find the response $f(t) * h(t)$ to the input

$$f(t) = \begin{cases} 1, & 0 < t < 10 \\ 0, & \text{elsewhere.} \end{cases}$$

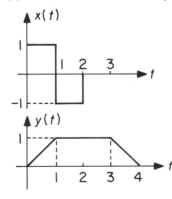

Problem 11.16

a) Show that the $\mathcal{L}$-transforms of the following functions are as given. (HINT: For $Y(s)$ it may be easier to find $\mathcal{L}[dy(t)/dt]$ first.)

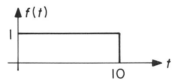

$$X(s) = \frac{(1 - e^{-s})^2}{s}$$

$$Y(s) = \frac{(1 - e^{-s})(1 - e^{-3s})}{s^2}$$

b) If $x(t)$ and $y(t)$ are the input and output respectively of an LTI system, find the system function $H(s)$ of this system.

c) Find and sketch the impulse response $h(t)$ of this system. Is the system BIBO stable?

d) Find and sketch the step response of the system.

Problem 11.17

The proper handling of unilateral transforms of functions containing singularities at $t = 0$ is not without its subtleties. This problem is intended to help you appreciate some of the implications of the convention described by (11.2–8), and the discussion in Section 11.4.

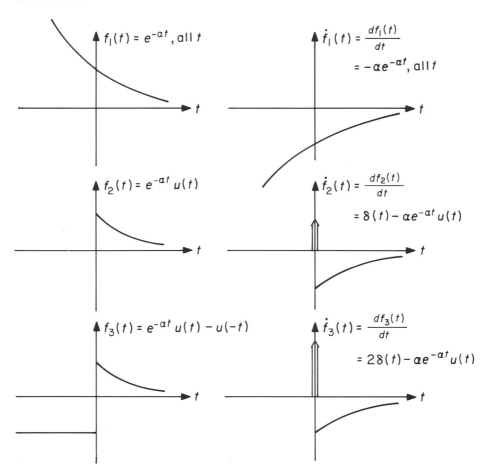

a) Find the Laplace transform, defined as

$$\mathcal{L}[f(t)] = \int_{0-}^{\infty} f(t)e^{-st}\, dt$$

for each of the six functions above.

b) Show that your results are consistent with the differentiation formula

$$\mathcal{L}[\dot{f}(t)] = s\mathcal{L}[f(t)] - \text{``}f(0)\text{''}, \quad \dot{f}(t) = \frac{df(t)}{dt}$$

provided that "$f(0)$" is interpreted as $f(0-)$.

c) On the other hand show that your results are consistent with the Initial Value
Theorem of Problem 2.4,

$$\lim_{s \to \infty} s \mathcal{L}[f(t)] = \text{``}f(0)\text{''}$$

provided that the limit exists and "$f(0)$" is interpreted as $f(0+)$.

Problem 11.18

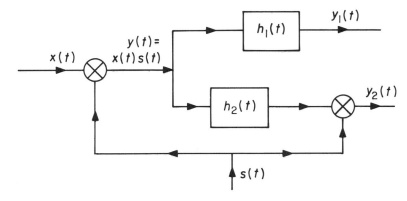

In the block diagram above $s(t)$ is a periodic train of unit impulses at times $t = nT$,
$n = \ldots, -2, -1, 0, 1, 2, \ldots$.

a) Find $h_1(t)$ (in terms of $h_2(t)$) such that $y_1(t) = y_2(t)$ for any $x(t)$.

b) Sketch $h_1(t)$ satisfying this condition if $h_2(t)$ is as shown below.

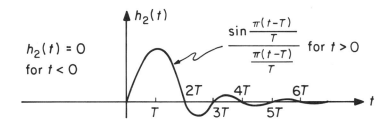

Problem 11.19

Approximations to ideal delay elements can be achieved in various ways—lumped
networks, distributed electrical or acoustic systems, tape recorders with multiple heads,
charge-coupled devices, etc. The inevitable effect of approximations, however, is to
restrict the class of inputs for which the output looks reasonably like a delayed copy
of the input. As an example, consider the all-pass circuit of Problem 4.2 with $R = 1\,\Omega$,
$C = 1\,\text{F}$, or equivalently the op-amp circuit on the next page. Both are intended to
approximate an ideal delay of two seconds.

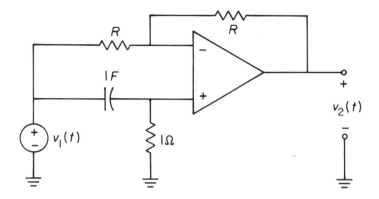

a) Show that the system function for this system is $H(s) = \dfrac{s-1}{s+1}$.

b) By inverse transforming, find the impulse response $h(t)$. (Be careful: $H(s)$ is an improper function.)

c) Find by convolution the response of this system to a unit step drive $v_1(t) = u(t)$.

d) Check your answer to (c) by $\mathcal{L}$-transforms. Check it also by showing that the derivative of the step response is $h(t)$.

e) Find by convolution the response of this system to a unit ramp drive $v_1(t) = tu(t)$.

f) Check your answer to (e) by $\mathcal{L}$-transforms. Check it also by showing that the derivative of the unit ramp response is the unit step response.

g) To test the extent to which this system approximates an ideal delay of two seconds, sketch carefully and approximately to scale on separate axes the impulse response, the step response, and the ramp response. Superimpose on each plot the ideal responses, that is, a delayed impulse, delayed step, and delayed ramp respectively.

h) What is the system function of an ideal delay of two seconds, that is, the system function corresponding to an impulse response $\delta(t - 2)$? Show that the $H(s)$ for the system above approximates the ideal system function for *small* values of s. (HINT: Expand $H(s)$ and the system function of the ideal delay in power series in s.)

i) Use the results of (h) to help explain why the system is a much better approximation to an ideal delay for the ramp input than for the step or impulse.

Problem 11.20

Linear system theory can be applied to the dynamics of simple automobile traffic situations.

a) Suppose first that the driver of the second of two cars attempts to keep his car a fixed distance, D, behind the first car. Let us suppose that to achieve this he senses the distance error

$$e_d(t) = x_1(t) - x_2(t) - D$$

(where $x_1(t)$ and $x_2(t)$ are the positions of the two cars as functions of time), and then accelerates (or decelerates) the second car in proportion. Of course, his reactions are not instantaneous. Assume that the acceleration $\ddot{x}_2(t) = d^2 x_2(t)/dt^2$ of the second car can be described as the output of an LTI system with impulse response $h_a(t)$ as shown below.

$$e_d(t) \longrightarrow \boxed{h_a(t) = \begin{cases} k_d\, e^{-at}, & t \ge 0 \\ 0, & t < 0 \\ \end{cases} \quad k_d > 0, \quad a > 0} \longrightarrow \ddot{x}_2(t)$$

Draw a block diagram that relates $x_2(t)$ to $x_1(t)$. For what values of k_d and α is the system stable?

b) Suppose next that the driver of the second car ignores the distance between cars and instead attempts to match his velocity $\dot{x}_2(t) = dx_2(t)/dt$ to that of the first car. Again his reactions are slow. Assume for this case that they can be described by the LTI system shown below.

$$e_v(t) = \dot{x}_1(t) - \dot{x}_2(t) \longrightarrow \boxed{h_b(t) = \begin{cases} k_v\, e^{-\beta t}, & t \ge 0 \\ 0, & t < 0 \\ \end{cases} \quad k_v > 0, \quad \beta > 0} \longrightarrow \ddot{x}_2(t)$$

Draw a block diagram relating $x_2(t)$ to $x_1(t)$ for this case. For what values of k_v and β is the system stable?

c) Consider a system of the type in (b), but where the second driver's reactions are so fast that we may approximate

$$h_b(t) = k_v \delta(t).$$

At $t = 0-$, both cars are stationary at $x = 0$. The velocity of the first car is shown below.

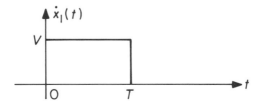

Sketch $x_2(t)$ as a function of time.

d) Prove or disprove the following statement with respect to the system in (c):

At $t = 0-$, both cars are stationary at $x = 0$. If $\dot{x}_1(t)$ is an arbitrary, bounded, non-negative function of time, then the second car will never hit the first. (Assume that the cars have zero length.)

Problem 11.21

One of the classical "paradoxes" of electrical circuit theory is the following.   An uncharged ideal capacitor of C farads is suddenly connected to an ideal battery of V volts; the voltage across the capacitor is thus

$$v(t) = V u(t).$$

The current that flows is

$$i(t) = C\frac{dv(t)}{dt} = CV\delta(t).$$

The "paradox" comes from the fact that the energy finally stored in the capacitor is clearly $\frac{1}{2}CV^2$ whereas the energy apparently supplied by the battery is

$$\int_{-\infty}^{\infty} V i(t)\, dt = V^2 \int_{-\infty}^{\infty} C\delta(t)\, dt = CV^2.$$

What has become of half of the energy? This is a puzzle, certainly, but the real nature of the "paradox" arises from the fact that several different explanations are possible.

Explanation 1:

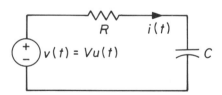

Every "real" capacitor must have some lead resistance and every "real" battery has a nonzero internal resistance. Thus suppose we model the situation more exactly by the circuit above.

a) Write a differential equation describing this circuit, and solve this equation to obtain $i(t)$, $t > 0$, assuming that the charge on C at $t = 0$ is zero.

b) Show that the energy lost in heating the resistor, $\int_0^{\infty} Ri^2(t)\, dt$, is equal for any $R > 0$ to the final energy stored in the capacitor, $\frac{1}{2}CV^2$, and that the sum of these energies is equal to the energy supplied by the battery, $\int_0^{\infty} V i(t)\, dt$. Thus one can argue that no matter how small R may be, the energy supplied by the battery is CV^2, and half of it is turned into heat.

Explanation 2:

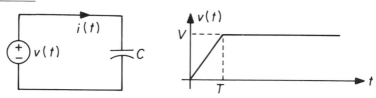

No "real" switch could in fact apply the battery voltage instantaneously to the capacitor. Suppose we model the actual behavior of the switch by the circuit and waveform shown above.

a) Determine and sketch the current $i(t)$ in the circuit.

b) Show that the energy supplied by the source, $\int_0^\infty v(t)i(t)\,dt$, is just equal to the energy finally stored in C, $\frac{1}{2}CV^2$, for any value of T, $0 < T < \infty$.

Thus one can argue that the energy supplied by the battery, if "correctly" computed, is simply $\frac{1}{2}CV^2$, so that it is unnecessary to postulate an energy loss in incidental resistance to achieve an energy balance.

Explanation 3:

It can be shown that the battery energy as computed in Explanation 2 is independent of the particular waveform $v(t)$ as long as $v(t)$ increases monotonically from 0 to V. To combine the effects of the slowly acting switch and the incidental resistance, let's re-solve the problem with

$$v(t) = V(1 - e^{-\alpha t})u(t), \quad \alpha > 0$$

which is somewhat simpler to analyze than the ramp-and-step.

a) Re-solve the differential equation of Explanation 1 to find $i(t)$ for this choice of $v(t)$ and the same initial condition as before.

b) Determine the energy supplied by the battery and show that the result is

$$\int_0^\infty e(t)i(t)\,dt = \frac{1}{2}CV^2\frac{1 + 2\alpha RC}{1 + \alpha RC}$$

which lies between $\frac{1}{2}CV^2$ and CV^2 depending on the value of αRC.

What, then, is the "true" energy supplied by the idealized battery? Is it CV^2, $\frac{1}{2}CV^2$, or something in between? There is no way to answer this question within the domain of circuit theory. To give a value for the energy supplied by the idealized battery implies giving a value for the symbolic integral

$$\int_{-\infty}^{\infty} \delta(t)u(t)\,dt$$

and, as we have emphasized in this chapter, this cannot be done in an unambiguous way. The energy supplied by the battery depends on the precise details of the modelling; different limiting arguments may give different results even if the "limits" appear to be the same.

Problem 11.22

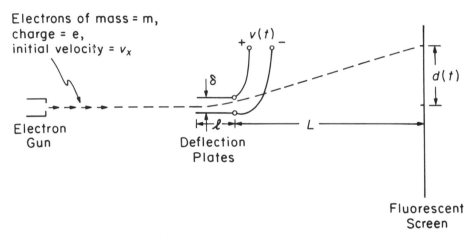

Electrons of mass = m, charge = e, initial velocity = v_x

Electron Gun

Deflection Plates

$v(t)$

$d(t)$

Fluorescent Screen

The figure above shows the essential features of the electrostatic deflection system of a cathode ray tube.

a) Assume that $\delta \ll \ell$ and that $v(t)$ changes slowly enough so that the electric field between the deflection plates is essentially the quasi-static value $v(t)/\delta$ and may be considered uniform throughout the space between the plates (i.e., neglect fringing). Find the response $d(t)$ to a unit step input $v(t) = u(t)$.

b) By differentiating, find the impulse response of this system. Evaluate the impulse response numerically for $\delta = 2$ mm, $\ell = 1$ cm, $L = 30$ cm, $e/m = 1.76 \times 10^5$ coul/kg, and assuming that the electron gun has accelerated the electrons through a potential of 2500 volts.

c) By combining step responses, sketch the responses to short pulses of several durations. What is the shortest pulse duration for the numbers in (b) that you feel would be reproduced by this oscilloscope with "acceptable" fidelity? How does this duration compare with the "duration" of the impulse response?

12

FREQUENCY-DOMAIN METHODS
FOR GENERAL LTI SYSTEMS

12.0 Introduction

The frequency-domain methods employed so far in this book were introduced in Chapters 2 and 8 through the unilateral $\mathcal{L}$-transform and the unilateral Z-transform. They are thus automatically restricted to causal systems. Moreover, only stimuli and responses for $t > 0$ have been explicitly considered; the effects of inputs for $t < 0$, if any, have been described in terms of the ZIR to an initial state.

The usefulness of frequency-domain methods is not dependent on state formulations and causality, but it is intimately tied to the concepts of linearity and time-invariance. The key to the description of linear systems and their signals is to recognize that if $\{\xi_i(t)\}$ is a set of inputs yielding known responses $\eta_i(t) = F[\xi_i(t)]$, then any input that can be represented as a weighted sum of these inputs, in the form

$$x(t) = \sum_i x_i \xi_i(t) \tag{12.0--1}$$

will produce the response

$$y(t) = F[x(t)] = \sum_i x_i \eta_i(t) . \tag{12.0--2}$$

Thus the principal problem in the functional description of LTI systems is to choose the set $\{\xi_i(t)\}$ so that

a) the set of waveforms that can be represented as in (12.0–1) is as large as possible, and

b) the description of the sets $\{\xi_i(t)\}$ and $\{\eta_i(t)\}$ is as simple as possible.

As we saw in Chapters 9 and 11, the set of delayed impulses is one appropriate choice for $\{\xi_i(t)\}$, but there are others. In particular, for every linear system (time-invariant or not) there is a special set of functions called *eigenfunctions* or *characteristic functions* having the property that

$$\eta_i(t) = F[\xi_i(t)] = \lambda_i \xi_i(t) \tag{12.0--3}$$

where the λ_i are constants called *eigenvalues* or *characteristic values*. For these special inputs *the output is identically the same waveform as the input* (except for a multiplying constant), and thus the difficulty of describing the sets $\{\xi_i(t)\}$ and $\{\eta_i(t)\}$ is cut in half. In general, the set of functions $\{\xi_i(t)\}$ that satisfy (12.0–3) depends upon the detailed properties of the system, as reflected in $F[\,\cdot\,]$. But, if the system is time-invariant as well as linear, then it is a remarkable fact that the eigenfunctions are almost independent of the specific system details: *The eigenfunctions of every LTI system are simply the complex exponentials* e^{st}, $-\infty < t < \infty$. Different values of the complex frequency s give different eigenfunctions.*

We have already essentially proved this result for lumped causal circuits in Section 3.3, where we demonstrated that the response to the input $e^{st}u(t)$ becomes predominantly $H(s)e^{st}$ as time passes (provided that s lies to the right of the rightmost pole of $H(s)$, that is, in the domain of convergence). Since the location of the time origin for time-invariant systems is arbitrary, this basically means that the response to e^{st}, $-\infty < t < \infty$ (that is, to e^{st} started "at" $t = -\infty$), will be $H(s)e^{st}$ at all *finite* times, $-\infty < t < \infty$, for such causal circuits and for s in the domain of convergence.

We can readily prove this eigenfunction property in general for an arbitrary (not necessarily causal) LTI system using the superposition integral (11.3–11), which was introduced to describe such systems:

$$y(t) = \int_{-\infty}^{\infty} x(\tau)h(t-\tau)\,d\tau = \int_{-\infty}^{\infty} h(\tau)x(t-\tau)\,d\tau\,. \qquad (12.0\text{–}4)$$

Setting $x(t) = e^{st}$, $-\infty < t < \infty$, in the second form we obtain

$$y(t) = \int_{-\infty}^{\infty} h(\tau)e^{s(t-\tau)}\,d\tau = \left[\int_{-\infty}^{\infty} h(\tau)e^{-s\tau}\,d\tau\right]e^{st} = H(s)e^{st} \qquad (12.0\text{–}5)$$

where

$$\boxed{H(s) = \int_{-\infty}^{\infty} h(t)e^{-st}\,dt} \qquad (12.0\text{–}6)$$

is a generalization of the system function defined for causal systems as $\mathcal{L}[h(t)]$ in Section 10.1. From (12.0–5), $H(s)$ is the eigenvalue for the eigenfunction e^{st}.

For an LTI system, the response to any member of the set of inputs $\{e^{s_i t}\}$, $-\infty < t < \infty$, where $\{s_i\}$ is some allowed set of complex numbers, is easy to describe—we need know only one function $H(s)$ since the set of responses is $\{H(s_i)e^{s_i t}\}$, $-\infty < t < \infty$. Moreover, we deduce immediately from linearity that the response to any input that we can write as

$$x(t) = \sum_i X_i e^{s_i t}\,, \quad -\infty < t < \infty \qquad (12.0\text{–}7)$$

---

*There are certain limitations on the values of s for which e^{st} is an eigenfunction, as we shall discuss. There are in addition special situations in which functions more complex than simple exponentials are also eigenfunctions of LTI systems. See Problem 14.14.

is just

$$y(t) = \sum_i X_i H(s_i) e^{s_i t}, \quad -\infty < t < \infty. \tag{12.0-8}$$

The idea behind equations (12.0–5) to (12.0–8) is one of the most significant in system theory and the main reason why the notion of frequency is so important in applications. (Indeed, practicing engineers sometimes become such slaves to the concept of frequency that they find it hard to think about signals as time functions!) From an analytical point of view, an outstanding advantage of the frequency-domain description of systems (as we have seen with L-transforms) is that the system functions of cascaded systems simply *multiply* (which is often much simpler both to visualize and to evaluate than the convolution of their impulse responses). Moreover, the eigenfunction property of e^{st} implies a type of isolation between signals that is fundamental to a wide variety of devices. For example, the whole idea behind assigning separate radio stations distinct *frequency* channels is that their signals will not become intermixed in passing through the (essentially LTI) transmission medium they share, and one of the problems with a satellite as a communication relay station to be used jointly by a number of users (*multiplexed*) is precisely that such a "transmission medium" is usually neither linear nor time-invariant. The frequency-domain approach to general LTI systems provides an alternate vantage point to the time-domain approach discussed in the last chapter; for purposes of both analysis and synthesis the view will often be clearer, as we shall see, from the frequency domain. And, of course, two ways of looking at the same situation are likely to provide more insight than either one alone.

There remain two general questions:

1. What restrictions, if any, are there on the values of s for which $x(t) = e^{st}$ $\implies y(t) = H(s)e^{st}$?

2. How large a class of time functions $x(t)$ can be represented as in (12.0–7) with the s_i restricted as required?

These are the basic topics of this chapter and the following one.

12.1 Strips of Convergence for $H(s)$

The region of the s-plane within which the input e^{st}, $-\infty < t < \infty$, yields the output $H(s)e^{st}$, $-\infty < t < \infty$, for a particular LTI system is that domain within which the integral

$$H(s) = \int_{-\infty}^{\infty} h(t)e^{-st}\, dt$$

exists or is convergent in some sense. It is easy to prove that this domain in general is a single strip parallel to the $j\omega$-axis, as shown in Figure 12.1–1. Either one or both of the boundaries may be at $\pm\infty$; it is also possible that $\sigma_1 = \sigma_2$, so that the strip becomes a single line.

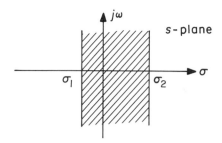

Figure 12.1–1. A general convergence strip.

Some examples illustrating these various possibilities are explored in Problem 12.2. Several important special cases deserve particular notice:

a) If $h(t)$ is *causal*—or even if $h(t)$ is simply *right-sided* (that is, if there exists a T, not necessarily positive, such that $h(t) \equiv 0$, $t < T$)—the domain of convergence of $H(s)$ is always a *right half-plane*.

b) Similarly if $h(t)$ is *left-sided*, the domain of convergence is a *left half-plane*.

c) If $h(t)$ is a *pulse* (that is, both right- and left-sided), the domain of convergence is the *entire s-plane*.

d) If $h(t)$ is *BIBO stable* (that is, if $\int_{-\infty}^{\infty} |h(t)| \, dt < \infty$), the domain of convergence always *includes the jω-axis*.

Generalizing the discussion of the causal case in Section 3.3, we note that attempts to drive a system with the input e^{st}, $-\infty < t < \infty$, with s *outside* the strip of convergence of $H(s)$ would induce "transients" resulting from the way the input is "started" and "stopped" at $t = \pm\infty$. These would be infinitely large at finite times and would thus dominate the response. Only for s in the domain of convergence is e^{st} an eigenfunction of the system.

12.2 The Fourier Integral

To complete our development of the frequency-domain analysis scheme of Section 12.0, we must still answer the questions:

a) How large a class of time functions can be represented as a weighted sum of exponentials with complex frequencies selected from the values lying solely in the strip of convergence as defined in the previous section?

b) How can we determine the appropriate weights?

Formulating these questions symbolically, we seek a representation for input time functions that has the form of equation (12.0–7), that is,

$$x(t) = \sum_i X_i e^{s_i t}, \quad -\infty < t < \infty \qquad (12.2\text{–}1)$$

where the s_i are restricted to the convergence strip for the system function $H(s)$; that is, in general, $\sigma_1 < \Re e[s_i] < \sigma_2$.

One might expect that the most general representations would result from exploiting all allowed values of s, such as representations of the form

$$x(t) = \int_{\sigma_1}^{\sigma_2} d\sigma \int_{-\infty}^{\infty} d\omega\, X(\sigma,\omega)e^{(\sigma+j\omega)t}.$$

But schemes of this sort have been little studied and even less used—largely because such generality is rarely needed.* For most purposes it is sufficient to restrict attention to the values of s along a single straight line parallel to the $j\omega$-axis, that is, to representations of the form

$$x(t) = \frac{1}{2\pi} \int_{-\infty}^{\infty} X(\sigma_0 + j\omega)e^{(\sigma_0+j\omega)t}\, d\omega. \tag{12.2--2}$$

(The $1/2\pi$ factor is introduced for later convenience.) Equation (12.2–2) is to be considered analogous to (12.2–1), that is, a representation of $x(t)$ in terms of a "sum" of characteristic inputs. If σ_0 can be chosen inside the convergence strip for $H(s)$, that is, if $\sigma_1 < \sigma_0 < \sigma_2$, then in analogy to (12.0–8) we have

$$y(t) = \frac{1}{2\pi} \int_{-\infty}^{\infty} X(\sigma_0 + j\omega)H(\sigma_0 + j\omega)e^{(\sigma_0+j\omega)t}\, d\omega. \tag{12.2--3}$$

In other words, $y(t)$ has the representation

$$y(t) = \frac{1}{2\pi} \int_{-\infty}^{\infty} Y(\sigma_0 + j\omega)e^{(\sigma_0+j\omega)t}\, d\omega \tag{12.2--4}$$

with

$$Y(\sigma_0 + j\omega) = X(\sigma_0 + j\omega)H(\sigma_0 + j\omega). \tag{12.2--5}$$

This illustrates the overall scheme we have in mind.

We shall return briefly to representations similar to (12.2–2) in the appendix to Chapter 13. For the moment, however, we shall further restrict attention to the case in which $h(t)$ is stable, so that the $j\omega$-axis is inside the convergence strip and we may choose $\sigma_0 = 0$. Then (12.2–2) becomes

$$x(t) = \frac{1}{2\pi} \int_{-\infty}^{\infty} X(j\omega)e^{j\omega t}\, d\omega \tag{12.2--6}$$

which is called the *Fourier integral representation* of $x(t)$. In manipulating Fourier integrals it turns out for most purposes to be convenient to choose $f = \omega/2\pi$, the angular frequency in Hertz (or cycles/sec), rather than ω, the radian frequency in radians/sec, as the variable of integration, so that (12.2–6) becomes

$$x(t) = \int_{-\infty}^{\infty} X(j2\pi f)e^{j2\pi f t}\, df. \tag{12.2--7}$$

---

*See, however, W. M. Brown, *Analysis of Linear Time-Invariant Systems* (New York, NY: McGraw-Hill, 1963).

Furthermore, since we have not as yet defined $X(j2\pi f)$, it is also convenient to suppress the $j2\pi$ factor and write simply

$$x(t) = \int_{-\infty}^{\infty} X(f)e^{j2\pi ft}\, df \qquad (12.2\text{–}8)$$

which we shall adopt as the basic form for the Fourier integral representation of $x(t)$. For the same reasons of convenience, we replace the system function $H(j\omega) = H(j2\pi f)$ by $H(f)$, suppressing the $j2\pi$ factor. $H(f)$ will be called the *frequency response* to distinguish it from the system function $H(s)$. It is related to the impulse response $h(t)$ by

$$H(f) = \int_{-\infty}^{\infty} h(t)e^{-j2\pi ft}\, dt \qquad (12.2\text{–}9)$$

which exists if the system is stable. In terms of $H(f)$ and $X(f)$ we can write

$$y(t) = \int_{-\infty}^{\infty} Y(f)e^{j2\pi ft}\, df$$

with

$$Y(f) = H(f)X(f). \qquad (12.2\text{–}10)$$

We are, of course, running some risk of confusion by using the same functional notation for the frequency response, $H(f)$, and the system function, $H(s)$—particularly since the frequency response is equal to the system function with s replaced by $j2\pi f$. A similar potential for confusion exists (if $x(t) = 0$ for $t < 0$) between the *Fourier transform*, $X(f)$, and the $\mathcal{L}$-transform, $X(s)$, of $x(t)$. In practice, however, this turns out to be less troublesome than one might anticipate. The gains in simplicity and symmetry seem generally to be well worth the risk. And on those comparatively rare occasions when we do wish to talk about frequency responses and system functions, Fourier and Laplace transforms, in the same problem we can simply revert to the notation $H(j2\pi f)$ and $X(j2\pi f)$.

Our fundamental problem in the next few chapters is to determine what limits if any must be placed on the classes of $x(t)$ that can be represented as in (12.2–8), and to learn how to select the appropriate corresponding Fourier transform $X(f)$.

12.3 A Special Case—Fourier Series

The most illuminating approach to an understanding of the Fourier integral seems to be to follow what is loosely the historical route, that is, to start with *Fourier series*. In some ways the idea behind Fourier series goes back to early Babylon,* and traces of the notion can be seen in the cycles and epicycles of Ptolemy's astronomical system and in much of the thinking on musical consonances starting perhaps with Pythagoras. In more modern times, a number of early 18th-century mathematicians (including Euler and D. Bernoulli) were aware that if it somehow were known that a waveform $x(t)$ could be expressed as a *finite* weighted sum of *harmonically* related sinusoids, that is, if it were *known* that

$$x(t) = a_0 + \sum_{n=1}^{N} a_n \cos \frac{2\pi nt}{T} + \sum_{n=1}^{N} b_n \sin \frac{2\pi nt}{T} \tag{12.3-1}$$

(where T is a constant whose significance we shall explore shortly), then the values of the coefficients a_n and b_n could be obtained from the formulas

$$a_0 = \frac{1}{T} \int_0^T x(t)\, dt \tag{12.3-2}$$

$$a_m = \frac{2}{T} \int_0^T x(t) \cos \frac{2\pi mt}{T}\, dt, \quad m \neq 0 \tag{12.3-3}$$

$$b_m = \frac{2}{T} \int_0^T x(t) \sin \frac{2\pi mt}{T}\, dt. \tag{12.3-4}$$

These results follow easily from the observation that sinusoids whose frequencies are integer multiples of some *fundamental frequency*, $f_0 = 1/T$, form an *orthogonal*[†] set of functions, that is,

$$\frac{2}{T} \int_0^T \sin \frac{2\pi nt}{T} \cos \frac{2\pi mt}{T}\, dt = 0, \quad \text{all } n, m \tag{12.3-5}$$

$$\frac{2}{T} \int_0^T \sin \frac{2\pi nt}{T} \sin \frac{2\pi mt}{T}\, dt = \frac{2}{T} \int_0^T \cos \frac{2\pi nt}{T} \cos \frac{2\pi mt}{T}\, dt$$

$$= \begin{cases} 0, & n \neq m \\ 1, & n = m \neq 0. \end{cases} \tag{12.3-6}$$

Thus substituting the expansion for $x(t)$ from (12.3–1) into, for example, the right-hand side of (12.3–3), and interchanging the order of integration and summation, all of the terms vanish by (12.3–5) or (12.3–6) except the one whose value is a_m.

Obviously only a restricted class of waveforms can be represented as in (12.3–1). In particular, since $\begin{Bmatrix} \sin \\ \cos \end{Bmatrix}(\alpha + 2\pi n) = \begin{Bmatrix} \sin \\ \cos \end{Bmatrix}(\alpha)$, it is evident that

$$x(t + T) = x(t) \tag{12.3-7}$$

---

*See G. de Santillana, *The Origins of Scientific Thought* (Chicago, IL: University of Chicago Press, 1961).

[†]The geometric implications of this name will be explored in the appendix to Chapter 14.

that is, $x(t)$ must be periodic with *period T*. (Obviously any multiple of T is also a period of $x(t)$.) But even among periodic functions the class of $x(t)$ consisting of a finite sum of sinusoids would seem to be a rather special one. Nevertheless, J. B. J. Fourier* rather audaciously proposed in 1807 that an infinite series of the form (12.3–1) could in fact represent an arbitrary periodic function, even one containing discontinuities. Moreover, he argued that the coefficients of the expansion could be determined from the same formulas, (12.3–2, 3, 4), as in the finite case. Fourier was more an engineer or physicist than a mathematician, and his proofs of this proposition did not satisfy even his contemporaries.[†] But it soon became clear that he was more nearly right than wrong, although many years passed before the problem was fully clarified.[‡] Indeed the study of Fourier series has turned out to be one of the most fruitful in the history of mathematics,[§] leading to a number of important results including major revisions in the notions of functions and convergence. But before we get further involved with such matters, we can easily illustrate with an example that at least in some cases the Fourier series does in fact seem to converge to the $x(t)$ from which it was derived, even though $x(t)$ is not at all as smooth as one would expect a sum of sinusoids to be.

Example 12.3–1

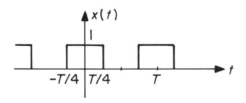

Figure 12.3–1. A periodic square wave.

---

*J. B. J. Fourier, *Théorie Analytique de la Chaleur*, originally published in 1822. A recent edition in English was published by Dover, New York.

[†]Fourier presented his ideas in a prize paper on the theory of heat submitted to the French Academy. The judges—including that "illustrious trio" Laplace, Lagrange, and Legendre— agreed to give Fourier the prize but expressed concern over the non-rigorous mathematical methods he had employed. It is easy to see with our modern understanding of relative orders of infinity that some sort of restriction on $x(t)$ is necessary to make Fourier's proposal correct since the number of points at which $x(t)$ can be independently specified in the period $0 < t < T$ is uncountably infinite, whereas the series (12.3–1) with $N \to \infty$ contains only a countably infinite number of independent coefficients. Incidentally, the founder of the theory of infinite sets, G. Cantor, was led to his studies by his earlier work on Fourier series.

[‡]The first rigorous proof of a version of Fourier's Theorem was given by Dirichlet in 1829. A complete discussion was not possible until the development of Lebesgue integration in the early years of this century—nearly a hundred years after the "theorem" was first proposed. The study of Fourier series remains an active branch of mathematics.

[§]See, for example, H. S. Carslaw, *Introduction to the Theory of Fourier's Series and Integrals* (London: Macmillan, 1930) pp. 1–30.

Consider the periodic square wave shown in Figure 12.3–1. The fundamental frequency is $f_0 = 1/T$ Hz. Since $\sin(2\pi nt/T)$ is an odd function and $x(t)$ is an even function, all of the b_m from (12.3–2) will be zero. However,

$$a_0 = \frac{1}{T} \int_{-T/2}^{T/2} x(t)\,dt = \frac{1}{T} \int_{-T/4}^{T/4} 1\,dt = \frac{1}{2}.$$

(Note that since $x(t)$ is periodic the limits on the integrals in (12.3–2, 3, 4) may be any convenient period such as $-T/2$ to $T/2$ rather than 0 to T.) Furthermore, for $m \neq 0$,

$$
\begin{aligned}
a_m &= \frac{2}{T} \int_{-T/2}^{T/2} x(t) \cos \frac{2\pi mt}{T}\,dt = \frac{2}{T} \int_{-T/4}^{T/4} \cos \frac{2\pi mt}{T}\,dt \\
&= \frac{2}{T} \frac{T}{2\pi m} \sin \frac{2\pi mt}{T} \Bigg|_{-T/4}^{T/4} = \frac{1}{\pi m}\left[\sin\left(\frac{\pi m}{2}\right) - \sin\left(-\frac{\pi m}{2}\right) \right] \\
&= \begin{cases} \dfrac{2(-1)^{(m-1)/2}}{\pi m}, & m = 1,\,3,\,5,\,\ldots \\ 0, & m = 2,\,4,\,6,\,\ldots. \end{cases}
\end{aligned}
$$

The Fourier series for this $x(t)$ is thus

$$
\begin{aligned}
x(t) &= \frac{1}{2} + \sum_{\substack{n=1 \\ n\ \text{odd}}}^{\infty} \frac{2(-1)^{(n-1)/2}}{\pi n} \cos \frac{2\pi nt}{T} \\
&= \frac{1}{2} + \frac{2}{\pi} \cos \frac{2\pi t}{T} - \frac{2}{3\pi} \cos \frac{6\pi t}{T} + \frac{2}{5\pi} \cos \frac{10\pi t}{T} - \cdots.
\end{aligned}
\tag{12.3–8}
$$

The way in which partial sums of this series converge to $x(t)$ is illustrated in Figure 12.3–2. In each sketch the preceding sum is shown dotted so that the effect of the added term can be assessed.

▶ ▶ ▶

12.4 Other Forms of Fourier Series; Spectra

The form of Fourier series described by

$$x(t) = a_0 + \sum_{n=1}^{\infty} a_n \cos \frac{2\pi nt}{T} + \sum_{n=1}^{\infty} b_n \sin \frac{2\pi nt}{T} \tag{12.4–1}$$

where

$$a_0 = \frac{1}{T} \int_0^T x(t)\,dt \tag{12.4–2}$$

and

$$\left\{ \begin{matrix} a_n \\ b_n \end{matrix} \right\} = \frac{2}{T} \int_0^T x(t) \left\{ \begin{matrix} \cos \dfrac{2\pi nt}{T} \\[2mm] \sin \dfrac{2\pi nt}{T} \end{matrix} \right\} dt, \quad n \geq 1 \tag{12.4–3}$$

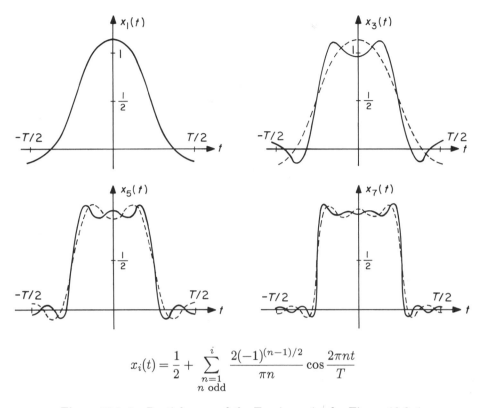

$$x_i(t) = \frac{1}{2} + \sum_{\substack{n=1 \\ n \text{ odd}}}^{i} \frac{2(-1)^{(n-1)/2}}{\pi n} \cos \frac{2\pi n t}{T}$$

Figure 12.3–2. Partial sums of the Fourier series for Figure 12.3–1.

is called the *sine-cosine form*. This is probably the most familiar form, but there are others that are often more useful. An important one is the *magnitude-angle form*, based on the observation that $a_n \cos \dfrac{2\pi n t}{T} + b_n \sin \dfrac{2\pi n t}{T}$ can also be written as $c_n \cos \left(\dfrac{2\pi n t}{T} - \theta_n \right)$, so that (12.4–1) becomes

$$x(t) = a_0 + \sum_{n=1}^{\infty} c_n \cos \left(\frac{2\pi n t}{T} - \theta_n \right) \tag{12.4–4}$$

where

$$c_n = \sqrt{a_n^2 + b_n^2} \tag{12.4–5}$$

$$\theta_n = \tan^{-1}(b_n / a_n) \tag{12.4–6}$$

or

$$a_n = c_n \cos \theta_n, \qquad b_n = c_n \sin \theta_n. \tag{12.4–7}$$

But perhaps the simplest and most useful form—and the one we shall employ almost exclusively in the sequel—is the *exponential form:*[*]

$$x(t) = \sum_{n=-\infty}^{\infty} X[n]e^{j2\pi nt/T} \tag{12.4-8}$$

$$X[n] = \frac{1}{T}\int_{-T/2}^{T/2} x(t)e^{-j2\pi nt/T}\, dt. \tag{12.4-9}$$

In the exponential form only a single integral is required to define the coefficients, but note that the sum in (12.4-8) extends over *negative* values of n (negative "frequencies"!) as well as positive values. However, if $x(t)$ is real (as we have previously tacitly assumed), then it follows at once from (12.4-9) that $X[-n]$ is the complex conjugate of $X[n]$:

$$X[-n] = X^*[n]. \tag{12.4-10}$$

Thus the complex amplitudes of the negative frequency components are entirely determined by those for positive frequencies. By comparison with previous results we can readily establish that, for $n \geq 0$,

$$\begin{aligned}
&a_0 = X[0] && c_n = 2|X[n]| \\
&a_n = 2\Re e[X[n]], \ n \neq 0 && \theta_n = -\angle X[n] \\
&b_n = -2\Im m[X[n]] && X[n] = \tfrac{1}{2}(a_n - jb_n) = \tfrac{1}{2}c_n e^{-j\theta_n}.
\end{aligned} \tag{12.4-11}$$

The sine-cosine and magnitude-angle forms are rarely used for complex $x(t)$, but the exponential form (12.4-8) can easily be applied to complex waveforms (although, of course, (12.4-10) will not then be correct).

Example 12.4–1

For the square wave of Example 12.3–1 we have, from (12.4–9),

$$\begin{aligned}
X[n] &= \frac{1}{T}\int_{-T/4}^{T/4} 1\, e^{-j2\pi nt/T}\, dt = \frac{1}{T}\left(\frac{T}{-j2\pi n}\right)e^{-j2\pi nt/T}\Big|_{-T/4}^{T/4} \\
&= \frac{1}{-j2\pi n}\left[e^{-j\pi n/2} - e^{j\pi n/2}\right] \\
&= \begin{cases}
\dfrac{(-1)^{(n-1)/2}}{\pi n}, & n \text{ odd} \\
0, & n \text{ even}, n \neq 0 \\
1/2, & n = 0.
\end{cases}
\end{aligned}$$

From (12.4–11) these results are in agreement with those of Example 12.3–1.
▶ ▶ ▶

---

[*]As before, the square brackets in $X[n]$ indicate that the variable n takes on only integer values.

The set of Fourier coefficients $\{X[n]\}$ of (12.4–9) collectively constitute the *spectrum* of $x(t)$, and the process of determining them is called *spectral analysis*.* In particular, $X[0]$ is the *average* or *d-c* (for "direct-current") value of $x(t)$, and $X[1]$ is usually called the complex amplitude of the *fundamental component*. The angular frequency of the fundamental component is the reciprocal of the *fundamental period*, T, which is the smallest nonzero number such that

$$x(t + T) = x(t). \tag{12.4–12}$$

For $n > 1$, $X[n]$ is often called the complex amplitude of the nth *harmonic* of $x(t)$. The process of determining the set $\{X[n]\}$ is thus sometimes described as *harmonic analysis* of $x(t)$—the word "harmonic," of course, coming from the relationship with musical consonances.[†]

A periodic signal has a *discrete spectrum* because only a discrete set of frequencies is required in the spectral synthesis of such waveforms—the Fourier series (12.4–8). Discrete spectra are also called *line spectra*—an appropriate label since a convenient way to represent discrete spectral information graphically is in the form of a line or bar graph such as those shown in Figure 12.4–1. (Historically, the name "line spectrum" came from the fact that, in the usual display of the output of an optical spectrometer, a single isolated frequency component in the source appears as a bright line.) Since the Fourier coefficients are complex, two graphs—either real and imaginary parts, or magnitude and phase angle—are required for a complete representation.

For real waveforms, the conjugate symmetry of the coefficients implies that the real part and the magnitude of the spectrum are *even* functions of frequency or index n, whereas the imaginary part and phase angle are *odd* functions. Thus, if it is known that $x(t)$ is real, only the positive-frequency part of the spectrum need be shown. If, in addition to being real, $x(t)$ is an *even* function, that is, if $x(-t) = x(t)$, then it follows at once from (12.4–3) and (12.4–9) that the coefficients $X[n]$ are all *real* and that the sine-cosine form contains *cosines only*. Conversely, if $x(t)$ is real and *odd*, that is, if $x(-t) = -x(t)$, then the coefficients $X[n]$ are all purely *imaginary* and the sine-cosine form contains *sines only*.

---

*The word "spectrum" was introduced into physics by Newton (1664) to describe the analysis of light by a prism into its component colors or frequencies. The word is variously used in the mathematical and physical literature to denote the sets $\{X[n]\}$, $\{|X[n]|\} = \{c_n\}$, $\{|X[n]|^2\}$, etc. In this book "spectrum," without any modifying adjectives, will generally refer to the set of complex amplitudes $\{X[n]\}$; other types of spectra will be identified by labels such as "magnitude spectrum" or "power spectrum."

[†]That the idea of spectral analysis of sounds (in at least a limited sense) antedated Fourier is illustrated by the fact that Mersenne suggested in 1636 that a vibrating string "struck and sounded freely makes at least five sounds at the same time, the first of which is the natural sound of the string and serves as the foundation for the rest—[which] follow the ratio of the numbers 1, 2, 3, 4, 5." The terms "fundamental" and "harmonic" were introduced by Sauveur in 1704—more than 100 years before Fourier's prize paper. (For an interesting discussion of the early history of sound analysis, see R. Plomp, *Experiments on Tone Perception* (Soesterberg, Netherlands: Institute for Perception RVO-TNO, 1966).)

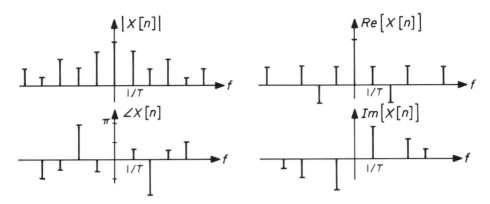

Figure 12.4–1. Representative discrete or line spectra, drawn for $x(t)$ real.

One final symmetry property requires a bit more discussion. A periodic function $x(t)$ is said to be *odd-harmonic* if its Fourier series is of the form

$$x(t) = \sum_{n \text{ odd}} X[n] e^{j2\pi nt/T}. \qquad (12.4\text{--}13)$$

That is, $x(t)$ is odd-harmonic if $X[n]$ vanishes for all even values of n. A necessary and sufficient condition for a periodic function to be odd-harmonic is that

$$x(t - T/2) = -x(t). \qquad (12.4\text{--}14)$$

To prove this, observe that if $X[n] = 0$ for n even, then

$$x(t - T/2) = \sum_{n \text{ odd}} X[n] e^{j2\pi nt/T} e^{-j\pi n} = -\sum_{n \text{ odd}} X[n] e^{j2\pi nt/T} = -x(t)$$

so (12.4–14) is necessary. Conversely, if (12.4–14) is satisfied, then for n even, so that we may define $m = n/2$,

$$X[2m] = \frac{1}{T} \int_{-T/2}^{T/2} x(t) e^{-j4\pi mt/T} \, dt$$

$$= \frac{1}{T} \int_0^{T/2} x(t) e^{-j4\pi mt/T} \, dt + \frac{1}{T} \int_{-T/2}^0 x(t) e^{-j4\pi mt/T} \, dt$$

$$= \frac{1}{T} \int_0^{T/2} x(t) e^{-j4\pi mt/T} \, dt + \frac{1}{T} \int_0^{T/2} x(t - T/2) e^{-j4\pi m(t-T/2)/T} \, dt$$

$$= \frac{1}{T} \int_0^{T/2} x(t) e^{-j4\pi mt/T} \, dt - \frac{1}{T} \int_0^{T/2} x(t) e^{-j4\pi mt/T} \, dt = 0$$

which proves the sufficiency of (12.4–14). In principle, it might also seem reasonable to define *even-harmonic* functions, but a moment's reflection will show that an even-harmonic function merely has a fundamental period half as long as what we have assumed, that is,

$$x(t - T/2) = x(t) \tag{12.4–15}$$

and hence is not especially interesting.

On the other hand, since we can write the sum for an arbitrary periodic $x(t)$ in the form

$$x(t) = \sum_{n=-\infty}^{\infty} X[n]e^{j2\pi nt/T} = \sum_{n \text{ odd}} X[n]e^{j2\pi nt/T} + \sum_{n \text{ even}} X[n]e^{j2\pi nt/T}$$

we can always represent an arbitrary $x(t)$ as the sum of even-harmonic and odd-harmonic components. To find these components, observe that

$$x(t) = \frac{1}{2}[x(t) + x(t - T/2)] + \frac{1}{2}[x(t) - x(t - T/2)].$$

The first bracket is easily shown from (12.4–15) to be even-harmonic and the second bracket from (12.4–14) to be odd-harmonic.

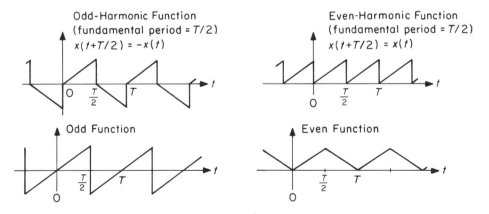

Figure 12.4–2. Some examples of symmetry conditions for periodic functions.

Examples of even and odd functions, as well as of even-harmonic and odd-harmonic functions, are shown in Figure 12.4–2. To illustrate some of these symmetry properties and to suggest how Fourier series can be applied to system analysis, consider the following example.

Example 12.4–2

Suppose that the frequency response, $H(f)$, of some real LTI system (called, for reasons that we shall clarify later, a *bandlimited differentiator*) has been measured experimentally and can be represented (perhaps somewhat idealized) as shown in Figure 12.4–3. By the frequency response, $H(f)$, we mean, of course, that if the input were $x(t) = e^{j2\pi ft}$, then the output would be $y(t) = H(f)e^{j2\pi ft}$. To be sure, the actual stimulus used to measure $H(f)$ is not a complex exponential but a real sinusoid, say $A\cos(2\pi ft + \theta)$. The steady-state response is then also a real sinusoid, say $B\cos(2\pi ft + \varphi)$, and we identify $|H(f)| = B/A$ and $\angle H(f) = \varphi - \theta$ for $f > 0$.

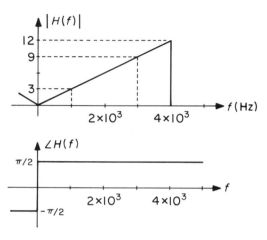

Figure 12.4–3. Frequency response, $H(f)$.

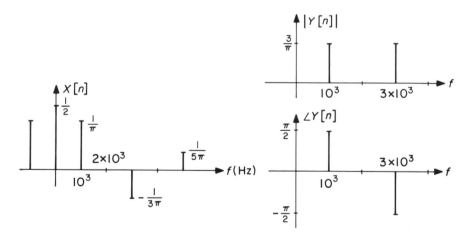

Figure 12.4–4. Square-wave spectrum.   **Figure 12.4–5.** Spectrum of $y(t)$.

We seek the response of this system to a 1000 Hz square wave, such as shown in Figure 12.3–1 with $T = 10^{-3}$ sec. From Example 12.4–1 we know that this waveform can be represented by the complex Fourier series,

$$x(t) = \sum_{n=-\infty}^{\infty} X[n]e^{j2\pi n10^3 t}$$

where the coefficients $X[n]$ were found in Example 12.4–1 and are plotted in Figure 12.4–4. Since the square wave of Figure 12.3–1 is an even function, the $X[n]$ are real, and only a single plot is required to show them. Note that in both the preceding figures and the present ones we show primarily positive frequencies; since both the system and $x(t)$ are real, both $H(f)$ and $X[n]$ have conjugate symmetry, that is,

$$X[-n] = X^*[n]$$

$$H(-f) = H^*(f).$$

(We have already discussed this result for $X[n]$; the conjugate symmetry of $H(f)$ follows in a similar way from (12.2–9), as we shall later explain.)

Since the response to $e^{j2\pi ft}$ is $H(f)e^{j2\pi ft}$ and since superposition holds, the response to $x(t)$ should be

$$y(t) = \sum_{n=-\infty}^{\infty} X[n]H(10^3 n)e^{j2\pi 10^3 nt} = \sum_{n=-\infty}^{\infty} Y[n]e^{j2\pi 10^3 nt}.$$

Comparing the sketches of $H(f)$ and $X[n]$, it should be clear that only a finite number of $Y[n]$ are nonzero. In fact

$$Y[0] = X[0]H(0) = (1/2) \cdot 0 = 0$$

$$Y[1] = Y^*[-1] = X[1]H(10^3) = \frac{1}{\pi}\left(3e^{j\pi/2}\right) = j\frac{3}{\pi}$$

$$Y[3] = Y^*[-3] = X[3]H(3 \times 10^3) = -\frac{1}{3\pi}\left(9e^{j\pi/2}\right) = -j\frac{3}{\pi}.$$

All other $Y[n]$ are zero; the magnitude and phase angle of $\{Y[n]\}$ are shown in Figure 12.4–5.

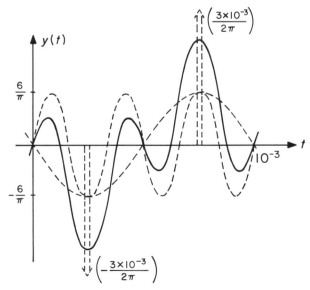

Figure 12.4–6. Response for Example 12.4–2.

Finally

$$y(t) = j\frac{3}{\pi}e^{-j6\pi 10^3 t} - j\frac{3}{\pi}e^{-j2\pi 10^3 t} + j\frac{3}{\pi}e^{j2\pi 10^3 t} - j\frac{3}{\pi}e^{j6\pi 10^3 t}$$

$$= -\frac{6}{\pi}\sin 2\pi 10^3 t + \frac{6}{\pi}\sin 6\pi 10^3 t\,.$$

This is sketched in Figure 12.4–6. Note that $y(t)$ is both odd and odd-harmonic. The dotted impulses correspond to $\dfrac{3 \times 10^{-3}}{2\pi}\dfrac{dx(t)}{dt}$; it is easy to see that the area of each impulse is roughly equal to the area under the peak in $y(t)$ at the same point in time, so that $y(t) \approx dx(t)/dt$.

▶ ▶ ▶

12.5 Averages of Periodic Functions; Parseval's Theorem

Since periodic functions extend with undiminished amplitude over $-\infty < t < \infty$, infinite-time integrals of such functions are generally undefined. But infinite-time *averages* of periodic functions are often interesting. The infinite-time average of a waveform $x(t)$ may be defined as

$$\langle x(t)\rangle \overset{\Delta}{=} \lim_{T_0 \to \infty} \frac{1}{2T_0}\int_{-T_0}^{T_0} x(t)\,dt\,. \tag{12.5–1}$$

If $x(t)$ is periodic with period T, then the infinite-time average (12.5–1) is equal to the average over a single period:

$$\langle x(t)\rangle = \frac{1}{T}\int_0^T x(t)\,dt\,.$$

This follows because for periodic $x(t)$ we can always write

$$\frac{1}{2T_0}\int_{-T_0}^{T_0} x(t)\,dt = \frac{1}{2T_0}\int_{-nT}^{nT} x(t)\,dt + \frac{1}{2T_0}\left[\int_0^\delta x(t)\,dt + \int_{-\delta}^0 x(t)\,dt\right]$$

where n is the largest number of periods contained in T_0 and δ is the leftover fraction of a period; that is, $T_0 = nT + \delta$ and $0 < \delta < T$. For large T_0 the second term vanishes and the first becomes

$$\lim_{T_0 \to \infty} \frac{2n}{2T_0}\int_0^T x(t)\,dt = \frac{1}{T}\int_0^T x(t)\,dt\,.$$

Of course, as we have already pointed out, the integral from 0 to T can be taken over any period such as $-T/2$ to $T/2$ or δ to $T + \delta$ without changing its value.

From (12.4–9) it should be evident that the average of $x(t)$ is simply $X[0]$, the zero-frequency or d-c Fourier coefficient:

$$\langle x(t)\rangle = X[0]\,.$$

But the idea of averaging can be extended to functions of $x(t)$. Thus the nth Fourier coefficient is the average of the product of $x(t)$ and $e^{-j2\pi nt/T}$:

$$X[n] = \langle x(t)e^{-j2\pi nt/T}\rangle. \tag{12.5-2}$$

This formula suggests a method for measuring experimentally the Fourier coefficients of an unknown periodic waveform by applying the product $x(t)e^{-j2\pi nt/T}$ (or, more realistically, by separately applying the real and imaginary parts of this product) to an LTI system with impulse response $h(t)$, as shown in Figure 12.5–1. If we examine the outputs $y_c(t)$ and $y_s(t)$ at any time $t > T_0$ after the input is switched on, they will approximate $\Re e[X[n]]$ and $\Im m[X[n]]$ respectively. Indeed, if T_0 is an integer multiple of the period T of $x(t)$, then the outputs for $t > T_0$ will be constants precisely equal to the desired averages. If T_0 is not an integer multiple of T but is large, $T_0 \gg T$, then the outputs will not be constants for $t > T_0$ but will vary periodically about the desired average, the variations tending to zero in amplitude as $T_0 \to \infty$. If T_0 is large enough, then the detailed shape of $h(t)$ is not very important; any $h(t)$ of long duration and unit area will perform about equally well. (Note the similarity of this discussion to that describing the unit impulse in Chapter 11. In a sense the impulse and the infinite-time average are dual generalized functions—one describing the limiting behavior of $h(t)$ in Figure 12.5–1 when T_0 is small, the other when T_0 is large.)

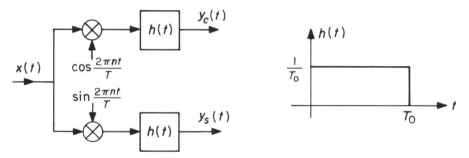

Figure 12.5–1. A method for measuring Fourier coefficients.

Another important average of a function of a real periodic waveform $x(t)$ is the *average power* in $x(t)$:

$$\langle x^2(t)\rangle = \frac{1}{T}\int_0^T x^2(t)\,dt. \tag{12.5-3}$$

The square root of the average power, called the *root-mean-square* (or *rms*) value of $x(t)$, is a useful measure of the amplitude of a complicated waveform.

Example 12.5–1

a) Let $x(t) = A\cos(\omega_0 t + \sigma)$. This $x(t)$ is periodic with period $T = 2\pi/\omega_0$; the average power in $x(t)$ is

$$\langle x^2(t)\rangle = \frac{\omega_0}{2\pi}\int_0^{2\pi/\omega_0} A^2\cos^2(\omega_0 t + \sigma)\,dt = \frac{A^2}{2}.$$

The rms amplitude of a sinusoid is thus $1/\sqrt{2} \approx 0.707$ times its peak value.

b) Let $x(t)$ be the square wave of Figure 12.3–1 or Figure 12.5–2:

$$x(t) = \begin{cases} 1, & -\dfrac{T}{4} < t < \dfrac{T}{4} \\[2mm] 0, & \dfrac{T}{4} < t < \dfrac{3T}{4} \end{cases}.$$

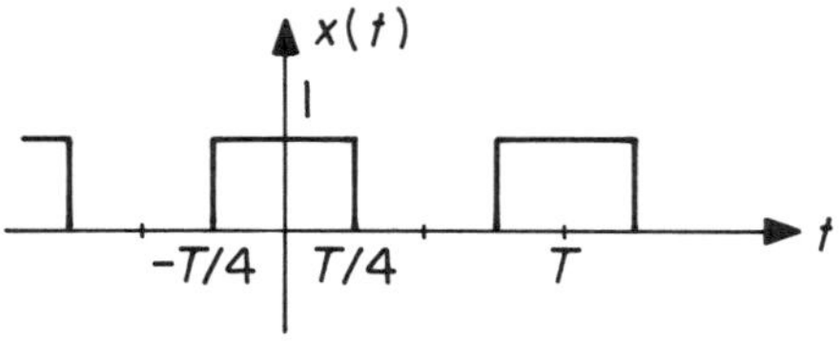

Figure 12.5–2. A periodic square wave.

Then the average power in $x(t)$ is

$$\langle x^2(t) \rangle = \frac{1}{T} \int_{-T/4}^{3T/4} x^2(t)\, dt$$

$$= \frac{1}{T} \int_{-T/4}^{T/4} 1^2\, dt = \frac{1}{2}.$$

Thus this square wave has the same power (or heating) effect as a steady d-c signal of amplitude 0.707.

▶ ▶ ▶

The average power in a periodic signal can also be written in another interesting way. Substituting the Fourier series expansion for $x(t)$ into the average power definition and interchanging summation and integration, we obtain

$$\frac{1}{T} \int_0^T x^2(t)\, dt = \frac{1}{T} \int_0^T x(t) \left[\sum_{n=-\infty}^{\infty} X[n] e^{j2\pi nt/T} \right] dt$$

$$= \sum_{n=-\infty}^{\infty} X[n] \underbrace{\left[\frac{1}{T} \int_0^T x(t) e^{j2\pi nt/T}\, dt \right]}_{X^*[n]}$$

or

$$\boxed{\langle x^2(t) \rangle = \frac{1}{T} \int_0^T x^2(t)\, dt = \sum_{n=-\infty}^{\infty} |X[n]|^2} \qquad (12.5\text{–}4)$$

which is *Parseval's Theorem* for real periodic functions. Note that the sinusoidal component of $x(t)$ constituting the nth harmonic is

$$X[n] e^{j2\pi nt/T} + X^*[n] e^{-j2\pi nt/T} = 2|X[n]| \cos\left(\frac{2\pi nt}{T} + \angle X[n] \right).$$

The average power in this sinusoid is (from Example 12.5–1a) just $2|X[n]|^2$. Hence Parseval's Theorem simply states that the average power in $x(t)$ is equal to the sum of the powers in its harmonic components.

Example 12.5–2

The Fourier coefficients of the square wave of Figure 12.5–2 were found in Example 12.4–1:

$$X[n] = \begin{cases} 1/2, & n = 0 \\ 0, & n \text{ even}, n \neq 0 \\ \dfrac{(-1)^{(n-1)/2}}{\pi n}, & n \text{ odd}. \end{cases}$$

From Example 12.5–1, $\langle x^2(t) \rangle = 1/2$. Hence by Parseval's Theorem we should have

$$\frac{1}{2} = \sum_{n=-\infty}^{\infty} |X[n]|^2 = \frac{1}{4} + 2 \sum_{\substack{n=1 \\ n \text{ odd}}}^{\infty} \frac{1}{n^2 \pi^2}.$$

This expression can be rearranged to give an interesting series for π^2:

$$\frac{\pi^2}{8} = 1 + \left(\frac{1}{3}\right)^2 + \left(\frac{1}{5}\right)^2 + \left(\frac{1}{7}\right)^2 + \cdots.$$

▶ ▶ ▶

12.6 Summary

Complex exponentials of the form e^{st}, $-\infty < t < \infty$, are (for appropriate values of s) eigenfunctions of a general LTI system, and thus an attractive basis set from which to build up functional descriptions of the input-output behavior of such systems. For stable systems (characterized by an absolutely integrable impulse response), the range of allowed values of s includes the $j\omega$-axis; inputs of the form $e^{j2\pi ft}$, $-\infty < t < \infty$, will yield responses of the form $H(f)e^{j2\pi ft}$, $-\infty < t < \infty$, where the frequency response, $H(f)$, is related to the impulse response, $h(t)$, by

$$H(f) = \int_{-\infty}^{\infty} h(t) e^{-j2\pi ft}\, dt. \qquad (12.6-1)$$

Furthermore, any input that we can write in the form of a Fourier integral,

$$x(t) = \int_{-\infty}^{\infty} X(f) e^{j2\pi ft}\, df \qquad (12.6-2)$$

will yield a response in Fourier integral form,

$$y(t) = \int_{-\infty}^{\infty} Y(f) e^{j2\pi ft}\, df \qquad (12.6-3)$$

where

$$Y(f) = H(f)X(f). \tag{12.6-4}$$

If the class of $x(t)$ that can be represented as in (12.6-2) is large (and we shall argue in the next chapter that it is very large indeed), then (12.6-4) provides a frequency-domain characterization of general CT LTI system behavior that is an important and useful alternative to the time-domain characterization given in the last chapter in terms of the superposition integral (11.3-11).

As a first step in studying the generality of (12.6-2), we considered the representation of a periodic $x(t)$ in the form of a Fourier series,

$$x(t) = \sum_{n=-\infty}^{\infty} X[n] e^{j2\pi nt/T} \tag{12.6-5}$$

where the $X[n]$ are given by Fourier's formula,

$$X[n] = \frac{1}{T} \int_0^T x(t) e^{-j2\pi nt/T} \, dt. \tag{12.6-6}$$

Although we made no effort to prove the generality of (12.6-5) and (12.6-6), we showed with an example that the class of $x(t)$ representable in this way might be larger than one might at first suspect. And we explored various properties of these formulas, including Parseval's Theorem, that will be useful in the following chapters.

In the next chapter, we shall continue to follow the historical path, deriving a statement of Fourier's Theorem for non-periodic $x(t)$. Without really proving this theorem either, we shall discuss the kind of conditions necessary to yield a rigorous statement. In a later chapter, we shall show that the general form of Fourier's Theorem includes the Fourier series as a special case—and thus justify the formal steps taken in this chapter.

EXERCISES FOR CHAPTER 12

Exercise 12.1

Without doing any integrations, show that the exponential Fourier series for

$$x(t) = (1 + \cos 2\pi t)\left[\cos(10\pi t + \pi/6)\right]$$

is

$$x(t) = \sum_{n=-\infty}^{\infty} X[n]e^{j2\pi nt}$$

with all $X[n] = 0$ except

$$X[4] = X^*[-4] = \frac{1}{2}X[5] = \frac{1}{2}X^*[-5] = X[6] = X^*[-6] = \frac{1}{4}e^{j\pi/6} \,.$$

Exercise 12.2

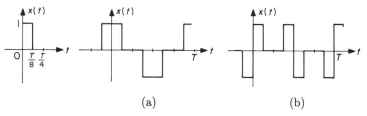

(a)   (b)

A periodic function $x(t)$ with period T is given in the interval $0 < t < T/4$ by the waveform shown to the left above. Show that the complete specification of $x(t)$ in the interval $0 < t < T$ must have the forms shown in (a) and (b) if it is known that $x(t)$ is

  a) an odd-harmonic, even function, or

  b) an odd-harmonic, odd function.

Exercise 12.3

The periodic waveform $x(t)$ is the input to an LTI system with frequency response $H(f)$ as shown. Show that the output waveform is $y(t) = \dfrac{4}{3\pi}\cos\left(2\pi t - \dfrac{\pi}{2}\right)$.

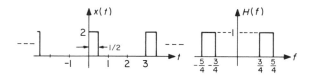

PROBLEMS FOR CHAPTER 12

Problem 12.1

Show that the following statement is false by constructing a counterexample:

> If two stable LTI networks have identically the same response
> (other than zero) to *any* one particular input, then they must
> have identically the same response to *every* input.

What kind of inputs must be excluded as test inputs if the statement is to have any chance of being true?

Problem 12.2

For each of the following impulse responses, show that the system function and domain of convergence are as given.

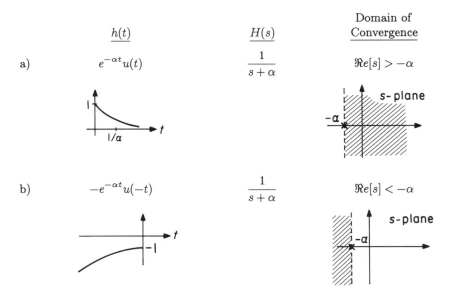

	$h(t)$	$H(s)$	Domain of Convergence
a)	$e^{-\alpha t}u(t)$	$\dfrac{1}{s+\alpha}$	$\Re e[s] > -\alpha$
b)	$-e^{-\alpha t}u(-t)$	$\dfrac{1}{s+\alpha}$	$\Re e[s] < -\alpha$

Note that the functional formulas for $H(s)$ in (a) and (b) are the same; only the domains of convergence differ. Hence in describing a system by giving its system function we must be careful to give, explicitly or implicitly, the domain of convergence as well as the formula. In earlier chapters we provided this information implicitly via the (usually unstated) assumption that we were dealing with causal systems.

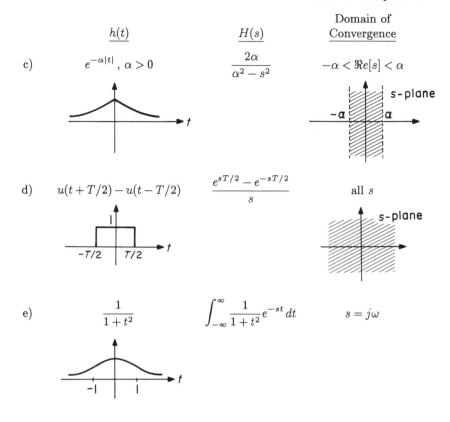

	$h(t)$	$H(s)$	Domain of Convergence
c)	$e^{-\alpha|t|}$, $\alpha > 0$	$\dfrac{2\alpha}{\alpha^2 - s^2}$	$-\alpha < \Re e[s] < \alpha$
d)	$u(t+T/2) - u(t-T/2)$	$\dfrac{e^{sT/2} - e^{-sT/2}}{s}$	all s
e)	$\dfrac{1}{1+t^2}$	$\displaystyle\int_{-\infty}^{\infty} \dfrac{1}{1+t^2} e^{-st}\, dt$	$s = j\omega$

This integral is not quite elementary. However, you should be able to show that the domain of convergence is the single line $\Re e[s] = 0$. In fact, for $s = j\omega$ we shall later show that $H(j\omega) = e^{-|\omega|}$ for $h(t) = 1/(1+t^2)$.

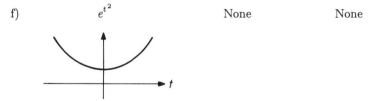

f) e^{t^2} None   None

You should be able to show that this system has no system function at all in any ordinary sense, that is, there is *no* domain of convergence.

Note in all the above that whenever $h(t)$ is right-sided (as in (a) and (d)) the domain of convergence is a right half-plane, whenever $h(t)$ is left-sided (as in (b) and (d)) the domain of convergence is a left half-plane, whenever $h(t)$ is both left- and right-sided (that is, a pulse as in (d)) the domain of convergence is the entire plane, and whenever $h(t)$ is absolutely integrable, that is, BIBO stable, as in (a) if $a > 0$, (b) if $a < 0$, (c), (d), and (e) the domain of convergence includes the $j\omega$-axis.

Problem 12.3

R. V. L. Hartley once proposed [*Proc. IRE, 30* (Mar. 1942): 144] to develop Fourier analysis in terms of the *cas functions* (an acronym for "cosine-and-sine"), where

$$\operatorname{cas}\frac{2\pi kt}{T} \equiv \cos\frac{2\pi kt}{T} + \sin\frac{2\pi kt}{T} .$$

a) Show that these functions are orthogonal over any interval of length T and normalized so that

$$\frac{1}{T}\int_0^T \operatorname{cas}\frac{2\pi kt}{T}\operatorname{cas}\frac{2\pi \ell t}{T}\,dt = \begin{cases} 1, & k = \ell \\ 0, & k \neq \ell . \end{cases}$$

b) Let $x(t)$ be a real periodic function. By manipulation of the Fourier series for $x(t)$, show that $x(t)$ may be written

$$x(t) = \sum_{k=-\infty}^{\infty} \Xi[k]\operatorname{cas}\frac{2\pi kt}{T} ,$$

where the coefficients $\Xi[k]$ are real and given by

$$\Xi[k] = \frac{1}{T}\int_0^T x(t)\operatorname{cas}\frac{2\pi kt}{T}\,dt .$$

Find a formula for $\Xi[k]$ in terms of the coefficients in the expansion

$$x(t) = \sum_{k=-\infty}^{\infty} X[k]e^{j2\pi kt/T}$$

and in terms of the coefficients in the expansion

$$x(t) = \sum_{k=0}^{\infty} c_k \cos\left(\frac{2\pi kt}{T} - \theta_k\right).$$

(Further ramifications of Hartley's proposal are considered in Problem 13.1.)

Problem 12.4

Let $x(t)$ be an arbitrary real periodic function with the exponential Fourier series representation

$$x(t) = \sum_{k=-\infty}^{\infty} X[k]e^{j2\pi kt/T} .$$

Describe the symmetry properties of each of the following partial sums (e.g., $y(t) = y(-t)$ or $y(t) = -y(t - T/2)$) and represent each by a formula in terms of $x(t)$ (e.g., $y(t) = \frac{1}{2}[x(t) + x(T/2 - t)]$):

a) $y_1(t) = \displaystyle\sum_{\text{all } k} \Re e[X[k]]e^{j2\pi kt/T}.$

b) $y_2(t) = \displaystyle\sum_{k\text{ odd}} \Re e[X[k]]e^{j2\pi kt/T}.$

c) $y_3(t) = \displaystyle\sum_{k\text{ odd}} \{\Re e[X[k]] - j\Im m[X[k]]\}e^{j2\pi kt/T}.$

Problem 12.5

Each of the following periodic functions is expanded in the exponential series

$$x(t) = \sum_{n=-\infty}^{\infty} X[n]e^{j2\pi nt/T}.$$

(A)
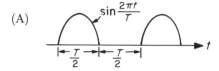

(C)
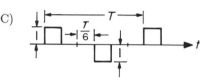

(B)
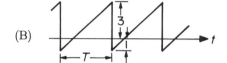

(D) $x(t) = \left[\cos \dfrac{2\pi(t - t_0)}{T}\right]^2$
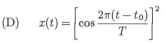

a) Find for each waveform the value of $X[0]$.

b) For which waveforms can the time origin (or the value of t_0 in (D)) be chosen such that all $X[n]$ are real? For each such waveform and with the time origin so chosen (the choice may not be unique), find $X[1]$.

c) For which waveforms can the time origin (or the value of t_0 in (D)) be chosen such that all $X[n]$ (except possibly $X[0]$) are purely imaginary? For each such waveform and with the time origin so chosen (the choice may not be unique), find $X[1]$.

d) For which waveforms does $X[n] = 0$ for $n = \pm 2, \pm 4, \pm 6, \ldots$?

Problem 12.6

a) Without evaluating the Fourier series coefficients, tell which of the four periodic waveforms on the next page have exponential Fourier series with the following properties, and justify your selections:

i) odd harmonics only

ii) coefficients purely real

iii) coefficients purely imaginary.

b) Find Fourier series representations for each of the following waveforms. Express your answers in three forms: sin and cos, magnitude and angle, and exponential.

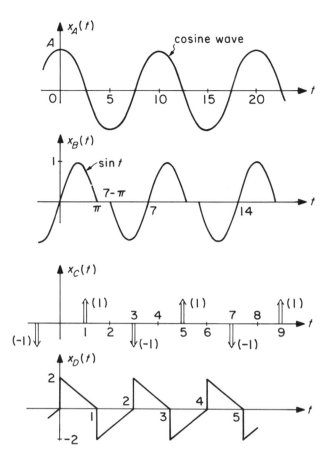

Problem 12.7

The periodic waveform $x(t)$ is the input to an LTI system with frequency response $H(f)$ as shown. Find the output waveform $y(t)$ in the form of a sum of cosines with appropriate amplitudes and phase angles.

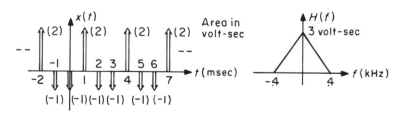

Problem 12.8

The input voltage $v_1(t)$ to the circuit below is the periodic train of positive and negative impulses on the right. Assume that the input impedance to $H(f)$ is infinite.

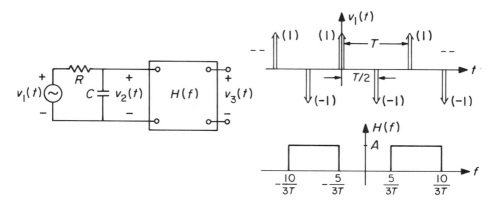

Find:

a) An expansion of $v_1(t)$ in the form $\displaystyle\sum_{n=-\infty}^{\infty} V_1[n]e^{j2\pi nt/T}$.

b) An expansion of $v_2(t)$ in the form $\displaystyle\sum_{n=-\infty}^{\infty} V_2[n]e^{j2\pi nt/T}$.

c) The average power in the output $v_3(t)$ if $H(f)$ is as shown.

d) The average power in the output if the input is $v_1(t) + v_1(t - T/3)$.

Problem 12.9

$H_1(f)$ and $H_2(f)$ are filters whose frequency responses are real and *periodic in frequency* as shown below.

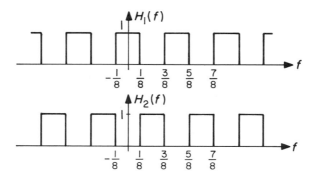

Find the rms value of the output of each of these filters for the periodic input $x(t)$ shown below.

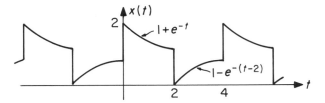

Problem 12.10

A practical op-amp integrator circuit includes a resistor shunted across the capacitor to limit the low-frequency gain, as shown in the figure. (Otherwise the small but inevitable unbalanced offset voltages and bias currents would be integrated, leading eventually to saturation of the op-amp.)

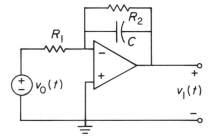

$$R_1 = 5 \text{ k}\Omega, \ R_2 = 0.5 \text{ M}\Omega, \ C = 0.01 \ \mu\text{F}$$

a) Determine the system function of this circuit for the parameter values given.

b) Use Bode's method to sketch the magnitude of the frequency response.

c) Sketch, approximately to scale, the square-wave response of this circuit for fundamental frequencies of 10, 100, and 1000 Hz. (HINT: Compare the spectra of the signals with the frequency response of the system.)

13

FOURIER TRANSFORMS AND FOURIER'S THEOREM

13.0 Introduction

As discussed in the last chapter, we seek to discover both what class of time functions can be represented as weighted sums of $e^{j\omega t}$'s, and how to choose the appropriate weights to represent a particular time function. We saw, informally, that a large class of periodic functions can apparently be represented as Fourier series,

$$x(t) = \sum_{n=-\infty}^{\infty} X[n]e^{j2\pi nt/T}$$

and that the Fourier coefficients $X[n]$ in such a representation can apparently be found from

$$X[n] = \frac{1}{T}\int_{-T/2}^{T/2} x(t)e^{-j2\pi nt/T}\,dt\,.$$

In this chapter we shall extend these formulas, again informally, to show that a large class of non-periodic functions can be represented as *Fourier integrals*,

$$x(t) = \int_{-\infty}^{\infty} X(f)e^{j2\pi ft}\,df$$

and that the *Fourier transform* $X(f)$ in such a representation can be found from

$$X(f) = \int_{-\infty}^{\infty} x(t)e^{-j2\pi ft}\,dt\,.$$

Together, these formulas constitute *Fourier's Theorem*. We shall explore (without, however, any real attempt at formal rigor) the kinds of conditions on $x(t)$ that are necessary for Fourier's Theorem to be true, we shall study a number of important examples, and we shall begin a discussion of the properties of the Fourier transform that make it useful in LTI system analysis.

13.1 Extension of the Fourier Series to the Fourier Integral

The generalization of Fourier series to non-periodic functions was suggested by Fourier himself and can be deduced from examination of the structure of the

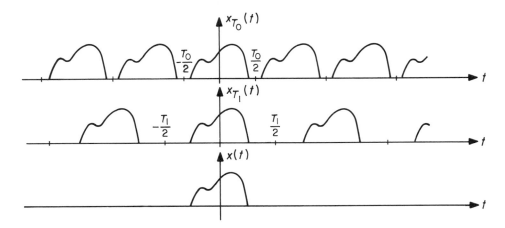

Figure 13.1–1. Enlarging the period of a periodic function.

Fourier series for a periodic function in the limit of a very large period. What we have in mind is illustrated in Figure 13.1–1. Holding the waveform near the origin fixed, we propose to let the period tend to infinity. In the limit we should obtain a representation for essentially the single pulse in the last figure.

In the beginning we have

$$x_T(t) = \sum_{n=-\infty}^{\infty} X_T[n]e^{j2\pi nt/T} \tag{13.1–1}$$

and

$$X_T[n] = \frac{1}{T}\int_{-T/2}^{T/2} x_T(t)e^{-j2\pi nt/T}\,dt. \tag{13.1–2}$$

The trick is to focus attention on a particular component in (13.1–1) at a frequency $f = n/T$, varying n as T varies to hold f fixed. Thus define

$$X_T(f) = TX_T[n] = \int_{-T/2}^{T/2} x_T(t)e^{-j2\pi ft}\,dt. \tag{13.1–3}$$

As we increase T, the lines in the spectrum $X_T(f)$ get closer together, as shown in Figure 13.1–2. Writing (13.1–1) formally in terms of $f = n/T$ gives

$$x_T(t) = \sum_{n=-\infty}^{\infty} X_T(f)e^{j2\pi ft}\frac{1}{T}. \tag{13.1–4}$$

So far we have done nothing but make some changes of notation. Now, however, let $T \to \infty$, holding $f = n/T$ fixed. The periodic function $x_T(t)$ approaches the

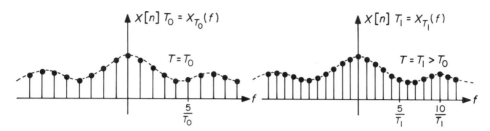

Figure 13.1–2. Spectrum of $x_T(t)$ for two values of T.

aperiodic function $x(t)$. Formally (13.1–3) tends to

$$X_T(f) \rightarrow X(f) = \int_{-\infty}^{\infty} x(t)e^{-j2\pi ft}\,dt \qquad (13.1-5)$$

and the sum (13.1–4) becomes an approximation to the integral

$$x(t) = \int_{-\infty}^{\infty} X(f)e^{j2\pi ft}\,df. \qquad (13.1-6)$$

The purely formal argument in the preceding paragraph must not be confused with a proof. It is exceedingly difficult to justify rigorously the operations necessary to pass from (13.1–4) to (13.1–6)—even assuming that we had proved the original Fourier series representation (13.1–1). But proving a theorem in mathematics is often easier than discovering a statement that is potentially a theorem. Our plausibility argument has provided us with a potential theorem: If we define the *Fourier transform* of $x(t)$ by the formula

$$\boxed{X(f) = \int_{-\infty}^{\infty} x(t)e^{-j2\pi ft}\,dt} \qquad (13.1-7)$$

then $x(t)$ has the *Fourier integral* (or *inverse Fourier transform*) representation

$$\boxed{x(t) = \int_{-\infty}^{\infty} X(f)e^{j2\pi ft}\,df} \qquad (13.1-8)$$

that is, $X(f)\,df$ is the "amount" of the complex exponential $e^{j2\pi ft}$ "contained in" $x(t)$ (consequently $X(f)$ is also called the *spectral density* of $x(t)$). Or, to use slightly different words, (13.1–7) *analyzes* $x(t)$ into its spectral components and (13.1–8) reconstructs or *synthesizes* $x(t)$ from these components; *Fourier's Theorem* states that this analysis-synthesis process can be carried out without loss—the reconstructed waveform is proclaimed to be identical to the waveform from which we started. It would be remarkable if this theorem were true without some sort of conditions or restrictions—and indeed appropriate restrictions are necessary—but before considering such matters let's look at several examples.*

Example 13.1–1

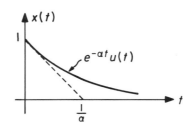

Figure 13.1–3. $x(t) = e^{-\alpha t} u(t)$.

Let $x(t) = e^{-\alpha t} u(t), \alpha > 0$, as shown in Figure 13.1–3. Then, from (13.1–7), the Fourier transform of $x(t)$ is

$$X(f) = \int_0^\infty e^{-\alpha t} e^{-j2\pi ft} \, dt = \frac{1}{-(\alpha + j2\pi f)} e^{-(\alpha + j2\pi f)t} \bigg|_0^\infty = \frac{1}{\alpha + j2\pi f}. \qquad (13.1\text{–}9)$$

According to (13.1–8), we thus should have the Fourier integral formula

$$x(t) = e^{-\alpha t} u(t) = \int_{-\infty}^\infty \frac{1}{\alpha + j2\pi f} e^{j2\pi ft} \, df \qquad (13.1\text{–}10)$$

which is to be considered as representing $x(t)$ as a weighted "sum" of complex exponentials. That (13.1–10) is valid can be shown by any of a number of procedures (such as contour integration) for evaluating the integral.

---

*The Fourier transform is variously defined in the literature. Two other common forms are:

$$X'(\omega) = \int_{-\infty}^\infty x(t) e^{-j\omega t} \, dt \quad \Longleftrightarrow \quad x(t) = \frac{1}{2\pi} \int_{-\infty}^\infty X'(\omega) e^{j\omega t} \, d\omega$$

$$X''(\omega) = \frac{1}{\sqrt{2\pi}} \int_{-\infty}^\infty x(t) e^{-j\omega t} \, dt \quad \Longleftrightarrow \quad x(t) = \frac{1}{\sqrt{2\pi}} \int_{-\infty}^\infty X''(\omega) e^{j\omega t} \, d\omega.$$

The form we have chosen seems to yield the greatest symmetry and simplicity in most applications. Care must be exercised in reading the literature since other choices, such as those described in this footnote, will lead to 2π factors in various theorems and formulas.

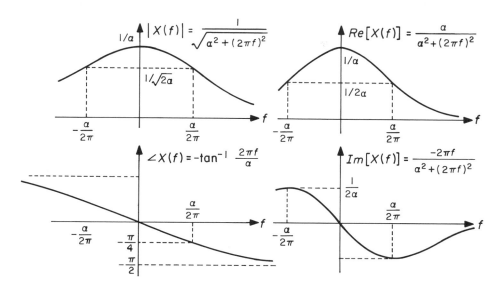

Figure 13.1–4. Real and imaginary parts, and magnitude and phase, of (13.1–9). The frequency at which $|X(f)|^2 = \frac{1}{2}|X(0)|^2$ is called the *half-power frequency*. Note that for this $X(f)$ the phase angle is $-45°$ at the half-power frequency.

Since $X(f)$ is complex, two plots—real and imaginary parts, or magnitude and phase—are required to describe it graphically, as shown in Figure 13.1–4. This figure illustrates a symmetry property that applies to the Fourier transform of any real $x(t)$:

CONJUGATE SYMMETRY PRINCIPLE:

    If $\Im m[x(t)] = 0$ (that is, if $x(t)$ is real), then

$$X(f) = X^*(-f).$$

    (That is, $\Re e[X(f)]$ and $|X(f)|$ are even functions, whereas $\Im m[X(f)]$ and $\angle X(f)$ are odd functions.)

▶ ▶ ▶

In Example 2.2–2 we showed that the unilateral $\mathcal{L}$-transform of $e^{-\alpha t}u(t)$ is $1/(s + \alpha)$. $X(f)$ as derived above, (13.1–9), is thus the $\mathcal{L}$-transform evaluated along the $j\omega$-axis, that is, for $s = j\omega = j2\pi f$. From definition (13.1–7), this is clearly a general result: For any time function that vanishes for $t < 0$, the Fourier transform is the unilateral $\mathcal{L}$-transform with $s = j2\pi f$. Indeed, we can now begin to see why the $\mathcal{L}$-transform worked so well in analyzing LTI system behavior.

The $\mathcal{L}$-transform of an input time function $x(t)$ measures the "amount" of e^{st} for each s "contained in" $x(t)$; multiplying by the system function determines the "amount" of each e^{st} "contained in" the output $y(t)$. This interpretation of the $\mathcal{L}$-transform is discussed more formally in Appendix B to this chapter.

Example 13.1–2

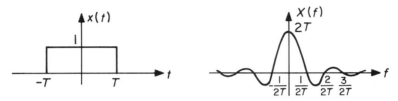

Figure 13.1–5. A symmetric square pulse and its transform.

Let

$$x(t) = \begin{cases} 1, & -T < t < T \\ 0, & \text{elsewhere}. \end{cases}$$

The Fourier transform of $x(t)$ is

$$X(f) = \int_{-T}^{T} 1\, e^{-j2\pi ft}\, dt = \frac{1}{-j2\pi f} e^{-j2\pi ft} \Big|_{-T}^{T}$$

$$= \frac{\sin 2\pi fT}{\pi f}. \tag{13.1–11}$$

Both $x(t)$ and $X(f)$ are sketched in Figure 13.1–5. From Fourier's Theorem we can write

$$x(t) = \int_{-\infty}^{\infty} \frac{\sin 2\pi fT}{\pi f} e^{j2\pi ft}\, df = \begin{cases} 1, & -T < t < T \\ 0, & \text{elsewhere}. \end{cases} \tag{13.1–12}$$

Again, the validity of this formula can be shown by various direct procedures.
► ► ►

Example 13.1–3

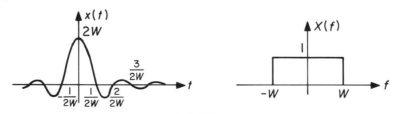

Figure 13.1–6. $\dfrac{\sin 2\pi Wt}{\pi t}$ and its transform.

Let $x(t) = \dfrac{\sin 2\pi W t}{\pi t}$ as sketched in Figure 13.1–6. The Fourier transform of $x(t)$ is

$$X(f) = \int_{-\infty}^{\infty} \frac{\sin 2\pi W t}{\pi t} e^{-j2\pi f t}\, dt.$$

Clearly, this integral is formally identical (except for a change of variable) to the inverse transform integral (13.1–12) in Example 13.1–2. Hence we expect

$$X(f) = \begin{cases} 1, & -W < f < W \\ 0, & \text{elsewhere} \end{cases} \qquad (13.1\text{–}13)$$

as shown in Figure 13.1–6. As a check we note that the inverse transform is

$$x(t) = \int_{-\infty}^{\infty} X(f) e^{j2\pi f t}\, df = \int_{-W}^{W} 1\, e^{j2\pi f t}\, df = \frac{\sin 2\pi W t}{\pi t}$$

which is the function with which we began.

▶ ▶ ▶

The symmetry of the Fourier transform relationships can often be exploited in this way. Formally we have the following principle:

> **TIME-FREQUENCY DUALITY PRINCIPLE:**
> If $X(f)$ is the Fourier transform of $x(t)$, then $x(-f)$ is the Fourier transform of $X(t)$.

The proof of this principle follows at once from appropriate changes of variables in the defining integrals. We can also see illustrated in these examples another general principle:

> **EVEN AND ODD SYMMETRY PRINCIPLES:**
> a) If $x(t) = x(-t)$ (that is, if $x(t)$ is an even function), then
>
> $$X(f) = X(-f).$$
>
> If $x(t)$ is both even and real, then $X(f)$ is both even and real.
>
> b) If $x(t) = -x(-t)$ (that is, if $x(t)$ is an odd function), then
>
> $$X(f) = -X(-f).$$
>
> If $x(t)$ is both odd and real, then $X(f)$ is both odd and purely imaginary.

The converses of all these statements are also true: If $X(f)$ has the properties stated, then $x(t)$ satisfies the conditions given. Again the proofs are straightforward from the defining integrals.

Example 13.1–4

Figure 13.1–7. A square pulse and its transform.

The two formulas

$$X(0) = \int_{-\infty}^{\infty} x(t)\,dt\,, \qquad x(0) = \int_{-\infty}^{\infty} X(f)\,df$$

are, of course, simply special cases of the general Fourier transform relations that are easily evaluated and are extremely useful for checking. Consider, for example, the transform pair derived in Example 13.1–2 and redrawn in Figure 13.1–7. Clearly

$$\int_{-\infty}^{\infty} x(t)\,dt = 2T$$

which is the value given for $X(0)$. (Note that $\displaystyle\lim_{f\to 0} \frac{\sin 2\pi fT}{\pi f}$ can be evaluated by L'Hôpital's Rule, or equivalently by remembering that $\sin x \approx x$ for small x.) On the other hand, we apparently must have

$$\int_{-\infty}^{\infty} \frac{\sin x}{\pi x}\,dx = \int_{-\infty}^{\infty} \frac{\sin 2\pi fT}{\pi f}\,df = 1$$

which is $x(0)$. Indeed, accepting the validity of Fourier's Theorem now provides an evaluation of this definite integral (which was used in Exercise 11.5).

It is helpful to note that the area under the function $\dfrac{\sin ax}{bx}$ is equal to the area of the inscribed triangle shown shaded in Figure 13.1–8. Knowledge of the two area formulas, and the general fact that the transform of $\dfrac{\sin ax}{bx}$ is a pulse and vice versa, are all you need to know to work out the constants in these transform pairs.

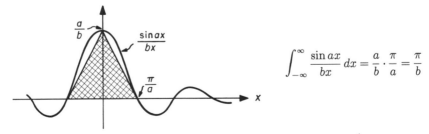

$$\int_{-\infty}^{\infty} \frac{\sin ax}{bx}\,dx = \frac{a}{b}\cdot\frac{\pi}{a} = \frac{\pi}{b}$$

Figure 13.1–8. Construction for finding the area of $\dfrac{\sin ax}{bx}$.

▶ ▶ ▶

Examples 13.1–1, 13.1–2, and 13.1–3 illustrate a common situation; one of the two integrals in Fourier's Theorem can be evaluated readily but the other is much less simple. A direct check of the theorem in such cases is thus difficult. The next example describes one of the few cases in which a direct check is feasible.

Example 13.1–5

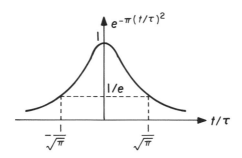

Figure 13.1–9. The Gaussian function.

From the fundamental definition (13.1–7), the Fourier transform of the *Gaussian function*, $x(t) = e^{-\pi(t/\tau)^2}$, is

$$X(f) = \int_{-\infty}^{\infty} e^{-\pi(t/\tau)^2} e^{-j2\pi ft}\, dt \,. \tag{13.1–14}$$

This $x(t)$ (as shown in Figure 13.1–9) tends to zero so rapidly as $|t| \to \infty$ that the integral clearly exists. We can be evaluate it by starting from the known integral*

$$\int_{-\infty}^{\infty} e^{-\pi t^2}\, dt = 1 \,. \tag{13.1–15}$$

By a change of variable this becomes

$$\frac{1}{\tau} \int_{-\infty}^{\infty} e^{-\pi\left(\frac{t-\alpha}{\tau}\right)^2}\, dt = 1 \,. \tag{13.1–16}$$

---

*Equation (13.1–15) can be derived by observing that

$$\left(\int_{-\infty}^{\infty} e^{-\pi x^2}\, dx \right)^2 = \int_{-\infty}^{\infty}\int_{-\infty}^{\infty} e^{-\pi(x^2+y^2)}\, dx\, dy = \int_{0}^{\infty} dr \int_{0}^{2\pi} d\theta r e^{-\pi r^2}$$

$$= \int_{0}^{\infty} 2\pi r e^{-\pi r^2}\, dr = \int_{0}^{\infty} e^{-u}\, du = 1$$

where the second equality implies a shift to polar coordinates and the next to last uses the change of variable $u = \pi r^2$.

Equation (13.1–16) is valid even when α is complex. Equation (13.1–14) can be put in the form of (13.1–16) by completing the square in the exponent:

$$
\begin{aligned}
X(f) &= \int_{-\infty}^{\infty} e^{-\pi[(t/\tau)^2 + j2ft - f^2\tau^2]} e^{-\pi f^2 \tau^2} \, dt \\
&= e^{-\pi f^2 \tau^2} \int_{-\infty}^{\infty} e^{-\pi[t/\tau + jf\tau]^2} \, dt \\
&= \tau e^{-\pi f^2 \tau^2}
\end{aligned}
\tag{13.1–17}
$$

which is the desired result. Since $X(f)$ and $x(t)$ have the same form, the inverse transform integral (13.1–8) can be evaluated by the same procedure to give

$$
\int_{-\infty}^{\infty} X(f) \, e^{j2\pi ft} \, df = e^{-\pi(t/\tau)^2}
\tag{13.1–18}
$$

which recovers $x(t)$ and directly demonstrates in this case the validity of Fourier's Theorem.

▶ ▶ ▶

13.2 More Careful Statements of Fourier's Theorem

The validity of Fourier's Theorem for the functions considered in the examples of the previous section could (perhaps with some difficulty) be justified in each case using appropriate ad hoc methods. Attempts to state and prove mathematically rigorous forms of Fourier's Theorem that apply to some general class of $x(t)$ must cope with a dual problem. On the one hand, the "tails" of $x(t)$ must be adequately constrained as $|t| \to \infty$ so that the infinite integral defining $X(f)$ exists in some appropriate sense; we shall call this a *global* condition. On the other hand, $x(t)$ must not be too "wiggly" or else the "tails" of $X(f)$ as $|f| \to \infty$ will be so badly behaved that the inverse transform integral will have no satisfactory meaning; we shall call this a *local* condition.

The first mathematically acceptable formulation of Fourier's Theorem (by Dirichlet in 1829) dealt with this dual problem by imposing two restrictions:

a) (Global condition) Requiring that $x(t)$ be absolutely integrable:

$$
\int_{-\infty}^{\infty} |x(t)| \, dt < \infty \, .
$$

b) (Local condition) Constraining $x(t)$ to have a finite number of maxima and minima and a finite number of discontinuities in every finite interval.*

---

*The simplest example of a function failing condition (b) is perhaps $\sin(1/t)$ near $t = 0$. For a proof of Dirichlet's Theorem, see, e.g., E. C. Titchmarch, *Introduction to the Theory of Fourier Integrals* (Oxford: Clarendon Press, 1937) p. 13.

Together, (a) and (b) ensure that $X(f)$ exists and that the tails of $X(f)$ vanish sufficiently rapidly that

$$\lim_{W \to \infty} \int_{-W}^{W} X(f)\, e^{j2\pi ft}\, df = \frac{x(t+) + x(t-)}{2}. \qquad (13.2\text{–}1)$$

The value of the synthesis integral is thus not precisely $x(t)$ but rather (for each t) the average of the values of $x(t)$ approaching t from the right and from the left. Hence the inverse transform integral (13.1–8) interpreted as in (13.2–1) converges to $x(t)$ at every point of continuity and to a value halfway between the right-hand and left-hand limits at a point of discontinuity.

Since Dirichlet's time, many variations of Fourier's Theorem have been proposed, differing in the conditions imposed on $x(t)$ and in the ways in which the infinite integrals are interpreted. However, most of these formulations (including Dirichlet's) are relatively unsatisfactory as a basis for LTI system analysis since they attempt to describe both $x(t)$ and $X(f)$ as ordinary functions, thus excluding as components of $x(t)$ not only singularity functions but even such well-behaved and essential time functions as $\sin t$ and $u(t)$ (whose transforms contain, as we shall see, singularity functions in frequency). Moreover, many of these formulations (like Dirichlet's) have a rather inconvenient lack of symmetry in that the mathematical properties required of $x(t)$ and of $X(f)$ may be quite different.

For our purposes the most useful form of Fourier's Theorem seems to result if *both* the time functions *and* their transforms are defined in terms of "what they do" rather than "what they are," that is, as generalized functions in the sense introduced in Chapter 11. In this formulation non-vanishing time functions such as $\sin t$ and $u(t)$ and even time functions that grow as fast as any finite power of t as $|t| \to \infty$ are acceptable; their Fourier transforms contain impulses or higher-order singularity functions in frequency. Conversely, singularity functions in time correspond to transforms that may grow as fast as a finite power of f as $|f| \to \infty$. The properties of time functions and their transforms in this form of the theorem are precisely dual; the theorem is completely symmetrical. Stated informally,* Fourier's Theorem in this interpretation implies that a time function containing singularity functions and growing no faster than a power of t (not, however, as fast as an exponential) has a unique transform (which may also contain singularity functions and may grow as fast as a power of f) from which the original generalized time function can be uniquely recovered. The direct and inverse integral formulas can be used for analysis and synthesis provided the values of improper integrals are interpreted as what they "do," that is, in terms of their effect inside integrals multiplying test functions as in Chapter 11.

---

*A more careful statement of the condition $x(t)$ must satisfy if it is to have a Fourier transform in the sense of generalized functions is that $\int_{-\infty}^{\infty} x(t)\phi(t)\, dt$ be well-defined for any $\phi(t)$ that is infinitely differentiable and that vanishes as $|t| \to \infty$ faster than any power of t. $X(f)$ will then satisfy the same condition in f.

A full development of the theory of Fourier transforms of generalized functions is regrettably a major undertaking, well beyond the scope of an introductory book such as this one (see the references cited in Chapter 11). Fortunately, a major consequence of this theory is to justify rigorously for the class of functions under consideration a variety of formal operations (such as integration by parts, interchange of orders of integration, differentiation of series term by term, and interchange of the order of operations and passing to limits) that we might carelessly be willing to accept without proof. Learning how to manipulate generalized functions and their transforms correctly and consistently, and how to apply the results to the understanding and design of important practical systems, is thus not difficult; such will be our goal in these notes. Rigorous justifications will come later in more advanced treatments.

13.3 Further Examples of Fourier's Theorem; Singularity Functions

Example 13.3–1

Consider $x(t) = \delta(t)$. Formally, the Fourier transform presents no difficulties:

$$X(f) = \int_{-\infty}^{\infty} x(t) e^{-j2\pi ft} \, dt = \int_{-\infty}^{\infty} \delta(t) e^{-j2\pi ft} \, dt = 1 \, .$$

The inverse transform leads, however, to a most improper-looking formula

$$\int_{-\infty}^{\infty} X(f) e^{j2\pi ft} \, df = \int_{-\infty}^{\infty} e^{j2\pi ft} \, df \, .$$

If Fourier's Theorem is valid under these conditions (as we have stated it is), then we must somehow accept the fact that

$$\int_{-\infty}^{\infty} e^{j2\pi ft} \, df = \delta(t) \, !$$

We can in fact easily show that $\int_{-\infty}^{\infty} e^{j2\pi ft} \, df$ "behaves" like the impulse $\delta(t)$ by putting it inside an integral,

$$\int_{-\infty}^{\infty} \left[\int_{-\infty}^{\infty} e^{j2\pi ft} \, df \right] \phi(t) \, dt$$

where $\phi(t)$ is some arbitrary well-behaved (for example, continuous at $t = 0$ and possessing a well-defined Fourier transform) time function. Interchanging orders of integration, we have

$$\int_{-\infty}^{\infty} \left[\int_{-\infty}^{\infty} \phi(t) e^{j2\pi ft} \, dt \right] df = \int_{-\infty}^{\infty} \Phi(-f) \, df$$

where $\Phi(f)$ is the Fourier transform of $\phi(t)$. But it follows at once from Fourier's Theorem that

$$\int_{-\infty}^{\infty} \Phi(-f) \, df = \int_{-\infty}^{\infty} \Phi(f) \, df = \phi(0) \, .$$

That is, $\int_{-\infty}^{\infty} e^{j2\pi ft} \, df$ as a time function "behaves" like an impulse at $t = 0$.

▶ ▶ ▶

Example 13.3–2

We can illustrate the meaning of the formal expression

$$\int_{-\infty}^{\infty} e^{j2\pi ft}\, df = \delta(t)$$

in several other ways:

a) Using Euler's relation,

$$e^{j2\pi ft} = \cos 2\pi ft + j\sin 2\pi ft$$

we can think of the integral above as the sum of a large number of cosines and sines of various frequencies. As Figure 13.3–1 suggests, it is not unreasonable to expect that at any t except $t = 0$ as many of these terms will be positive as will be negative, and hence their contributions may destructively interfere or cancel. At $t = 0$, however, all the cosine terms are $+1$; their contributions add, producing an infinite spike at the origin.

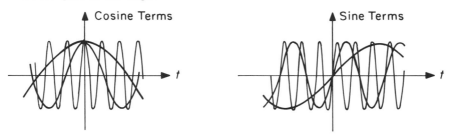

Figure 13.3–1. Interpretation of $\int_{-\infty}^{\infty} e^{j2\pi ft}\, df$.

b) Let us consider a sequence of functions approaching $\delta(t)$ in the limit as discussed in Exercise 11.5. For example, consider the sequence of Gaussian functions

$$\xi_n(t) = n e^{-\pi t^2 n^2}$$

in the limit $n \to \infty$. It seems reasonable that the sequence of transforms

$$\Xi_n(f) = \int_{-\infty}^{\infty} \xi_n(t) e^{j2\pi ft}\, dt = e^{-\pi(f/n)^2}$$

should in some sense* approach in the limit the transform of $\delta(t)$. Clearly

$$\lim_{n\to\infty} \Xi_n(f) = 1$$

---

*It is not true in the ordinary sense that if some sequence of functions $x_n(t)$ converges to a limit $x(t)$, then the sequence of transforms $X_n(f)$ necessarily converges to the transform of the limit. Thus consider the sequence $x_n(t) = \frac{1}{n}e^{-t/n}u(t)$, which converges uniformly to $x(t) = 0$ as $n \to \infty$. The sequence of transforms is, from Example 13.1–1, $X_n(f) = 1/(1 + j2\pi nf)$, which converges to

$$\lim_{n\to\infty} X_n(f) = \begin{cases} 1, & f = 0 \\ 0, & f \neq 0 \end{cases} \neq 0 = X(f).$$

Clearly, however, in this case (and it can be shown in general), local averages of $X_n(f)$ converge to the local average of $X(f)$. If we define our functions in terms of what they "do," then $\lim\limits_{n\to\infty} x_n(t) = x(t)$ implies $\lim\limits_{n\to\infty} X_n(f) = X(f)$.

as expected. The sequences $\xi_n(t)$ and $\Xi_n(f)$ are shown in Figure 13.3–2.

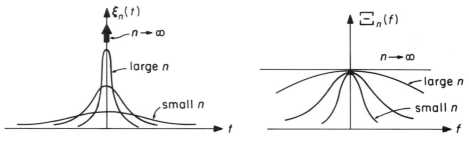

Figure 13.3–2. The sequences $\xi_n(t)$ and $\Xi_n(f)$.

Other sequences approaching an impulse give identical results, for example, those based on square pulses or $\sin t/t$ functions whose transforms were derived in Examples 13.1–2 and 13.1–3.

▶ ▶ ▶

Example 13.3–3

Example 13.3–1 led to the Fourier transform pair

$$x(t) = \delta(t) \quad \Longleftrightarrow \quad X(f) = 1. \tag{13.3–1}$$

The symmetry of the defining integrals, or the duality principle, then implies another pair:

$$x(t) = 1 \quad \Longleftrightarrow \quad X(f) = \delta(f). \tag{13.3–2}$$

That is, a steady constant d-c waveform

$$x(t) = 1, \quad -\infty < t < \infty$$

corresponds to an impulse in frequency at $f = 0$. To say this another way, the only frequency present in $x(t)$ is $f = 0$, so that

$$x(t) = e^{j2\pi 0 t} = 1.$$

In this case, it is the analysis integral

$$X(f) = \int_{-\infty}^{\infty} 1 \, e^{-j2\pi ft} \, dt$$

that presents difficulties, while the synthesis integral

$$x(t) = \int_{-\infty}^{\infty} \delta(f) \, e^{j2\pi ft} \, df = e^{j2\pi ft} \bigg|_{f=0} = 1$$

is more straightforward, at least operationally.

▶ ▶ ▶

Example 13.3–4

The technique, illustrated in Example 13.3–2b, of studying the limiting behavior of sequences of functions and their Fourier transforms is broadly useful when singularity functions of one kind or another are likely to be involved. As another example, let's explore the Fourier transform of the unit step function, $x(t) = u(t)$. A direct approach clearly leads to the improper integral

$$X(f) = \int_{-\infty}^{\infty} u(t)e^{-j2\pi ft}\,dt = \int_{0}^{\infty} e^{-j2\pi ft}\,dt\,.$$

The meaning of this expression can be obtained, as in Example 13.3–1, by studying what it "does" inside an integral. However, it is more illuminating in this case to consider the transforms of a sequence of functions such as

$$\xi_n(t) = e^{-t/n}u(t)$$

that tends to $x(t) = u(t)$ as $n \to \infty$.

Proceeding formally, the Fourier transform of $\xi_n(t)$ is (from Example 13.1–1)

$$\Xi_n(f) = \frac{1}{(1/n) + j2\pi f}\,.$$

In the limit $n \to \infty$, $\Xi_n(f)$ apparently tends to $1/(j2\pi f)$. But we must proceed carefully; from the appearance of the improper integral in the direct effort to determine the transform of $u(t)$, we suspect that singularity functions may be present. Furthermore $1/(j2\pi f)$ cannot be the complete Fourier transform of $u(t)$—it is an odd, purely imaginary frequency function and thus must correspond (from the odd-symmetry principle of Example 13.1–2) to an odd, purely real time function, whereas $u(t)$ is real but neither even nor odd. The correct transform of $u(t)$ will thus have both real and imaginary parts that are nonzero.

The key to the complete Fourier transform of $u(t)$ is to observe that $\lim_{n\to\infty} \Xi_n(f) = 0$ for all f except $f = 0$. Indeed, for $f = 0$, $\Xi_n(f) = n$, which is real and becomes arbitrarily large as $n \to \infty$. This suggests that $\Re[\Xi_n(f)]$ may contain a singularity function at $f = 0$ as $n \to \infty$. Writing $\Xi_n(f)$ in terms of its real and imaginary parts, we obtain

$$\Xi_n(f) = \frac{1/n}{(1/n)^2 + (2\pi f)^2} - j\frac{2\pi f}{(1/n)^2 + (2\pi f)^2}\,.$$

The imaginary part tends to $1/(j2\pi f)$ as $n \to \infty$, but the real part is a sequence of functions that behaves in the limit like an impulse of area $1/2$. This is easily seen because we may write

$$\Re[\Xi_n(f)] = \frac{n}{2}G(nf)$$

where $G(f)$ is the absolutely integrable function

$$G(f) = \frac{2}{1 + (2\pi f)^2}\,.$$

The sequence $nG(nf)$ is sketched in Figure 13.3–3. The area under $G(f)$ is unity,

$$\int_{-\infty}^{\infty} G(f)\,df = 1$$

as may be proved in several ways (see, for example, Problem 13.6). Hence $nG(nf)$ behaves as a unit impulse* in frequency:

$$\lim_{n \to \infty} nG(nf) = \delta(f).$$

Thus we obtain the complete Fourier transform of the unit step as

$$\int_{-\infty}^{\infty} u(t)e^{-j2\pi ft}\,dt = \frac{1}{2}\delta(f) + \frac{1}{j2\pi f}$$

which is sketched in Figure 13.3–4.

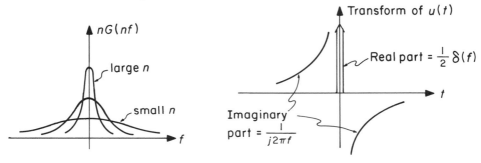

Figure 13.3–3.   The sequence $nG(nf)$.   Figure 13.3–4. Transform of $u(t)$.

As we have already seen in Example 13.3–1, the transform $\frac{1}{2}\delta(f)$ corresponds to the time function $x_1(t)$ shown in Figure 13.3–5, that is, to the constant $1/2$ for all time. Thus (as also shown in Figure 13.3–5) the second, purely imaginary, term in the transform of $u(t)$, that is, $1/(j2\pi f)$, must be the transform of the time function

$$x_2(t) = \frac{1}{2}\operatorname{sgn}t = \begin{cases} \dfrac{1}{2}, & t > 0 \\[2mm] -\dfrac{1}{2}, & t < 0. \end{cases} \tag{13.3–4}$$

We note with satisfaction that $\frac{1}{2}\operatorname{sgn}t$ is an odd real function of time and thus is an appropriate time function to correspond to the odd, purely imaginary transform $1/(j2\pi f)$.

---

*That $\lim_{n \to \infty} nG(nt) = \delta(t)$ was in effect known to Cauchy (indeed $G(t)$ is sometimes called *Cauchy's function* in his honor) and was used both by him and by Poisson to deduce Fourier's Theorem in the form

$$\lim_{n \to \infty} \int_{-\infty}^{\infty} e^{-|f|/n}\, e^{j2\pi ft}\left[\int_{-\infty}^{\infty} x(\tau)e^{-j2\pi f\tau}\,d\tau\right]df = \lim_{n \to \infty} \int_{-\infty}^{\infty} x(\tau)nG[n(t-\tau)]\,d\tau = x(t).$$

[See Problem 13.6, and B. Van der Pol and H. Bremmer, *Operational Calculus* (Cambridge, UK: Cambridge University Press, 1950) p. 63.]

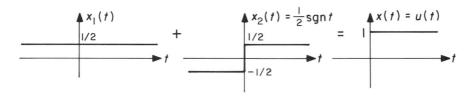

Figure 13.3–5. Construction of $u(t)$ from even and odd components.

▶ ▶ ▶

13.4 The Convolution Property of Fourier Transforms

Our interest in Fourier transforms arose initially from the observation that $e^{j2\pi ft}$ was an eigenfunction of a stable LTI system, that is, if the input were

$$x(t) = e^{j2\pi ft}, \quad -\infty < t < \infty$$

for some fixed frequency f, then the output would be

$$y(t) = H(f)e^{j2\pi ft}, \quad -\infty < t < \infty$$

where we now recognize (12.2–9) as defining the frequency response $H(f)$ as the Fourier transform of the impulse response:

$$H(f) = \int_{-\infty}^{\infty} h(t)e^{-j2\pi ft}\, dt.$$

If, furthermore, we can describe some other input as a weighted sum of eigenfunctions,

$$x(t) = \int_{-\infty}^{\infty} X(f)e^{j2\pi ft}\, df$$

then the response is similarly a weighted sum of the eigenfunctions,

$$y(t) = \int_{-\infty}^{\infty} X(f)H(f)e^{j2\pi ft}\, df.$$

Thus, the Fourier transform of the output $Y(f)$ is related to the Fourier transform of the input by

$$Y(f) = X(f)H(f).$$

Since we can also relate the output to the input by

$$y(t) = x(t) * h(t)$$

the preceding argument suggests what is in fact a general Fourier transform property:

CONVOLUTION PROPERTY:

The Fourier transform of the convolution of two functions, $w(t) * v(t)$, is the product of their Fourier transforms; that is,

$$w(t) * v(t) \quad \Longleftrightarrow \quad W(f)V(f).$$

The conditions given in Chapter 10 are sufficient but not necessary for convolution to be well-defined. Loosely, the convolution property applies whenever the product of the transforms is well-defined, by which we mean, for example, that the product of transforms does not contain products of singularity functions such as $\delta(f)u(f)$ or $\delta^2(f)$ that have no satisfactory meaning (as pointed out in Chapter 11). Hence Fourier's methods can be extended to such not strictly stable systems as lossless networks, ideal differentiators, and ideal integrators (indeed, even to higher-order integrators having impulse responses $t^n u(t)$ that grow without bound). To be sure such systems, if they actually existed, might not be very useful since small incidental input disturbances due to thermal agitation, unbalanced bias effects, minor non-linearities, etc., could induce large uncontrolled output components. But as idealized analytical approximations to systems that actually exist they are extremely useful.

The dual of the convolution property is the product property:

PRODUCT PROPERTY:

The Fourier transform of the product of two functions, $w(t)v(t)$, is the convolution of their Fourier transforms; that is,

$$w(t)v(t) \quad \Longleftrightarrow \quad W(f) * V(f).$$

This is also a most important theorem, as we shall amply illustrate in the chapters to follow. The convolution of two functions of frequency is defined in precisely the same way as the convolution of two functions of time:

$$W(f) * V(f) = \int_{-\infty}^{\infty} W(v)V(f - v)\, dv = \int_{-\infty}^{\infty} W(f - v)V(v)\, dv .$$

Another important formula can be derived as a special case of the above theorems. Thus let

$$
\begin{aligned}
w(t) &= x(t) & \Longleftrightarrow & \quad W(f) = X(f) \\
v(t) &= y^*(t + \tau) & \Longleftrightarrow & \quad V(f) = Y^*(-f)e^{j2\pi f\tau}
\end{aligned}
$$

(where we have allowed for the possibility that $x(t)$ and $y(t)$ in applications of the formula to be derived might be complex, and where we have used the time-delay property in Table XIII.2 of Appendix A to this chapter). Applying the product property above, the transform of the product $w(t)v(t)$ is

$$\int_{-\infty}^{\infty} x(t)y^*(t + \tau)e^{-j2\pi ft}\, dt = \int_{-\infty}^{\infty} X(f - v)Y^*(-v)e^{j2\pi v\tau}\, dv$$

or, changing variable to $\mu = -v$,

$$\left| \int_{-\infty}^{\infty} x(t)y^*(t+\tau)e^{-j2\pi ft}\, dt = \int_{-\infty}^{\infty} X(f+\mu)Y^*(\mu)e^{-j2\pi\mu\tau}\, d\mu \right| \qquad (13.4\text{--}1)$$

which we shall often find useful. The special case $f = \tau = 0$ and $y(t) = x(t)$ gives

$$\left| \int_{-\infty}^{\infty} |x(t)|^2\, dt = \int_{-\infty}^{\infty} |X(f)|^2\, df \right| \qquad (13.4\text{--}2)$$

which is usually called *Parseval's Theorem* (although in fact Parseval formulated only the analogous theorem given in Section 12.5 for Fourier series—the extension to Fourier integrals is apparently due to Rayleigh). The quantity $\int_{-\infty}^{\infty} |x(t)|^2\, dt$ is often called the *energy* in $x(t)$. (Of course, if $x(t)$ is a voltage or a current, $\int_{-\infty}^{\infty} |x(t)|^2\, dt$ will not have quite the dimensions of physical energy, but the name is useful and conventional nonetheless.) Correspondingly, $|X(f)|^2$ is called the *energy density spectrum* of $x(t)$. Parseval's Theorem can thus be interpreted as stating that the total energy in a waveform is equal to the sum of the energies in each frequency component. As will be shown in the appendix to Chapter 14, this has a most interesting vectorial interpretation.

Example 13.4–1

Parseval's Theorem in the form of (13.4–2) does not apply to waveforms composed of a discrete sum of one or more sinusoids of various frequencies over the interval $-\infty < t < \infty$. Both of the integrals in (13.4–2) are infinite in such a case. The time integral is unbounded because $|x(t)|^2$ extends with undiminished amplitude forever. Over an infinite time a sinusoid conveys an infinite energy. The transform of a sum of sinusoids is a set of impulses; the frequency integral is unbounded because it contains squares of impulses.

However, the average *power* in a sum of sinusoids is finite. It was in this sense that we derived a form of Parseval's Theorem in Chapter 12 for periodic functions and Fourier series. Thus, in general, if

$$x(t) = \sum_{m=-M}^{M} X[m]e^{j2\pi f_m t} \qquad (13.4\text{--}3)$$

then

$$\langle x^2(t) \rangle = \lim_{t \to \infty} \frac{1}{2T} \int_{-T}^{T} x^2(t)\, dt = \sum_{m=-M}^{M} |X[m]|^2 \qquad (13.4\text{--}4)$$

which is readily shown to be valid whether or not the frequencies f_m are multiples of a fundamental frequency as they are in the Fourier series representation of periodic functions. Incidentally, even if the frequencies f_m are not harmonically related, $x(t)$ as in (13.4–3) is *almost periodic* in that, if one waits an appropriately long time, all of the exponential terms in (13.4–3) can simultaneously be brought as close as one chooses to the value unity that they all had at $t = 0$, so that $x(t)$ almost repeats with this period.

▶ ▶ ▶

Example 13.4–2

As a final example, we illustrate an important application of Parseval's Theorem. Suppose we must pass a signal $x(t)$ through a *bandlimiting filter*, for which $H(f) \equiv 0$ except for a limited band of frequencies. We seek to determine how $H(f)$ should be shaped within the band so as to minimize the *integrated squared error* between output and input,

$$\mathcal{E} = \int_{-\infty}^{\infty} [x(t) - y(t)]^2 \, dt \, .$$

Conceivably, one might suspect that enhancing the transmission near the edge of the band could compensate to some extent for those frequency components just outside that are rejected by the filter—or, alternately, that a smooth gentle transition from *pass* to *stop* bands could reduce the error (for example, by eliminating the overshoot at a discontinuity of $x(t)$ that, as we shall show in Chapter 16, is produced by an abrupt transition in $H(f)$). Either of these suspicions may be correct for certain categories of signals and error criteria. However, as we shall now prove, to minimize the integrated squared error one should choose $H(f) \equiv 1$ in the pass band, so that $Y(f)$ is identically equal to $X(f)$ wherever possible.

This result is easily proved using Parseval's Theorem. Suppose, to be specific, that we choose the pass band of the filter to be $-W < f < W$, although any other interval or combination of intervals would do as well. In other words, we require that $Y(f) \equiv 0$ for all $|f| > W$. We then seek to find that $Y(f)$ in the range $|f| < W$ that will make $y(t)$ a best approximation to $x(t)$ in the sense of minimizing the integrated squared error. Since by linearity the time function $x(t) - y(t)$ has the Fourier transform $X(f) - Y(f)$, Parseval's Theorem gives

$$\mathcal{E} = \int_{-\infty}^{\infty} |X(f) - Y(f)|^2 \, df$$

$$= \int_{-W}^{W} |X(f) - Y(f)|^2 \, df + \int_{|f|>W} |X(f)|^2 \, df$$

where we have used the fact that $Y(f) \equiv 0$ for $|f| > W$. No choice of $Y(f)$ can influence the second integral above, and both integrals are positive. Thus the smallest value of $\mathcal{E}$ results if we set $Y(f) = X(f)$ for $|f| < W$, which makes the first integral zero. But this is just what we wished to show. In the appendix to Chapter 14 we shall demonstrate that this minimum-squared-error property is by no means restricted to Fourier integrals, but indeed applies to any *orthogonal expansion*, of which Fourier series and integrals are merely the most familiar examples.

▶ ▶ ▶

13.5 Summary

In the limit as the period becomes very large, the Fourier series representation for a periodic function becomes the Fourier integral representation for an aperiodic function:

$$x(t) = \int_{-\infty}^{\infty} X(f) e^{j2\pi ft} \, df \, . \tag{13.5–1}$$

Equation (13.5–1) describes $x(t)$ as a "sum" of eigenfunctions of the form $e^{j2\pi ft}$ with "weights" given by the Fourier transform,

$$X(f) = \int_{-\infty}^{\infty} x(t)e^{-j2\pi ft}\, dt. \tag{13.5-2}$$

The representation (13.5–1) with $X(f)$ given by (13.5–2) is valid provided $x(t)$ is defined operationally (by what it "does") and provided $x(t)$ grows no faster than a finite power of $|t|$. These are examples of the local smoothness conditions and global integrability conditions that any rigorous statement of Fourier's Theorem must contain.

The bulk of the chapter consisted of a series of examples illustrating important Fourier transform pairs and properties. The results are summarized in Appendix A to this chapter.

APPENDIX A TO CHAPTER 13

Tables of Fourier Transforms and Their Properties

Most applications of Fourier methods do not directly involve such fundamental manipulations as have occupied us in this chapter. A more common procedure is to derive the desired transforms and inverse transforms by applying one or more of 8 to 10 basic theorems and properties to one or more of a dozen or so simple transform pairs. Table XIII.1 gives some of the more useful Fourier transform pairs; much more extensive lists have been compiled by Campbell and Foster, and Erdelyi et al., among others.* Many of the pairs in Table XIII.1 have been derived as examples earlier in this chapter. Most of the remaining entries are discussed in the exercises or problems, or follow trivially from other pairs in the table.

Table XIII.2 is a list of the more important properties, theorems, and formulas concerning Fourier integrals. Some of these have already been introduced. The others follow almost immediately from appropriate manipulations of the defining integrals. The proofs are left as exercises or problems.

---

*G. A. Campbell and R. M. Foster, *Fourier Integrals for Practical Applications* (New York, NY: Van Nostrand, 1948); A. Erdelyi (Ed.), W. Magnus, F. Oberhettinger, F. G. Tricom, *Tables of Integral Transforms* (New York, NY: McGraw-Hill, 1954).

Table XIII.1—Short Table of Fourier Transforms

$$x(t) = \int_{-\infty}^{\infty} X(f)e^{j2\pi ft}\,df \qquad\qquad X(f) = \int_{-\infty}^{\infty} x(t)e^{-j2\pi ft}\,dt$$

a) $\delta(t) \qquad\Longleftrightarrow\qquad 1$

b) $1 \qquad\Longleftrightarrow\qquad \delta(f)$

c) $e^{-\pi(t/\tau)^2} \qquad\Longleftrightarrow\qquad \tau e^{-\pi(f\tau)^2}$

d) $e^{-\alpha t}u(t) \qquad\Longleftrightarrow\qquad \dfrac{1}{\alpha + j2\pi f},\quad \alpha > 0$

e) $u(t) \qquad\Longleftrightarrow\qquad \dfrac{\delta(f)}{2} + \dfrac{1}{j2\pi f}$

f) $\operatorname{sgn} t = \begin{cases} 1, & t > 0 \\ -1, & t < 0 \end{cases} \qquad\Longleftrightarrow\qquad \dfrac{1}{j\pi f}$

g) $\dfrac{1}{\pi t} \qquad\Longleftrightarrow\qquad -j\operatorname{sgn} f$

h) $e^{-\alpha|t|} \qquad\Longleftrightarrow\qquad \dfrac{2\alpha}{\alpha^2 + (2\pi f)^2}$

i) $e^{j2\pi f_0 t} \qquad\Longleftrightarrow\qquad \delta(f - f_0)$

j) $\sin 2\pi f_0 t \qquad\Longleftrightarrow\qquad \dfrac{\delta(f - f_0) - \delta(f + f_0)}{2j}$

k) $\cos 2\pi f_0 t \qquad\Longleftrightarrow\qquad \dfrac{\delta(f - f_0) + \delta(f + f_0)}{2}$

l) $e^{j2\pi f_0 t} u(t) \qquad\Longleftrightarrow\qquad \dfrac{\delta(f - f_0)}{2} + \dfrac{1}{j2\pi}\left[\dfrac{1}{f - f_0}\right]$

m) $\sin 2\pi f_0 t\, u(t) \qquad\Longleftrightarrow\qquad \dfrac{\delta(f - f_0) - \delta(f + f_0)}{4j} + \dfrac{1}{2\pi}\left[\dfrac{f_0}{f_0^2 - f^2}\right]$

n) $\cos 2\pi f_0 t\, u(t) \qquad\Longleftrightarrow\qquad \dfrac{\delta(f - f_0) + \delta(f + f_0)}{4} + \dfrac{1}{j2\pi}\left[\dfrac{f}{f^2 - f_0^2}\right]$

o) $\dot\delta(t) \qquad\Longleftrightarrow\qquad j2\pi f$

p) $te^{-\alpha t} u(t),\quad a > 0 \qquad\Longleftrightarrow\qquad \dfrac{1}{(\alpha + j2\pi f)^2}$

q) $t \qquad\Longleftrightarrow\qquad \dfrac{j\dot\delta(f)}{2\pi}$

r) $\qquad\Longleftrightarrow\qquad$ $\dfrac{\sin 2\pi fT}{\pi f}$

s) $\dfrac{\sin 2\pi Wt}{\pi t} \qquad\Longleftrightarrow\qquad$

t) $\displaystyle\sum_{n=-\infty}^{\infty} \delta(t - nT) \qquad\Longleftrightarrow\qquad \dfrac{1}{T}\sum_{n=-\infty}^{\infty} \delta\left(f - \dfrac{n}{T}\right)$

Table XIII.2—Properties of Fourier Transforms

	$x(t) = \int_{-\infty}^{\infty} X(f)e^{j2\pi ft}\,df$	$X(f) = \int_{-\infty}^{\infty} x(t)e^{-j2\pi ft}\,dt$
Conjugate Symmetry	$\Im m[x(t)] = 0$ (i.e., $x(t)$ is real)	$X(f) = X^*(-f)$ (i.e., $\Re e[X(f)] = \Re e[X(-f)]$, $\Im m[X(f)] = -\Im m[X(-f)]$)
Even Symmetry	$x(t) = x(-t)$	$X(f) = X(-f)$
Odd Symmetry	$x(t) = -x(-t)$	$X(f) = -X(-f)$
Linearity	$ax_1(t) + bx_2(t)$	$aX_1(f) + bX_2(f)$
Duality	$X(t)$	$x(-f)$
Scale Change	$x(at)$	$\dfrac{1}{|a|}X(f/a)$
Time Delay	$x(t - t_0)$	$e^{-j2\pi ft_0}X(f)$
Times $e^{j2\pi f_0 t}$	$e^{j2\pi f_0 t}x(t)$	$X(f - f_0)$
Differentiation	$\dfrac{dx(t)}{dt}$	$j2\pi f X(f)$
Times t	$tx(t)$	$\dfrac{1}{-j2\pi}\dfrac{dX(f)}{df}$
Convolution	$\displaystyle\int_{-\infty}^{\infty} w(\tau)v(t - \tau)\,d\tau$	$W(f)V(f)$
Product	$w(t)v(t)$	$\displaystyle\int_{-\infty}^{\infty} W(\nu)V(f - \nu)\,d\nu$
Integration	$\displaystyle\int_{-\infty}^{t} x(\tau)\,d\tau$	$\dfrac{X(f)}{j2\pi f} + \dfrac{X(0)\delta(f)}{2}$

Other formulas:

$$X(0) = \int_{-\infty}^{\infty} x(t)\,dt\,; \qquad x(0) = \int_{-\infty}^{\infty} X(f)\,df$$

$$\int_{-\infty}^{\infty} |x(t)|^2\,dt = \int_{-\infty}^{\infty} |X(f)|^2\,df \qquad \text{(Parseval)}$$

$$\int_{-\infty}^{\infty} x(t)y^*(t + \tau)e^{-j2\pi\nu t}\,dt = \int_{-\infty}^{\infty} X(f + \nu)Y^*(f)e^{-j2\pi f\tau}\,df$$

APPENDIX B TO CHAPTER 13

The Bilateral Laplace Transform

The versions of Fourier's Theorem thus far discussed still exclude an important and useful class of time functions, namely those that blow up exponentially as $t \to +\infty$ or $-\infty$. Typical examples are $x(t) = e^{-\alpha t}$, $-\infty < t < \infty$, or $x(t) = e^{\alpha t} u(t)$, $\alpha > 0$. Such functions often become Fourier transformable if they are first multiplied by an exponential convergence factor. For example, $e^{-\sigma_0 t} x(t) = e^{-\sigma_0 t} e^{\alpha t} u(t)$ satisfies both Dirichlet's conditions and the generalized function conditions for any $\sigma_0 > \alpha$. The Fourier transform of $e^{-\sigma_0 t} x(t)$ is

$$\int_{-\infty}^{\infty} \left[e^{-\sigma_0 t} x(t) \right] e^{-j2\pi f t} \, dt = \int_{0}^{\infty} e^{\alpha t} e^{-(\sigma_0 + j2\pi f)t} \, dt$$

$$= \frac{1}{\sigma_0 + j2\pi f - \alpha}, \quad \sigma_0 > \alpha.$$

Hence we may write from Fourier's Theorem

$$e^{-\sigma_0 t} x(t) = \int_{-\infty}^{\infty} \frac{1}{\sigma_0 + j2\pi f - \alpha} e^{j2\pi f t} \, df$$

or, multiplying both sides by $e^{\sigma_0 t}$,

$$x(t) = e^{\alpha t} u(t) = \int_{-\infty}^{\infty} \frac{1}{\sigma_0 + j2\pi f - \alpha} e^{(j2\pi f + \sigma_0)t} \, df.$$

Note that this formula represents $x(t)$ as a "sum" of weighted e^{st} terms with values of s selected from the region of the s-plane to the right of the line $\Re e[s] = \alpha$, that is, not including the $j\omega$-axis if $\alpha > 0$. However, if this region of the s-plane overlaps the domain of convergence for the system function of the system to which $x(t)$ is the input (as will, for example, always be the case for this $x(t)$ if the system is causal), then the general frequency-domain scheme suggested at the beginning of Chapter 12 remains possible.

To generalize, suppose that $e^{-\sigma t} x(t)$ is Fourier transformable for some range of values of σ; in general, this range will be a strip, $\sigma_1 < \sigma < \sigma_2$. Define the *bilateral Laplace transform*

$$\boxed{X(s) = \int_{-\infty}^{\infty} x(t) e^{-st} \, dt, \quad \sigma_1 < \Re e[s] < \sigma_2} \tag{13.B-1}$$

where $s = \sigma + j2\pi f$ and $X(s)$ is to be considered the Fourier transform* (vs. f) of $e^{-\sigma t} x(t)$ for $\sigma_1 < \sigma < \sigma_2$. Then, from Fourier's Theorem, $x(t)$ has the *Laplace representation*

$$x(t) = \int_{-\infty}^{\infty} X(s) e^{st} \, df, \quad s = \sigma + j2\pi f, \quad \sigma_1 < \Re e[s] < \sigma_2.$$

---

*Note that we have here reverted to the previous meaning of the symbol $X(\ )$, with a $j2\pi$ included in the argument, rather than the meaning corresponding to $X(f)$ as recently employed. In the literature on Laplace transforms, unilateral or bilateral, this choice is conventional and appropriate.

More commonly, this is written as a line or contour integral in the complex variable s,

$$\boxed{x(t) = \frac{1}{2\pi j} \int_C X(s) e^{st} \, ds}$$ (13.B–2)

where C is the contour shown in Figure 13.B–1 within the convergence strip for $X(s)$.

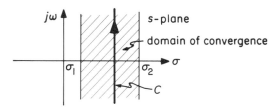

Figure 13.B–1. Contour for Laplace integral representation.

In a very real sense the bilateral Laplace transform reflects the fullest power of the frequency-domain approach:

a) If the domain of convergence of the system function $H(s)$ (which we now identify as the bilateral Laplace transform of $h(t)$) overlaps the domain of convergence of $X(s)$, then each input e^{st}, $-\infty < t < \infty$, with s in the overlap region yields $H(s)e^{st}$, $-\infty < t < \infty$, as an output. The total response is thus

$$y(t) = \frac{1}{2\pi j} \int_C Y(s) e^{st} \, ds$$

with

$$Y(s) = X(s)H(s).$$

b) If the system is causal and $x(t) = 0$, $t < 0$, then the domains of convergence of $H(s)$ and $X(s)$ are right half-planes and must overlap; the overlap region is to the right of the rightmost pole of $X(s)$ or $H(s)$. The procedure above then becomes identical with the ZSR obtained by the unilateral Laplace transform described earlier, but now seen in a new light—not as simply a powerful operational tool that mysteriously "works" but rather as an exploitation of the eigenfunction property of e^{st} and the Fourier representation of signals as sums of sinusoids.

Nevertheless, in spite of its elegance and generality, we shall not pursue the properties of the bilateral Laplace transform further in this book. The full power of the bilateral Laplace transform is rarely needed; most problems in circuit analysis or control applications, for example, fit comfortably within the more limited domain of the unilateral Laplace transform.* Moreover, the power of the bilateral transform depends critically on the domain of convergence being of nonzero width, so that the elegant methods (such as contour integration) of the theory of functions of a complex variable can be fully exploited. Continuous sinusoids, $\sin t/t$, and other on-going waveforms over $-\infty < t < \infty$ that regularly arise in communication problems imply convergence domains of zero width and are thus better handled with Fourier transforms, as we shall illustrate in the chapters that follow.

---

*The bilateral transform is, however, extremely successful in handling certain boundary-value problems that arise in the study of, for example, electromagnetic fields.

EXERCISES FOR CHAPTER 13

Exercise 13.1

a) Show by means of a counterexample that in general the fact that $X(f)$ is the transform of the (possibly complex) time function $x(t)$ does *not* imply that $\Re e[X(f)]$ is the transform of $\Re e[x(t)]$.

b) Construct, however, a particular example of a complex time function $x(t)$ with transform $X(f)$ for which $\Re e[X(f)]$ is in fact the transform of $\Re e[x(t)]$.

Exercise 13.2

The Gaussian function $e^{-\pi t^2}$ of Example 13.1–5 is an example of a time function whose transform $e^{-\pi f^2}$ has precisely the same functional form; thus the Gaussian pulse "is its own Fourier transform." Show that this property is not unique to the Gaussian function by proving that if $x(t)$ is any *even* (possibly complex) time function with Fourier transform $X(f)$, then the time function $x(t)+X(t)$ is its own Fourier transform. Illustrate this result with an example.

Exercise 13.3

Show that the phase of the Fourier transform $H(f)$ of $h(t)$ below is $\angle H(f) = -j2\pi f$. (HINT: Exploit symmetry.)

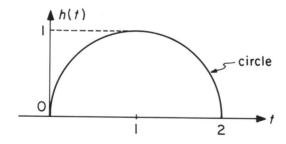

PROBLEMS FOR CHAPTER 13

Problem 13.1

Continuing the discussion of Hartley's proposal for Fourier analysis in terms of the cas ("cosine-and-sine") functions begun in Problem 12.3, let $x(t)$ be a real Fourier transformable function and define the "Hartley transform"

$$X(f) = \int_{-\infty}^{\infty} x(t) \operatorname{cas} 2\pi f t \, dt$$

where

$$\operatorname{cas} 2\pi f t = \cos 2\pi f t + \sin 2\pi f t.$$

Then according to Hartley, $x(t)$ should be given by the "inverse Hartley transform"

$$x(t) = \int_{-\infty}^{\infty} X(f) \operatorname{cas} 2\pi f t \, df$$

which has identically the same form as the transform. Moreover, all quantities involved are real.

a) Prove Hartley's formulas by relating $X(f)$ to the real and imaginary parts of $X(f)$, the ordinary Fourier transform of $x(t)$.

b) Find and sketch $X(f)$ for

   i) $x(t) = \delta(t - \tau)$;

   ii) $x(t) = e^{-\alpha t} u(t)$.

c) Let $X(f)$ and $H(f)$ be the Hartley transforms for $x(t)$ and $h(t)$, respectively. Find a formula for $Y(f)$, the Hartley transform of $y(t) = x(t) * h(t)$, in terms of $X(f)$ and $H(f)$.

d) Would Hartley's formulas—in spite of their symmetry and the fact that no complex quantities are involved—be as useful for the analysis of LTI systems as the ordinary Fourier integral? Explain.

Problem 13.2

If the Fourier transform of a (generally complex) function $x(t)$ is $X(f)$, express the transforms of the following in terms of $X(f)$:

   a) $x^*(t)$.

   b) $\Re[x(t)]$.

   c) $x(t - 2) + x^*(-t - 2)$.

   d) $e^{-j4\pi t} x(t/3)$.

Problem 13.3

Suppose that a certain telephone channel can be characterized by the frequency response

$$H(f) = e^{j\varphi(f)}, \quad \varphi(-f) = -\varphi(f)$$

where $\varphi(f)$ is some complicated function of frequency. Such a channel might very well have serious distortion, such as "hollow barrel" effects. To correct for these, it has been suggested (rather impractically) that the signal at the receiving end be recorded on tape, flown by fast jet back to the transmitting end, played *backwards* through the original channel a second time, and recorded again on tape. The signal on this second tape, if played *backwards*, should (it is claimed) be the original signal, independent of $\varphi(f)$. Would the scheme work? Explain your answer.

Problem 13.4

A Gaussian-shaped pulse signal

$$v_1(t) = 3e^{-2\pi t^2}$$

is applied at the input of a square-law circuit whose instantaneous output $v_2(t)$ is the square of $v_1(t)$:

$$v_2(t) = v_1^2(t).$$

The square-law circuit is followed by a (non-causal) linear filter whose impulse response is a Gaussian pulse,

$$h(t) = e^{-\pi t^2}, \quad -\infty < t < \infty.$$

The filter output is $v_3(t)$.

a) Describe $v_3(t)$ and its spectrum.

b) How are the results modified if the squaring operation is preceded by the filter instead of being followed by the filter?

Problem 13.5

a) Find directly by integration the Fourier transforms of $e^{-\alpha t}\cos\omega_0 t\, u(t)$, $\alpha > 0$, and $e^{-\alpha t}\sin\omega_0 t\, u(t)$, $\alpha > 0$.

b) The frequency response of an ideal tuned circuit can be put in the standardized form

$$H(f) = \frac{j(f/f_0)}{j(f/f_0) + Q(1 - (f/f_0)^2)}.$$

Find the impulse response $h(t)$ of this tuned circuit, where

$$h(t) = \int_{-\infty}^{\infty} H(f)e^{j2\pi ft}\, df.$$

Problem 13.6

a) Let $x(t) = e^{-\alpha t} u(t)$. Observe that $e^{-\alpha |t|}$ can be written

$$e^{-\alpha |t|} = x(t) + x(-t).$$

Observe also that the scale-change property implies that $x(-t) \iff X(-f)$ are a Fourier transform pair. Using these facts and the linearity property, derive pair (h) of Table XIII.1. Check your work by direct integration.

b) Use the duality principle and pair (h) of Table XIII.1 to show that

$$\frac{1}{1+t^2} \quad \iff \quad \pi e^{-2\pi |f|}$$

are a Fourier transform pair. Check your result by direct evaluation of the inverse transform.

c) Use the scale-change property and the fact that

$$\int_{-\infty}^{\infty} x(t)\, dt = X(0)$$

to deduce the fact used in Example 13.3–4 that

$$\int_{-\infty}^{\infty} \frac{2}{1+(2\pi t)^2}\, dt = 1.$$

d) Consider the transform pair

$$x(t) = e^{-\alpha t} u(t) \quad \iff \quad \frac{1}{\alpha + j2\pi f} = X(f).$$

Observe that since $x(t)$ is real, the real and imaginary parts of $X(f)$ are even and odd functions of frequency, respectively. Hence the inverse transform integral can be written

$$\int_{-\infty}^{\infty} \frac{1}{\alpha + j2\pi f} e^{j2\pi ft}\, df = \int_{-\infty}^{\infty} \Re\left[\frac{1}{\alpha + j2\pi f}\right] \cos 2\pi ft\, df$$

$$- \int_{-\infty}^{\infty} \Im\left[\frac{1}{\alpha + j2\pi f}\right] \sin 2\pi ft\, df.$$

The integrals on the right are even and odd functions of time, respectively; call them $x_e(t)$ and $x_o(t)$, respectively. Show that

$$x_e(t) = \frac{1}{2}[x(t) + x(-t)] = \frac{1}{2} e^{-\alpha |t|}$$

$$x_o(t) = \frac{1}{2}[x(-t) - x(t)] = -\operatorname{sgn} t\, e^{-\alpha |t|}.$$

Problem 13.7

a) Derive the times-$e^{j2\pi f_0 t}$ property of Table XIII.2,

$$e^{j2\pi f_0 t} x(t) \iff X(f - f_0).$$

b) Derive the Fourier transform of $e^{j2\pi f_0 t}$ (entry (i) of Table XIII.1) from the transform of the constant $x(t) = 1$ and the times-$e^{j2\pi f_0 t}$ property. Check your result by directly evaluating the inverse transform.

c) Use this result and the linearity property to derive the Fourier transforms of $\sin 2\pi f_0 t$ and $\cos 2\pi f_0 t$. Sketch the real and imaginary parts of the resulting spectra, including an appropriate range of both positive and negative frequencies.

d) Similarly, derive the Fourier transforms of the functions $e^{j2\pi f_0 t} u(t)$, $\cos 2\pi f_0 t\, u(t)$, and $\sin 2\pi f_0 t\, u(t)$.

e) Use the result of (b) to argue that the response of an LTI system with frequency response $H(f)$ to the input $e^{j2\pi f_0 t}$ is $H(f_0)e^{j2\pi f_0 t}$.

Problem 13.8

a) Derive the differentiation property of Table XIII.2,

$$\frac{dx(t)}{dt} \iff j2\pi f\, X(f)$$

indirectly by differentiating both sides of the inverse transform integral

$$x(t) = \int_{-\infty}^{\infty} X(f)e^{j2\pi ft}\, df$$

and drawing appropriate conclusions.

b) Derive the times-t property of Table XIII.2,

$$tx(t) \iff \frac{1}{-j2\pi}\frac{dX(f)}{df}$$

by a procedure analogous to (a) but applied to the direct transform integral.

c) Apply the differentiation or times-t properties to appropriate transform pairs of Table XIII.1 to establish the following transform pairs:

i) $\cos 2\pi f_0 t \iff \dfrac{\delta(f - f_0) + \delta(f + f_0)}{2}$.

ii) $\dot{\delta}(t) \iff j2\pi f$.

iii) $te^{-\alpha t} u(t) \iff \dfrac{1}{(\alpha + j2\pi f)^2}$, $\quad \alpha > 0$.

iv) $t \iff \dfrac{j\dot{\delta}(f)}{2\pi}$.

Problem 13.9

Use the times-t property of Table XIII.2 to find the Fourier transform of $x(t) = te^{-\pi t^2}$.
Check your answer by using the differentiation property.

Problem 13.10

a) Derive the time-delay property of Table XIII.2,

$$x(t - t_0) \quad \Longleftrightarrow \quad e^{-j2\pi f t_0} X(f)$$

by a simple change of variable in the defining Fourier-transform integral.

b) Derive the Fourier transform of the pulse

$$x(t) = \begin{cases} 1, & -T < t < T \\ 0, & \text{elsewhere} \end{cases}$$

by writing

$$x(t) = u(t + T) - u(t - T)$$

and exploiting the delay and linearity properties and the transform of $u(t)$ from
entry (e) of Table XIII.1.

Problem 13.11

Check the validity of the formulas $X(0) = \int_{-\infty}^{\infty} x(t)\, dt$ and $x(0) = \int_{-\infty}^{\infty} X(f)\, df$ by
applying them to a number of transform pairs in Table XIII.1—for example, pairs (a),
(b), (c), (h), and (s).

Problem 13.12

a) Let $y(t)$ be the integral of $x(t)$:

$$y(t) = \int_{-\infty}^{t} x(\tau)\, d\tau .$$

Show that $Y(f) = \dfrac{X(f)}{j2\pi f} + \dfrac{X(0)\delta(f)}{2}$ by constructing $y(t)$ to be the convolution of
$x(t)$ and a unit step, $y(t) = x(t) * u(t)$, and exploiting the convolution property of
Table XIII.2.

b) Use the integration property to evaluate the Fourier transform of the integral of
each of the following functions. Compare in each case with the result obtained by
first integrating and then evaluating the transform.

    i) $x(t) = \delta(t)$.   iii) $x(t) = e^{-\alpha t} u(t)$.

    ii) $x(t) = \dot{\delta}(t)$.   iv)

Problem 13.13

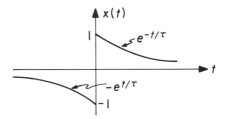

a) Use the linearity and scale-change properties from Table XIII.2 to derive the Fourier transform of $x(t)$ in the figure from the basic pair

$$e^{-\alpha t} u(t) \quad \Longleftrightarrow \quad \frac{1}{\alpha + j2\pi f}.$$

b) From the result of (a), derive the transform of $\operatorname{sgn} t$ by passing to an appropriate limit. Check with entry (f) of Table XIII.1.

c) Combine the result of (b) with the basic pair $1 \Longleftrightarrow \delta(f)$ to derive the transform of $u(t)$. Check with entry (e) of Table XIII.1.

Problem 13.14

In the following diagrams, the blocks represent ideal integrators, delay lines, amplifiers, and summation elements.

a) Sketch the impulse response of each block diagram. As a partial check on your work, each impulse response should turn out to be strictly a pulse: $h(t) = 0$ except for a finite interval of time, $t_1 < t < t_2$.

b) Find and sketch (magnitude and phase) the frequency response of each block diagram.

(HINT: It may be easier to return to the block diagrams and use certain properties of Fourier transforms than to proceed directly. As an aid to sketching, note that

$$-1 + e^{ja} = e^{ja/2}(e^{ja/2} - e^{-ja/2}) = 2je^{ja/2}\sin(a/2).$$

Similarly

$$1 - 2e^{ja} + e^{j2a} = -4e^{ja}(\sin(a/2))^2.)$$

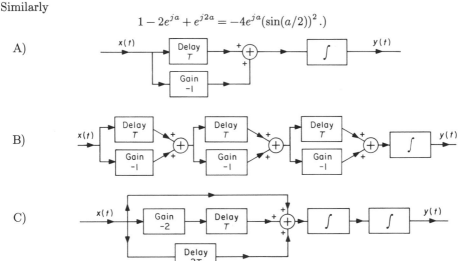

Problem 13.15

A real symmetric (even) function $x(t)$ is said to *reproduce itself* under *convolution* if

$$x(t/a) * x(t/b) = kx(t/c)$$

where k and c in general depend on a and b but the form of the relation is to be valid for any a and b. In words, if such a function is convolved with a similar function having the same *shape* (i.e., identical except for time scale and possibly amplitude), then the result also has the same shape.

a) Find an equivalent condition in terms of $X(f)$, the transform of $x(t)$.

b) Assuming that $\int_{-\infty}^{\infty} x(t)\, dt = 1$ and $\int_{-\infty}^{\infty} t^2 x(t)\, dt$ is finite but not zero, show that $c^2 = a^2 + b^2$ and $k = |ab/c|$. (HINT: The assumed integrability conditions guarantee that $X(0) = 1$ and that $\ddot{X}(0)$ exists but is not zero. Symmetry implies that $\dot{X}(0) = 0$.)

c) Show that both of the following symmetric time functions reproduce themselves under convolution:

$$x_1(t) = e^{-\pi t^2}$$

$$x_2(t) = \frac{1}{\pi(1 + t^2)}\,.$$

Are the relationships derived in (b) satisfied for both $x_1(t)$ and $x_2(t)$?

The class of functions that reproduce themselves under convolution is a rather narrow one. This problem was first studied by P. Lévy in the 1930s and is of considerable importance in probability theory. For a readable discussion, see J. Lamperti, *Probability* (New York, NY: W. A. Benjamin, 1966).

Problem 13.16

The impulse function $\delta(t)$ has a *sifting property*: If $f(t)$ is continuous at t,

$$f(t) = \int_{-\infty}^{\infty} f(\tau)\delta(t - \tau)\, d\tau\,.$$

Show that the function $\dfrac{\sin \pi t}{\pi t}$ also has the sifting property:

$$f(t) = \int_{-\infty}^{\infty} f(\tau)\frac{\sin \pi(t - \tau)}{\pi(t - \tau)}\, d\tau$$

if the spectrum of $f(t)$ is suitably restricted.

Problem 13.17

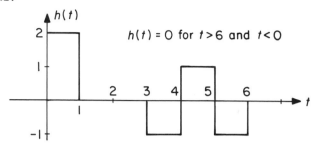

$h(t) = 0$ for $t > 6$ and $t < 0$

The impulse response $h(t)$ of a certain filter is shown above. Determine the *equivalent noise bandwidth* of this filter, defined as

$$BW = \int_{-\infty}^{\infty} \frac{|H(f)|^2}{|H(0)|^2}\, df$$

where

$$H(f) = \int_{-\infty}^{\infty} h(t) e^{-j2\pi ft}\, dt\,.$$

(HINT: This is an easy problem! The significance of the equivalent noise bandwidth will be discussed in Chapter 19.)

Problem 13.18

Use (13.4–1) to prove that

$$x(t) = \frac{\sin 2\pi W t}{\pi t}$$

and

$$x\!\left(t - \frac{n}{2W}\right) = \frac{\sin 2\pi W\!\left(t - \dfrac{n}{2W}\right)}{\pi\!\left(t - \dfrac{n}{2W}\right)}$$

are *orthogonal* for any $n \neq 0$; that is,

$$\int_{-\infty}^{\infty} x(t)x\!\left(t - \frac{n}{2W}\right) dt = 0\,.$$

Problem 13.19

Suppose that we want to approximate a periodic waveform $x(t)$ by a finite sum of harmonically related sinusoids,

$$\hat{x}(t) = \sum_{k=-N}^{N} a_k e^{j2\pi kt/T}\,.$$

Prove by a method paralleling that of Example 13.4–2 that if $\hat{x}(t)$ is to be a *minimum mean-square-error approximation* to $x(t)$, that is, if $(1/T)\int_0^T |x(t) - \hat{x}(t)|^2\, dt$ is to be a minimum, then one should choose the a_k to be the Fourier coefficients $X[k]$.

Problem 13.20

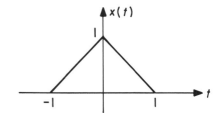

a) Find the Fourier transform of the time function to the right. (HINT: Consider $x(t)$ as the convolution of two pulses.)

b) Evaluate the integral

$$\int_{-\infty}^{\infty} \left(\frac{\sin t}{t}\right)^4 dt.$$

14

SAMPLING IN TIME
AND FREQUENCY

14.0 Introduction

We shall open this chapter by developing an extremely useful Fourier transform pair—a *periodic impulse train* in time and its transform, which is a periodic impulse train in frequency. Using this pair as a tool, we shall study two important ideas that turn out to be time-frequency duals—the Fourier transforms of arbitrary periodic time functions (which, of course, will reconnect us with the Fourier series notions from which we began) and the properties of time-sampled signals (which will lead us both to the justly celebrated Sampling Theorem and to formulas called the discrete-time Fourier transform). Finally, we shall consider time functions that are both sampled and periodic and thus derive the discrete Fourier transform which is the basis for the Fast Fourier Transform (FFT) algorithm that underlies much of modern digital signal processing. In an appendix, many of the ideas and results of this and earlier chapters will be combined and reinterpreted in terms of a vector-space representation of signals.

14.1 The Periodic Impulse Train

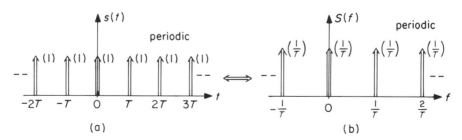

Figure 14.1–1. The periodic unit impulse train and its transform.

We seek the Fourier transform of the periodic unit impulse train $s(t)$ shown in Figure 14.1–1a. As we shall show, the transform $S(f)$ of $s(t)$ is the periodic impulse train in frequency illustrated in Figure 14.1–1b. To derive this result, truncate $s(t)$ to $2N + 1$ impulses centered on the origin (as shown in Figure

14.1–2a) and consider the transform $S_N(f)$ of the truncated time function $s_N(t)$:

$$S_N(f) = \int_{-\infty}^{\infty} s_N(t)e^{-j2\pi ft}\, dt = \int_{-\infty}^{\infty} \left[\sum_{n=-N}^{N} \delta(t-nT)\right] e^{-j2\pi ft}\, dt$$

$$= \sum_{n=-N}^{N} \int_{-\infty}^{\infty} \delta(t-nT)e^{-j2\pi ft}\, dt = \sum_{n=-N}^{N} e^{-j2\pi nfT}. \qquad (14.1-1)$$

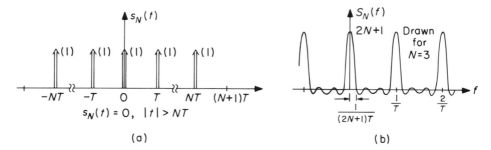

(a)   (b)

Figure 14.1–2. A truncated impulse train and its transform.

To determine the shape of $S_N(f)$, recognize the sum in (14.1–1) as the first $2N+1$ terms of a geometric series and then put it in closed form:

$$\sum_{n=-N}^{N} e^{-j2\pi nfT} = e^{j2\pi NfT} + e^{j2\pi(N-1)fT} + \cdots + 1 + \cdots + e^{-j2\pi NfT}$$

$$= e^{j2\pi NfT}\left[1 + e^{-j2\pi fT} + (e^{-j2\pi fT})^2 + \cdots + (e^{-j2\pi fT})^{2N}\right]$$

$$= \frac{e^{j2\pi NfT}(1 - e^{-j2\pi(2N+1)fT})}{1 - e^{-j2\pi fT}}$$

$$= \frac{e^{j2\pi NfT}e^{-j2\pi(N+1/2)fT}\left[e^{j2\pi(N+1/2)fT} - e^{-j2\pi(N+1/2)fT}\right]}{e^{-j2\pi fT/2}\left[e^{j2\pi fT/2} - e^{-j2\pi fT/2}\right]}.$$

Hence

$$S_N(f) = \frac{\sin \pi(2N+1)fT}{\sin \pi fT}. \qquad (14.1-2)$$

This has the behavior shown in Figure 14.1–2b—peaks of magnitude $(2N+1)$ at frequencies that are multiples of $1/T$, with zeros at multiples of $1/(2N+1)T$ in between.

As $N \to \infty$, $s_N(t) \to s(t)$. The peaks in $S_N(f)$ get progressively higher and proportionately narrower as $N \to \infty$, so that they approach impulses. To show that their area is $1/T$, note that the area under each lobe of $S_N(f)$ is

$$\int_{-1/2T}^{1/2T} S_N(f)\,df = \int_{-1/2T}^{1/2T} \sum_{n=-N}^{N} e^{-j2\pi nfT}\,df = \sum_{n=-N}^{N} \underbrace{\int_{-1/2T}^{1/2T} e^{-j2\pi nfT}\,df}_{=\,0 \text{ unless } n=0}$$

$$= \frac{1}{T}$$

for any N. Thus we have demonstrated that the transform of a "picket fence" in time is a "picket fence" in frequency, as indicated in Figure 14.1–1 and described symbolically by the transform pair

$$\boxed{\sum_{n=-\infty}^{\infty} \delta(t - nT) \Longleftrightarrow \frac{1}{T}\sum_{n=-\infty}^{\infty} \delta\left(f - \frac{n}{T}\right).} \qquad (14.1\text{–}3)$$

14.2 Fourier Transforms of Periodic Functions; Fourier Series Revisited

Any periodic function $x(t)$ may be considered as the convolution

$$x(t) = x_T(t) * s(t) \qquad (14.2\text{–}1)$$

where $x_T(t)$ is the time function describing a single period,

$$x_T(t) = \begin{cases} x(t), & 0 < t < T \\ 0, & \text{elsewhere} \end{cases}$$

and $s(t)$ is a periodic unit impulse train. Graphically, this construction is illustrated in Figure 14.2–1.

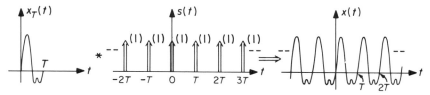

Figure 14.2–1. Graphical representation of (14.2–1).

In the frequency domain convolution becomes multiplication, so that the same construction can be interpreted as

$$X(f) = X_T(f)S(f) \qquad (14.2\text{–}2)$$

where $X_T(f)$ is the transform of $x_T(t)$. This formula describes the situation represented in Figure 14.2–2; since $S(f)$ is an impulse train, $X(f)$ is also an impulse train with impulse areas proportional to the heights of $X_T(f)$ at the corresponding frequencies. Formally,

$$X(f) = \frac{1}{T} \sum_{n=-\infty}^{\infty} X_T\left(\frac{n}{T}\right) \delta\left(f - \frac{n}{T}\right). \tag{14.2–3}$$

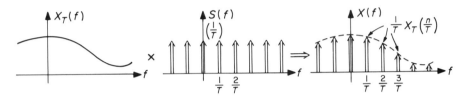

Figure 14.2–2.   Graphical representation of (14.2–2).

Thus we should be able to reconstruct $x(t)$ from the inverse Fourier transform formula

$$
\begin{aligned}
x(t) &= \int_{-\infty}^{\infty} X(f) e^{j2\pi ft} \, df \\
&= \int_{-\infty}^{\infty} \left[\frac{1}{T} \sum_{n=-\infty}^{\infty} X_T\left(\frac{n}{T}\right) \delta\left(f - \frac{n}{T}\right) \right] e^{j2\pi ft} \, df \\
&= \frac{1}{T} \sum_{n=-\infty}^{\infty} X_T\left(\frac{n}{T}\right) \int_{-\infty}^{\infty} \delta\left(f - \frac{n}{T}\right) e^{j2\pi ft} \, df \\
&= \frac{1}{T} \sum_{n=-\infty}^{\infty} X_T\left(\frac{n}{T}\right) e^{j2\pi nt/T}.
\end{aligned}
\tag{14.2–4}
$$

That is, a periodic time function can represented as a discrete sum of harmonically related sinusoids or exponentials. This, of course, is the Fourier series representation introduced in Chapter 12,

$$\boxed{x(t) = \sum_{n=-\infty}^{\infty} X[n] e^{j2\pi nt/T}} \tag{14.2–5}$$

where we now interpret the Fourier series coefficients $X[n]$ as proportional to values at the harmonic frequencies of the Fourier transform $X_T(f)$ of a single period of $x(t)$:

$$X[n] = \frac{1}{T} X_T\left(\frac{n}{T}\right). \tag{14.2–6}$$

Since

$$X_T(f) = \int_{-\infty}^{\infty} x_T(t)e^{-j2\pi ft}\, dt$$

$$= \int_{0}^{T} x(t)e^{-j2\pi ft}\, dt$$

we obtain the same formula as before for the Fourier coefficients:

$$\boxed{X[n] = \frac{1}{T}\int_{0}^{T} x(t)e^{-j2\pi nt/T}\, dt\,.} \qquad (14.2\text{--}7)$$

We are thus now entitled to consider the Fourier series formulas of Chapter 12 as formally established—a special case of the Fourier transform formulas. Note that the inverse Fourier transform describes a periodic function as an integral of a series of impulses in frequency; the Fourier series represents such a function as a discrete sum of finite-amplitude sinusoids. The two representations are related through the formula

$$\boxed{X(f) = \sum_{n=-\infty}^{\infty} X[n]\delta\!\left(f - \frac{n}{T}\right).} \qquad (14.2\text{--}8)$$

Example 14.2–1

Consider the periodic square wave shown in Figure 14.2–3 on the next page. Here it is convenient to take $x_T(t)$ as the period of $x(t)$ from $-T/2 < t < T/2$ rather than from $0 < t < T$. (It should be obvious on reviewing the previous argument that $x_T(t)$ may be taken as any convenient period with no change in the final results.) From Chapter 13, $X_T(f)$ then has the form shown in Figure 14.2–3. Multiplying by $S(f)$ gives $X(f)$; note that the even harmonics are cancelled by the zeros of $X_T(f)$, as we expect from the odd-harmonic-only symmetry properties of $x(t)$. The present argument makes it clear why even-harmonic cancellation requires the *duty cycle* (or fraction of the period during which $x(t)$ has its higher value) to be exactly $1/2$.

To reconstruct $x(t)$ we can either integrate the impulse train $X(f)$ or first determine the Fourier coefficients

$$X[n] = \frac{1}{T}X_T\!\left(\frac{n}{T}\right) = \begin{cases} 1/2, & n = 0 \\ 0, & n \text{ even, } n \neq 0 \\ \dfrac{(-1)^{(n-1)/2}}{n\pi}, & n \text{ odd} \end{cases}$$

and then sum to determine the Fourier series already given in Chapter 12:

$$x(t) = \sum_{n=-\infty}^{\infty} X[n]e^{j2\pi nt/T} = \frac{1}{2} + \sum_{\substack{n=-\infty \\ n \text{ odd}}}^{\infty} \frac{(-1)^{(n-1)/2}}{n\pi}e^{j2\pi nt/T}\,.$$

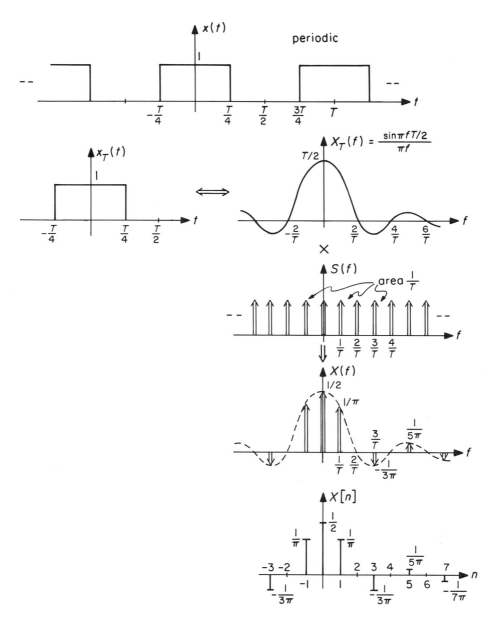

Figure 14.2–3. Analysis of a square wave.

▶ ▶ ▶

14.3 The Sampling Theorem

Sampling theorems formalize an intuitively reasonable idea. If the spectrum $X(f)$ of $x(t)$ contains primarily lower frequencies—that is, if $|X(f)|$ is relatively small for $|f| > W$, say—one would not expect $x(t)$ to change significantly in a time interval short compared with the period $1/W$ of the highest frequency that $x(t)$ does contain in significant amplitude. Thus sample values of $x(t)$ taken at intervals a fraction of $1/W$ seconds apart should describe $x(t)$ rather well; a smooth curve passing through these sample points should not differ greatly from $x(t)$ anywhere. The simplest and most famous form of theorem embodying this idea can be stated as follows.

SAMPLING THEOREM:

Suppose that $X(f)$, the Fourier transform of $x(t)$, is identically zero for all $|f| > W$ and has no singularities at $|f| = W$. Then $x(t)$ has exactly the representation

$$x(t) = \sum_{n=-\infty}^{\infty} x\left(\frac{n}{2W}\right)\frac{\sin 2\pi W\left(t - \dfrac{n}{2W}\right)}{2\pi W\left(t - \dfrac{n}{2W}\right)}. \qquad (14.3\text{--}1)$$

That is, samples of $x(t)$ at points $t = n/2W$ (samples that are $1/2W$ apart in time) completely determine $x(t)$ at all points in time.

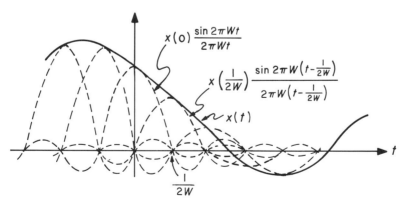

Figure 14.3–1. The Sampling Theorem as an interpolation formula.

In the time domain the Sampling Theorem implies that the values of a bandlimited function between the sample points can be exactly found from the sample values by interpolating with $\frac{\sin 2\pi Wt}{2\pi Wt}$ functions as in Figure 14.3–1. As

an interpolation formula, (14.3–1) has long been familiar to mathematicians.*
Indeed, Black[†] credits one rather general formulation of the Sampling Theorem
to Cauchy in 1841. In the 1920s the essential idea of the theorem was rediscovered by Carson, Nyquist, and Hartley and became in their hands one of the
cornerstones of the modern theory of communication.[‡] (As a result $1/2W$ is often
called the *Nyquist interval*.) More recently the Sampling Theorem has been associated with the names of Gabor, Shannon, and Kotel'nikov.[§]

Example 14.3–1

As an illustration of the simple but effective ways in which Hartley and Nyquist used
the Sampling Theorem, consider the problem of computing the minimum bandwidth
required for standard black-and-white television. Suppose we want to "resolve" 500
(lines) × 650 (elements per line) in each picture or "frame" at a rate of 30 frames/second.
Through scanning, the brightness of each point is to be sent in turn as the amplitude
of the video waveform. If the maximum frequency in the video waveform is W Hz,
then from the Sampling Theorem the video waveform can take on independent values
no oftener than every $1/2W$ seconds. Hence we must have

$$2W > 500 \times 650 \times 30$$

or

$$W > 4.875 \text{ MHz.}$$

In fact a bandwidth of about 6 MHz is used for a variety of reasons, some of which will
be explained in later sections.

▶ ▶ ▶

To prove the Sampling Theorem in the form of (14.3–1), observe that the
system in Figure 14.3–2 implements the right-hand side of (14.3–1). That is, the
output of the multiplier is the impulse train $\sum_{n=-\infty}^{\infty} x(n/2W)\delta(t - n/2W)$, and
each impulse applied to the LTI system $H(f)$ triggers the impulse response

$$h(t) = \frac{\sin 2\pi Wt}{2\pi Wt}$$

appropriately weighted and delayed. To see that this output is in fact $x(t)$,
consider the operation of the same system in the frequency domain. The result

---

*See, for example, E. T. Whittaker, *Proc. Roy. Soc. Edin.*, *35*, (1915): 181–194. See also J.
M. Whittaker, *Interpolation Function Theory*, Camb. Tracts in Math. and Math. Phys., *33*
(Cambridge, UK: Cambridge Univ. Press, 1935).

[†]H. S. Black, *Modulation Theory* (New York, NY: Van Nostrand, 1953).

[‡]Some early references are H. Nyquist, *Trans. AIEE*, *47*, (1928): 617–644; H. Nyquist, *Bell Sys.
Tech. J.*, *3*, 2, (1924): 324–346; and R. V. L. Hartley, *Bell Sys. Tech. J.*, *7*, 3, (1928): 535–563.

[§]D. Gabor, *J. IEE*, *93*, pt. III, (1946): 429–457; C. E. Shannon, *Proc. IRE 37*, (1949): 10–21; V.
A. Kotel'nikov, "On the capacity of the 'Ether' and wires in electrical communication" (1933).

of multiplying $x(t)$ by the impulse train $\sum_{n=-\infty}^{\infty} \delta(t - n/2W)$ is to convolve $X(f)$ with the impulse train $2W \sum_{n=-\infty}^{\infty} \delta(f - 2Wn)$. Thus, the spectrum of the output of the multiplier consists of $2WX(f)$ reproduced periodically in frequency every $2W$ Hz as shown. If the spectrum $X(f)$ is restricted as assumed, then the output of the LTI system $H(f)$ has precisely the same spectrum $X(f)$ as the original input $x(t)$, which proves the theorem.

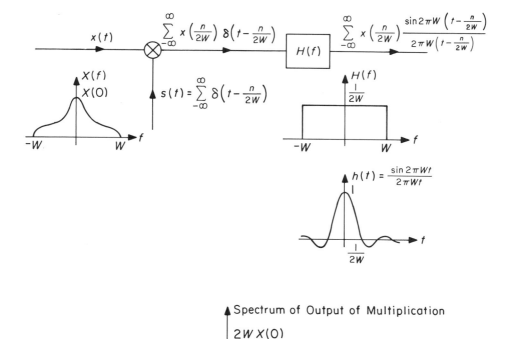

Figure 14.3–2. Construction demonstrating validity of the Sampling Theorem.

Conversely, note that if $x(t)$ has spectral energy outside the band $|f| < W$, the result of sampling at the rate $2W$ samples/sec and then lowpass filtering to the band $|f| < W$ is not the waveform that would result if $x(t)$ were simply lowpass-filtered to $|f| < W$. In particular, if $x(t)$ consists of a single sinusoid at a frequency $f > W$, the output of the system shown in Figure 14.3–2 is a sinusoid of a different frequency in the band $|f| < W$. A frequency outside the band $|f| < W$ thus "impersonates" or "passes itself off" as a sinusoid of lower frequency; in consequence, this phenomenon is called *aliasing*.

A number of generalizations and variations of the Sampling Theorem have been devised. Some special attributes of the bandpass case are discussed in Problem 17.16. For the lowpass case, it can be shown that a pair of samples taken close to each other every $1/W$ seconds (instead of a single sample every $1/2W$ seconds) will also suffice. Indeed, it is possible to prove that samples can be taken almost at random so long as, on average, $2W$ samples are taken per second (and the times at which they are taken are recorded).[*] In addition, samples of $x(t)$ at a rate slower than $2W$/sec can be combined with samples of various transformations of $x(t)$ to give a combined set of $2W$ samples/sec from which $x(t)$ can be recovered; Problem 14.12 gives two examples. These generalizations of the Sampling Theorem have some theoretical interest but only limited practical utility, since schemes in which any significant fraction of the samples are spaced further apart than $1/2W$ tend to be very sensitive to errors in sample values and timing.

A segment of a bandlimited waveform is often said to have $2TW$ "degrees of freedom." To see what this means, suppose that a long segment of duration $T \gg 1/2W$ is chopped out of a waveform bandlimited to $|f| < W$. The segment contains $2TW$ sampling points. The sample values at these $2TW$ points do not precisely specify the segment of the waveform, since the "tails" of the interpolating functions centered on sample instants outside the the segment will influence the values of $x(t)$ at points other than sample points inside the segment. But if $2TW \gg 1$, the error caused by ignoring these tails will be largely confined to small intervals near the beginning and end of the segment, and the energy in the error will be small relative to the energy in the whole waveform segment. Thus for $2TW \gg 1$, a segment of a bandlimited waveform is approximately specified by $2TW$ numbers. Indeed, if the numbers are not restricted to be sample values, a quite accurate description of a segment of a bandlimited waveform can be conveyed by about $2TW$ numbers. A precise and important theorem to this effect, including a bound on the error, has been proved by Landau and Pollak.[†]

The effect inside a segment of a bandlimited waveform of the "tails" corresponding to sample points outside suggests an interesting conjecture. Might it

---

[*]A specific theorem (J. L. Yen, *IRE PGCT-3*, (1956): 251) is the following: Pick a finite integer N and divide time into successive intervals of length $N/2W$. Pick N sampling instants in any manner from each interval. Knowledge of the sampling times and of the values of $x(t)$ at those times uniquely specifies $x(t)$ if $X(f) \equiv 0$, $|f| > W$.

[†]H. L. Landau and H. O. Pollak, *Bell Sys. Tech. J.*, *41*, (1962): 1295–1336. See also, J. M. Wozencraft and I. M. Jacobs, *Principles of Communication Engineering* (New York, NY: John Wiley, 1965) Appendix 5A.

be possible by examining a segment of a bandlimited waveform at each instant—not just at sample points—to estimate the amplitude of the waveform at sample points outside the segment, and thus to determine the entire waveform, past and future, given only a finite segment? The answer, in principle, is "yes." It can in fact be shown that any bandlimited waveform is an analytic function. Thus a Taylor's series can be formed, with coefficients obtained from the derivatives of the waveform inside the segment, that will converge in at least some interval outside the segment. By successive operations of this sort, the waveform can, in principle, be determined everywhere. The possibility of extrapolating a bandlimited waveform in this way is intimately related to the fact that LTI systems with frequency response identically zero over a band of frequencies such as $|f| > W$ are inherently non-causal, as we shall discuss in Chapter 16.

Bandlimited waveforms are often used as models for signals in communication and control systems; the fact that the entire past and future of such waveforms are uniquely specified by a small segment of the present is thus slightly embarrassing. A strictly bandlimited waveform could never provide new and unexpected information. But although these statements are true in principle, they have little practical significance. It can be shown, for example, that extrapolation without substantial error for even a short distance $(< 1/2W)$ requires unreasonable precision of measurement and computation. And experience clearly suggests that systems designed or analyzed on the basis of bandlimited idealizations usually perform more or less as expected. Nevertheless, one must proceed with caution, and "unexpected" results derived from such idealizations should be examined with considerable suspicion.

14.4 Pulse Modulation Systems

In addition to its general analytic usefulness, the Sampling Theorem is closely related to a number of practical applications such as pulse modulation systems, sampled-data systems, and digital data processing. One feature of pulse modulation systems is that they permit a number of independent signals to be sent simultaneously over the same channel by *time-division multiplexing*.* For example, three pulse signals derived as shown in Figure 14.4–1 by sampling separate bandlimited sources can be interleaved as shown and transmitted together. By appropriate synchronization of gating and smoothing operations at the receiver, the three signals can readily be separated and recovered.

---

*This idea goes back at least to Moses B. Farmer, who proposed in 1852 the use of synchronous commutators to transmit several telegraph messages over the same wire. Time-division multiplexing was first applied to continuous signals such as speech around 1900.

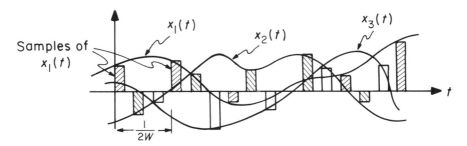

Figure 14.4-1. An example of time-division multiplexing.

Example 14.4-1

A simple commercial eight-voice-channel time-division-multiplexed system might have the following characteristics. Each of the eight voices is lowpass-filtered to the band $|f| < 3.3$ kHz. Preserving all the information in these signals would require sampling them at least as fast as the Nyquist rate, $2 \times 3.3 = 6.6$ kHz. In fact, the common practice is to *oversample* at a rate of 8 kHz, that is, once every 125 μsec; as a result (as shown in Figure 14.4-2) a *guard band* is created between the energy at low frequencies in the sampled signal and the energy in the band around 8 kHz. This greatly simplifies the design of the smoothing or interpolating filters in the receiver since they can use this guard band as a transition region from the pass to the stop band.

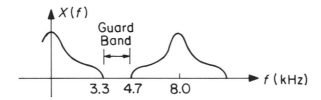

Figure 14.4-2. Spectrum of an oversampled voice sample.

Each sample is converted into a pulse of appropriate height and a few microseconds in duration. An electronic switch or commutator is arranged to select samples from each of the eight speech waveforms in succession, switching every $125/8 = 15.6$ μsec. The resulting waveform is a train of about 5 μsec duration pulses 15.6 μsec apart, every eighth of which comes from a particular speech waveform. Before transmission (which we might do by using the multiplexed signal to frequency modulate a carrier, as we shall discuss in Chapter 17) a synchronizing signal, such as a carefully phased 64 kHz sinusoid (which has a period of 15.6 μsec), would probably be added and the whole waveform passed through a *pulse-shaping filter* designed to restrict the bandwidth as much as possible without so spreading out each pulse as to cause significant overlapping of successive pulses (*interpulse interference*). The design of such a filter requires interesting compromises, as we shall discuss in Chapter 15.

▶ ▶ ▶

Another advantage of pulse modulation systems is that they readily and flexibly permit matching the properties of the message source to the bandwidth, power, and noise properties of a communication channel in such a way as to improve efficiency and performance. For example, the *pulse-position-modulated* (PPM) waveform shown in Figure 14.4–3 (in which the position of each narrow pulse relative to a uniformly spaced reference grid specifies the amplitude of the sample) has a much wider bandwidth (on the order of $1/\delta$) than the original waveform. But this wider bandwidth, as we shall illustrate in Chapter 20, implies a reduced vulnerability to certain types of wideband noise and interference, and thus may either give improved performance (such as a better signal-to-noise ratio or longer range) or permit a lowering of transmitter power.

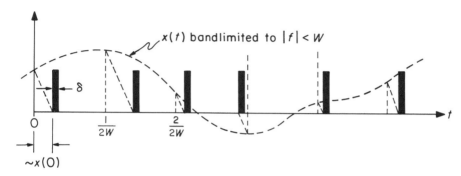

Figure 14.4–3. Pulse-position modulation.

This advantage of PPM is also shared by *pulse-code modulation* (PCM) illustrated in Figure 14.4–4. In PCM each sample value is first *quantized*, that is, replaced by the closest one of a finite number of discrete levels (8 in the illustration). With each number is associated a characteristic waveform of duration $1/2W$ seconds; commonly (as shown) this waveform is a sequence of 1 and 0 or + and − pulses corresponding, for example, to the binary representation of the level number. As with PPM, the PCM waveform obtains improved performance at the expense of a wider bandwidth. But PCM has other advantages, chief of which are the ease with which error-correcting codes and various special characteristics of the message source can be exploited and the opportunity to reshape the pulses at repeater stations in a cascaded communication system, so that the effects of noise do not accumulate. PCM also shares with other digital communication systems the advantage of flexibility. Once a speech wave, for example, has been converted into a sequence of digits by sampling and quantizing, it loses its identity as a speech wave. Strings of digits from many different sources—speech, pictures, data—can be flexibly interleaved for transmission over the same facility. Some of the major performance attributes of such pulse modulation systems as PPM and PCM will be discussed in Chapter 20.

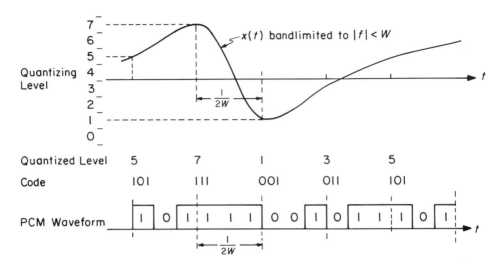

Figure 14.4-4. Pulse-code modulation.

14.5 The Discrete-Time Fourier Transform

The situation considered in the Sampling Theorem is the dual of the problem analyzed in connection with Fourier series. A duration-limited time function $x_T(t)$ (describing a single period of a periodic function) has a transform $X_T(f)$ that is completely described by samples at frequencies $1/T$ apart (because these samples describe the Fourier series from which the entire periodic function, of which $x_T(t)$ is one period, can be reconstructed). Alternately, a time function that is an impulse train (for example, the sequence of samples $x(t)s(t)$) has a transform that is periodic in frequency. The dual relationship can be made even more obvious by deriving a pair of formulas that are precisely the duals of the Fourier series formulas; these formulas will prove extremely useful in later chapters.

Paralleling our earlier development, let $x(t)$ be a time function whose transform $X(f)$ is periodic in frequency with period $2W$. $X(f)$ can be considered the convolution in frequency of the impulse train $S(f)/2W$ and the frequency function $X_W(f)$ describing a single period. That is, if we let

$$X_W(f) = \begin{cases} X(f), & -W < f < W \\ 0, & \text{elsewhere} \end{cases}$$

then

$$X(f) = X_W(f) * S(f)/2W. \tag{14.5-1}$$

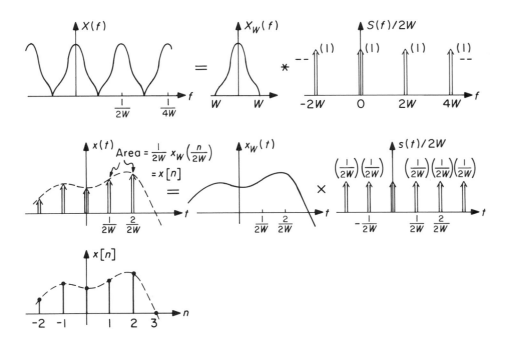

Figure 14.5–1. A time function whose transform is periodic.

This situation is represented diagrammatically in Figure 14.5–1. (In general $X_W(f)$ and $X(f)$ will be complex functions of frequency; for pictorial simplicity we have represented them as real.) In the time domain the corresponding representation is

$$x(t) = \frac{x_W(t)s(t)}{2W} = \sum_{n=-\infty}^{\infty} \frac{1}{2W} x_W\left(\frac{n}{2W}\right) \delta\left(t - \frac{n}{2W}\right). \tag{14.5–2}$$

Thus, a time function $x(t)$ whose transform is periodic must be an *impulse train* in time whose amplitudes are proportional to samples of $x_W(t)$, the inverse transform of one period of $X(f)$. It is convenient to describe the amplitude of these impulses by a function of the discrete variable n. Thus let

$$x[n] \overset{\Delta}{=} \frac{1}{2W} x_W\left(\frac{n}{2W}\right) \tag{14.5–3}$$

so that (14.5-2) becomes

$$x(t) = \sum_{n=-\infty}^{\infty} x[n]\delta\left(t - \frac{n}{2W}\right). \qquad (14.5\text{-}4)$$

We can derive $x[n]$ directly from $X(f)$ since

$$2Wx[n] = x_W\left(\frac{n}{2W}\right) = \int_{-\infty}^{\infty} X_W(f)e^{j2\pi nf/2W}\, df = \int_{-W}^{W} X(f)e^{j2\pi nf/2W}\, df.$$

Thus

$$\boxed{x[n] = \frac{1}{2W}\int_{-W}^{W} X(f)e^{j2\pi nf/2W}\, df.} \qquad (14.5\text{-}5)$$

On the other hand, we also have directly

$$X(f) = \int_{-\infty}^{\infty} x(t)e^{-j2\pi ft}\, dt = \int_{-\infty}^{\infty} \sum_{n=-\infty}^{\infty} x[n]\delta\left(t - \frac{n}{2W}\right)e^{-j2\pi ft}\, dt$$

which yields, on interchanging summation and integration,

$$\boxed{X(f) = \sum_{n=-\infty}^{\infty} x[n]e^{-j2\pi nf/2W}.} \qquad (14.5\text{-}6)$$

This is clearly periodic in frequency with period $2W$.

The dual relationship of these formulas to the usual Fourier series formulas is best illustrated by displaying them side-by-side.

Periodic $x(t)$	Periodic $X(f)$
$x(t) = x(t+T)$	$X(f) = X(f+2W)$
$x(t) = \displaystyle\sum_{n=-\infty}^{\infty} X[n]e^{j2\pi nt/T}$	$X(f) = \displaystyle\sum_{n=-\infty}^{\infty} x[n]e^{-j2\pi nf/2W}$
$X[n] = \displaystyle\frac{1}{T}\int_{-T/2}^{T/2} x(t)e^{-j2\pi nt/T}\, dt$	$x[n] = \displaystyle\frac{1}{2W}\int_{-W}^{W} X(f)e^{j2\pi nf/2W}\, df$
$X(f) = \displaystyle\sum_{n=-\infty}^{\infty} X[n]\delta\left(f - \frac{n}{T}\right)$	$x(t) = \displaystyle\sum_{n=-\infty}^{\infty} x[n]\delta\left(t - \frac{n}{2W}\right)$

The principal usefulness of the periodic-$X(f)$ formulas derives from the fact that $x[n]$ is a function of discrete time (an indexed sequence of numbers) rather than a function of continuous time like $x(t)$. Hence the synthesis formula (14.5-5) can be interpreted as representing an arbitrary discrete-time function as a "sum" (integral) of exponentials. The periodic-$X(f)$ formulas (14.5-5) and (14.5-6) thus play the same role with respect to discrete-time LTI systems as the usual Fourier transform formulas play with respect to continuous-time LTI sustems. Further properties and applications of these formulas will be discussed in Chapter 18.

14.6 Summary

One of the most useful Fourier transform pairs relates a periodic impulse train in time to a periodic impulse train in frequency:

$$s(t) = \sum_{n=-\infty}^{\infty} \delta(t - nT) \Longleftrightarrow \frac{1}{T} \sum_{n=-\infty}^{\infty} \delta\left(f - \frac{n}{T}\right) = S(f).$$

Thus if we convolve a waveform $x_T(t)$ of duration less than T with $s(t)$, the result is a periodic waveform. The transform of this periodic waveform is the product of the impulse train $S(f)$ and the transform $X_T(f)$ of the single-period waveform $x_T(t)$. The result is an impulse train whose areas are the Fourier series coefficients; the Fourier integral becomes in this case a sum—the Fourier series for the periodic waveform.

On the other hand, if we multiply a waveform $x(t)$ by $s(t)$, the result is an impulse train whose areas are samples of $x(t)$ at points $t = nT$. The transform of the impulse train of samples is the convolution of $S(f)$ and the transform $X(f)$ of $x(t)$. The result of this convolution is periodic in frequency. If the bandwidth W of $X(f)$ is small enough $(< 1/2T)$, then $X(f)$ can be recovered from $X(f) * S(f)$ without any of the distorting effects of aliasing; in this case the samples of $x(t)$ accurately describe $x(t)$ at all times.

These two arguments describe precisely dual situations. Note in particular:

Periodic function in time $\Longleftrightarrow$ Impulse train in frequency

Impulse train in time $\Longleftrightarrow$ Periodic function in frequency.

If the areas of the impulse train in time are considered to define a function in discrete time, the Fourier integral describes such a DT function as a weighted sum of sinusoids. This representation will be exploited in Chapter 18 to extend Fourier methods to discrete-time LTI systems.

APPENDIX TO CHAPTER 14

Vector-Space Representations of Signals

Many of the ideas developed in this and earlier chapters apply to expansions in series of functions other than complex exponentials. Many of these series are useful for system analysis. For example, the eigenfunctions of time-varying linear systems have much the same advantages as a basis for representing signals in such systems as complex exponentials have for signals in LTI systems. And even for LTI systems, certain special cases (such as bandlimited systems) or restricted classes of signals can be most conveniently discussed in terms of series composed of functions other than sinusoids. Most importantly, however, a general treatment of series representations suggests a geometric analog of our analytical procedures that is exceedingly valuable.

In this appendix we shall study some of the general properties of functional representations of the form

$$x(t) = \sum_k x_k \phi_k(t).$$

It is convenient to begin by assuming that the (possibly complex) functions $\phi_k(t)$ are *orthogonal* and *normalized* (or *orthonormal* to include both properties), although as we shall see these assumptions impose no serious restrictions. Formally, these properties state that, on some interval $t_0 < t < t_1$,

$$\text{Orthogonality} \Rightarrow \int_{t_0}^{t_1} \phi_k(t) \phi_m^*(t)\, dt = 0, \qquad k \neq m$$

$$\text{Normalization} \Rightarrow \int_{t_0}^{t_1} |\phi_k(t)|^2\, dt = 1.$$

Equivalently, these conditions can be combined in the single equation*

$$\int_{t_0}^{t_1} \phi_k(t) \phi_m^*(t)\, dt = \delta[k - m] = \begin{cases} 1, & k = m \\ 0, & k \neq m. \end{cases}$$

Example 14.A–1

Some examples of orthonormal functions and their ranges that have already been introduced in the previous chapters or problems are:

a) $\phi_k(t) = \dfrac{e^{j 2\pi k t/T}}{\sqrt{T}}, \qquad t_1 = t_0 + T.$

---

*More commonly (as mentioned in Chapter 8), $\delta[k - m]$ is denoted δ_{km} and called Kronecker's delta.

b) $\quad \phi_k(t) = \begin{cases} \dfrac{1}{\sqrt{\delta}}, & k\delta < t < (k+1)\delta \\ 0, & \text{elsewhere} \end{cases}$, $\qquad t_0 = -\infty,\ t_1 = \infty$

as shown in Figure 14.A–1.

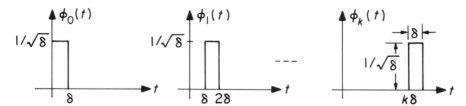

Figure 14.A–1. An orthonormal set of pulse functions.

c) $\quad \phi_k(t) = \dfrac{1}{\sqrt{T}}\left[\cos\dfrac{2\pi kt}{T} + \sin\dfrac{2\pi kt}{T}\right]$, $\qquad t_1 = t_0 + T$.

d) $\quad \phi_k(t) = \dfrac{\sin 2\pi W\left(t - \dfrac{k}{2W}\right)}{\sqrt{2W}\,\pi\left(t - \dfrac{k}{2W}\right)}$, $\qquad t_0 = -\infty,\ t_1 = \infty$.

▶ ▶ ▶

Of course, many more sets of orthonormal functions can be defined. In fact, any set of linearly independent* square-integrable functions $\xi_k(t)$ can be combined by a scheme called the *Gram-Schmidt*[†] procedure to generate a set of orthonormal functions $\phi_k(t)$. This procedure is elementary in principle and generates the $\phi_k(t)$ one at a time as follows:

i) Let

$$\phi_0(t) = a_0\xi_0(t)$$

and choose a_0 so that $\phi_0(t)$ is normalized.

ii) Let

$$\phi_1(t) = a_1[b_0\phi_0(t) + \xi_1(t)]$$

---

*A set of functions is linearly independent if no one of the functions can be constructed as a weighted sum of the remainder.

[†]See, for example, B. Friedman, *Principles and Techniques of Applied Mathematics* (New York, NY: John Wiley, 1956) p. 16.

and choose b_0 so that the bracketed expression is orthogonal to $\phi_0(t)$. This leads immediately to the value

$$b_0 = -\int_{t_0}^{t_1} \phi_0^*(t)\xi_1(t)\, dt.$$

Then a_1 can be selected to normalize $\phi_1(t)$.

iii) Similarly let

$$\phi_2(t) = a_2[c_0\phi_0(t) + c_1\phi_1(t) + \xi_2(t)].$$

Forcing the bracketed expression to be orthogonal to both $\phi_0(t)$ and $\phi_1(t)$ leads to values for c_0 and c_1:

$$c_0 = -\int_{t_0}^{t_1} \phi_0^*(t)\xi_2(t)\, dt$$

$$c_1 = -\int_{t_0}^{t_1} \phi_1^*(t)\xi_2(t)\, dt$$

and a_2 can then be selected to normalize $\phi_2(t)$.

iv) Continuing in this way, find successively $\phi_3(t)$, $\phi_4(t)$, etc.

Like many of our results in this book, the Gram-Schmidt procedure is more important as an existence argument than as a computational procedure. It shows, for example, that restricting our attention to series of orthonormal functions is not a serious limitation. And it also shows that, given any set of M functions, $s_i(t)$, there exists a set of no more than M orthonormal functions $\phi_i(t)$ such that $s_i(t) = \sum_{j=1}^{M} s_{ij}\phi_i(t)$.

Suppose that we somehow know that a function $x(t)$ can be represented by a finite series of M orthonormal functions:

$$x(t) = \sum_{k=1}^{M} x_k\phi_k(t), \quad t_0 < t < t_1. \tag{14.A--1}$$

Then (as discussed in Chapter 12 for series of sinusoids), it is easy to deduce a formula from which we can determine the coefficients by multiplying both sides of (14.A–1) by $\phi_m^*(t)$ and integrating:

$$\int_{t_0}^{t_1} x(t)\phi_m^*(t)\, dt = \int_{t_0}^{t_1} \left[\sum_{k=1}^{M} x_k\phi_k(t)\right]\phi_m^*(t)\, dt$$

$$= \sum_{k=1}^{M} x_k \int_{t_0}^{t_1} \phi_k(t)\phi_m^*(t)\, dt$$

$$= \sum_{k=1}^{M} x_k\delta[k - m]$$

$$= x_m. \tag{14.A--2}$$

For example, using $\phi_k(t)$ from Example 14.A–1a, we easily verify from (14.A–2) that

$$x_m = \sqrt{T}\left[\frac{1}{T}\int_{t_0}^{t_0+T} x(t)e^{-j2\pi mt/T}\,dt\right] = \sqrt{T}X[m]$$

where $X[m]$ is the usual complex Fourier series coefficent, and the series becomes

$$x(t) = \sum_{k=-\infty}^{\infty} \sqrt{T}X[k]\frac{1}{\sqrt{T}}e^{j2\pi kt/T} = \sum_{k=-\infty}^{\infty} X[k]e^{j2\pi kt/T}$$

which is the ordinary Fourier series, as it should be. By analogy, a series of orthonormal functions (even if they are not sinusoids) whose coefficients are determined as in (14.A–2) is called a *generalized Fourier series*, and the x_m are called *generalized Fourier coefficients*.

Even if we do not know that $x(t)$ can be represented as in (14.A–1), we can still show, as we did in Example 13.4–1 and Problem 13.9 for the ordinary Fourier series and integrals,* that the squared error

$$\xi_M^2 = \int_{t_0}^{t_1} \left| x(t) - \sum_{k=1}^{M} x_k\phi_k(t) \right|^2 dt$$

is minimized for fixed M by choosing the x_k to be generalized Fourier coefficients given by (14.A–2). Moreover, for x_k given by (14.A–2), the minimum value of ξ_M^2 is

$$\xi_{M_{\min}}^2 = \int_{t_0}^{t_1} |x(t)|^2\,dt - \sum_{k=1}^{M} |x_k|^2 \geq 0 \qquad (14.A-3)$$

which is called *Bessel's inequality*. If for some particular value of M (which may be infinite) and for every member of some particular class of functions $x(t)$, it is true that $\xi_M^2 = 0$, then the set of M functions $\phi_k(t)$ is said to be *complete* for that class of $x(t)$, and equation (14.A–3) becomes

$$\int_{t_0}^{t_1} |x(t)|^2\,dt = \sum_{k=1}^{M} |x_k|^2 \qquad (14.A-4)$$

which is Parseval's Theorem.

The results that we have been discussing can be given a most illuminating interpretation by pointing out a formal analogy that exists between signals represented as in (14.A–1) and multidimensional vectors. Some such analogy, of course, has already been implied by our choice of the geometric word

---

*The method of proof must, however, be somewhat different since—as we shall shortly see—the generalizaton of Parseval's Theorem is not valid unless $x(t)$ can in fact be represented as in (14.A–1). However, it is easy to minimize ξ_M^2 in the ordinary way by setting to zero the partial derivatives of ξ_M^2 with respect to each x_k and solving the resulting equations for x_k. Another proof will be given in Example 14.A–3.

"orthogonal" for an important relationship between waveforms. Indeed, the parallel between signals and vectors is so close that much of the material of this book is discussed in the more rigorous mathematical literature under the heading of *linear vector spaces*.

The analogy is based on the observation that any signal that can be represented in the form

$$x(t) = \sum_{k=1}^{M} x_k \phi_k(t)$$

can be completely described for any fixed set of $\phi_k(t)$ by specifying the ordered set of M quantities x_k—that is, by specifying an M-dimensional *vector* $\mathbf{x}$ with *components* x_k. Moreover, if $y(t)$ is another waveform that can be represented in terms of the same *basis set* $\phi_k(t)$, that is, if

$$y(t) = \sum_{k=0}^{N} y_k \phi_k(t)$$

then the vector corresponding to the waveform $Ax(t) + By(t)$ (where A and B are constants) is simply $A\mathbf{x} + B\mathbf{y}$. But the real power of the analogy stems from the fact that if the $\phi_k(t)$ are orthonormal over the interval $t_0 < t < t_1$, then*

$$\int_{t_0}^{t_1} x(t)y(t)\,dt = \int_{t_0}^{t_1} \left[\sum_{m=1}^{M} x_m \phi_m(t)\right]\left[\sum_{n=1}^{M} y_n \phi_n(t)\right] dt$$

$$= \sum_{n=1}^{M} x_n y_n = \mathbf{x} \cdot \mathbf{y}. \tag{14.A–5}$$

The *scalar, dot,* or *inner product* of the two vectors is thus equal to the integral of the product of the two waveforms. We note three particular consequences:

a) The condition that $y(t)$ and $x(t)$ are orthogonal,

$$\int_{t_0}^{t_1} x(t)y(t)\,dt = 0$$

is equivalent to the condition that the dot product of $\mathbf{x}$ and $\mathbf{y}$ must vanish, so that $\mathbf{x}$ and $\mathbf{y}$ are perpendicular. Thus, in particular, the vectors ϕ_k corresponding to the functions $\phi_k(t)$ (the components of ϕ_k are all zero except for the kth, which is 1) form a mutually perpendicular set of M vectors (in M dimensions, of course!).

b) If $y(t) = x(t)$, then

$$\int_{t_0}^{t_1} x^2(t)\,dt = \sum_{n=1}^{M} x_n^2 = \mathbf{x} \cdot \mathbf{x} = ||\mathbf{x}||^2 \tag{14.A–6}$$

---

*For simplicity, we shall assume in the following paragraphs that $x(t)$, x_k, $\phi_k(t)$, and all similar quantities are real.

where $||\mathbf{x}||^2$ is the square of the *magnitude* or *length* of the vector $\mathbf{x}$ and is thus equal to the energy in $x(t)$. In particular, since the functions $\phi_k(t)$ are normalized to have unit energy, the vectors $\boldsymbol{\phi}_k$ are mutually perpendicular *unit* vectors and thus determine a Cartesian coordinate system in an M-dimensional Euclidean space. Each x_k is then the component of $\mathbf{x}$ along the kth coordinate, and (14.A–6) is simply the M-dimensional extension of the Pythagorean Theorem.

c) Since $\boldsymbol{\phi}_k$ is a unit vector in the direction of the kth coordinate, x_k is the *projection* of $\mathbf{x}$ along $\boldsymbol{\phi}_k$:

$$x_k = \mathbf{x} \cdot \boldsymbol{\phi}_k = \int_{t_0}^{t_1} x(t)\phi_k(t)\,dt\,.$$

But this is just the formula (14.A–2) for the Fourier coefficients that we previously obtained from an algebraic argument.

The following examples illustrate some of the advantages of this vectorial representation of waveforms.

Example 14.A–1

The dot product of two vectors can also be written in the form

$$\mathbf{x} \cdot \mathbf{y} = ||\mathbf{x}||\,||\mathbf{y}||\cos\theta$$

where θ is the angle between $\mathbf{x}$ and $\mathbf{y}$. Since $|\cos\theta| \leq 1$, it must be true that

$$|\mathbf{x} \cdot \mathbf{y}|^2 \leq ||\mathbf{x}||^2\,||\mathbf{y}||^2$$

or

$$\left[\int_a^b x(t)y(t)\,dt\right]^2 \leq \int_a^b x^2(t)\,dt \int_a^b y^2(t)\,dt$$

which is called the *Schwarz inequality*. This most useful formula can, of course, be proved directly, but its demonstration as above is an excellent example of the insight to be gained from the vectorial point of view.

▶ ▶ ▶

Example 14.A–2

Consider a set of M vectors $\mathbf{x}_n$, $n = 1, 2, \ldots, M$. These span at most M dimensions, in that we can always find a set of M mutually perpendicular unit vectors $\boldsymbol{\phi}_m$ that determine an M-dimensional coordinate system in which all of the $\mathbf{x}_n$ can be represented. Or to put this another way, we can always find a set of $\boldsymbol{\phi}_m$ that will allow us to represent any of the $\mathbf{x}_n$ in the form

$$\mathbf{x}_n = \sum_{m=1}^{M} \alpha_{nm}\boldsymbol{\phi}_m\,.$$

One method for finding an appropriate set of $\boldsymbol{\phi}_m$ is the Gram-Schmidt procedure—pick $\boldsymbol{\phi}_1$ along $\mathbf{x}_1$; pick $\boldsymbol{\phi}_2$ perpendicular to $\boldsymbol{\phi}_1$, in the plane determined by $\mathbf{x}_1$ and $\mathbf{x}_2$; pick $\boldsymbol{\phi}_3$ perpendicular to both $\boldsymbol{\phi}_1$ and $\boldsymbol{\phi}_2$; etc. Of course, if more than two of the vectors $\mathbf{x}_n$ lie in the same plane, then the vectors are not linearly independent and they span less than M dimensions. But note that if only these M vectors (or waveforms) are important in a given problem, we can restrict our entire analysis to a finite-dimensional space in spite of the fact that each waveform might be quite complicated. Shannon used this point of view to great effect in the development and proof of the fundamental theorems of information theory (see Chapter 20).* As another application, the observation made earlier in this chapter that a set of duration- and bandwidth-limited waveforms has only about $2TW$ "degrees of freedom" is equivalent to the fact that there are only about $2TW$ orthogonal waveforms of duration T and bandwidth W.

Of course, the M-dimensional coordinate system obtained above is not unique; any *rotated* orthogonal coordinate system will do as well. If $\mathbf{u}_\ell$ represents a set of unit vectors along the new coordinates, then the rotation can be described by specifying each new unit vector in terms of the old unit vectors, that is,

$$\mathbf{u}_\ell = \sum_{n=1}^{M} \lambda_{\ell n} \boldsymbol{\phi}_n$$

where $\lambda_{\ell n}$ is the component of $\mathbf{u}_\ell$ along $\boldsymbol{\phi}_n$,

$$\lambda_{\ell n} = \mathbf{u}_\ell \cdot \boldsymbol{\phi}_n$$

and is called a *direction cosine* (since $\|\mathbf{u}_\ell\| = \|\boldsymbol{\phi}_n\| = 1$, $\lambda_{\ell n}$ is simply the cosine of the angle between $\mathbf{u}_\ell$ and $\boldsymbol{\phi}_n$).

Obviously the set of direction cosines must satisfy certain conditions if the $\mathbf{u}_\ell$ are to be a set of mutually orthogonal unit vectors. Indeed, since

$$\mathbf{u}_\ell \cdot \mathbf{u}_m = \delta[\ell - m]$$

we must have

$$\sum_{n=1}^{M} \lambda_{\ell n} \lambda_{mn} = \delta[\ell - m] = \begin{cases} 1, & \ell = m \\ 0, & \ell \neq m. \end{cases}$$

The coefficients in the expansion of a given vector $\mathbf{x}_n$ in terms of $\mathbf{u}_\ell$,

$$\mathbf{x}_n = \sum_{\ell=1}^{M} \gamma_{n\ell} \mathbf{u}_\ell$$

are simply the projections of $\mathbf{x}_n$ along the new coordinates:

$$\gamma_{n\ell} = \mathbf{x}_n \cdot \mathbf{u}_\ell.$$

Since $\mathbf{x}_n$ as a vector is independent of the coordinate system in which it is represented, it must be true that

$$\|\mathbf{x}_n\|^2 = \sum_{\ell=1}^{M} \gamma_{n\ell}^2 = \sum_{m=1}^{M} \alpha_{nm}^2.$$

---

*C. E. Shannon, *Bell Sys. Tech. J.* (July and October, 1948). Reprinted as *The Mathematical Theory of Communication* (Urbana, IL: Univ. of Illinois Press, 1949).

As a loose but graphic application of these notions, consider the two principal waveform representation schemes of this and the preceding chapter:

$$x(t) = \int_{-\infty}^{\infty} x(\tau)\delta(t-\tau)\,d\tau$$

$$x(t) = \int_{-\infty}^{\infty} X(f)e^{j2\pi ft}\,df.$$

If we consider the set of delayed impulses as determining one set of orthogonal vectors and the set of complex exponentials as determining another set, then $x(\tau)d\tau$ and $X(f)df$ are the components of $x(t)$ along the corresponding coordinates. The frequency-domain representation of $x(t)$ thus amounts to picking a coordinate system for describing $x(t)$ that is simply *rotated* from the time-domain coordinate system. And Parseval's Theorem,

$$\int_{-\infty}^{\infty} x^2(t)\,dt = \int_{-\infty}^{\infty} |X(f)|^2\,df$$

is just a statement of the fact that the length of a vector is independent of the coordinate system in which it is described.

Example 14.A–3

As a final example, let's reconsider a problem we discussed in Example 13.4–2 and Problem 13.19 in terms of waveforms. We want to choose coefficients x_k in the finite sum

$$\sum_{k=1}^{M} x_k \phi_k(t)$$

that will best approximate some waveform $x(t)$ in the sense of minimum squared error,

$$\xi_M^2 = \int_{t_0}^{t_1} \left| x(t) - \sum_{k=1}^{M} x_k \phi_k(t) \right|^2 dt\,.$$

The solution obtained algebraically in (14.A–2) was that the x_k should be the Fourier coefficients

$$x_k = \int_{t_0}^{t_1} x(t)\phi_k(t)\,dt\,.$$

The nature of this problem and its solution becomes strikingly clear when expressed in vectorial terms. Suppose that we seek the "best approximation" $\mathbf{x}'$ to a vector $\mathbf{x}$, where $\mathbf{x}'$ is restricted to have nonzero components only in a fixed subset (such as the first M) of the coordinates required to describe $\mathbf{x}$, and where "best approximation" is interpreted to mean that the distance between the tips of the vectors $\mathbf{x}$ and $\mathbf{x}'$, that is, the length $\|\mathbf{x} - \mathbf{x}'\|$ of the vector $\mathbf{x} - \mathbf{x}'$, is to be minimized. Clearly, minimum distance between vectors in this sense corresponds to minimum squared error between waveforms, since

$$\|\mathbf{x} - \mathbf{x}'\|^2 = \left\| \mathbf{x} - \sum_{k=0}^{M} \alpha_k \boldsymbol{\phi}_k \right\|^2 = \xi_M^2\,.$$

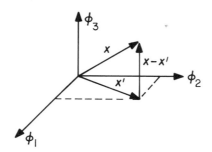

Figure 14.A–2. The minimum squared error problem in three dimensions.

Specifically, suppose we seek the best approximation lying in the ϕ_1-ϕ_2-plane to the three-dimensional vector $\mathbf{x}$ shown in Figure 14.A–2. Geometrically, it is clear that to minimize $\|\mathbf{x} - \mathbf{x}'\|$, we should have $\mathbf{x} - \mathbf{x}'$ *perpendicular* to the ϕ_1-ϕ_2-plane, which implies that the components of $\mathbf{x}'$ should be simply equal to the components of $\mathbf{x}$ along ϕ_1 and ϕ_2, that is, they should be the Fourier coefficients, as we saw earlier. Clearly, this conclusion generalizes to more than three dimensions. Moreover, from the properties of right triangles, we have

$$\|\mathbf{x}\| \geq \|\mathbf{x}'\|$$

which is Bessel's inequality.

▶ ▶ ▶

Further applications of the vectorial representation of waveforms will be found in the problems and in succeeding chapters.

EXERCISES FOR CHAPTER 14

Exercise 14.1

The signal $x(t)$ with Fourier transform $X(f)$ specified in the figure below is sampled by the system shown. Show that the output is $y(t) = \delta(t)$.

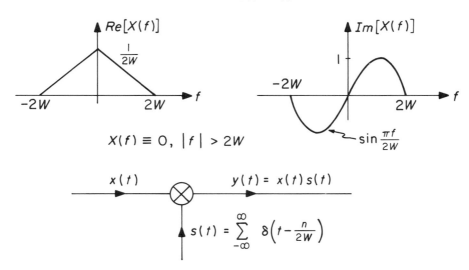

Exercise 14.2

Suppose that the Fourier transform of $x(t)$ is zero outside the band $|f| < 100$ Hz and that that of $y(t)$ is zero outside the band $|f| < 200$ Hz. What is the maximum time between samples that will completely characterize the following time signals:

     a) $x(t)$ b) $x(10t)$ c) $x(t) + y(t)$ d) $x(t) * y(t)$ e) $x(t)y(t)$

Answers: (a) 5 msec (b) 0.5 msec (c) 2.5 msec (d) 5 msec (e) 1.67 msec.

Exercise 14.3

a) Starting from the Fourier series representation for an impulse train, derive the formula for a train of doublets:

$$\sum_{k=-\infty}^{\infty} \dot{\delta}\left(t - \frac{k}{T}\right) = \sum_{k=-\infty}^{\infty} \frac{j2\pi k}{T^2} e^{j2\pi kt/T}$$

$$= -\sum_{k=1}^{\infty} \frac{4\pi k}{T^2} \sin \frac{2\pi kt}{T}.$$

b) In traditional developments of Fourier series it is sometimes argued that

i) $\displaystyle\lim_{N\to\infty}\sum_{k=1}^{N}\frac{4\pi k}{T^2}\sin\frac{2\pi kt}{T}=0,\quad t\neq nT$ (which is true).

ii) For $t=nT$, every term in the sum is zero (which is true).

iii) Therefore

$$\lim_{N\to\infty}\sum_{k=1}^{N}\frac{4\pi k}{T^2}\sin\frac{2\pi kt}{T}=0,\quad\text{all }t.$$

Hence the Fourier series representation of a periodic function is *not unique*: An arbitrary multiple of $j2\pi k/T^2$ could be added to every coefficient without altering the synthesis result.

Discuss.

Exercise 14.4

A signal $x(t)$ with the Fourier transform $X(f)=\int_{-\infty}^{\infty}x(t)e^{-j2\pi ft}\,dt$ shown below is passed through the non-causal "comb" filter with periodic impulse response $h(t)$.

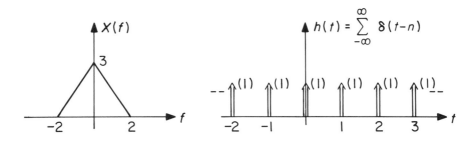

a) Show that the energy in the input $x(t)$ is $\int_{-\infty}^{\infty}x^2(t)\,dt=12$, whereas the average power is $\lim_{T\to\infty}\frac{1}{2T}\int_{-T}^{T}x^2(t)\,dt=0$.

b) Show that the energy in the output $y(t)=x(t)*h(t)$ is infinite, whereas the average power is 13.5.

PROBLEMS FOR CHAPTER 14

Problem 14.1

Let $x(t)$ be the periodic signal shown below.

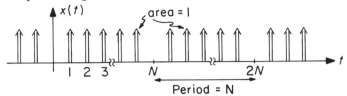

a) By representing $x(t)$ as a difference of two impulse trains, one with period 1 and the other with period N, find the Fourier integral representation of $x(t)$.

b) Find the Fourier series for $x(t)$ directly from the formula for Fourier coefficients. Exercise caution for harmonics that are multiples of N. Compare your results with part (a).

c) $x(t)$ is fed into an LTI system with impulse response shown below. What is the frequency response of this system?

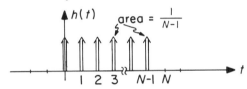

d) What is the Fourier series for the output $y(t)$ of the system?

e) What is the time function $y(t)$ having the Fourier series computed in part (d)?

f) Check your answer to part (e) by a time-domain analysis of the situation.

Problem 14.2

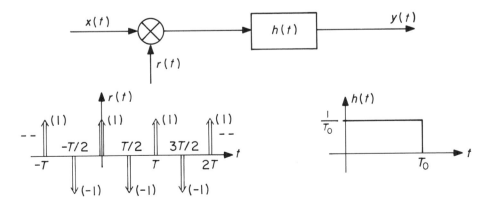

The purpose of the system shown on the preceding page is to provide a large output when the input is a sinusoid of frequency near $1/T$ Hz and a small output otherwise. That is, the system behaves in some respects as a narrowband filter. The frequency to which the "filter" is "tuned" can be adjusted by changing the period T of the impulse train.

a) Find and sketch the Fourier transform $R(f)$ of the periodic impulse train $r(t)$.

b) Find and sketch the magnitude of the Fourier transform $|H(f)|$ of the finite-time integrator with impulse response $h(t)$.

c) Suppose $x(t) = A\cos(2\pi f_0 t + \theta)$ with f_0 a frequency near $1/T$. Sketch the rms value of $y(t)$ as a function of $f_0 - 1/T$ for $T_0 = 100T$. (Make reasonable approximations.)

d) For what frequencies other than $f_0 = 1/T$ will this system have a large rms output? Suggest a scheme for reducing the response to these other frequencies, keeping in mind that we would like to be able to "tune" the system over some range of frequencies by varying T.

Problem 14.3

A half-wave rectifier with condenser-input filter is shown schematically below. For small load currents, at least, the various waveforms will appear as shown. In particular, the diode conducts only for a very small part of each cycle.

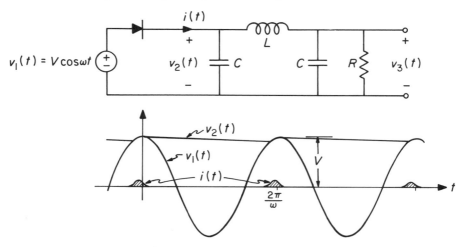

a) What is $v_3(t)$ if $i(t) = e^{j\omega t}$?

b) What is $v_3(t)$ if $i(t) = \displaystyle\sum_{n=-\infty}^{\infty} a_n e^{jn\omega_0 t}$?

c) Suppose that $i(t)$, $-\infty < t < \infty$, is a periodic train of impulses of area T, with period T; that is,

$$i(t) = \sum_{n=-\infty}^{\infty} T\delta(t - nT).$$

Find a Fourier series or Fourier integral representation of $v_3(t)$.

d) If I_{dc} is the average value of $i(t)$ and I_f is the rms value of the fundamental component of $i(t)$, show that

$$I_f \approx \sqrt{2} I_{dc}$$

and so, if $\dfrac{1}{\omega C} \ll \omega L$,

$$V_{3f}/V_{3dc} \approx \frac{\sqrt{2}}{\omega^3 LC^2 R}$$

where V_{3f} and V_{3dc} are the rms fundamental component and average value of $v_3(t)$ respectively. (HINT: Consider $i(t)$ as being approximately a periodic impulse train.)*

Problem 14.4

Repeated tone bursts are frequently used as test sounds in psychoacoustic experiments. Consider two possible ways of generating such sounds:

a) The tone is generated with the same starting phase at the beginning of each burst, as depicted in (A). T_2 is not necessarily a multiple of $1/f_0$.

b) A continuous tone is "blanked" or "gated" by a rectangular wave as shown in (B). The frequency of the tone is not necessarily a multiple of the blanking frequency.

Discuss the differences between the time functions and the spectra of these two signals. Are either or both of these signals periodic?

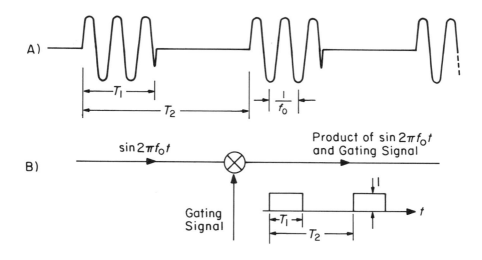

---

*L. B. Arguimbau, *Vacuum Tube Circuits* (New York, NY: John Wiley, 1948) pp. 27–28.

Problem 14.5

In any real communication or control system employing sampled waveforms, finite
pulses must be used instead of ideal impulses. Two ways of generating finite pulses are
illustrated below.

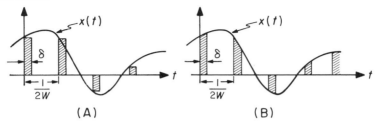

(A)   (B)

In (A) the value of $x(n/2W)$ is sampled and held for the pulse duration δ; in (B) sections
of $x(t)$ of duration δ are gated out every $1/2W$ seconds. The pulse waveform in (A) can
be considered the result of ideal impulse sampling followed by a filter whose impulse
response is a short square pulse of duration δ. The pulse waveform in (B) is simply
the result of multiplying $x(t)$ by a periodic rectangular wave. Assuming that $X(f)$ is
bandlimited to $|f| < W$, describe in detail the spectra of these two pulse waveforms for
$|f| < W$ and show how in each case $x(t)$ can be precisely recovered.

Problem 14.6

Show that a bandlimited $x(t)$ can be recovered from a gated pulse waveform such as
(B) of Problem 14.5 even if pulses are spaced as much as $1/W$ apart, provided that
δW is a *rational* number. (Indeed, $x(t)$ can in theory be recovered from such a gating
process under much weaker conditions, as discussed in the chapter.)

Problem 14.7

a) In the system below find the maximum value of T and the values of the constants
   A, f_1, and f_2 such that $z(t) = x(t)$.

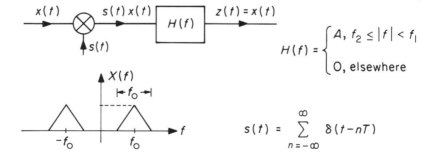

$$H(f) = \begin{cases} A, f_2 \le |f| < f_1 \\ 0, \text{ elsewhere} \end{cases}$$

$$s(t) = \sum_{n=-\infty}^{\infty} \delta(t - nT)$$

b) Repeat part (a) for $s(t)$ replaced by $s(t - T_0)$, where T_0 is a constant in the range
   $0 < T_0 < T$.

Problem 14.8

In theory, the correct way to recover a continuous bandlimited signal from its samples is to pass the sample pulses (ideally, impulses) through an ideal lowpass filter. In practice, other smoothing or interpolating schemes are often employed instead. One of the simplest is the *boxcar* or *zero-order hold circuit* shown in Figure 1 below. The switch is closed briefly every T seconds to charge the capacitor C to the value of the input pulse at that moment. Between pulses, the switch is open and the capacitor simply holds or "remembers" the last pulse amplitude. The output $y(t)$ is then a "staircase" approximation to the original input $x(t)$ from which the samples were derived.

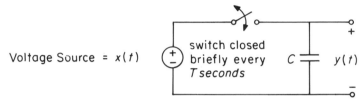

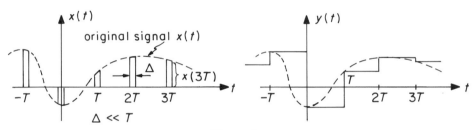

Figure 1.

The boxcar circuit is a time-varying linear system, but as long as inputs are restricted to narrow pulse trains such as $\hat{x}(t)$ the effect is approximately the same as that of an appropriate LTI system (see Figure 2 below).

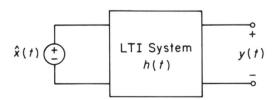

Figure 2.

a) Assuming that $\hat{x}(t)$ is an impulse train,

$$\hat{x}(t) = \sum_{n=-\infty}^{\infty} \Delta x(nT)\delta(t - nT)$$

find $h(t)$ such that $y(t)$ will be the same in Figure 2 as it is in Figure 1 when the input is the corresponding pulse train.

b) Determine the relationship between $Y(f)$ and the spectrum of the original band-limited signal $x(t)$ from which $\hat{x}(t)$ was derived. Under what circumstances will $y(t)$ be a very good approximation to this original signal (discuss in both the time domain and the frequency domain)?

Problem 14.9

As explained in the chapter, a signal limited in bandwidth to $|f| < W$ can be recovered from non-uniformly spaced samples as long as the average sample density is $2W$ per second. This problem illustrates a particular example of non-uniform sampling.

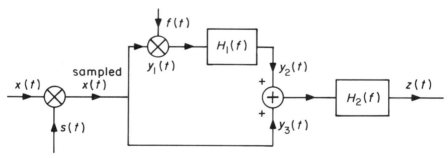

Figure 1.

Assume that

    i) $x(t)$ is bandlimited: $X(f) = 0$, $|f| > W$.

    ii) $s(t)$ is a non-uniformly spaced periodic unit impulse train as shown in Figure 2.

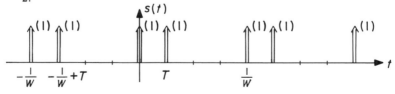

Figure 2.

    iii) $f(t)$ is a periodic waveform with period $1/W$. Since $f(t)$ multiplies an impulse train, only its values at $t = 0$ and $t = T$, $f(0) = a$ and $f(T) = b$, are significant.

iv) $H_1(f)$ is a $90°$ phase-shifter:

$$H_1(f) = \begin{cases} j, & f > 0 \\ -j, & f < 0. \end{cases}$$

v) $H_2(f)$ is an ideal lowpass filter:

$$H_2(f) = \begin{cases} K, & 0 < f < W \\ K^*, & -W < f < 0 \\ 0, & |f| > W \end{cases}$$

where K is a (possibly complex) constant.

Find (and sketch where appropriate) the following:

a) The Fourier transform $S(f)$ of $s(t)$.

· b) The Fourier transform of the product $s(t)f(t)$ in terms of the as yet unspecified parameters a and b.

c) An expression for the Fourier transform $Y_1(f)$ of $y_1(t)$ that is valid in the interval $0 < f < W$.

d) An expression for the Fourier transform $Y_2(f)$ of $y_2(t)$ that is valid in the interval $0 < f < W$.

e) An expression for the Fourier transform $Y_3(f)$ of $y_3(t)$ that is valid in the interval $0 < f < W$.

f) Values of the real parameters a and b and the complex gain K as functions of T, such that $z(t) = x(t)$ for any bandlimited $x(t)$ and any T in the range $0 < T < 1/2W$.

Problem 14.10

It is frequently helpful to display on an oscilloscope screen waveforms having very rapid time structure—for example, showing important changes occurring in a fraction of a nanosecond. Since the rise time of even the fastest oscilloscopes is much longer than this, such displays cannot be achieved directly. But if the waveform can be repeated periodically, the desired result can be obtained indirectly by a device called a *sampling oscilloscope*.

The idea, as shown in Figure 1, is to sample the fast waveform $x(t)$ once each period but at successively later points in successive periods. The increment Δ should be an appropriate sampling interval for the bandwidth of $x(t)$. If the resulting impulse train is then passed through an appropriate lowpass interpolating filter, the output $y(t)$ should be proportional to the original fast waveform slowed down or stretched out in time, that is, $y(t) \sim x(at)$, where $a < 1$.

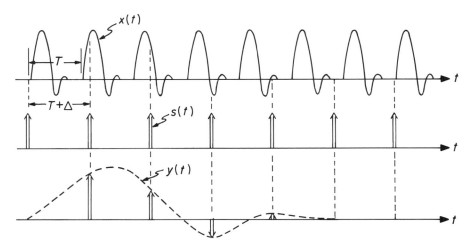

Figure 1.

a) If $X(f)$ is zero outside the band $|f| < W_x$ and if $Y(f)$ is restricted to the band $|f| < W_y$, what is the *smallest* period T and the *largest* value of Δ that can be employed?

b) Let T and Δ be equal to the extreme values determined in (a). Find an explicit formula for $y(t)$ in Figure 2 in terms of $x(t)$, W_x, and W_y.

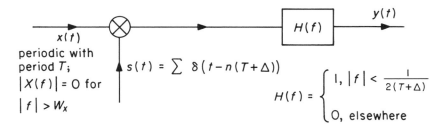

Figure 2.

Problem 14.11

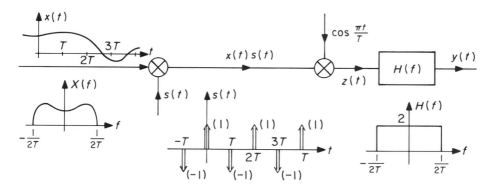

a) Determine and give a labelled sketch of $S(f)$.
b) Determine and give a labelled sketch of $h(t)$.
c) Explain with appropriate sketches why $y(t) = Kx(t)$ if $x(t)$ is bandlimited as shown. Determine the value of K.

Problem 14.12

In the system below* samples are taken of both the bandlimited functions $x(t)$ and $x(t) * h(t)$ at the rate of W samples/second, half the rate that would be required to specify $x(t)$ if $x(t)$ alone were sampled. The objective of this problem is to show that, for many choices of $H(f)$, it is possible to find LTI systems $G_1(f)$ and $G_2(f)$ such that $y(t) = x(t)$, so that the samples of $x(t)$ and $x(t) * h(t)$ jointly specify $x(t)$ uniquely.

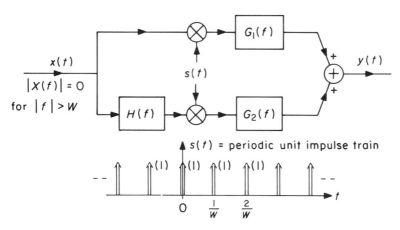

a) Show that $Y(f)$ in the interval $0 < f < W$ is given by
$$Y(f) = WG_1(f)[X(f) + X(f - W)]$$
$$+ WG_2(f)[X(f)H(f) + X(f - W)H(f - W)].$$

*D. A. Linden, *Proc. IRE*, 47 (July 1959): 1219–1226.

b) Show that if the output of $H(f)$ is the derivative of $x(t)$ and if $G_1(f)$ and $G_2(f)$ have the forms shown below, then $y(t) = x(t)$; that is, samples at half the normal rate of a bandlimited function and its derivative specify the function.

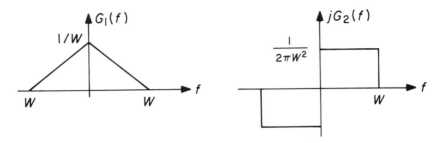

c) Show that if the output of $H(f)$ is the Hilbert transform of $x(t)$ (from Problem 15–3),

$$H(f) = \begin{cases} -j, & f > 0 \\ j, & f < 0 \end{cases}$$

then it is possible to choose $G_1(f)$ and $G_2(f)$ so that $y(t) = x(t)$; that is, samples at half the normal rate of a bandlimited function and its Hilbert transform specify the function.

Problem 14.13

To avoid intersymbol interference in a modem (modulator/demodulator) intended to transmit binary digits over a telephone line, the following scheme was proposed long ago:[*]

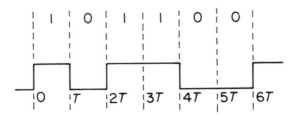

i) Represent the string of binary digits at the modulator by an on-off pulse waveform as shown above.

ii) Let the combined effects of various filters in the modulator, demodulator, telephone line, etc., be described by an LTI system with frequency response $H(f)$. Let $p(t)$ represent the response of this system to a single unit pulse of duration T.

---

[*]H. Nyquist, Trans. AIEE, 47 (April 1928): 617–644.

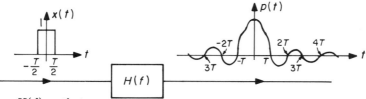

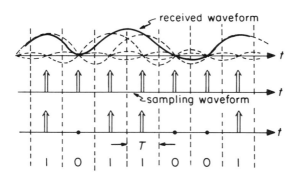

iii) Choose $H(f)$ so that
  1) $p(0) = 1$;
  2) $p(nT) = 0$ for n any positive or negative integer except 0;
  3) $P(f) \approx 0$ for $|f| > W$, where W is the usable bandwidth of the phone line.

Then the on-off pulse waveform in (i) will be received as the waveform shown below. Sampling at the points shown will then yield the original binary digit string without intersymbol interference.

This problem illustrates various examples of this scheme.

a) Choose $p(t)$ to have the form

$$p(t) = \frac{\sin \dfrac{\pi t}{T}}{\dfrac{\pi t}{T}}.$$

What is the minimum value of T (that is, the maximum bit rate) if $P(f)$ is exactly to satisfy condition (3) above?

b) For $p(t)$ as in (a) and with T equal to its minimum allowed value, determine a formula for $H(f)$ and sketch its shape.

c) Another choice for $p(t)$ is almost any waveform of the form

$$p(t) = g(t)\frac{\sin \dfrac{\pi t}{T}}{\dfrac{\pi t}{T}}$$

where $g(0) = 1$ and T is suitably constrained. Specifically, let

$$g(t) = \frac{\sin \dfrac{\pi t}{4T}}{\dfrac{\pi t}{4T}}.$$

Sketch $p(t)$ and $P(f)$. Describe how you would compute $H(f)$ (an explicit formula is not required) and determine the minimum value of T if condition (3) is to be satisfied.

d) Argue that a general class of spectra $P(f)$ meeting the conditions for zero inter-symbol interference stated earlier have a roll-off characteristic at the band edge that is anti-symmetric as shown below.

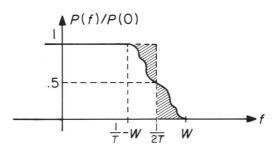

Problem 14.14

Show that the time functions

$$x_i(t) = \frac{\sin 2\pi W(t - \tau_i)}{2\pi W(t - \tau_i)}$$

are eigenfunctions of the system

$$H(f) = \begin{cases} 1, & |f| < W \\ 0, & \text{elsewhere} \end{cases}$$

for any choice of the constants τ_i (see Section 12.0). Discuss the relationship of this fact to the Sampling Theorem.

Problem 14.15

Let $\xi_k(t) = t^k e^{-t/2} u(t)$. Use the Gram-Schmidt procedure to derive the first few functions of the following series of orthonormal functions over the interval $0 < t < \infty$. These are called the *Laguerre functions*. (HINT: You may find it helpful to know that $\int_0^\infty t^k e^{-t}\, dt = k!$.)

$$\phi_0(t) = e^{-t/2}\, u(t)$$
$$\phi_1(t) = (-t + 1)e^{-t/2}\, u(t)$$
$$\phi_2(t) = \left(\frac{t^2}{2} - 2t + 1\right)e^{-t/2}\, u(t)$$
$$\vdots \qquad\qquad \vdots$$
$$\phi_k(t) = \frac{1}{k!}e^{t/2}\frac{d^k}{dt^k}\left[t^k e^{-t}\right]u(t).$$

Problem 14.16

a) Let $H(f)$ be the frequency response of an LTI system. Suppose that $H(f)$ is periodic in frequency: $H(f + f_0) = H(f)$ for some f_0 and all f. Describe as accurately as you can the character of the impulse response $h(t)$ for this system, and give an expression for $h(t)$ in terms of $H(f)$ over a single period $0 < f < f_0$.

b) In general, time-*varying* linear operators do not commute, so that the effect of cascading two linear time-varying systems (or one time-varying and one time-invariant system) depends on the order. Consider, however, the particular systems shown below—an amplifier (multiplier) with a time-varying gain and an LTI system with a periodic frequency response. Discuss the commutability of the cascade of these systems: Prove that the two cascade systems shown are equivalent for all inputs or demonstrate a particular input for which they give different responses. (HINT: Consider the responses to short pulses applied at various times.)

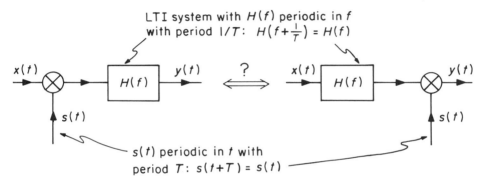

LTI system with $H(f)$ periodic in f
with period $1/T$: $H\left(f + \frac{1}{T}\right) = H(f)$

$s(t)$ periodic in t with
period T: $s(t+T) = s(t)$

Problem 14.17

a) Use vector-space arguments to prove that the Schwarz inequality becomes an equality if and only if $x(t) = ky(t)$.

b) Use the Schwarz inequality to prove that the average value of a real periodic waveform is always less than or equal to its rms value:

$$\langle x(t) \rangle \overset{\Delta}{=} \frac{1}{T} \int_0^T x(t)\, dt \leq \sqrt{\frac{1}{T} \int_0^T x^2(t)\, dt} \overset{\Delta}{=} \sqrt{\langle x^2(t) \rangle}.$$

c) Derive the result of (b) from Parseval's Theorem.

d) The *autocorrelation function* of a real periodic waveform will be defined in Chapter 19 as

$$R_x(\tau) \overset{\Delta}{=} \langle x(t)x(t - \tau) \rangle = \frac{1}{T} \int_0^T x(t)x(t - \tau)\, dt.$$

Use the Schwarz inequality to prove that

$$|R_x(\tau)| \leq R_x(0).$$

Problem 14.18

a) Use formula (c) of Table XIII.2 to prove that if a set of functions $\phi_k(t)$ are orthonormal over the infinite interval $-\infty < t < \infty$, then their Fourier transforms $\Phi_k(f)$ are also orthonormal over the infinite interval $-\infty < f < \infty$.

b) Apply this result to the functions of Problem 14.15 to prove that the functions

$$\Phi_k(f) = \left(\frac{-\dfrac{1}{2} + j2\pi f}{\dfrac{1}{2} + j2\pi f} \right)^k \frac{1}{\dfrac{1}{2} + j2\pi f}$$

are orthonormal over $-\infty < f < \infty$.

15

FILTERS, REAL AND IDEAL

15.0 Introduction

A simple narrow pulse-like waveform—an ideal example would be an impulse such as $A\delta(t-T)$—is specified just as completely by sketching its Fourier transform as by describing it as a function of time. But for most of us the specification of impulse-like waveforms directly in the time domain is much more graphic—closer, somehow, to their essential "impulsive" character, their amplitude and time of occurrence—than a description in the frequency domain. And, for similar reasons no doubt, the effects of the operations of *gating* or *sampling* (multiplication by a narrow pulse) and *delay* (convolution with a narrow pulse) seem easiest to appreciate in the time domain. The frequency domain, on the other hand, is well suited for describing the duals of these operations and waveforms. The dual of a time pulse is a "pulse" in frequency—a sinusoid or narrowband waveform whose Fourier transform is zero or very small for all frequencies except those in the vicinity of some particular frequency. And the duals of gating and delay are *filtering* (convolution with a narrowband waveform) and *frequency shifting* (multiplication by a sinusoid). Presumably these should be studied as operations on the transforms of waveforms if we expect to understand their essential character. Such a study will be a principal objective of the next few chapters.

The four elementary operations—gating and delay, filtering and frequency shifting—in various combinations and approximations account for a major fraction of the component functions in even the most sophisticated communication, control, and signal-processing systems. Thus an ability to shift rapidly and easily from signal descriptions in the time domain to corresponding descriptions in the frequency domain is most helpful for an understanding of these complex systems. Moreover, a number of simple examples, idealizations, and theorems relating to pulses in time and frequency and to the corresponding elementary operations constitute in substantial degree the metaphorical basis for the *language* of system characterization—the idealizations or abstractions that help us explain complicated systems by saying that certain parts are similar to various well-known simple systems. The broader objectives of this book are to study some of these simple examples and theorems, and to provide practice in shifting flexibly from time to frequency descriptions. In Chapter 17 in particular we shall illustrate the usefulness of this language in the description of several important modulation and detection systems.

15.1 Ideal Filters

In ordinary usage, a *filter* is a device for separating an aggregate into two classes—those that *pass through* and those that are *stopped*—depending on some attribute of the objects such as size, mass, or color. An *ideal* (frequency) *filter* can thus be defined as an LTI system characterized by one or more *pass bands* (sets of frequencies for which $|H(f)| = 1$) and one or more *stop bands* (the complementary sets of frequencies for which $|H(f)| = 0$). Simple ideal filters are frequently described as *lowpass, highpass,* or *bandpass* as shown in Figure 15.1–1.

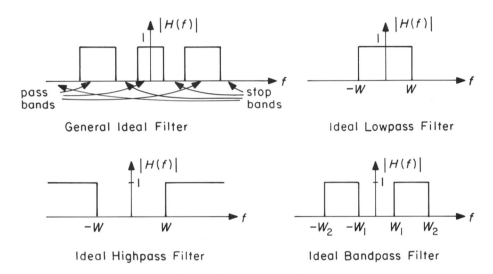

Figure 15.1–1. Ideal filters.

The magnitude of the frequency response of an ideal filter clearly conveys the essence of its filtering action. But the phase angle must be specified as well to complete the description of the filter as an LTI system. Often we shall assume that $\angle H(f)$ for an ideal filter is either zero or proportional to frequency,

$$\angle H(f) = -\alpha f, \quad \alpha \geq 0 \tag{15.1–1}$$

since under such circumstances a signal $x(t)$ whose spectrum is entirely restricted to the pass band of the filter is reproduced at the output without any alteration other than a simple delay, $T = \alpha/2\pi$ (see Figure 15.1–2). Such an ideal filter is said to be *distortionless*.

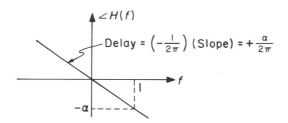

Figure 15.1–2. Phase characteristic of ideal distortionless filter.

For a number of reasons—to be explored in the following sections—the frequency response of an actual physical LTI system can only approach that of an ideal filter. Some of the ways in which the frequency response of a real system will differ from the ideal are illustrated in Figure 15.1–3. First, we observe that $|H(f)|$ for an actual physical filter cannot change suddenly from 1 to 0 at the *cutoff frequency*, but requires a *transition band* of nonzero width. Second, $|H(f)|$ cannot be precisely zero for all frequencies in the stop band; at most it is possible to have $|H(f)| = 0$ at isolated frequencies such as f_3 in Figure 15.1–3. The relative effectiveness of the filter in the stop band is often

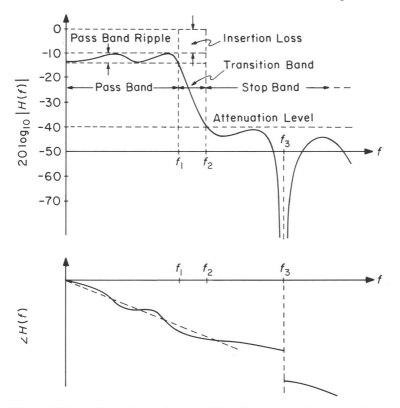

Figure 15.1–3. Deviations of actual filters from ideal characteristics.

partially described by the *attentuation level*, corresponding to an upper bound on $|H(f)|$ in this frequency region. Finally, it is not usually possible to achieve precisely either $|H(f)| = 1$ or $\angle H(f) = -\alpha f$ in the pass band. Typically, the average gain is less than 1, described as an *insertion loss*. And some deviations from an ideal constant response will generally occur; such deviations will produce *amplitude* and *phase distortion* of the transmitted signal. Often the amplitude deviations have the appearance sketched in Figure 15.1–3 and thus are called *ripples*. Even small distortions may be significant if hundreds of filters are cascaded—as they are in long-distance telephone lines—and it may become necessary to insert *equalizers* in such systems in an attempt to compensate. (Of course, such linear amplitude and phase distortions must not be confused with the non-linear distortions that are present at least to some degree in every real physical system and whose effects cannot usually be so readily corrected.)

Since the earliest days of radio, practical techniques for the design and fabrication of improved filters have been intensively sought. Some of the important achievements in this effort were the "electric wave filters" of Campbell,* Zobel, Bode, Foster, and others at the Bell Telephone Laboratories in the 1920s and 30s, and the "modern" theories of network synthesis developed by Cauer, Guillemin,† and many others in the 1930s and 40s. Most of the early studies concentrated on realizing precise specifications on $|H(f)|$, with little attention paid to the phase characteristic. Such an approach was acceptable in large part because most of the early applications were to telephony or radio, and the ear is remarkably insensitive to many types of phase distortion. But the rapid growth of radar, television, and pulse circuitry during and after World War II introduced more serious requirements; in particular, it became necessary to control certain aspects of the impulse or step response as well as the frequency response.

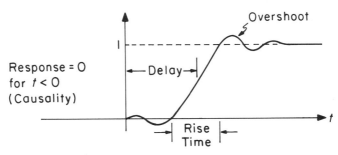

Figure 15.1–4. Step response of lowpass filter.

Obviously, independent specification of both the complete impulse or step response of a filter and the complete frequency response is impossible—either response alone entirely determines the other. But partial specification of the characteristics of each response is permissible, provided that certain compromises are made and certain constraints satisfied. Some of the important characteristics

---

*G. A. Campbell, U.S. Patent 1,227,113, May 22, 1917.

†E. A. Guillemin, *Synthesis of Passive Networks* (New York, NY: John Wiley, 1957).

that one might seek to control in the typical step response of a real approximation to a lowpass filter are illustrated in Figure 15.1–4. In the next few sections and the following chapter we shall explore the principal ways in which the values of delay and rise time and the desire to control overshoot and satisfy causality interact with such attributes of the frequency response as bandwidths and attentuation levels. Most of our discussion will be directed toward lowpass filters since, as we shall later show, the properties of bandpass filters can for the most part be discussed in terms of related lowpass filters.

15.2 Causality Conditions and Hilbert Transforms

The impulse response of an ideal lowpass filter characterized by

$$|H(f)| = \begin{cases} 1, & |f| < W \\ 0, & |f| > W \end{cases}$$

and

$$\angle H(f) = 0$$

is given by pair (s) of Table XIII.1 as

$$h(t) = \frac{\sin 2\pi W t}{\pi t}, \qquad -\infty < t < \infty \tag{15.2-1}$$

and is sketched in Figure 15.2–1.

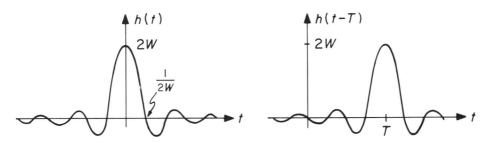

Figure 15.2–1. $h(t)$ for an ideal filter, with and without delay.

Since $h(t)$ does not vanish for $t < 0$, an ideal lowpass filter is non-causal. And simply adding a linear phase function cannot overcome this difficulty; there is no value of T such that $h(t-T)$ for an ideal lowpass filter vanishes for $t < 0$. Of course, for a sufficiently large delay T, the value of $h(t-T)$ for negative t can be made as small as we wish, so that it seems plausible that with a small amount of amplitude or phase distortion of the right sort plus a large delay it might be possible to make an otherwise ideal filter causal. But this hope turns out to be largely false; the difficulty with any ideal filter is that we seek to achieve $|H(f)| \equiv 0$ over a *range* of frequencies, such as $|f| > W$. As Norbert Wiener put it, "No [causal] filter can have infinite attenuation in any finite [nonzero] band. The perfect filter is physically unrealizable by its nature, not merely because of the paucity of means at our disposal. No instrument acting solely on the past has a sufficiently sharp discrimination to separate one [band of frequencies] from another [band of frequencies] with unfailing accuracy."[*] This conclusion is the consequence of a celebrated theorem due to Wiener and Paley,[†] which states that if $h(t)$ is square-integrable and causal, then

$$\int_{-\infty}^{\infty} \frac{|\ln |H(f)||}{1 + f^2} \, df < \infty. \qquad (15.2-2)$$

Conversely, if $|H(f)|$ is square-integrable and if the integral in (15.2–2) is unbounded, then we cannot make $h(t) \equiv 0$ for all $t < 0$, no matter what $\angle H(f)$ we associate with $|H(f)|$.

We shall make no attempt to prove (15.2–2), but some of the consequences are interesting.[‡] Thus, as we have already suggested, no filter that absolutely rejects some *band* of frequencies while passing others (isolated zeros are allowed, however) can be realized exactly, no matter what shape we assume for $|H(f)|$ in the pass band and no matter what phase function $\angle H(f)$ we select to complete the specification of $H(f)$. Moreover, the Paley-Wiener Theorem restricts the rate at which the frequency response of a causal LTI system can vanish as $|f| \to \infty$. No causal LTI sytem can have

$$|H(f)| = e^{-|f|} \qquad (15.2-3)$$

or

$$|H(f)| = e^{-\pi f^2} \qquad (15.2-4)$$

that is, $|H(f)|$ vanishing exponentially or faster as $|f| \to \infty$. The Paley-Wiener Theorem also contains a sufficiency statement: Given any square-integrable function $|H(f)|$ for which (15.2–2) is satisfied, there exists an $\angle H(f)$ such that $H(f) = |H(f)|e^{j\angle H(f)}$ is the Fourier transform of a causal $h(t)$. But this part of the theorem is of somewhat less interest than the necessity of (15.2–2).

---

[*]N. Wiener, *The Interpolation, Extrapolation and Smoothing of Stationary Time Series* (Cambridge, MA: Technology Press, 1949) p. 37.

[†]N. Wiener and R. E. A. C. Paley, *Fourier Transforms in the Complex Domain* (Providence, RI: Am. Math. Soc. Coll. Pub., 1934) p. 16.

[‡]See, for example, A. Papoulis, *The Fourier Integral and Its Applications* (New York, NY: McGraw-Hill, 1962) p. 215, for a more detailed discussion. The theorem can be extended to $h(t)$ satisfying less severe conditions than square-integrability; see E. Pfaffelhuber, *IEEE Trans. Cir. Thy.*, *CT-18* (March 1971): 218–223.

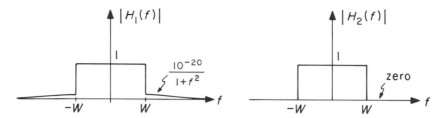

Figure 15.2–2. An approximation to a lowpass filter.

The Paley-Wiener Theorem is less important for applications than it might at first appear. Whether a particular $|H(f)|$ does or does not satisfy (15.2–2) unfortunately says very little about the relative difficulty of *approximating* $|H(f)|$ with a causal physical device. Thus in Figure 15.2–2 $|H_1(f)|$ is no easier to approximate physically than $|H_2(f)|$, but $|H_1(f)|$ satisfies the Paley-Wiener condition and $|H_2(f)|$ does not. On the other hand, a number of classes of causal LTI systems are known that contain members whose frequency response magnitudes approach the ideal lowpass characteristic as closely as may be desired. An example is the class of *Butterworth filters* discussed earlier, for which

$$|H_n(f)|^2 = \frac{1}{1 + (f/W)^{2n}}. \qquad (15.2-5)$$

The way in which $|H_n(f)|$ approximates an ideal lowpass filter for large n is illustrated in Figure 15.2–3. Some further properties of Butterworth filters will be discussed in Chapter 16.

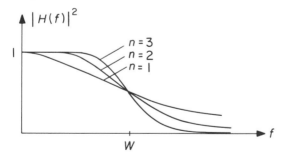

Figure 15.2–3. Frequency responses of Butterworth filters.

Another procedure often suggested for specifying a causal approximation to a non-causal filter is simply to delay the non-causal impulse response such as (15.2–1) until the "tail" for $t < 0$ is so small that it may presumably be chopped off and ignored. Thus the LTI system corresponding to the impulse response shown in Figure 15.2–4 can presumably be made to approximate an ideal lowpass filter as closely as desired by taking T large enough. As pointed out in Chapter 13, however, we must not cavalierly assume that the transforms of a convergent

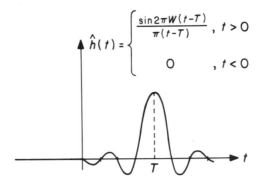

Figure 15.2–4. A causal approximation to the impulse response of an ideal lowpass filter.

sequence of time functions necessarily converge (in the ordinary sense) to the transform of the limit. In fact, for large finite values of T it is easy to show that the Fourier transform of $h(t)$ above approaches an ideal lowpass filter response except for infinite "ears" or "horns" at the band edge, as shown in Figure 15.2–5. Since the area under these "ears" tends to zero for large T, the sequence of frequency responses does converge in a local-average sense, but the presence of the "ears" might cause trouble in some applications. Analytically, the "ears" are a consequence of the unbalanced, slowly decreasing tail of $h(t)$, and they can be largely eliminated by the simple artifice of chopping off $h(t)$ for large *positive* t as well as for negative t. A particular example is the symmetrically chopped waveform

$$h(t) = \begin{cases} \dfrac{\sin 2\pi W(t-T)}{\pi(t-T)}, & 0 < t < 2T \\ 0, & \text{elsewhere} \end{cases} \qquad (15.2-6)$$

which, together with other schemes, is explored in Problem 15.6.

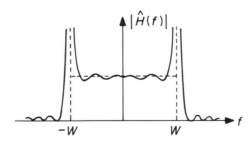

Figure 15.2–5. Frequency response corresponding to Figure 15.2–4.

Causality is a tight constraint on an LTI system not only because $|H(f)|$ is then required to satisfy the Paley-Wiener Theorem but also because causality implies a strong relationship between the real and imaginary parts of $H(f)$, as well as a less strong but still significant interdependence of $\log|H(f)|$ and $\angle H(f)$. These relationships can be derived as follows. If $h(t)$ is causal (and free from singularities at $t = 0$), then apparently

$$h(t) = h(t)u(t). \tag{15.2-7}$$

Taking Fourier transforms of both sides yields the equivalent condition

$$H(f) = H(f) * \left[\frac{1}{2}\delta(f) + \frac{1}{j2\pi f}\right] \tag{15.2-8}$$

or, carrying out the convolution with the impulse and cancelling a factor of 2,

$$H(f) = H(f) * \frac{1}{j\pi f}. \tag{15.2-9}$$

Writing out (15.2-9) in terms of real and imaginary parts, we obtain

$$H_r(f) + jH_i(f) = \int_{-\infty}^{\infty} [H_r(\eta) + jH_i(\eta)]\frac{1}{j\pi(f-\eta)}\,d\eta. \tag{15.2-10}$$

Equating real and imaginary parts separately gives

$$H_r(f) = \frac{1}{\pi}\int_{-\infty}^{\infty}\frac{H_i(\eta)}{f-\eta}\,d\eta \tag{15.2-11}$$

$$H_i(f) = -\frac{1}{\pi}\int_{-\infty}^{\infty}\frac{H_r(\eta)}{f-\eta}\,d\eta \tag{15.2-12}$$

so that $H_r(f)$ and $H_i(f)$ are *completely interdependent* in that, given either, the other can be found.* Both of these formulas are of the form

$$\boxed{y(t) = \frac{1}{\pi}\int_{-\infty}^{\infty}\frac{x(\tau)}{t-\tau}\,d\tau} \tag{15.2-13}$$

which is called the *Hilbert transform* of $x(t)$ and has further important applications in modulation theory, as we shall see. For some further examples and properties of Hilbert transforms, see Exercise 15.3 and Problems 15.3–15.5.

---

*Impulses and other singularity functions at $t = 0$ are excluded from $h(t)$ by (15.2-7). These correspond to powers of f that must be subtracted from $H_r(f)$ and $H_i(f)$ before (15.2-11) and (15.2-12) are applied. It can also be shown that $\log|H(f)|$ and $\angle H(f)$ are Hilbert transforms of one another if the system is minimum-phase (see Problem 4.12).

15.3 The Ideal Filter Step Response and Gibbs' Phenomenon

In the time domain, the operation of an ideal (zero-phase) lowpass filter can be described by the convolution

$$y(t) = \int_{-\infty}^{\infty} x(t-\tau)\frac{\sin 2\pi W\tau}{\pi\tau}\,d\tau. \tag{15.3--1}$$

If, in particular, the input is a unit step, $x(t) = u(t)$, then

$$y(t) = \int_{-\infty}^{t} \frac{\sin 2\pi W\tau}{\pi\tau}\,d\tau \tag{15.3--2}$$

which is the integral of the impulse response. The integral in (15.3–2) cannot be evaluated in terms of elementary functions but can be written

$$y(t) = \text{Si}\,[2\pi Wt] \tag{15.3--3}$$

where

$$\text{Si}\,[t] = \frac{1}{\pi}\int_{-\infty}^{t} \frac{\sin\tau}{\tau}\,d\tau \tag{15.3--4}$$

is called the *sine-integral function* and is widely tabulated.* The general shape of $\text{Si}\,[2\pi Wt]$ is shown in Figure 15.3–1. It differs from a step in at least two important ways: The step response takes a finite time (on the order of $1/2W$ seconds) to go from near zero to near its final value (as contrasted with the zero rise time of the input step), and the response oscillates about the final value with an initial overshoot of about 9% of the discontinuity and a settling time equal to several times $1/2W$ seconds. The duration of the rise time of any lowpass filter, ideal or not, is limited primarily by its bandwidth, as we shall show in the next chapter. But whether or not the step response of a lowpass filter overshoots and oscillates about the final value is a critical function of the detailed shape of the frequency response in the pass band, and indeed for a given shape is independent of the bandwidth (as should be obvious on dimensional grounds).

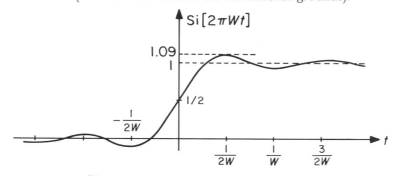

Figure 15.3–1. The sine-integral function.

---

*For example, E. Jahnke and F. Emde, *Tables of Functions* (New York, NY: Dover, 1945).

We have already encountered an equation essentially identical to (15.3–1) in our discussion of the validity of Fourier's Theorem in Section 13.2. Thus Dirichlet's version of Fourier's Theorem, equation (13.2–1), states that if $x(t)$ is absolutely integrable and smooth enough, then the inverse transform can be written in the form

$$\lim_{W\to\infty} \int_{-W}^{W} X(f)e^{j2\pi ft}\,df = \lim_{W\to\infty} \int_{-\infty}^{\infty} x(\tau)\frac{\sin 2\pi W(t-\tau)}{\pi(t-\tau)}\,d\tau$$

$$= \frac{x(t+)+x(t-)}{2} \qquad (15.3{-}5)$$

where the second integral follows from the first since "chopping off" $X(f)$ at $\pm W$ is equivalent to passing $x(t)$ through an ideal lowpass filter. Now, if $x(t)$ is continuous, then $x(t-) = x(t+)$ and the limit in (15.3-5) simply states that $\frac{\sin 2\pi Wt}{\pi t}$ (which is often called *Dirichlet's kernel* because of its appearance in this argument) is a basis function for an impulse, in the sense discussed in Exercise 11.5. But if t_0 is a point of discontinuity of $x(t)$, then the factors in the integrand of the convolution integral in (15.3–5) appear as shown in Figure 15.3–2. The results of the convolution for two different values of W are also shown in the lower part of the figure. Note first that for any value of W, the value of the convolution at $t = t_0$ is precisely $[x(t_0+) + x(t_0-)]/2$, consistent with (15.3–5). But note also that, just as with the step response of the ideal lowpass filter, there is an overshoot at the discontinuity, amounting to about 9% of the discontinuity independent of the value of W. For larger values of W the overshoot occurs closer to the discontinuity and the oscillations die away more rapidly, but the peak amplitude of the overshoot does not get smaller. This peculiar behavior was first clarified by J. Willard Gibbs in 1898 and is accordingly called *Gibbs' phenomenon.*

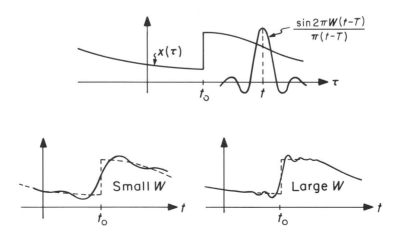

Figure 15.3–2. Schematic explanation of Gibbs' phenomenon.

Mathematically, Gibbs' phenomenon implies that even though the limit in Dirichlet's Theorem (15.3–5) converges to the average of $x(t+)$ and $x(t-)$, the convergence is not uniform in any interval containing a discontinuity in $x(t)$. For many analytical purposes such non-uniform convergence is undesirable. Moreover, physically an overshoot in the system response to a unit step can be quite objectionable. "Underdamped" automobile suspensions and elevator control systems are familiar examples of this sort of response to a discontinuity. In audio systems the corresponding phenomenon is usually called "ringing"; it sounds like a short whistle accompanying each sharp consonant or other sudden sound. In television, if the video amplifiers were ideal lowpass filters with an overshoot in their response to a step discontinuity, then a vertical black-white boundary in the picture would be fringed with narrow bands of gray, which would be visually quite unacceptable. Thus for both mathematical and physical reasons the design of lowpass filters without overshoot is an important problem.

To avoid overshoot, it is evidently sufficient to require that the impulse response of the lowpass filter be entirely positive; the step response will then be montonically increasing (or non-decreasing), and the sort of situation that led to Gibbs' phenomenon will not arise. The frequency response of such a filter must satisfy a number of conditions, as we shall discuss in Chapter 19. In particular, a necessary condition is that $H(0)$ must be strictly greater than $|H(f)|$ for $f \neq 0$; that is, $|H(f)|$ cannot be a constant in the pass band, so that no truly distortionless filter can have a monotonic step response. Various kinds of approximations are possible, of course, but there is no general solution to the problem of finding a best approximation to an ideal lowpass filter that has a monotonic step response.

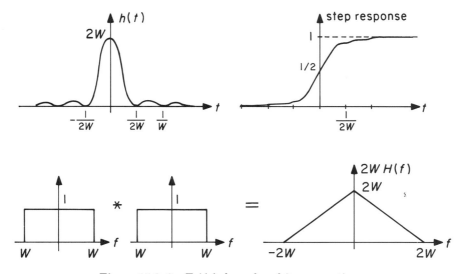

Figure 15.3–3. Fejér's kernel and its properties.

One important example of a filter with a monotonic step response is the filter with impulse response

$$h(t) = \frac{1}{2W}\left(\frac{\sin 2\pi Wt}{\pi t}\right)^2 \tag{15.3–6}$$

which is called *Fejér's kernel*. This impulse response and the corresponding step response are sketched in Figure 15.3–3. The frequency response of the filter can be found from the product formula of Table XIII.2 as the convolution of the ideal lowpass filter response of Dirichlet's kernel with itself; this is the triangle-shaped frequency response shown in the bottom part of Figure 15.3–3.

Since the Fourier transform of $h(t)$ of (15.3–6) is 1 at $f = 0$, the area under $h(t)$ is 1 (which is consistent with the fact that the step response $\to 1$ as $t \to \infty$). Thus $h(t)$ can be the basis for an impulse (see Exercise 11.5), and we should expect that

$$\lim_{W\to\infty} \int_{-2W}^{2W}\left(1 - \frac{|f|}{2W}\right)X(f)e^{j2\pi ft}\,df = \frac{x(t+) + x(t-)}{2} \tag{15.3–7}$$

where the convergence should now be smoother than for Dirichlet's Theorem and Gibbs' phenomenon will be avoided. The limit in (15.3–7) is said to interpret the improper integral $\int_{-\infty}^{\infty}X(f)e^{j2\pi ft}\,df$ *in the sense of Césaro*, and (15.3–7) is a statement of *Fejér's version* of Fourier's Theorem.

15.4 Summary

The impulse response and the frequency response of any LTI system are Fourier transforms of one another. The step response and the frequency response are also directly related. Desirable properties in one domain may thus impose constraints that are unacceptable in the other. In this chapter we have explored several such time-frequency interactions:

a) Any system that has infinite rejection for signal frequencies in some stop band of nonzero width—including, in particular, any ideal filter—is inherently non-causal.

b) Causality also implies a strong relationship between the real and imaginary parts (or the magnitude and the phase) of the frequency response—indeed one implies the other.

c) Attempts to make the transition from pass to stop frequency bands too precipitous lead to ringing (Gibbs' phenomenon).

Other time-frequency interactions—in particular, the reciprocal relation between duration (or rise time) and bandwidth—will be discussed in the next chapter.

EXERCISES FOR CHAPTER 15

Exercise 15.1

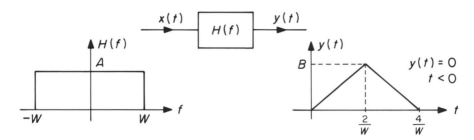

Explain why there is no input $x(t)$, $-\infty < t < \infty$, that will yield the output $y(t)$ when applied to the ideal lowpass filter shown above.

Exercise 15.2

An ideal amplifier or differentiator cannot in general be represented by the superposition integral with an impulse response that is an ordinary function (as contrasted with a singularity function such as $\delta(t)$ or $\dot{\delta}(t)$). Show, however, that if $x(t)$ is restricted to be a bandlimited function, $X(f) = 0$ for $|f| > W$, then ordinary functions $h_1(t)$ and $h_2(t)$ can be found such that $Kx(t) = x(t) * h_1(t)$ and $\dot{x}(t) = x(t) * h_2(t)$.

Exercise 15.3

Let $\hat{x}(t)$ be the Hilbert transform of $x(t)$:

$$\hat{x}(t) = \frac{1}{\pi} \int_{-\infty}^{\infty} \frac{x(\tau)}{t - \tau} \, d\tau \,.$$

Show that $\hat{x}(t)$ and $x(t)$ are orthogonal:

$$\int_{-\infty}^{\infty} x(t)\hat{x}^*(t) \, dt = 0 \,.$$

PROBLEMS FOR CHAPTER 15

Problem 15.1

Let $H(f)$ be the frequency response of an ideal lowpass filter with delay:

$$H(f) = \begin{cases} e^{-j2\pi fT}, & |f| < W \\ 0, & \text{elsewhere.} \end{cases}$$

Let the input be a symmetric pulse, $x(t) = x(-t)$. Show that the effects of amplitude distortion alone are symmetric, whereas the effects of phase distortion alone are anti-symmetric.

Problem 15.2

Determine and sketch the impulse response and step response of the ideal highpass filter shown below. (HINT: Consider $1 - H(f)$.)

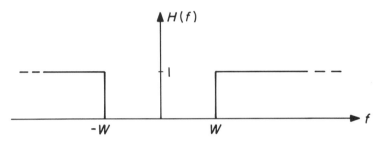

Problem 15.3

Let $y(t)$ be the Hilbert transform of $x(t)$ as defined by (15.2–13). By interpreting (15.2–13) as a convolution, show that

$$Y(f) = -jX(f)\,\text{sgn}\,f = \begin{cases} -jX(f), & f > 0 \\ +jX(f), & f < 0 \end{cases}$$

that is, the Hilbert transform of a waveform can be obtained by passing it through an LTI system with frequency response

$$H(f) = \begin{cases} -j, & f > 0 \\ j, & f < 0. \end{cases}$$

Show also that the complex time function

$$z(t) = x(t) + jy(t)$$

has a Fourier transform $Z(f)$ that is identically zero for all $f < 0$. And show further that the Hilbert transform of the Hilbert transform of $x(t)$ is $-x(t)$.

Problem 15.4

a) Using the formulas of Problem 15.3, show that the Hilbert transform of $x(t) = \cos 2\pi W t$ is $y(t) = \sin 2\pi W t$.

b) Similarly, show that the Hilbert transform of

$$x(t) = \sum_{n=1}^{\infty} a_n \cos \frac{2\pi nt}{T} + \sum_{n=1}^{\infty} b_n \sin \frac{2\pi nT}{T}$$

is

$$y(t) = \sum_{n=1}^{\infty} a_n \sin \frac{2\pi nt}{T} - \sum_{n=1}^{\infty} b_n \cos \frac{2\pi nT}{T}.$$

Problem 15.5

a) Use equation (15.2–12) and the duals of the formulas of Problem 15.3 to find the (negative) Hilbert transform, $H_i(f)$, of the function

$$H_r(f) = \frac{2}{1 + (2\pi f)^2}.$$

b) Find $h(t)$, the inverse transform of $H(f) = H_r(f) + jH_i(f)$, and thus demonstrate directly that $h(t)$ is causal.

Problem 15.6

By the dual of the arguments of section 15.3 show that:

a) The filter with impulse response

$$h(t) = \begin{cases} \dfrac{\sin 2\pi W(t - T)}{\pi(t - T)}, & 0 < t < 2T \\ 0, & \text{elsewhere} \end{cases}$$

is a causal filter whose frequency response for large T approximates that of an ideal lowpass filter except for finite "ears" at the edge of the band.

b) The filter with impulse response

$$h(t) = \begin{cases} \left[1 - \dfrac{|t - T|}{T}\right] \dfrac{\sin 2\pi W(t - T)}{\pi(t - T)}, & 0 < t < 2T \\ 0, & \text{elsewhere} \end{cases}$$

is a causal filter approximating an ideal lowpass filter for large T without the "ears" of the impulse response of (a).

Problem 15.7

Another way to show that $H_r(f)$ and $H_i(f)$ are interdependent is to show that if $h(t)$ is real and causal, then $h(t)$ can be derived given $H_r(f)$ alone (this is sometimes called *real-part sufficiency*). To do this, proceed as follows:

a) Define for an arbitrary (not necessarily causal) $h(t)$

$$h_e(t) = \frac{1}{2}[h(t) + h(-t)] = \text{even part of } h(t)$$

$$h_o(t) = \frac{1}{2}[h(t) - h(-t)] = \text{odd part of } h(t).$$

Let $H(f) = H_r(f) + jH_i(f)$. Let $H_e(f)$ and $H_o(f)$ be the Fourier transforms of $h_e(t)$ and $h_o(t)$ respectively. Relate $H_e(f)$ and $H_o(f)$ to $H_r(f)$ and $H_i(f)$.

b) Show that if $h(t)$ is causal, then $h(t)$ is completely determined by $h_e(t)$, and thus using the results of (a), derive a formula for a causal $h(t)$ in terms of $H_r(f)$ alone.

16

DURATION-BANDWIDTH RELATIONSHIPS
AND THE UNCERTAINTY PRINCIPLE

16.0 Introduction

The step response of a typical lowpass filter will have nonzero *rise time* and nonzero *delay*, as illustrated in Figure 16.0–1. Since the step response is the integral of the impulse response, it should be clear that any measure of step response rise time is also a measure of impulse response *duration*, and vice versa. The principal purpose of this chapter is to show that the impulse response duration (or step response rise time) is inversely related to the frequency response bandwidth. For any particular shape of impulse response, it is obvious from dimensional considerations or the scale-change property of Table XIII.2 that such an inverse relationship should exist. The remarkable thing is that—independent of shape—there is a lower bound on the duration-bandwidth product; this is the formal content of the *Uncertainty Principle*, which will be our main analytical result.

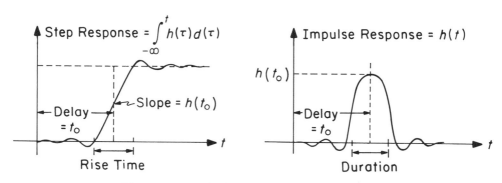

Figure 16.0–1. Delay, rise time, and duration.

16.1 Definitions of Delay, Rise Time, Duration, and Bandwidth

The quantities of interest to us can be formally defined in an unlimited number of ways. No one way, or set of ways, is best for all purposes. For example, a direct, easily interpreted measure of rise time is the time it takes the step response to go from 10% to 90% of its final value. But such a measure is tedious to compute and clumsy to manipulate. On the other hand, a very simple and convenient definition of rise time would be the ratio of the final value of the step response to the slope of the step response at some appropriate point along the rise (such as the 50% point). If we call the time at this point t_0, this definition of rise time can be expressed in a simple formula,

$$\text{Rise time} = \frac{\int_{-\infty}^{\infty} h(t)\, dt}{h(t_0)} \tag{16.1-1}$$

where $h(t)$ as usual represents the impulse response and we have exploited the fact that the step response is the integral of the impulse response. We can also understand this formula as defining the "width" of $h(t)$ as the ratio of its "area" to its "height." But (16.1–1) can give unreasonable values for the rise time if $h(t)$ is not a single, narrow, largely positive pulse-like waveform such as that shown in Figure 16.0–1. (See, for example, Problem 16.1.)

If we are to relate the rise time to the bandwidth, we must also come up with a way of specifying the effective bandwidth of an arbitrary lowpass frequency response $H(f)$. Except for the fact that $H(f)$ is in general complex, this problem is mathematically identical with specifying the duration of $h(t)$. A potential ambiguity is introduced by the conjugate symmetry of $H(f)$ (assuming that $h(t)$ is real). Thus, when we specify the bandwidth of $H(f)$, do we wish to include all frequencies, $-\infty < f < \infty$ (the bandwidth of the *double-sided* spectrum) or simply the positive frequencies, $f > 0$ (the bandwidth of the *single-sided* spectrum)? For example, should we specify the bandwidth of the ideal lowpass filter

$$H(f) = \begin{cases} 1, & |f| < W \\ 0, & \text{elsewhere} \end{cases}$$

as W or $2W$? Since little more than a factor of 2 is involved, this would seem to be largely a matter of style and convenience. Unfortunately, there is no uniform approach to this matter in the literature; considerable care (and sometimes a Ouija board) may be necessary to discover each author's intent. We shall not be dogmatic in this book either, but shall endeavor to make our intentions clear whenever ambiguity might arise.

Examples of various ways to measure durations and bandwidths—some fairly generally applicable and others restricted to the particular example in which they are applied—are shown in Figure 16.1–1. Note that in each case the product of duration and bandwidth is on the order of unity (within a factor of 2 or so). This is the observation we wish to generalize and make more precise.

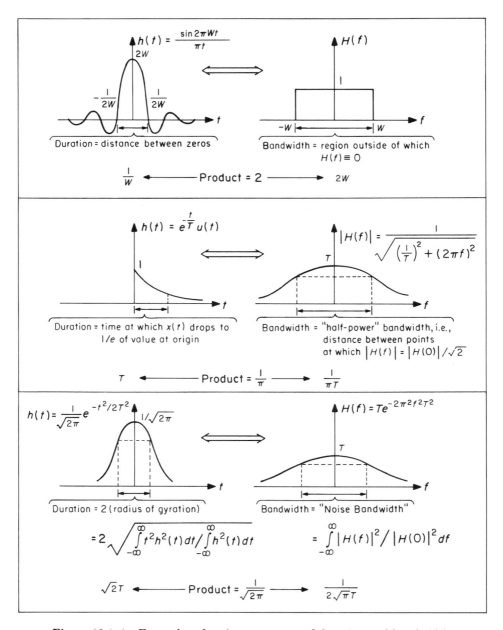

Figure 16.1–1. Examples of various measures of duration and bandwidth.

Probably the most analytically useful definitions of the delay, duration, and bandwidth are given by various moments of $h(t)$ and $H(f)$, or even better of $h^2(t)$ and $|H(f)|^2$. The definitions and some important properties of these moments are given in the following examples.*

Example 16.1–1

Consider an amplifier made up of cascaded stages. The stages need not be identical. Let $h_i(t)$ be the impulse response of the ith stage and assume that $h_i(t) \geq 0$, so that the step response of each stage (or indeed of any number of cascaded stages) is monotonic. Assuming that both integrals exist, define

$$T_i = \frac{\int_{-\infty}^{\infty} t h_i(t)\, dt}{\int_{-\infty}^{\infty} h_i(t)\, dt}. \qquad (16.1-2)$$

T_i, the (normalized) first moment of $h_i(t)$, can be interpreted as the center of gravity of a mass distributed along the t-axis with density $h_i(t)$. Thus we may consider T_i as measuring the delay in the impulse or step response of the ith stage.

It is then easy to show that the delay resulting from a cascade of n stages is the sum of the delays of each stage; that is, if

$$h(t) = h_1(t) * h_2(t) * \cdots * h_n(t) \qquad (16.1-3)$$

then

$$T = T_1 + T_2 + \cdots + T_n. \qquad (16.1-4)$$

In particular, if the stages are identical, the delay is directly proportional to the number of stages. We shall prove this for a cascade of two stages; the general result then follows by induction. Combining several formulas from Table XIII.2, we deduce that

$$\int_{-\infty}^{\infty} t h(t)\, dt = \frac{-1}{j 2\pi} \left. \frac{dH(f)}{df} \right|_{f=0}. \qquad (16.1-5)$$

If

$$h(t) = h_1(t) * h_2(t) \qquad (16.1-6)$$

then

$$H(f) = H_1(f) H_2(f) \qquad (16.1-7)$$

and

$$\int_{-\infty}^{\infty} t h(t)\, dt = \frac{-1}{j 2\pi} \left[H_2(0) \left. \frac{dH_1(f)}{df} \right|_{f=0} + H_1(0) \left. \frac{dH_2(f)}{df} \right|_{f=0} \right]. \qquad (16.1-8)$$

Moreover,

$$\int_{-\infty}^{\infty} h(t)\, dt = H(0) = H_1(0) H_2(0). \qquad (16.1-9)$$

---

*It is perhaps worth pointing out that, although we shall state all the results of this chapter in terms of system impulse responses and their transforms, $h(t)$ and $H(f)$, the results obviously apply broadly to any time function and its transform.

Finally,

$$T = \frac{\int_{-\infty}^{\infty} t h(t) \, dt}{\int_{-\infty}^{\infty} h(t) \, dt} = -\frac{1}{j2\pi} \left[\frac{dH_1(f)/df|_{f=0}}{H_1(0)} + \frac{dH_2(f)/df|_{f=0}}{H_2(0)} \right]$$
$$= T_1 + T_2 \tag{16.1-10}$$

as was to be shown.

Similarly, assuming that the integrals exist, we may define

$$(\Delta T_i)^2 = 4 \left[\frac{\int_{-\infty}^{\infty} t^2 h_i(t) \, dt}{\int_{-\infty}^{\infty} h_i(t) \, dt} - T_i^2 \right]$$
$$= 4 \frac{\int_{-\infty}^{\infty} (t - T_i)^2 h_i(t) \, dt}{\int_{-\infty}^{\infty} h_i(t) \, dt}. \tag{16.1-11}$$

$(\Delta T_i/2)^2$ is just the (normalized) moment of inertia about the center of gravity of a mass distribution $h_i(t)$.* That is, ΔT_i is twice the radius of gyration of the mass distribution, and thus a measure of the duration of $h_i(t)$ or of the rise time of the step response of the ith stage. Again we can prove (see Problem 16.3) that for a cascade of n stages,

$$(\Delta T)^2 = (\Delta T_1)^2 + (\Delta T_2)^2 + \cdots + (\Delta T_n)^2. \tag{16.1-12}$$

Thus, in particular, for identical stages the rise time is proportional to the *square root* of the number of stages.

▶ ▶ ▶

If $h_i(t)$ is not positive, the duration measure of (16.1-11) becomes somewhat dubious. A better measure is obtained by first squaring $h(t)$:

$$(\Delta T)^2 = 4 \left[\frac{\int_{-\infty}^{\infty} t^2 h^2(t) \, dt}{\int_{-\infty}^{\infty} h^2(t) \, dt} - \left(\frac{\int_{-\infty}^{\infty} t h^2(t) \, dt}{\int_{-\infty}^{\infty} h^2(t) \, dt} \right)^2 \right]. \tag{16.1-13}$$

Indeed, in many ways ΔT of (16.1-13) is the most analytically satisfactory simple general measure of duration; for virtually any $h(t)$ for which the integrals exist, (16.1-13) will give a not unreasonable estimate of the duration. Equivalently,

$$(\Delta W)^2 = 4 \frac{\int_{-\infty}^{\infty} f^2 |H(f)|^2 \, df}{\int_{-\infty}^{\infty} |H(f)|^2 \, df} \tag{16.1-14}$$

is probably the best simple measure of bandwidth for real lowpass waveforms. These definitions take on added significance because for ΔT and ΔW so defined it is possible to prove the following celebrated principle:

---

*T_i and $(\Delta T_i/2)^2$ also have analogs in statistics (where $h_i(t)$, if positive, is analogous to a probability density) and are thus sometimes called the *mean* and *variance* (or *dispersion*) of $h_i(t)$.

UNCERTAINTY PRINCIPLE:

For any real waveform for which ΔT and ΔW of (16.1–13) and (16.1–14) exist,

$$\boxed{\Delta T \, \Delta W \geq \frac{1}{\pi}.} \qquad (16.1-15)$$

In words, ΔT and ΔW cannot simultaneously be arbitrarily *small*: A short duration implies a large bandwidth, and a small-bandwidth waveform must last a long time. In relation to measurements, (16.1–15) is often interpreted to imply that the uncertainty in the determination of a frequency is on the order of magnitude of the reciprocal of the time taken to measure it. Indeed, using the equivalences of relativity theory, (16.1–15) can be transformed into the *Heisenberg Uncertainty Principle* of wave mechanics, which asserts the impossibility of simultaneously specifying the precise position and conjugate momentum of a particle.

The proof of (16.1–15) is an interesting exercise in Fourier manipulations. In outline, the argument is as follows. Assume for simplicity (and without loss of generality, since the moment of inertia is a minimum about an axis through the center of gravity) that the time origin is chosen at the center of gravity of $h^2(t)$, so that the second term in (16.1–13) is zero. Then, since the Fourier transform of $dh(t)/dt = \dot{h}(t)$ is (from Table XIII.2) $j2\pi f\, H(f)$, we have, applying Parseval's Theorem,

$$(2\pi)^2 \int_{-\infty}^{\infty} f^2 |H(f)|^2 \, df = \int_{-\infty}^{\infty} [\dot{h}(t)]^2 \, dt. \qquad (16.1-16)$$

Combining this formula with (16.1–13) and (16.1–14) gives

$$(\pi \Delta T \Delta W)^2 = 4 \frac{\int_{-\infty}^{\infty} t^2 h^2(t)\, dt \int_{-\infty}^{\infty} [\dot{h}(t)]^2 \, dt}{\left(\int_{-\infty}^{\infty} h^2(t)\, dt \right)^2}. \qquad (16.1-17)$$

Next apply the Schwarz inequality (Example 14.A–1) to the numerator of (16.1–17) to obtain

$$\pi \Delta T \Delta W \geq \frac{2 \left| \int_{-\infty}^{\infty} t h(t) \dot{h}(t)\, dt \right|}{\int_{-\infty}^{\infty} h^2(t)\, dt}. \qquad (16.1-18)$$

But the fraction on the right in (16.1–18) is identically equal to 1 (as can be seen by integrating either the numerator or the denominator by parts and noting that $h(t)$ must vanish rapidly as $t \to \pm\infty$ if the integrals in (16.1–13) and (16.1–14) are to exist as assumed), which gives the desired result.

It is interesting to explore the conditions under which the lower limit on the duration-bandwidth product can be achieved. To obtain equality in the Schwarz inequality as applied to (16.1–17) (see Problem 14.17) we must have

$$k t h(t) = \dot{h}(t) \qquad (16.1-19)$$

or

$$\frac{\dot{h}(t)}{h(t)} = kt.\tag{16.1-20}$$

Integrating, we have

$$\ln h(t) = \frac{kt^2}{2} + \text{constant}\tag{16.1-21}$$

or

$$h(t) \sim e^{kt^2/2}.\tag{16.1-22}$$

For k negative this is an acceptable pulse-like waveform—the Gaussian function already discussed. Among all waveforms, then, *the Gaussian pulse has the smallest duration-bandwidth product* in the sense of (16.1–13) and (16.1–14).

Example 16.1–2

As an example of the Uncertainty Principle, consider the 3rd-order Butterworth filter examined extensively in earlier chapters. The magnitude of the frequency response and the impulse response are

$$|H(f)|^2 = \frac{1}{1 + (f/f_0)^6}$$

$$h(t) = 2\pi f_0 \left\{ e^{-2\pi f_0 t} + e^{-\pi f_0 t} \left[\frac{1}{\sqrt{3}} \sin \sqrt{3}\pi f_0 t - \cos \sqrt{3}\pi f_0 t \right] \right\} u(t).$$

These functions are sketched in Figure 16.1–2. We readily compute

$$\int_{-\infty}^{\infty} |H(f)|^2 \, df = \int_{0}^{\infty} h^2(t) \, dt = \frac{2\pi f_0}{3}$$

$$\int_{-\infty}^{\infty} f^2 |H(f)|^2 \, df = \frac{\pi f_0^3}{3}, \qquad \int_{0}^{\infty} t^2 h^2(t) \, dt = \frac{35}{36\pi f_0}$$

$$\int_{0}^{\infty} t h^2(t) \, dt = \frac{3}{4}$$

from which, by (16.1–13) and (16.1–14),

$$(\Delta W)^2 = 2f_0^2, \qquad (\Delta T)^2 = \frac{37}{48(\pi f_0)^2}$$

and

$$(\Delta T \Delta W)^2 = \frac{37}{24} \frac{1}{\pi^2} > \frac{1}{\pi^2}$$

as required by the Uncertainty Principle. The relationships of ΔT and ΔW as measures of width to the actual impulse, step, and frequency response are shown in Figure 16.1–2.

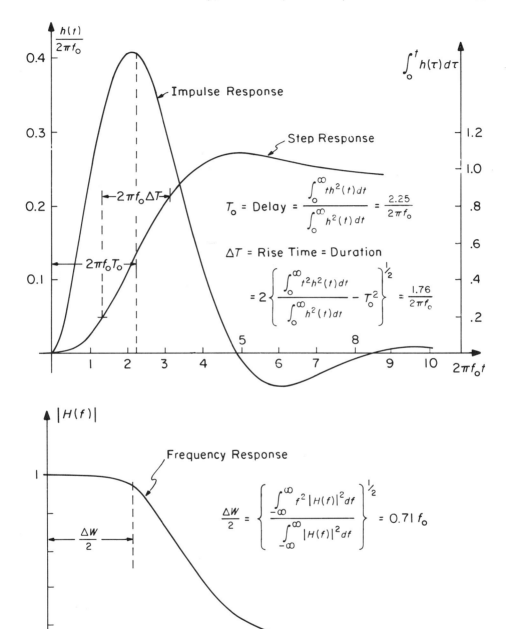

Figure 16.1–2. Impulse, step and frequency response of a 3rd-order Butterworth lowpass filter.

▶ ▶ ▶

Example 16.1–3

The Gaussian pulse cannot be the impulse response of a causal system, even with substantial delay (see Section 15.2). But it is easy to construct systems whose impulse responses approximate delayed Gaussian pulses. Consider, for example, an N-stage amplifier with the impulse response of each stage equal to

$$h(t) = \alpha\sqrt{N}\,e^{-\alpha\sqrt{N}t}u(t). \tag{16.1–23}$$

The frequency response of the cascade of N such stages is

$$H_N(f) = [H(f)]^N = \left(\frac{1}{1+j2\pi f/\alpha\sqrt{N}}\right)^N. \tag{16.1–24}$$

We seek to determine the shape of $H_N(f)$ for large N. By taking logarithms,

$$\ln H_N(f) = -N\ln\left(1+j\frac{2\pi f}{\alpha\sqrt{N}}\right) \tag{16.1–25}$$

and using the power series expansion $\ln(1+x) = x - \dfrac{x^2}{2} + \dfrac{x^3}{3}\cdots$, we have

$$\ln H_N = -N\left[\frac{j2\pi f}{\alpha\sqrt{N}} - \frac{1}{2}\left(\frac{j2\pi f}{\alpha\sqrt{N}}\right)^2 + \frac{1}{3}\left(\frac{j2\pi f}{\alpha\sqrt{N}}\right)^3\cdots\right]$$

$$\approx -j\frac{2\pi f}{\alpha}\sqrt{N} - \frac{1}{2}\left(\frac{2\pi f}{\alpha}\right)^2 \tag{16.1–26}$$

where the remaining terms vanish at least as fast as $1/\sqrt{N}$ for large N. Thus the frequency response of the cascade tends to

$$H_N(f) \approx e^{-\frac{1}{2}(2\pi f/\alpha)^2}e^{-j\sqrt{N}(2\pi f/\alpha)} \tag{16.1–27}$$

for large N, and the impulse response of the cascade tends to

$$h_N(t) \approx \frac{\alpha}{\sqrt{2\pi}}e^{-\frac{(t-\sqrt{N}/\alpha)^2}{2/\alpha^2}} \tag{16.1–28}$$

that is, a Gaussian pulse delayed by $\sqrt{N}/\alpha$.

This result is a very special case of a remarkable theorem*—the *Central Limit Theorem* of probability theory—which states in effect that, under very general conditions, the cascade of a large number of LTI systems will tend to have a Gaussian impulse response, almost independent of the characteristics of the systems cascaded! Sufficient conditions are that

  1. the absolute third moments, $\int_{-\infty}^{\infty}|t|^3h_i(t)\,dt$, exist for all the component systems and are uniformly bounded;

---

*See, for example, M. Fisz, *Probability Theory and Mathematical Statistics* (New York, NY: John Wiley, 1963). The version stated is a special case of Lyapunov's Theorem.

2. the durations, ΔT_i, of the component systems in the sense of (16.1–11) satisfy the relation $\lim\limits_{N\to\infty} \dfrac{1}{N} \sum\limits_{i=1}^{N} (\Delta T_i)^2 \neq 0$.

The first condition allows us to ignore for large N higher-order terms in an expansion such as (16.1–26), and the second guarantees that no finite subset of the component systems will dominate the result because the remainder all have relatively wide bandwidths. Given this theorem (which we shall not prove), it follows from Example 16.1–1 that the overall impulse response of N cascaded stages is approximately

$$h(t) \approx \frac{k}{\sqrt{2\pi}\,\Delta T}\, e^{-\frac{(t-T)^2}{2(\Delta T)^2}} \tag{16.1–29}$$

where T and ΔT are given by (16.1–4) and (16.1–12) respectively and

$$k = \prod_{i=1}^{N} \int_{-\infty}^{\infty} h_i(t)\,dt. \tag{16.1–30}$$

16.2 The Significance of the Uncertainty Principle; Pulse Resolution

The constraint imposed by the Uncertainty Principle on time waveforms and their transforms has an astonishingly wide domain of applications in addition to the rise-time/bandwidth context in which it was introduced—as the following examples illustrate.

Example 16.2–1

As we suggested in Section 14.4, the information waveform at an appropriate spot in a *pulse communication system* might appear as shown in Figure 16.2–1. In each successive time interval of duration T seconds a pulse of standardized shape and amplitude, such as the triangle shown below, is either presented or not presented accordingly as the symbol to be communicated in that interval is a "1" or a "0." The resulting waveform is then modulated on a carrier for transmission. Since the available bandwidth is usually limited for technical, economic, or legal reasons, the Uncertainty Principle sets a limit on the *bit rate* the system can achieve.

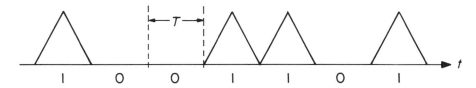

Figure 16.2–1. The information waveform in a pulse communication system.

▶ ▶ ▶

Example 16.2-2

A speech waveform contains quasi-periodicities, introduced by the glottal excitation and by resonances in the vocal tract, that are most readily studied in terms of the spectrum of the speech waveform. Of course, the waveform (and hence its quasi-periodic structure) is rapidly changing as the speaker moves from syllable to syllable (or more fundamentally from phoneme to phoneme)—indeed, it is precisely these changes that convey the meaning in the utterance and are thus of greatest interest. To follow these changes, we must compute the spectrum only over a time interval less than the phoneme duration. But the Uncertainty Principle tells us that our ability to resolve two frequency components will deteriorate inversely with the time taken to analyze them. Thus a compromise must be reached between our desires to follow rapid time changes and to explore fine details in the spectrum. Similar considerations apply to the spectral analysis of many phenomena—seismic records, weather patterns, ocean waves, econometric data, electroencephalograms and cardiograms, etc.

▶ ▶ ▶

Example 16.2-3

Designing a transmitting antenna for radio or radar systems usually involves either trying to maximize the field strength at some receiving site or trying to minimize the angular dimensions of a "pencil" or "fan" beam in which the radiated energy is to be concentrated. Since an increase in radiated intensity in one direction can be achieved only by reducing it in other directions, these goals are essentially the same. The maximization generally must be carried out subject to a constraint on the overall size (cost) of the antenna structure. On the other hand, the designer of a receiving antenna usually seeks either to maximize the effective cross-sectional area of the antenna for signals arriving from a given direction, or to narrow as much as possible the acceptance angle of the antenna so as to locate the direction of a source as precisely as possible. Again these goals are to be achieved for a fixed maximum aperture size of the actual antenna, and again these goals are essentially the same. Indeed, because a reciprocity principle applies to antennas that is similar to the reciprocity principle for circuits (in both cases, the principle follows directly from Maxwell's equations), the design problems for transmitting and receiving antennas are essentially identical—in both cases we seek to minimize a beam width subject to a constraint on aperture size. And since light is also an electromagnetic phenomenon, the same physical situation arises in the design of optical systems, such as microscopes or telescopes, to achieve maximum resolution. The connection between all of these problems and the Uncertainty Principle should be apparent once it is recognized that the distribution of energy in angle from a radiating antenna is essentially the magnitude of the spatial Fourier transform of the field strength across the aperture.* Reducing the aperture width necessarily broadens the beam angle, and the Uncertainty Principle sets a minimum on the product of beam angle and aperture width. (The value of this minimum does depend, however, on the wavelength or frequency of the radiated waveform.)

▶ ▶ ▶

---

*See, for example, R. Bracewell, *The Fourier Transform and Its Applications* (New York, NY: McGraw-Hill, 1965) Chap. 13.

Besides illustrating the wide significance of the Uncertainty Principle, these examples share another feature in common. In all of them a key problem is to design a pulse shape (or its dual, a spectral band) that meets three criteria simultaneously:

1. The spectrum of the pulse should essentially vanish for frequencies greater than some frequency W.

2. The pulse duration is to be reasonably close to the Uncertainty Principle limit.

3. The "tails" of the pulse must die away sufficiently rapidly that the tail of a large pulse will not "mask" or seriously distort another smaller pulse at an adjacent time instant; that is, we seek to be able to *resolve* two adjacent pulses even if one is much larger than the other. We must minimize as much as possible the effects of *interpulse* (or, in the dual situation, *interchannel*) *interference*.

No universally satisfactory solution to this problem is possible; the requirements are fundamentally contradictory. But there is one general principle that helps to suggest the nature of the compromises involved—the more slowly and smoothly a function changes, the more rapidly and precipitously its transform changes, and vice versa.

A formula that expresses this principle in a somewhat more quantitative fashion can be derived from Parseval's Theorem and the differentiation or the multiply-by-t properties:

$$(2\pi)^{2k} \int_{-\infty}^{\infty} |t|^{2k} |x(t)|^2 \, dt = \int_{-\infty}^{\infty} \left| \frac{d^k X(f)}{df^k} \right|^2 \, df. \qquad (16.2\text{--}1)$$

Thus if all the derivatives of $X(f)$ through the $(n-1)^{\text{st}}$ are square-integrable but the nth is not, we may in general conclude that $x(t)$ vanishes faster than $|t|^{-n+1/2}$ but no faster than $|t|^{-n-1/2}$. Indeed, for most ordinary square-integrable spectra, if $X(f)$ contains discontinuities but no worse singularities, then $n = 1$ and $x(t)$ will typically vanish as $|t|^{-1}$. For example, the frequency response of an ideal lowpass filter contains such discontinuities, and the impulse response is $\frac{\sin 2\pi Wt}{\pi t}$, which vanishes as $|t|^{-1}$. Continuous spectra with discontinuous derivatives, such as $e^{-|f|}$ or the triangular spectrum of Fejér's kernel, imply $n = 2$ and thus correspond to time functions that vanish more rapidly, in fact as $|t|^{-2}$. Other examples may be taken from Table XIII.1; thus, by the dual of this argument, the time function $e^{-\alpha t} u(t)$, being discontinuous, should have a spectrum falling off as $|f|^{-1}$, as indeed $(\alpha + j2\pi f)^{-1}$ does. Still other examples are provided by the following example.

Example 16.2–4

It is instructive to compare the spectra corresponding to each of the four pulses of Figure 16.2–2. The spectra may be computed in a straightforward manner; the results are plotted in Figure 16.2–3. The square pulse, $x_1(t)$, and the triangular pulse, $x_2(t)$, have spectral tails falling off at 6 dB/octave (that is, as f^{-1}) and 12 dB/octave (that is, as f^{-2}) as expected; for the same value of T, the effective width of the triangular pulse is somewhat narrower, and hence the bandwidth of $X_2(f)$ is somewhat wider than that of $X_1(f)$. The very useful waveform $x_3(t)$ is called the *raised-cosine* (or *Hanning*) *pulse*. The tails of its Fourier transform decay at 18 dB/octave because the second derivative of $x_3(t)$ is square-integrable but discontinuous. Since $x_3(t)$ is even narrower than the triangular pulse $x_2(t)$, $X_3(f)$ has a still wider bandwidth. This illustrates a general principle: If the interval in which $x(t)$ is nonzero is restricted, more rapidly decaying tails on $X(f)$ by and large imply a corresponding increase in bandwidth. The fourth pulse, $x_4(t)$ (called the *Hamming pulse**), illustrates one of the many sorts of compromises that are often useful. In this case, by combining $x_3(t)$ and $x_1(t)$ with appropriate weights, the height of those *side lobes* of $X_4(f)$ immediately adjacent to the *main lobe* can be substantially reduced at the expense of a slower rate of decay further out; since $x_4(t)$ is discontinuous, the tails of $X_4(f)$ must ultimately vanish as $|f|^{-1}$. The particular weighting in $x_4(t)$ was chosen to minimize the maximum side-lobe amplitude.

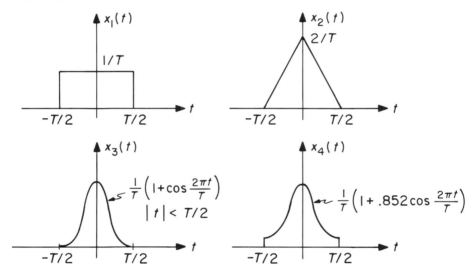

Figure 16.2–2. Four different pulse shapes.

▶ ▶ ▶

One application of the pulse-shape–spectral-shape relationships of Example 16.2–4 is provided by the pulse communication system described in Example 16.2–1. A succession of raised-cosine pulses of various amplitudes would have a slightly larger effective bandwidth than a similar train of square pulses or triangle pulses, but might be preferable in practice since the low tails of the raised-cosine spectrum would cause much less interference with other communication services occupying adjacent frequencies.

---

*R. B. Blackman and J. W. Tukey, *The Measurement of Power Spectra* (New York, NY: Dover, 1958).

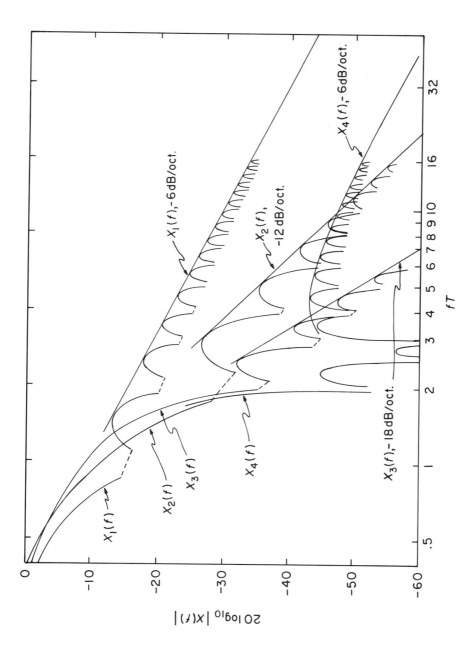

Figure 16.2-3. Spectra of the pulses of Figure 16.2-2.

A quite different application suggested by Example 16.2–3 is to the design of the large antennas used in radar or radio astronomy. Such antennas often employ parabolic reflectors to collect energy from a wide but finite aperture and focus it onto a single detector at the focal point of the "dish." If there is only a single active source (or radar reflector) in front of the antenna, then in general the output of the detector will be largest if this source is directly along the axis of the antenna, and it will fall off rapidly as the angle between the source and the antenna axis increases. The precise manner in which this *beam pattern* varies with the angle off axis depends on the weighting that the detector (for example, because of its shape) assigns to the signal coming from each element of the reflector surface. Often, areas near the edge of the reflector are intentionally less heavily weighted than areas near the center. The reason for such non-uniform weighting follows from the fact, noted in Example 16.2–3, that the beam pattern as a function of the angle, θ, is (for narrow beams) essentially the magnitude of the Fourier transform of the detector weighting as a function of position across the aperture. Thus, if the weighting is uniform across a finite aperature, the beam pattern is of the form $\frac{\sin k\theta}{k\theta}$ and the side lobes fall off only as $|\theta|^{-1}$. Lower side lobes can be obtained, at the expense of some increase in the width of the main beam, by an appropriate *tapered* weighting, which might be triangular or raised-cosine. The importance of such tapering is obvious once it is appreciated that in a radar, for example, the echo from a large nearby target may be more than 100 dB greater than that from a small remote target. Thus, even with a smooth raised-cosine tapering two targets might have to be separated by many beam widths before the echo from the weaker one would even be comparable with the side-lobe level of the stronger one.

Other applications will be discussed in the following chapters.

16.3 Summary

The Uncertainty Principle gives a lower bound on the product of the duration and bandwidth of any waveform. A related and almost equally powerful idea is that a rapidly decaying characteristic in one domain implies a high degree of smoothness in the other. Together, these concepts govern the design of waveforms and systems for an extraordinary range of applications—including the communication systems that will be the topic of our next chapter.

EXERCISES FOR CHAPTER 16

Exercise 16.1

For an ideal filter with impulse response $h(t) = \dfrac{\sin 2\pi W t}{\pi t}$, show that:

a) The rise time of the step response as defined by (16.1–1) is $1/2W$.

b) The 10%-to-90% rise time is $0.446/W$.

Exercise 16.2

For the simple RC circuit shown below, prove that the product of the 10%-to-90% rise time and the 3-dB-down single-sided bandwidth is approximately 0.35. (Since the performance of many electronic amplifiers and other systems is often dominated by a single pole, this result provides an approximate rule-of-thumb for arbitrary lowpass systems that is often surprisingly accurate. For example, the 741 op-amp data sheet gives a unity-gain bandwidth of 10^6 Hz and a unity-gain rise time of $0.3\ \mu\text{sec.}$)

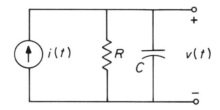

Exercise 16.3

Argue from the Paley-Wiener Theorem that no time function can be zero outside some finite time interval and also have a transform that is zero outside some finite frequency interval. Thus "duration" and "bandwidth" cannot simultaneously be interpreted in a strict sense.

PROBLEMS FOR CHAPTER 16

Problem 16.1

Choose $t_0 = 0$ in (16.1–1) and define

$$\Delta T_1 = \frac{\int_{-\infty}^{\infty} h(t)\, dt}{h(0)} = \text{duration of } h(t).$$

Similarly let

$$\Delta W_1 = \frac{\int_{-\infty}^{\infty} H(f)\, df}{H(0)} = \text{bandwidth of } H(f).$$

a) Show that

$$\Delta T_1 \Delta W_1 = 1$$

for any $h(t)$ whose Fourier transform is $H(f)$.

b) Evaluate ΔT_1, ΔW_1, and their product for each of the waveforms below and discuss the extent to which they represent "reasonable" measures of "duration" and "bandwidth," respectively, in each case.

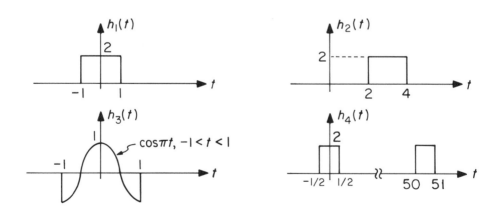

c) These definitions may not be very satisfactory if $h(t)$ or $H(f)$ is complex or changes sign. Instead, define

$$\Delta T_2 = \frac{\int_{-\infty}^{\infty} |h(t)|\, dt}{|h(0)|} \qquad \Delta W_2 = \frac{\int_{-\infty}^{\infty} |H(f)|\, df}{|H(0)|}$$

and show that

$$\Delta T_2 \Delta W_2 \geq 1.$$

Problem 16.2

Some of the difficulties with ΔT_1 and ΔW_1, the duration and bandwidth measures defined in Problem 16.1, arise from the arbitrary choice $t_0 = 0$. These can be circumvented to a degree by working with the *autocorrelation function**

$$R_h(\tau) = \int_{-\infty}^{\infty} h(t)h(t+\tau)\,dt$$

rather than $h(t)$.

a) Show that the Fourier transform of $R_h(\tau)$ is $|H(f)|^2$.

b) Define

$$\Delta T_2 = \frac{\int_{-\infty}^{\infty} R_h(\tau)\,d\tau}{R_h(0)}$$

$$= \frac{\left(\int_{-\infty}^{\infty} h(t)\,dt\right)^2}{\int_{-\infty}^{\infty} h^2(t)\,dt} = \text{duration of } h(t)$$

$$\Delta W_2 = \frac{\int_{-\infty}^{\infty} |H(f)|^2\,df}{|H(0)|^2} = \text{bandwidth of } H(f)$$

and argue[†] that

$$\Delta T_2 \Delta W_2 = 1\,.$$

For reasons that will become more apparent in Chapter 19, ΔW_2 is frequently called the *equivalent noise bandwidth* of $H(f)$.

c) Evaluate ΔT_2 and ΔW_2 for each of the waveforms of Problem 16.1 and discuss the extent to which they represent "reasonable" measures of "duration" and "bandwidth," respectively, in each case.

d) Show that if $h(t)$ is delayed, or more generally if $h(t)$ is chopped up into pieces of various lengths and the pieces are rearranged in any order, interspersed with intervals of zero height, and some or all of the intervals are reversed in time direction (see below), both the "autocorrelation duration" ΔT_2 of the new waveform and the bandwidth ΔW_2 are the same as those of the original waveform.

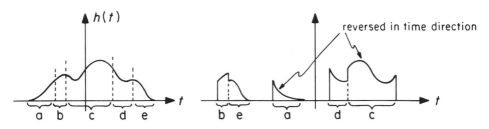

*Note that, in contrast with Problem 14.17, the autocorrelation function is here defined as an integral instead of an average. An average would, of course, be zero for the non-periodic waveforms $h(t)$ being considered.

[†]See D. G. Lampard, *IRE PGCT-3*, 4 (Dec. 1956): 286.

e) The process of *Steiner symmetrization* constructs an even function from an arbitrary function as shown below. Argue that ΔT_2 and ΔW_2 are unaltered by this process.*

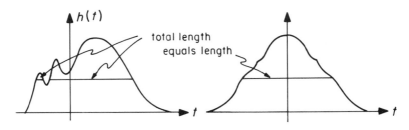

Problem 16.3

Derive equation (16.1–12) by a method paralleling that used to prove (16.1–4), that is, by exploiting the fact that

$$\int_{-\infty}^{\infty} t^2 H(t)\, dt = \frac{-1}{(2\pi)^2} \frac{d^2 H(f)}{df^2}\bigg|_{f=0}.$$

Problem 16.4

On a dB-vs.-log f plot, compare graphically the approximation obtained by transforming the Gaussian impulse response (16.1–29) with the exact frequency response of the cascade of four RC lowpass filters, each characterized by $H(f) = 1/(1 + j2\pi fRC)$.

Problem 16.5

In general, convolving two "pulse-like" waveforms (waveforms that are small outside some finite time interval) gives a result that is "wider" or has a longer effective duration than either component. This is a consequence of the theorem stated in equation (16.1–12) and proved in Problem 16.3. But for this theorem to be valid the waveforms must meet certain conditions. Thus show that

$$\frac{1}{t} * \frac{1}{t} = k\delta(t)$$

(a result much shorter than the "durations" of the $1/t$ "pulses" from which it is composed!) and find the constant k. Why is this example an exception to the general principle?

---

*See R. Bracewell, *The Fourier Transform and Its Applications* (New York, NY: McGraw-Hill, 1965) p. 175.

17

BANDPASS OPERATIONS AND
ANALOG MODULATION SYSTEMS

17.0 Introduction

For the most part, we have so far focused our attention on the properties of lowpass filters and on the special features of waveforms that have been passed through such filters. Much of what we have learned can be directly extended to apply to bandpass problems as well. The key to such an extension is the frequency-shift property of Table XIII.2,

$$e^{j2\pi f_c t} x_m(t) \iff X_m(f - f_c). \qquad (17.0\text{--}1)$$

(The reasons for the subscript choices will appear shortly.) The basic idea is to consider bandpass problems in terms of related lowpass problems plus a translation in frequency. Indeed, we shall show later that any *bandpass waveform* (that is, one whose spectrum is largely or entirely restricted to some pair of conjugate bands $W_1 < |f| < W_2$) can be written in the form

$$x(t) = \Re e[x_m(t)e^{j2\pi f_c t}] = \frac{1}{2}\big(x_m(t)e^{j2\pi f_c t} + x_m^*(t)e^{-j2\pi f_c t}\big) \qquad (17.0\text{--}2)$$

so that

$$X(f) = \frac{1}{2}\big(X_m(f - f_c) + X_m^*(-f - f_c)\big) \qquad (17.0\text{--}3)$$

where $x_m(t)$ is a (generally complex) lowpass waveform. And we shall show that the result of passing a bandpass waveform through a bandpass filter can be readily deduced from the result of passing the corresponding lowpass waveform through a lowpass filter. Moreover, there are a variety of important devices variously called (depending on their design and specific purpose) *mixers, modulators, demodulators,* and *detectors* that broadly are intended to convert lowpass waveforms into bandpass waveforms (or vice versa) and hence represent a direct application of these mathematical ideas. Thus, studying bandpass problems as transformed lowpass ones is not simply a mathematical artifice. A discussion of these analytical notions and physical devices, and of their applications in analog communication systems, is the ambitious objective of this chapter.

17.1 Amplitude Modulation

It is convenient to begin with the simple *modulator* shown in Figure 17.1–1 whose spectral properties follow almost immediately from the multiply-by-$e^{j2\pi f_c t}$ property of Table XIII.2.

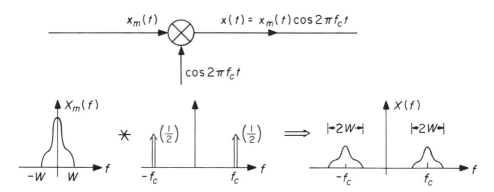

Figure 17.1–1. A simple modulator.

Here we assume that $x_m(t)$ is real, so that (17.0–2) and (17.0–3) become

$$x(t) = x_m(t)\cos 2\pi f_c t = x_m(t)\left[\frac{e^{j2\pi f_c t} + e^{-j2\pi f_c t}}{2}\right]$$
$$\Longleftrightarrow \quad \frac{1}{2}X_m(f - f_c) + \frac{1}{2}X_m(f + f_c). \qquad (17.1\text{–}1)$$

If we also assume that $x_m(t)$ has its spectrum strictly limited to the band $|f| \le W$, that is,

$$|X_m(f)| \equiv 0, \qquad |f| > W \qquad (17.1\text{–}2)$$

and if we further restrict W to be less than or equal to f_c, then $X_m(f - f_c)$ and $X_m(f + f_c)$ will not overlap and $x(t)$ will be a bandpass waveform with its spectrum (as shown in Figure 17.1–1) restricted to the bands

$$(f_c - W) < |f| < (f_c + W). \qquad (17.1\text{–}3)$$

We note, however, that because $x_m(t)$ is real, $X_m(f)$ has conjugate symmetry about $f = 0$, and so does each piece of $X(f)$ about $f = \pm f_c$; $X(f)$ is thus more structured than the most general bandpass spectrum needs to be—a point to which we shall return.

It is frequently useful to distinguish two classes of bandpass waveforms depending on whether the bandwidth is very much smaller than, or comparable with, the band-center frequency; the former are called *narrowband* (NB) and the latter *wideband* (WB) waveforms. Sometimes a signal is said to be narrowband if the width of the occupied frequency band is less than one octave; for the situation described above this would correspond to

$$(f_c + W) < 2(f_c - W) \quad \text{or} \quad W < f_c/3. \tag{17.1-4}$$

But the most characteristic properties of narrowband waveforms result if the bandwidth is in fact much smaller than the band-center frequency, say $W < f_c/10$. In that case $x(t) = x_m(t) \cos 2\pi f_c t$ (with $x_m(t)$ real) has the characteristic appearance of a sinusoid at the *carrier* frequency f_c (hence the subscript) whose amplitude is *modulated* or slowly changing in accordance with the *envelope* or *modulating signal* $x_m(t)$ (hence the subscript) as shown in Figure 17.1–2. An *amplitude modulator**** is thus a device for constructing a bandpass waveform of the form $x(t) = x_m(t) \cos 2\pi f_c t$ from a given lowpass waveform $x_m(t)$. An ideal amplitude modulator is in effect simply a multiplier.

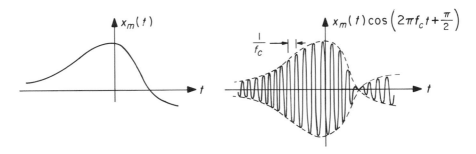

Figure 17.1–2. An amplitude-modulated carrier.

One of the most important properties of an amplitude-modulated (AM) waveform is that amplitude modulators preserve all of the details of the low-pass modulating waveform $x_m(t)$; that is, it is possible to reverse the operation generating $x(t)$ and recover $x_m(t)$ in complete detail. One simple *demodulator* or *detector* is the *synchronous* (synchrodyne, homodyne) *detector* or *product demodulator* shown in Figure 17.1–3.

---

*There are several forms of amplitude modulators; later we shall call this simple type a suppressed-carrier amplitude modulator.

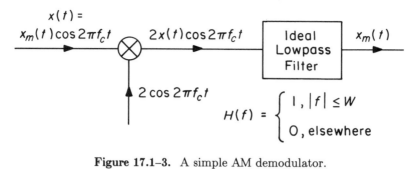

Figure 17.1-3. A simple AM demodulator.

That the output of this system is $x_m(t)$ follows from the fact that the output of the multiplier can be written

$$2x(t)\cos 2\pi f_c t = 2x_m(t)\cos^2 2\pi f_c t$$
$$= x_m(t) + x_m(t)\cos 2\pi 2 f_c t \qquad (17.1-5)$$

where we have used the trigonometric identity

$$\cos^2\phi = \frac{1}{2} + \frac{1}{2}\cos 2\phi. \qquad (17.1-6)$$

The spectrum of the second term in (17.1–5) is, from (17.1–1), centered around the frequencies $\pm 2 f_c$ and thus is not passed by the lowpass filter. Hence the output is just $x_m(t)$ as desired. (Of course, the same result could be obtained, as in the discussion of the modulator, from convolving the spectra of $x(t)$ and $2\cos 2\pi f_c t$; it is instructive to work out the details by this alternate route for practice.)

Although the bandpass AM waveform $x(t) = x_m\cos 2\pi f_c t$ is thus entirely equivalent to the lowpass waveform $x_m(t)$ in terms of the information it conveys, it is made up of a different range of frequencies. This difference has a number of practical consequences, of which perhaps the most important are:

a) By *frequency-division multiplexing* it becomes possible for a single communication link to serve a number of independent customers. In such a system each user modulates his lowpass information waveform onto a different carrier frequency so spaced that the resulting frequency bands or *channels* do not overlap and thus can be separated by filters at the receiver. Historically, frequency-division multiplexing developed more slowly than the time-division multiplexing discussed in Chapter 14 because it required more complicated filters and electronic circuits, but it has a substantial advantage in that it does not require accurate synchronization of the separate users (as time-division multiplexing does). It is also relatively insensitive to various forms of amplitude and phase distortion that (by causing pulse interference) can seriously disrupt a time-division system. On the other hand, frequency-division multiplexing is more sensitive than time-division multiplexing to various non-linear and time-varying distortions that can cause *cross-talk* between the channels. Frequency-division multiplexing is, of course, the primary means whereby the available spectrum of radio frequencies is shared among a number of services—see, for example, Table XVII.1. (Of course, many schemes of multiplexing are used in addition to time and frequency division; these include the phase or quadrature multiplexing discussed below and in Exercise 17.4, the use of polarization of electromagnetic radiation or different modes in waveguides, and the "phantom circuit" in telephony whereby two copper-wire pairs carry three conversations.)

b) Many of the factors that influence the cost, performance, or even the feasibility of a communication, control, or measurement system are likely to be strong functions of the channel frequency; the selection of an appropriate frequency is thus one of the more important steps in the system design. The factors that can be significant in this regard are too diverse and numerous for any simple listing, but—to pick the example of radio communications—can include such attributes as propagation mode (line of sight or ionospheric reflection), attenuation, dispersion, and distortion in the medium; interference and noise; available transmitter power and efficiency; antenna size and directivity; and in general the cost, size, weight, efficiency, and power handling capacity of the components. In particular, commercial radio communication would be virtually impossible if speech and music had to be transmitted as electromagnetic signals at the original frequencies of a few kilohertz, because of the size of the antennas that would be required. On the other hand, the obvious military advantage in being able to communicate with submerged submarines has led to the construction of radio stations operating at frequencies that are nearly this low—in spite of the size of antennas and the immense power that must be generated—because electromagnetic waves of higher frequency are rapidly attenuated in sea water.

The (suppressed-carrier) amplitude modulation scheme discussed above has one serious weakness—the *local oscillator* signal, $2\cos 2\pi f_c t$, in the demodulator must be accurately synchronized in phase with the carrier signal, $\cos 2\pi f_c t$, in the modulator. To explore this difficulty, suppose that the demodulator local oscillator signal were $2\cos(2\pi f_c t + \theta)$, so that the input to the lowpass filter in the product demodulator of Figure 17.1–3 became

$$2x(t)\cos(2\pi f_c t + \theta) = 2x_m(t)\cos 2\pi f_c t \cos(2\pi f_c t + \theta)$$
$$= x_m(t)\cos\theta + x_m(t)\cos(4\pi f_c t + \theta) \qquad (17.1-7)$$

where we have used the trigonometric identity

$$\cos\phi\cos\psi = \frac{1}{2}\cos(\phi + \psi) + \frac{1}{2}\cos(\phi - \psi). \qquad (17.1-8)$$

Table XVII.1—Typical Communication Uses of Frequency Bands*

Frequency Band	Name	Typical Uses
3–30 kHz	Very Low Frequency (VLF)	Long-range navigation; sonar
30–300 kHz	Low Frequency (LF)	Navigational aids; radio beacons
300–3000 kHz	Medium Frequency (MF)	Maritime radio; direction finding; distress and calling; commercial AM radio
3–30 MHz	High Frequency (HF)	Search and rescue; ship-to-shore and ship-to-aircraft; telegraph, telephone, and facsimile
30–300 MHz	Very High Frequency (VHF)	VHF television channels; FM radio; land transportation; private aircraft; air traffic control; taxicab; police; navigational aids
0.3–3 GHz	Ultra High Frequency (UHF)	UHF television channels; radio-sonde; navigational aids; surveillance radar; satellite communication; radio altimeters
3–30 GHz	Super High Frequency (SHF)	Microwave links; airborne radar; approach radar; weather radar; common carrier land mobile
30-300 GHz	Extremely High Frequency (EHF)	Railroad service; radar landing systems; experimental

---

*Adapted from R. E. Ziemer and W. H. Tronter, *Principles of Communications* (Boston, MA: Houghton Mifflin, 1976) by permission.

The second term in (17.1–7) has a spectrum centered on $\pm 2f_c$ and would be rejected by the lowpass filter, so the final output would be $x_m(t)\cos\theta$. The output level would thus depend on θ, and could even become zero if $\theta = \pi/2$, that is, if the local oscillator were $90°$ out of phase with the carrier. Moreover, if θ slowly changed with time (that is, if the frequency of the local oscillator were not quite correct), then the output would wax and wane, grow and fade, in a manner that might be quite undesirable.

Many applications of suppressed-carrier AM are in control systems where the modulator and demodulator are geographically close; the same oscillator can be used for modulation and demodulation so that no problems arise. But in suppressed-carrier AM communication systems, where the transmitter and receiver may be many miles apart, there can be severe synchronizing difficulties. One solution is to add to the transmitted signal a small-amplitude sinusoid at the carrier frequency that can be used to synchronize the local oscillator. Another way is to square the received signal in a separate channel of the receiver, thereby generating a true carrier at frequency $2f_c$ since

$$x_m^2(t)\cos^2 2\pi f_c t = \frac{1}{2}x_m^2(t) + \frac{1}{2}x_m^2(t)\cos 4\pi f_c t \qquad (17.1-9)$$

and $x_m^2(t)$ is always positive so that it has a nonzero average value; this double-frequency carrier can then be used to synchronize the local oscillator. In either of these schemes, the actual synchronization of the local oscillator usually involves a *phase-locked loop* (see Problem 17.12, which also describes a variant of the squaring system).

The way in which the performance of a synchronous detector changes with the phase relationship between the carrier and the local oscillator suggests an interesting possibility. If synchronization can be achieved between the transmitter and receiver, then it is possible to transmit two independent signals simultaneously over the same channel by what is called *quadrature multiplexing*. The details are explored in Exercise 17.4.

The synchronization problem discussed above disappears for narrowband waveforms if the modulating waveform $x_m(t)$ is itself always positive,

$$x_m(t) \geq 0 \qquad (17.1-10)$$

since the demodulator can then be constructed as a non-linear *envelope detector*. The operation of the simplest type of envelope detector should be qualitatively evident from Figure 17.1–4. The diode conducts only on the rising part of the carrier peaks, charging the capacitor rapidly to the peak value. During the rest of each carrier cycle the capacitor discharges slowly through the resistor. Further lowpass filtering or smoothing will remove the residual "ripples" at the carrier frequency.

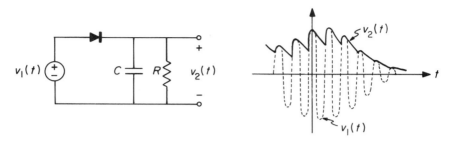

Figure 17.1–4. A simple envelope detector.

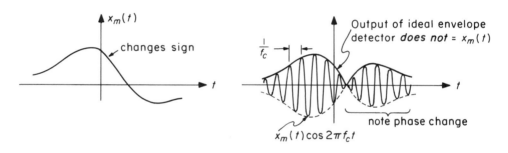

(a) $x_m(t)$ both positive and negative

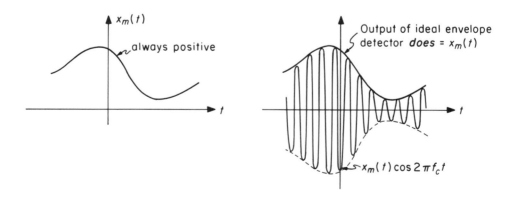

(b) $x_m(t)$ always positive

Figure 17.1–5. Output of an ideal envelope detector.

If $x_m(t) \geq 0$, then an envelope detector of this type will essentially recover $x_m(t)$; if $x_m(t)$ changes sign, however, then the envelope detector has no carrier reference to detect the resulting phase change and gives the wrong output—as shown in Figure 17.1–5. Some modulating waveforms, such as a video signal, are inherently positive. Others, such as speech and music signals, are usually bipolar; the ear, for example, does not respond to the low frequencies or d-c component that a unidirectional signal inherently contains (see Chapter 20). Moreover, the human voice and musical instruments do not generate such frequencies in the acoustic signal. To make a bipolar signal unidirectional it is sufficient to add a positive constant larger than the largest negative peak of the modulating signal, so that the modulator output becomes

$$x(t) = (A + x_m(t))\cos 2\pi f_c t$$
$$= A\cos 2\pi f_c t + x_m(t)\cos 2\pi f_c t. \tag{17.1–11}$$

This waveform differs from that obtained previously by the addition of the term $A\cos 2\pi f_c t$ at the carrier frequency (accordingly this term is called *the carrier* and the previously studied second term, $x_m(t)\cos 2\pi f_c t$, is called *suppressed-carrier* AM). In consequence of the carrier term, a pair of impulses are added to the spectrum of $x(t)$, as shown in Figure 17.1–6.* The amplitude of these impulses is independent of $x_m(t)$; all of the useful information in the waveform $x(t)$ is contained in the *side bands*, the spectral bands on either side of the carrier impulses. The only role of the carrier term is to permit a simpler demodulator.[†]

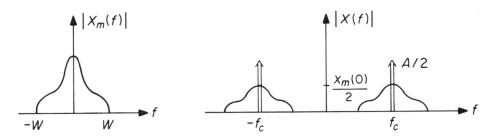

Figure 17.1–6. Spectrum of an AM signal.

---

*This is the conventional, albeit slightly dishonest, spectral picture of AM. We have assumed that the carrier term is of indefinite duration (yielding the impulse) whereas the signal is of finite duration, or at least integrable, so that it has a finite Fourier transform. A nonperiodic signal of infinite duration requires the power spectral density tools to be discussed in Chapter 19.

[†]The presence of a carrier does not require the use of a non-linear envelope detector such as the one illustrated in Figure 17.1–4. The synchronous detector of Figure 17.1–3 remains an effective alternative, particularly since the presence of the carrier makes it easy to employ a phase-locked loop to synthesize the local oscillator signal.

Example 17.1–1

The case of sinusoidal AM is easy to study in detail and provides important insights. It is conventional to write in this case

$$x(t) = A(1 + m\cos 2\pi f_m t)\cos 2\pi f_c t \qquad (17.1-12)$$

where f_m is the modulating frequency ($f_m \ll f_c$) and m, the *modulation index* or *factor*, measures the fractional "depth" of modulation. To avoid *overmodulation*, we must restrict $|m| \leq 1$. The index m is frequently expressed in percent; the waveforms for $m = 50\%$ and $m = 100\%$ are sketched in Figure 17.1–7.

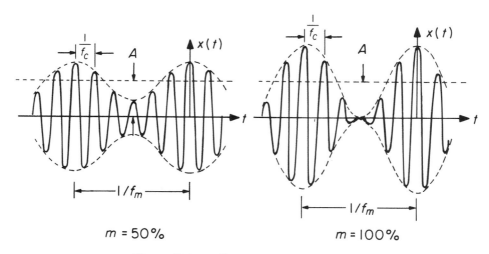

$m = 50\%$ $m = 100\%$

Figure 17.1–7. Sinusoidal AM waveforms.

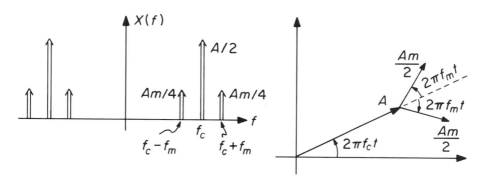

Figure 17.1–8. Spectrum of sinusoidal AM.

The spectrum of $x(t)$ of (17.1–12), shown in Figure 17.1–8, consists entirely of impulses, although $x(t)$ may or may not be truly periodic depending on whether f_c/f_m is or is not an integer. The character of $X(f)$ becomes obvious if $x(t)$ of (17.1–12) is written in the form

$$x(t) = A\cos 2\pi f_c t + \frac{Am}{2}\cos 2\pi(f_c + f_m)t + \frac{Am}{2}\cos 2\pi(f_c - f_m)t. \qquad (17.1\text{--}13)$$

This equation can also be interpreted in terms of the projections of a set of rotating vectors or phasors, as shown to the right in Figure 17.1–8. The large vector (of length A) rotates about the origin at a uniform rate f_c cycles/sec; the smaller vectors (of length $Am/2$) rotate in opposite directions about the tip of the larger vector at the rate f_m cycles/sec. Their overall rates of rotation are thus $f_c + f_m$ and $f_c - f_m$ cycles/sec. The projection of the vector sum on the horizontal axis is $x(t)$. Note that the resultant of the two side-band vectors always lies along the vector A, so that the overall vector rotates at constant speed while varying slowly in amplitude.

▶ ▶ ▶

Example 17.1–2

Amplitude modulation and synchronous detection are (ideally) linear (time-varying) operations on the modulating signal $x_m(t)$ (and consequently amplitude modulation is sometimes called *linear modulation*). Nevertheless, most practical amplitude modulators as well as detectors use non-linear devices, such as rectifiers, rather than linear multipliers. One practical modulator circuit is shown, somewhat idealized, in Figure 17.1–9. (Some other important modulator circuits are discussed in Problems 17.4 and 17.5.) An exact analysis of even such a simple non-linear circuit as this presents extreme difficulties, but an approximate steady-state analysis is easy if we further idealize the system, replacing the tuned circuit by an impedance that is infinite for frequencies in the vicinity of $\pm f_c$ and zero at all other frequencies. Suppose also that the voltage $x_m(t)$ varies so slowly compared to the bandwidth of the tuned circuit that we may analyze the system as if it were always in the steady state in response to a constant voltage x_m.

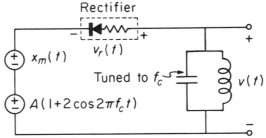

Figure 17.1–9. A simple circuit for AM modulation.

From the properties assumed for the tuned circuit, it follows that the voltage $v(t)$ must be a sinusoid of frequency f_c, that is, $v(t) = V \cos 2\pi f_c t$, where V is to be found. Hence the voltage across the rectifier has the form

$$v_r(t) = x_m - A + (V - 2A) \cos 2\pi f_c t.$$

We now identify three regions of operation:

    i) If $x_m < -A$, the diode is back-biased into the non-conducting state throughout the cycle. Thus $V = 0$.

    ii) If $x_m > A$, the diode is forward-biased into the conducting state throughout the cycle. Since the impedance of the tuned circuit is very large compared to the forward resistance of the rectifier, $V = 2A$ independent of x_m.

    iii) If $|x_m| < A$, the diode conducts during part of each cycle. But the tuned circuit impedance at frequency f_c is large, so that the circulating current component at this frequency is almost zero and the amount of conduction must be very small. Hence the peak positive value of $v_r(t)$ is approximately zero, or

$$x_m - A = V - 2A$$

which implies

$$V = x_m + A, \qquad -A \leq x_m \leq A. \qquad (17.1\text{–}14)$$

Letting $x_m(t)$ be slowly varying gives finally

$$v(t) = [A + x_m(t)] \cos 2\pi f_c t, \qquad |x_m(t)| < A \qquad (17.1\text{–}15)$$

which is identical with (17.1–11) subject to the peak-amplitude limits. Typical waveforms are shown in Figure 17.1–10. Note the distorting effects of *overmodulation* ($|x_m(t)| > A$), which in this type of modulator are equivalent to *amplitude limiting* or *peak clipping* of $x_m(t)$.

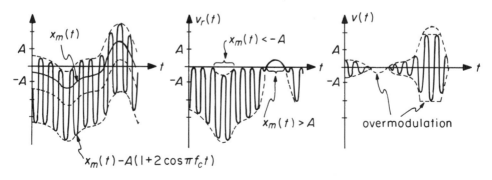

Figure 17.1–10. Waveforms in the modulator of Figure 17.1–9.

▶ ▶ ▶

An envelope detector allows a much simpler receiver than a synchronous detector since it does not require an oscillator accurately synchronized to the carrier frequency. But this simplicity is achieved at a price; in order to prevent distortion in the envelope detector resulting from overmodulation, the carrier amplitude A in (17.1–11) must always be greater than the maximum amplitude of the modulating signal $x_m(t)$. This carrier component in the transmitted signal conveys no information and will be filtered out in the receiver, but the carrier power must nevertheless be generated and radiated by the transmitter. (This can be a serious limitation in an environment, such as a satellite, where available power is limited.) Thus ordinary amplitude modulation with a substantial carrier is efficient for *broadcast* situations in which there is one transmitter but many receivers, whereas suppressed-carrier amplitude modulation (SC-AM) finds its principal communications use in *point-to-point* systems in which the number of receivers is small enough that the added cost and complexity of synchronous detectors is not serious.

The *efficiency* of amplitude modulation can be defined as the ratio of the average power in the side-band components of the modulated waveform to the total average power. From Example 13.4–1, the average power in a spectrum composed of impulses is the sum of the squares of the impulse areas. Hence the efficiency of a sinusoidally amplitude-modulated waveform such as (17.1–13) is (see Figure 17.1–8)

$$\frac{4\left[\dfrac{Am}{4}\right]^2}{2\left[\dfrac{A}{2}\right]^2 + 4\left[\dfrac{Am}{4}\right]^2} = \frac{m^2}{2+m^2} \tag{17.1–16}$$

which has a maximum value (at $m = 1$) of 33.3%.

But even this rather modest figure does not truly reflect the very low efficiency of ordinary AM in practical situations. The *peak factor* (pf) of a complex waveform is defined as the ratio of the peak or maximum value to the rms value. Measurements show, for example, that the peak factor for "uncontrolled" speech is about 35. It can be shown that the efficiency of an AM signal cannot exceed $1/[1+(pf)^2]$ if overmodulation is not permitted (see Exercise 17.2 for a discussion of this formula for periodic signals); thus the AM efficiency for "uncontrolled" speech is less than 0.1%!

To improve efficiency, commercial radio stations use automatic gain controls to reduce the gain during moments of high sound intensity (thus reducing pf for speech to about 8) and permit some overmodulation distortion to occur. In addition it is sometimes feasible (as illustrated in Figure 17.1–11) to pass the modulating signal through a resistive non-linearity that *compresses* the dynamic range before modulating it onto the carrier; a compensating *expanding* non-linearity can then be inserted in the receiver after demodulation. This combined process is called *companding*.

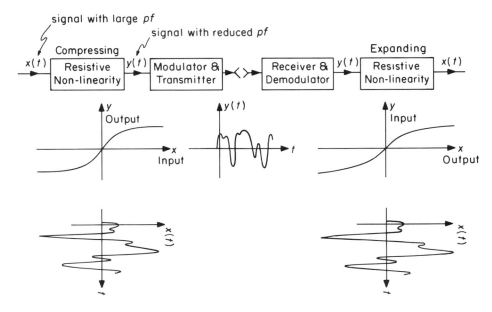

Figure 17.1–11. Waveforms in a companding system.

17.2 Mixers and Superheterodyne Receivers

The multiplier as an amplitude modulator and the synchronous detector analyzed in the previous section are both special cases of a general class of devices called *mixers* or *frequency converters* that are of great practical importance and whose properties we shall now explore more generally. The purpose of an ideal mixer is to shift the center frequency of a narrowband signal. The block diagram of an ideal mixer and the spectra explaining its operation are shown in Figure 17.2–1. Multiplying $x(t)$ and $\cos 2\pi f_0 t$ is equivalent to convolving their spectra; the result is to shift the spectrum of $x(t)$ to the right and to the left, creating side bands at $f_0 \pm f_1$ as shown. The bandpass filter selects one of these side bands. If the upper side band is selected (as shown), the overall effect is simply to move the center frequency of $x(t)$ from f_1 to $f_0 + f_1$. If the lower side band is selected (as shown dotted), the effect is not only to shift the center frequency of $x(t)$ to $f_0 - f_1$ but also to invert the spectrum. (Of course, if $X(f)$ is symmetrical about the center frequency f_1, as will be the case if $x(t)$ is an AM signal, an inversion has no effect. But in other cases it may be important.)

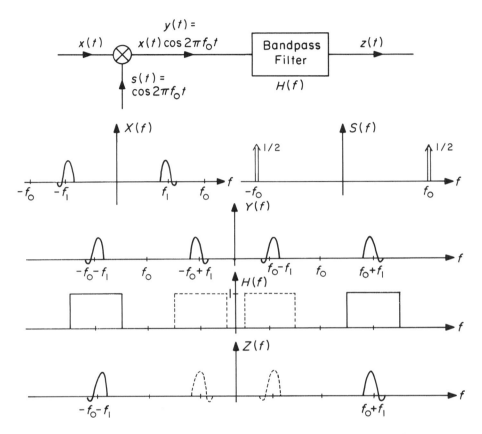

Figure 17.2–1. Mixer block diagram and spectra. (The spectrum of $x(t)$ is shown as real for pictorial convenience; it would, of course, be complex in general.)

An important example of the use of a mixer is the *superheterodyne receiver*.* The function of a radio receiver can be conveniently broken down into two steps. The first is to provide a highly selective narrowband filter with considerable gain that will amplify the signal on the desired frequency channel while suppressing interfering signals on other channels. Often we want to be able to *tune* the receiver, changing the filter from one channel to another over some band of frequencies that might be many channel-bandwidths wide. The second step is to recover or detect the modulation on the output signal of the filter. The superheterodyne principle is intended to assist in the solution of the first step in the receiver.

Early vacuum-tube radio receivers were of the *tuned-radio-frequency* (TRF)

Heterodyning is the process of mixing two signals at different frequencies to obtain a third frequency. Mixing two signals at the same frequency (as in synchronous detection) is sometimes called *homodyning*.

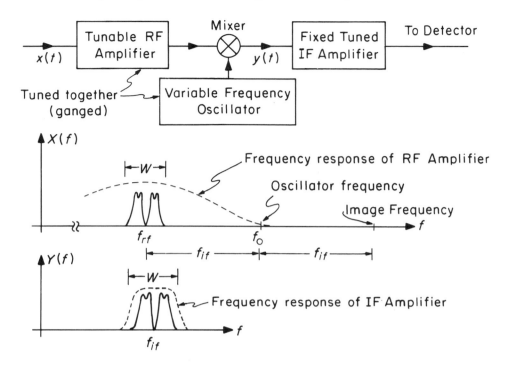

Figure 17.2-2. Superheterodyne receiver.

type; they consisted of a sequence of amplifier stages with sharply tuned RLC circuits between them, the C's or L's of which could be adjusted for tuning. The trouble with this type of arrangement is that it is difficult to keep the selectivity and gain of the various stages fairly constant over the entire tuning range. Moreover, if a number of stages are cascaded to give a large gain, there is a danger that feedback from the output to the input will lead to instabilities or oscillations (as we discussed in Chapter 6). The superheterodyne receiver gets around these problems by realizing most of the selectivity at a single fixed frequency while splitting up the gain between several frequencies. The method for accomplishing this is shown in Figure 17.2–2. The RF (radio-frequency) amplifier is now one or two stages and does not have to be very selective. The IF (intermediate-frequency) amplifier is where most of the selectivity is accomplished; its bandwidth W must be the same as the bandwidth of the received signal (in the case of AM, twice the audio bandwidth), but the center frequency f_{if} of this band is held fixed by changing the oscillator frequency f_0 when the RF amplifier is retuned. Normally, the oscillator is kept f_{if} Hz above the RF frequency. Note, however, that an RF signal f_{if} Hz above f_0 will also produce a difference frequency of f_{if} Hz; the RF amplifier must be selective enough to eliminate this unwanted *image* signal. This is quite easy if the IF frequency is reasonably high. Most modern receiver "front ends" operate on the superheterodyne principle, which was discovered by the American inventor Edwin H. Armstrong in 1917. For this and many other contributions to the practical art of radio communications (including in particular the invention of wideband FM) Armstrong even more than Marconi deserves the title "Father of Radio." *

17.3 Single-Side-Band Modulation; General Narrowband Representations

Because of the conjugate symmetry of the side bands, the spectrum of either an ordinary or a suppressed-carrier AM signal occupies twice as much bandwidth as is necessary to transmit the modulation signal. This should be clear from Figure 17.3–1: The original signal can be recovered just as well from the upper channel as from the lower even though the signal in the upper channel has only half the bandwidth of the lower. (No effort has been made to draw the spectra of this figure to a common vertical scale.) The signal $x_3(t)$ in the upper branch of the figure is called a *single-side-band AM* (SSB-AM) signal and is clearly more economical of bandwidth than an ordinary or suppressed-carrier AM signal.[†] Indeed, two independent single-side-band signals separated by filters at the receiver require precisely the same channel bandwidth and yield the same overall effect as the two quadrature multiplexed signals discussed in Exercise 17.4. In addition to its bandwidth advantage, single-side-band modulation is

---

*For a fascinating biography of Armstrong, see L. Lessing, *Man of High Fidelity: Edwin Howard Armstrong* (Philadelphia, PA: Lippincott, 1956).

[†]The discovery of SSB-AM is attributed to J. R. Carson in the early 1920s.

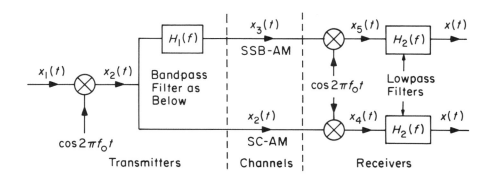

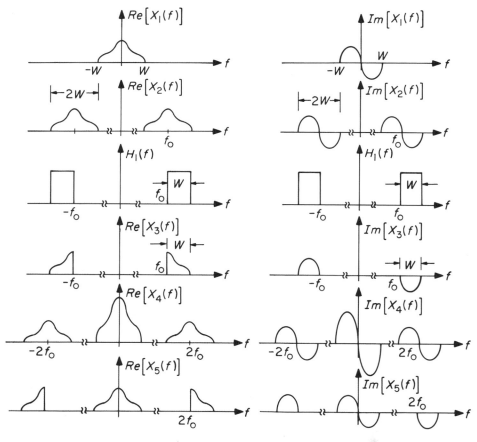

Figure 17.3–1. Comparison of SSB-AM and SC-AM.

less sensitive to certain kinds of interference and distortion than other kinds of amplitude modulation. Of course, SSB-AM shares with SC-AM the need for a complex receiver including a synchronized local oscillator, a requirement that limits its usefulness in broadcast situations. (SSB-AM used for transmitting speech or music is, however, somewhat less sensitive than SC-AM to frequency offset of the local oscillator.)

The bandpass filter $H_1(f)$ in the single-side-band modulator of Figure 17.3–1 requires a very sharp cutoff characteristic, at least at the edge of the pass band adjacent to the carrier. An ingenious alternative scheme (discovered by Hartley in 1928), partially circumventing these stringent filtering requirements, is shown in Figure 17.3–2. (Still another approach is described in Problem 17.8.) In this system, $H(f)$ is a 90°-phase-shift filter,

$$H(f) = \begin{cases} -j, & f > 0 \\ j, & f < 0. \end{cases} \qquad (17.3\text{--}1)$$

As discussed in Problem 15.3, this filter has the property that its output $x_2(t)$ is the Hilbert transform of its input $x_1(t)$.

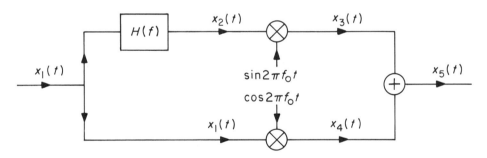

Figure 17.3–2. Another scheme for achieving SSB modulation.

The various spectra sketched in Figure 17.3–3 show clearly how the desired result is obtained. The impulse response of $H(f)$ is, from Table XIII.2, $h(t) = 1/\pi t$, $-\infty < t < \infty$, which is both rather badly behaved at $t = 0$ and non-causal. But satisfactory approximations effectively achieving the desired result can be built in various ways. For example, if $x_1(t)$ is speech or music, neither very low frequencies nor moderate alterations in the phase characteristics of the signal are noticed by the ear; the desired effect can then be achieved as shown in Figure 17.3–4, by passing $x_1(t)$ separately through a pair of filters that, over a band from, say, 50 Hz to 10 kHz, have unit magnitude and a 90° phase difference. Causal filters with such a property are relatively easy to design.[*]

*See S. Darlington, *Bell Sys. Tech. J.*, 29 (1950): 94.

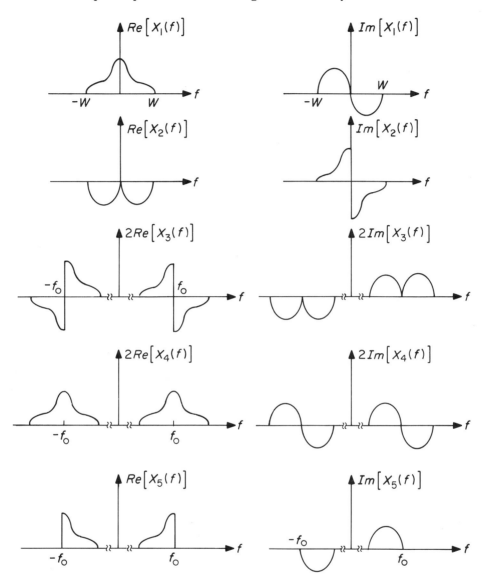

Figure 17.3–3. Spectra at various points in SSB system of Figure 17.3–2.

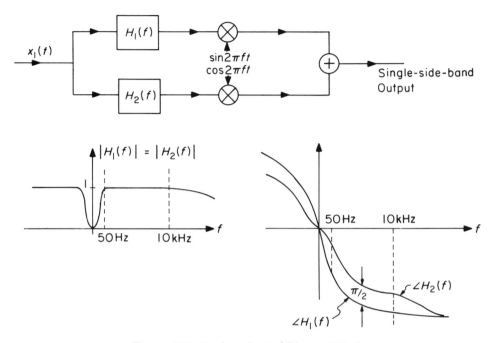

Figure 17.3–4. A variant of Figure 17.3–2.

Single-side-band AM waveforms are examples of general narrowband wave-forms in that, unlike conventional or suppressed-carrier AM waveforms, they need have no symmetry properties within the band. An arbitrary narrowband waveform cannot be represented simply as a product of a single real lowpass waveform and a cosine function, but not much more complication is needed. Specifically we now show that any narrowband waveform $x(t)$ can be represented in the form

$$x(t) = x_c(t) \cos 2\pi f_0 t + x_s(t) \sin 2\pi f_0 t. \tag{17.3-2}$$

This representation is not unique in that f_0 may be chosen somewhat freely to be any frequency in the vicinity of the band of $x(t)$. $x_c(t)$ and $x_s(t)$ are lowpass waveforms called the *quadrature components* of $x(t)$. The easiest way to show the validity of (17.3–2) is to prove that the output waveform from the block diagram in Figure 17.3–5 is simply the input narrowband waveform $x(t)$. This is done graphically in the spectra in Figure 17.3–6. A little time spent in relating each of the waveforms in Figure 17.3–6 to the operations in Figure 17.3–5 will prove rewarding.

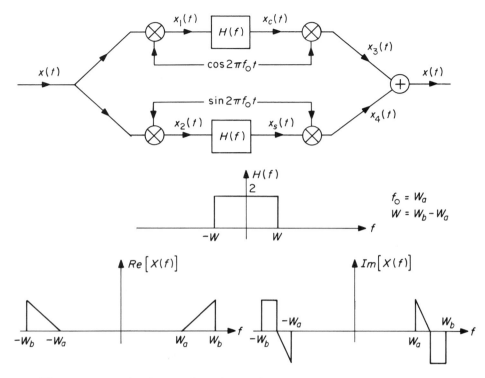

Figure 17.3–5. Analysis and synthesis of a general narrowband waveform.

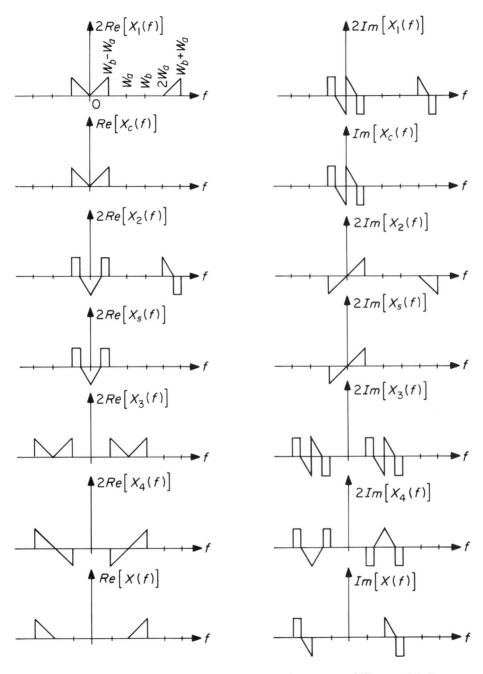

Figure 17.3–6. Spectra at various points in the system of Figure 17.3–5.

It is sometimes convenient in analytical manipulations of narrowband signals to define a *complex envelope*

$$\xi(t) = x_c(t) - jx_s(t) \tag{17.3--3}$$

so that (17.3–2) can be written as the real part of an *analytic signal*

$$x(t) = \Re[\xi(t)e^{j2\pi f_0 t}]. \tag{17.3--4}$$

We can also describe a general narrowband signal in the form

$$x(t) = E(t)\cos[2\pi f_0 t + \psi(t)] \tag{17.3--5}$$

where the *envelope* $E(t)$ is related to $\xi(t)$ by

$$E(t) = |\xi(t)| = \sqrt{x_c^2(t) + x_s^2(t)} \tag{17.3--6}$$

and the *phase* $\psi(t)$ is related to $\xi(t)$ by

$$\psi(t) = \angle\xi(t) = \tan^{-1}\frac{-x_s(t)}{x_c(t)}. \tag{17.3--7}$$

Equation (17.3–5) shows that we can consider a general narrowband signal as being simultaneously modulated in both amplitude and phase. Some properties of purely phase-modulated signals will be discussed in the next section.

Analytic signals have many of the same advantages over real signals that complex exponentials have over sinusoids. For example, as shown in Problem 17.10, successive bandpass operations on a narrowband signal $x(t)$ can be described in terms of equivalent lowpass operations on $\xi(t)$—thus justifying the detailed study of lowpass operations sketched out earlier. Since we shall have little formal need for analytic signals in this book, however, we shall not carry this discussion further.

17.4 Phase and Frequency Modulation

A purely *phase-modulated* (PM) signal results if in (17.3–5) we hold the envelope fixed and vary the phase $\psi(t)$ in proportion to the modulating signal $x_m(t)$, so that

$$x(t) = A\cos[2\pi f_c t + mx_m(t)]. \tag{17.4--1}$$

The proportionality factor m is called the *phase modulation index*. Since it is the phase angle of the cosine, rather than its amplitude, that is modulated, phase and frequency modulation are sometimes called collectively *angle modulation*.

To explore the spectral properties of a PM signal, we shall study first the case in which $x_m(t) = \cos 2\pi f_m t$ and assume that $m \ll 1$ so that

$$\begin{aligned}
x(t) &= A\cos[2\pi f_c t + m\cos 2\pi f_m t] \\
&= A\cos 2\pi f_c t \cos[m\cos 2\pi f_m t] - A\sin 2\pi f_c t \sin[m\cos 2\pi f_m t] \\
&\approx A\cos 2\pi f_c t - Am\sin 2\pi f_c t \cos 2\pi f_m t
\end{aligned} \qquad (17.4\text{--}2)$$

where we have used the approximations that, for small x, $\cos x \approx 1$ and $\sin x \approx x$. The last line in (17.4–2) can be written

$$x(t) \approx A\cos 2\pi f_c t - \frac{Am}{2}\sin 2\pi(f_c + f_m)t - \frac{Am}{2}\sin 2\pi(f_c - f_m)t. \qquad (17.4\text{--}3)$$

If each term is written in exponential form, the spectrum of a small-modulation-index PM signal is easily seen to be (approximately) as shown to the left in Figure 17.4–1. Note that the magnitude of this spectrum is identical with that for a sinusoidal AM signal, but the phase relations of the carrier and side bands are different. This phase relationship is more graphically illustrated by the phasor diagram on the right in Figure 17.4–1. As in the similar vector diagram for AM, the smaller vectors rotate slowly in opposite directions about the tip of the rapidly rotating large vector and $x(t)$ is the projection of the resultant on the horizontal axis. But unlike the case of AM, the resultant of the smaller vectors is perpendicular to the larger vector; if the side-band vectors are small as assumed ($m \ll 1$), then the overall resultant is nearly the same length, A, as the carrier vector but rotates at a non-uniform rate.

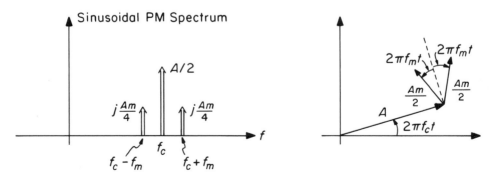

Figure 17.4–1. Spectrum and phasor diagram for a PM signal with $m \ll 1$.

The phase relationships in this vector diagram suggest a simple way to generate a small-modulation-index PM signal. The scheme is shown in Figure

17.4–2 for an arbitrary modulating signal $x_m(t)$ rather than a sinusoid. The modulating signal is first SC-AM-modulated on a carrier, $A\sin 2\pi f_c t$, by the mixer or balanced modulator. Then a large carrier, 90° out of phase with that used in the mixer, is added. The result—assuming $|x_m(t)| \ll 1$—is approximately a PM signal as desired.

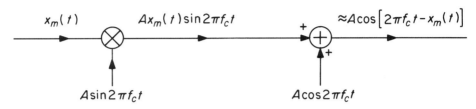

Figure 17.4–2. A PM modulator for $m \ll 1$.

For reasons that are in large part historical and conventional, phase modulation is normally discussed in the literature under the heading of *frequency modulation*. The formal distinction is simple enough. The waveform

$$x(t) = A\cos[2\pi f_c t + \psi(t)] \qquad (17.4-4)$$

is a carrier *phase*-modulated by $\psi(t)$. On the other hand, we may plausibly define the *instantaneous* (radian) *frequency* of $x(t)$ as the derivative of the argument of the cosine, that is, as

$$\frac{d}{dt}[2\pi f_c t + \psi(t)] = 2\pi f_c + \frac{d\psi(t)}{dt} . \qquad (17.4-5)$$

Setting $\dfrac{d\psi(t)}{dt} = \phi(t)$ or $\psi(t) = \displaystyle\int_{-\infty}^{t} \phi(\tau)\,d\tau$, we can say that

$$x(t) = A\cos\left[2\pi f_c t + \int_{-\infty}^{t} \phi(\tau)d\tau\right] \qquad (17.4-6)$$

is a carrier *frequency*-modulated by $\phi(t)$. If, then, we integrate the basic modulation waveform before phase-modulating, the result is frequency modulation. But since such an integration is an LTI operation that can be undone by differentiation at the receiver, the distinction, at least in the noise-free case, is not a particularly fundamental one. As a matter of fact, for reasons of noise reduction most practical so-called frequency-modulation (FM) communication systems pre-emphasize higher frequencies and might better be called PM systems anyway.

Let us consider in more detail the case of sinusoidal frequency modulation, that is,

$$\phi(t) = -k\sin 2\pi f_m t . \qquad (17.4-7)$$

Integrating, we have

$$\psi(t) = \int_{-\infty}^{t} \phi(\tau) \, d\tau = \frac{k}{2\pi f_m} \cos 2\pi f_m t. \qquad (17.4\text{--}8)$$

(The improper integral is handled as in earlier chapters; $\psi(t)$ is simply the response to a sinusoid of an LTI system with $h(t) = u(t)$.) Comparing (17.4–8) with (17.4–2), sinusoidal FM and PM are indistinguishable, and we can identify the modulation index m as

$$|m| = \frac{k}{2\pi f_m}. \qquad (17.4\text{--}9)$$

From (17.4–5) the maximum value of $\phi(t)$—that is, k—is the maximum *devia-tion* of the instantaneous (radian) frequency from $2\pi f_c$. Thus for FM, $|m|$ is usually called the *deviation ratio*, the ratio of the maximum deviation of the instantaneous frequency to the modulating frequency. (If the modulation is not a unit-amplitude sinusoid, the deviation ratio is usually defined as the ratio of the maximum value of $\phi(t)$ to the maximum frequency in the modulation signal.) Loosely, it is obvious that—at least if f_m is very small—the radian frequency of $x(t)$ varies from $2\pi f_c - k$ to $2\pi f_c + k$; that is, k would seem to be a measure of the bandwidth of the FM signal. Continuing such a line of reasoning, early workers on radio communication systems concluded that if k were very small—much less than $2\pi f_m$, for example—the bandwidth occupied by such a *narrowband frequency modulation* (NBFM) signal would be much *less* than that required to transmit the same modulation by AM. But "instantaneous frequency" is a treacherous concept, and this argument is totally fallacious: We have already seen that $k \ll 2\pi f_m$ implies $|m| \ll 1$, and the NBFM signal is thus identical with the small-modulation-index PM signal considered earlier, which has almost precisely the *same* bandwidth requirements as an AM signal with the same modulation. The use of FM to reduce the required transmission bandwidth is thus doomed to failure.

This fact led some early analysts—J. R. Carson in particular[*] —to conclude that frequency modulation was worthless, or at least had no advantages over amplitude modulation. Armstrong,[†] however, showed in a dramatic demonstration in 1936 that *wideband*[‡] frequency modulation ($|m| \gg 1$), while not reducing the bandwidth required for transmission (in fact, the bandwidth is substantially increased), nevertheless offered some surprising noise reduction features over conventional amplitude modulation. We shall suggest how this comes about in Chapter 20; for the moment we shall simply study some of the spectral characteristics of FM signals when $|m| \gg 1$.

---

[*]J. R. Carson, *Proc. IRE, 10* (Feb. 1922): 47.

[†]E. H. Armstrong, *Proc. IRE, 24* (May 1936).

[‡]Note that "wideband" is being used here in a different sense than as defined at the beginning of this chapter; a wideband FM signal is usually still a narrowband signal in the sense that the bandwidth occupied is much less than the carrier frequency.

The main result is that the bandwidth required to transmit a wideband FM (WBFM) signal is on the order of $2k$ rad/sec when $|m| \gg 1$. It is possible although difficult to show this for a general modulating signal, but we shall settle for a simple example. The simplest example, however, is not sinusoidal WBFM described by

$$x(t) = A\cos\left[2\pi f_c t + \frac{k}{2\pi f_m}\cos 2\pi f_r t\right] \qquad (17.4\text{--}10)$$

(although the spectrum of such a signal can in fact be found in terms of Bessel functions[*]) but rather square-wave modulation as shown in Figure 17.4–3.

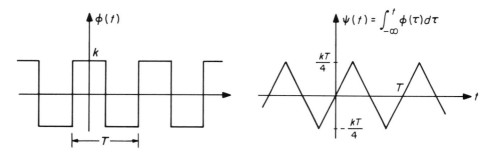

Figure 17.4–3. Modulating waveform for WBFM spectrum analysis.

For simplicity we shall assume that the carrier frequency f_c is an integer multiple of $2/T$. From (17.4–6) we may write for square-wave FM

$$x(t) = \begin{cases} A\cos 2\pi\left(f_c + \dfrac{k}{2\pi}\right)t, & -\dfrac{T}{4} < t < \dfrac{T}{4} \\[3mm] A\cos 2\pi\left(f_c - \dfrac{k}{2\pi}\right)\left(t - \dfrac{T}{2}\right), & \dfrac{T}{4} < t < \dfrac{3T}{4} \end{cases} \qquad (17.4\text{--}11)$$

continuing periodically in alternate intervals. That is, the frequency shifts suddenly back and forth between two values. Signals similar to this (although, of course, not periodic) are used in the *frequency-shift-key* (FSK) system of radio-telegraphy, in which the two frequencies represent *mark* and *space*, 0 and 1, etc. In the periodic case, $x(t)$ may be considered to be the superposition of two square-wave-modulated sinusoids of different frequencies, as shown in Figure 17.4–4. The spectrum of each component waveform is simply that derived earlier for a square wave shifted to the corresponding carrier frequency, as shown in Figure 17.4–5.

---

[*]See, for example, M. Schwartz, *Information Transmission, Modulation and Noise*, 3rd ed. (New York, NY: McGraw-Hill, 1980).

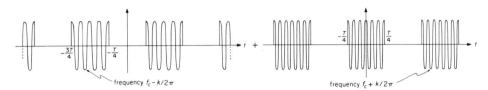

Figure 17.4–4. Analysis of square-wave FM into two components.

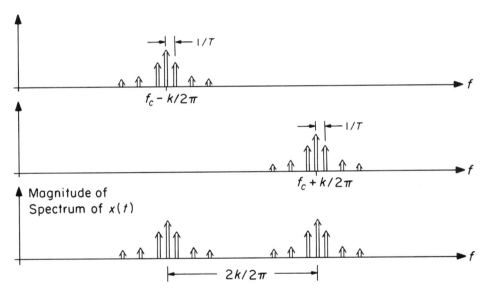

Figure 17.4–5. Spectrum of the components of Figure 17.4–4 and their superposition.

It is obvious that for k large compared with the modulation bandwidth (that is, for $k \gg 2\pi/T$, which implies a large deviation ratio), the bandwidth of $x(t)$ is roughly $2k/2\pi$ Hz, whereas for a smaller deviation ratio (k comparable with or smaller than $2\pi/T$) the overall bandwidth is essentially the bandwidth of the modulating signal. (This is sometimes called *Carson's Rule.*) In commercial broadcast FM, the value of $k/2\pi$, the frequency deviation corresponding to the maximum modulating signal, is 75 kHz. If the maximum frequency in the audio signal is taken as 15 kHz, this gives a deviation ratio of 5. The overall bandwidth of such signals is thus somewhat more than 150 kHz; indeed, adjacent FM channels are assigned 200 kHz apart.

Narrowband FM signals can be readily generated by the scheme illustrated in Figure 17.4–2, but the generation of WBFM requires other means. One method (due to Armstrong) starts by generating a narrowband FM signal as in Figure 17.4–2, then passing it through a succession of frequency multipliers and mixers. Analytically this scheme operates as follows. The output of the NBFM modulator can be written as $\cos[2\pi f_c t + \epsilon\theta(t)]$, where ϵ is small. If this signal is passed through a square-law device, the output is

$$\cos^2[2\pi f_c t + \epsilon\theta(t)] = \frac{1}{2} + \frac{1}{2}\cos[2\pi(2f_c)t + 2\epsilon\theta(t)]. \qquad (17.4-12)$$

Now if the band around $2f_c$ is filtered out and mixed in a balanced modulator with $\cos 2\pi f_c t$, the output of the mixer in the band around the original carrier, f_c, will be proportional to $\cos[2\pi f_c t + 2\epsilon\theta(t)]$, so that the deviation ratio will have been multiplied by 2. This procedure can be extended or repeated to achieve whatever deviation ratio is necessary.

A more direct method of generating WBFM might use electronic means to vary the capacitance in the tuned circuit of an oscillator in accordance with the modulating signal. In general, an oscillator whose frequency is determined by an input voltage is called a *voltage-controlled oscillator* (VCO). Several types of VCOs are now available in integrated-circuit form, and they have many uses in addition to generating WBFM; an example was described in Problem 1.14.

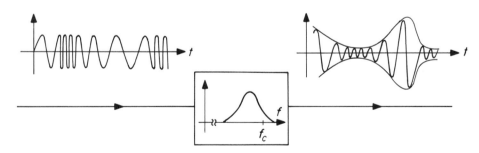

Figure 17.4–6. Tuned-circuit response to an FM signal.

The simplest demodulator, or *frequency discriminator*, for FM signals is a circuit tuned below (say) the carrier frequency (see Figure 17.4–6). Variations in the instantaneous frequency of the input carrier modulation are translated into variations in the output amplitude of the tuned circuit. These amplitude variations can be detected with an ordinary envelope detector. (Indeed NBFM can be received on a slightly detuned AM receiver.) The limited linear range of such a discriminator can be extended by using a pair of circuits, one tuned above and the other below the carrier frequency. The outputs are separately detected and the detected signals subtracted to give the overall characteristic shown in Figure 17.4–7. These discriminators will, of course, also have a change

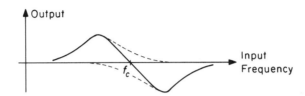

Figure 17.4–7. Discrimination response derived from a pair of tuned circuits.

in their output corresponding to amplitude changes, if any, in the input. In practical systems, amplitude variations are present on the FM signal as a result of noise, fading, interference, and the like. Hence discriminators of the type shown in Figure 17.4–6 and Figure 17.4–7 must be preceded by a *limiter*, which is (ideally) a non-linear resistive device having the characteristics shown in Figure 17.4–8. When followed by a tuned amplifier, the limiter approximately removes the amplitude variations in the envelope of a narrowband signal while preserving the phase variations.

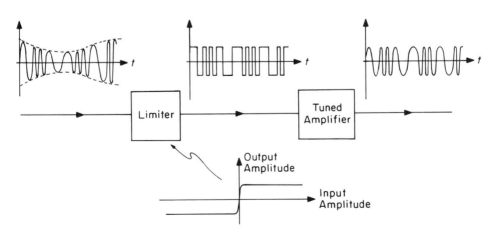

Figure 17.4–8. Combined operation of a limiter and a tuned amplifier.

The overall block diagram for a typical FM receiver is shown in Figure 17.4–9. Modern high-performance FM receivers often employ a phase-locked loop detector rather than a discriminator (see Problem 17.12). The construction of FM receivers has been much simplified in recent years by the availability of many of the required operations combined in a single integrated circuit.

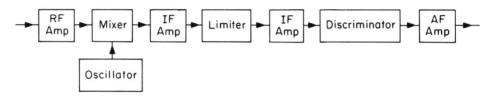

Figure 17.4–9. FM receiver.

17.5 Summary

Shifting the frequency content of a signal is useful for many reasons and has led to the development of a variety of modulators, mixers, detectors, and other devices. The modulation schemes discussed in this chapter—AM, PM, and FM— were invented 50 years ago or more but continue to occupy a central position in most communication systems. However, increasingly these older ideas are being employed in combination with pulse or "digital" techniques that have a variety of theoretical and practical advantages; we briefly mentioned some of these in Chapter 14 and shall explore them further in Chapter 20. The next chapter will complete our introduction to the tools of Fourier analysis by focusing in some detail on applications of the discrete-time Fourier transform developed as the dual of the Fourier series formulas in Chapter 14.

EXERCISES FOR CHAPTER 17

Exercise 17.1

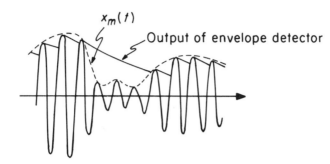

As shown in the figure, an envelope detector of the type illustrated in Figure 17.1–4 may fail to function properly if the modulating envelope changes too rapidly. Argue that a necessary condition to avoid this form of distortion is that the slope of the envelope $x_m(t)$ be bounded by

$$\frac{1}{x_m(t)}\left|\frac{dx_m(t)}{dt}\right| < \frac{1}{RC}.$$

For a given $x_m(t)$, this places an upper bound on the RC product. What, if anything, sets a lower bound?

Exercise 17.2

The peak factor (pf) of a waveform was defined in Section 17.1 as the ratio of the peak or maximum value to the rms value.

a) Show that $pf = 1$ for a symmetrical square wave with zero average value. (Since the rms value of a waveform cannot exceed the peak value, pf cannot be less than 1.)

b) Show that $pf = \sqrt{2}$ for a sinusoid.

c) Show that pf may be as large as $\sqrt{2N}$ for a waveform that is the sum of N equal-amplitude sinusoids of different frequencies.

d) If $x_m(t)$ is a periodic signal with zero average value and peak factor pf, show that the efficiency of the amplitude-modulated signal

$$x(t) = (A + x_m(t))\cos 2\pi f_c t$$

cannot exceed $1/[1 + (pf)^2]$ if overmodulation is not permitted.

Exercise 17.3

Find directly the impulse response $h(t)$ of the ideal bandpass filter on the left below and show that it can be written in the form $h(t) = h_m(t) \cos \pi(f_1 + f_2)t$, where $h_m(t)$ is the impulse response of the ideal lowpass filter on the right. Sketch $h(t)$ for the case $f_1 = 4$ kHz, $f_2 = 6$ kHz.

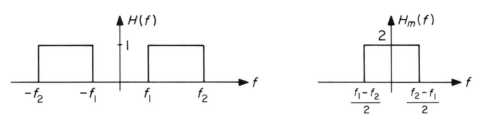

Exercise 17.4

Quadrature modulation is a scheme for sending two independent signals simultaneously over the same frequency channel. A simple block diagram for this process is shown below.

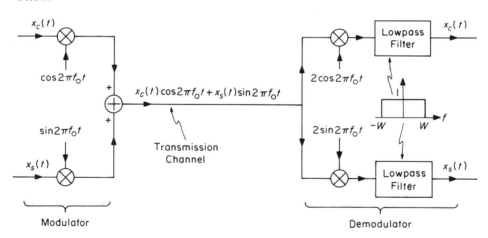

Let $x_c(t)$ and $x_s(t)$ be arbitrary signals with spectra restricted to the band $|f| < W \ll f_0$. Sketch "typical" spectra for the signals in the transmission channel and at the inputs to the lowpass filters in the demodulator and thus demonstrate that:

a) The bandwidth occupied by the superimposed signals in the transmission channel is no greater than it would be for either one alone.

b) Nevertheless, the original signals can be independently recovered; indeed, $\hat{x}_c(t) = x_c(t)$, $\hat{x}_s(t) = x_s(t)$.

PROBLEMS FOR CHAPTER 17

Problem 17.1

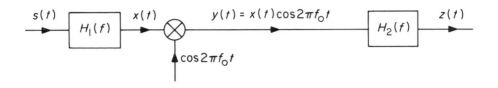

In the system above assume that

 i) $s(t)$ is a periodic chain of unit-area impulses spaced T seconds apart;

 ii) the impulse response $h_1(t)$ corresponding to $H_1(f)$ is zero everywhere except in the interval $0 < t < T$;

 iii) $H_2(f)$ is zero for $|f| \geq 1/2T$.

a) It is easy to see that $x(t)$ is periodic with period T and hence can be expanded in the form

$$x(t) = \sum_{n=-\infty}^{\infty} X[n] e^{j2\pi nt/T}.$$

Find $\sum_{n=-\infty}^{\infty} |X[n]|^2$ in terms of an integral involving $H_1(f)$. Do not leave your answer in a form containing an infinite sum.

b) It is easy to see that $y(t)$ has a line spectrum for any choice of f_0. For $f_0 = \dfrac{9}{4T}$, find the amplitude and location of all lines with frequencies in the range

$$-\frac{1}{2T} \leq f \leq \frac{1}{2T}$$

in terms of values of $H_1(f)$ at specific fequencies.

c) Is $y(t)$ necessarily strictly periodic for an arbitrary choice of f_0? Why?

d) Is $z(t)$ necessarily strictly periodic for an arbitrary choice of f_0? Why?

Problem 17.2

In the system shown below, $x(t)$ is a signal whose spectrum is zero for frequencies greater than $1/2T$ Hz and $s(t)$ is a periodic impulse function with period $2T$. The filter has a frequency response that is real and bandlimited. Suggest a scheme for recovering $x(t)$ from $y(t)$ and prove that it works. (All constants should be evaluated.)

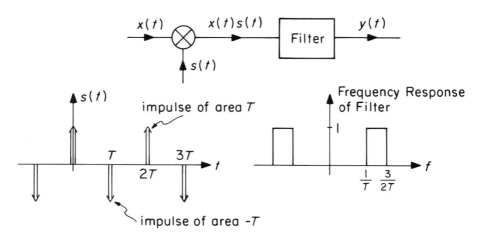

Problem 17.3

The design of high-gain d-c amplifiers (that is, amplifiers whose pass band includes $f = 0$) frequently presents difficulties because small slow changes (due to aging, temperature fluctuations, supply voltage changes, etc.) in the quiescent operating points of the active elements can produce responses indistinguishable from those due to small desired signals. One way to circumvent these difficulties is to use a *chopper* to modulate the signal onto a carrier so that an a-c coupled amplifier can be used instead. Another chopper is then used as a synchronous detector to restore the signal to its original frequency range. The scheme is illustrated on the next page. Assume that the spectrum of the input signal $x_1(t)$ is restricted to $|f| < f_1$. The periodic time functions $v(t)$ and $v(t - \tau)$ describe the chopping action. $H_0(f)$ is an ideal lowpass filter with gain 1 over the pass band $|f| < f_1$; and $H_1(f)$ is a bandpass high-gain amplifier with characteristics as shown.

a) Find the first few terms of the Fourier series for $v(t)$ and $v(t - \tau)$.

b) Sketch the spectra of $y_1(t)$ and $y_2(t)$ and label them accurately in terms of an assumed spectral shape for $X_1(f)$.

c) Find an expression for $x_2(t)$ in terms of $x_1(t)$ and the system parameters.

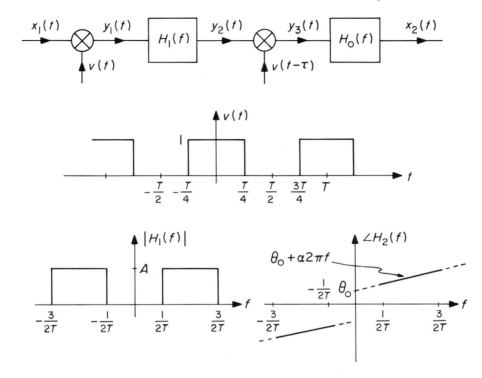

Problem 17.4

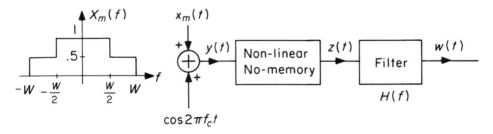

In the modulator above (which is a variant of the scheme discussed in Example 17.1–2), the modulating signal $x_m(t)$ and a sinusoid at the intended carrier frequency are added to produce $y(t) = x_m(t) + \cos 2\pi f_c t$, which is then passed through a non-linear device described by $z(t) = 5y(t) + y^2(t)$.

a) Assume that $x_m(t)$ is a real, even function having the spectrum shown to the right above. Make a carefully labelled sketch of $Z(f)$ over the range $-3f_c < f < 3f_c$.

b) Describe the frequency response $H(f)$ of the filter such that $w(t)$ has the form of $x_m(t)$ double-side-band amplitude-modulated (with carrier) on a carrier at f_c.

Problem 17.5

This problem describes the performance of a *balanced modulator*, which is another variant of the circuit described in Example 17.1–2 and one of the most common practical schemes for multiplying a sinusoidal carrier by a low-frequency signal.

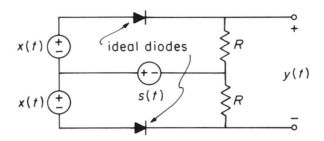

a) Show that at any instant t, if $s(t) > |x(t)|$, then $y(t) = 2x(t)$; on the other hand, if $s(t) < -|x(t)|$, then $y(t) = 0$.

b) In practice, $x(t)$ is a low-frequency signal and $s(t)$ is a sinusoid $A \cos 2\pi f_0 t$, where A is very large compared with the peak value of $x(t)$. Consequently almost no time is spent with $|s(t)| < |x(t)|$, and so, approximately,

$$y(t) = \begin{cases} 2x(t), & \cos 2\pi f_0 t > 0 \\ 0, & \cos 2\pi f_0 t < 0. \end{cases}$$

Make a sketch showing how the spectrum of $y(t)$ is related to the spectrum of $x(t)$.

c) To obtain a modulated carrier, we pass the output of the preceding system through a bandpass filter that lets through only the frequencies near $\pm f_0$. If the filter has unit gain and no phase shift, how is the output related to $x(t)$?

Problem 17.6

One of the disadvantages of a compander is that compressing a bandlimited signal introduces energy at frequencies outside the band; that is, the compressed signal has a smaller peak factor but requires a larger bandwidth to transmit without distortion.* A much simpler procedure is to sample the output of the compressor at the Nyquist rate before passing it through the ideal bandlimiting filter. (Alternately, the compressor can be designed to operate on the amplitudes of the samples rather than on the original continuous signal.) In this case the recovery process is straightforward.

Draw a block diagram (with each block described in detail) that starts with a bandlimited waveform, compresses it, samples it, and passes the result through an ideal lowpass filter to yield a compressed signal occupying the original bandwidth. Devise and discuss in detail a scheme for recovering the original signal.

---

*Actually it can be shown (see the theorem of Buerling mentioned in H. J. Landau, *Bell Sys. Tech. J.*, 35 (1960): 351–356) that restricting the band of the compressed signal by passing it through an ideal lowpass filter does not destroy any information, in that the original signal can still in principle be recovered exactly from the output. The procedure is, however, complicated.

Problem 17.7

The superheterodyne receiver below employs synchronous detection. The antenna signal contains the desired amplitude-modulated signal $x_m(t)\cos(2\pi f_c t + \theta)$ as well as many interfering signals at other frequencies. Assume that $f_c = 10^6$ Hz and that the spectrum of $x_m(t)$ is zero outside the band $|f| < 5\times10^3$ Hz. The frequency responses of the various amplifiers are shown; assume for simplicity that all the frequency responses are real, that is, the phase angles of $H_{RF}(f)$, $H_{IF}(f)$, and $H_{AF}(f)$ are all zero.

a) Explain how this receiver is intended to function, using carefully drawn spectra where appropriate. Model the mixer and the synchronous detector as ideal multipliers; assume that $f_m > f_c$ and that $\theta = \phi = \psi = 0$.

b) In order to prevent interfering signals at certain "image" frequencies (different from those in the desired signal but still passed by the RF amplifier) from also being shifted to the IF amplifier band by the mixer, we must not pick a local oscillator frequency in a certain range. Explain, and determine the excluded range. (Consider the possibility of using any $f_m > 0$, not just the values $f_m > f_c$.)

c) If the transmitter oscillator at frequency f_c and the two receiver oscillators are independent, then the phase angles are unlikely all to be (or stay) zero. What relationship must be maintained among these phase angles if the system is to function as desired? Suggest a scheme employing a phase-locked loop for ensuring the desired relationship. Are any special properties or conditions on the transmitted waveform necessary for your scheme to function successfully?

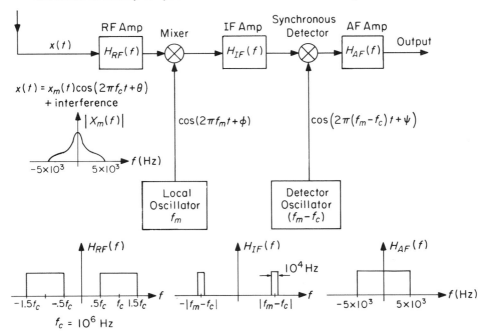

Problem 17.8

Another SSB generation scheme, due to D. K. Weaver, Jr.,* is illustrated below. Assume that $x(t)$ has the (real) Fourier transform shown. Sketch the Fourier transforms of $x_1(t)$, $x_2(t)$, $x_3(t)$, $x_4(t)$, and $y(t)$, thus demonstrating that $y(t)$ is in fact $x(t)$ single-side-band-modulated on the carrier f_c as desired. (This system depends for its success on a careful balance of the gains between the upper and lower paths—a condition hard to meet with analog circuits but easy to achieve with digital signal processing of the kind to be discussed in Chapter 18.[†])

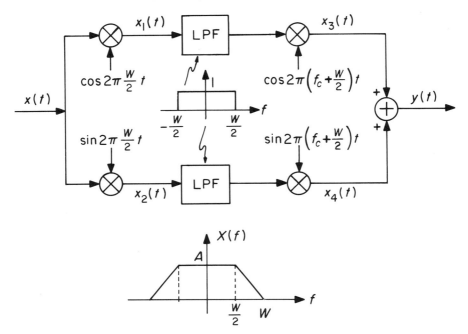

Problem 17.9

In the following assume that $H(f) \approx 0$ for all $|f| > W$ and that $W \ll f_0$.

a) One simple form of *lowpass-bandpass transformation* replaces f by $(f^2 - f_0^2)/2f$ wherever it appears. Argue that, under the assumed conditions,

$$H\left(\frac{f^2 - f_0^2}{2f}\right) \approx H(f - f_0) + H(f + f_0)$$

and sketch the magnitude and phase of a simple example of $H(f)$ and the corresponding $H\big((f^2 - f_0^2)/2f\big)$.

---

*D. K. Weaver, Jr., *Proc. IRE, 44* (Dec. 1956): 1703–1705.

[†]See A. V. Oppenheim (ed.), *Applications of Digital Signal Processing* (New York, NY: Prentice-Hall, 1978) pp. 8–11.

b) Evaluate approximately the corresponding time function

$$\int_{-\infty}^{\infty} H\left(\frac{f^2 - f_0^2}{2f}\right) e^{j2\pi ft}\, df$$

in terms of $h(t)$.

c) Suppose that $H(f)$ is the transfer function of an RLC network. Argue that a network with transfer function $H\big((f^2 - f_0^2)/2f\big)$ can be obtained from that for $H(f)$ by replacing each L and C by the tuned circuits shown below, and find the values of C_1, L_1, C_2, and L_2 in terms of L, C, and f_0.

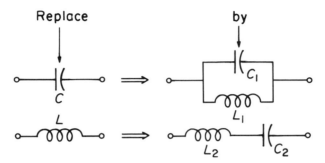

Problem 17.10

Let $x(t)$ and $h(t)$ be respectively a waveform and impulse response whose transforms are both zero except in narrow bands near $f = \pm f_0$. Define complex envelopes $\xi(t)$ and $\eta(t)$ such that

$$x(t) = \Re[\xi(t)e^{j2\pi f_0 t}]$$

$$h(t) = \Re[\eta(t)e^{j2\pi f_0 t}].$$

a) Let $y(t) = x(t) * h(t)$. Show that $y(t)$ can be written in the form

$$y(t) = \Re[\mu(t)e^{j2\pi f_0 t}]$$

with

$$2\mu(t) = \xi(t) * \eta(t)$$

or

$$2\mathcal{M}(f) = \Xi(f)\mathcal{H}(f)$$

where $\mathcal{M}(f)$, $\Xi(f)$, and $\mathcal{H}(f)$ are the Fourier transforms of $\mu(t)$, $\xi(t)$, and $\eta(t)$ respectively. In other words, the effect of a bandpass filter on a bandpass waveform can be analyzed in terms of an equivalent lowpass problem (involving, to be sure, complex time functions in general). CAUTION: Note that if u and v are complex numbers, then $\Re[u]\Re[v] \neq \Re[uv]$ in general.

b) The above result can be approximately valid even if both $x(t)$ and $h(t)$ are not strictly bandlimited as assumed. As an example, show that the response of an ideal narrow bandpass filter to a step-sinusoid at the band-center frequency is approximately a sinusoid modulated by the step response of an ideal lowpass filter.

Problem 17.11

One source of radio "static" is lightning. The physics of the generation and propagation of lightning effects is complex; we shall assume that the net effect is to induce at the receiver input a short pulse waveform shaped as shown to the left below. We shall model the receiver as in the block diagram on the right.

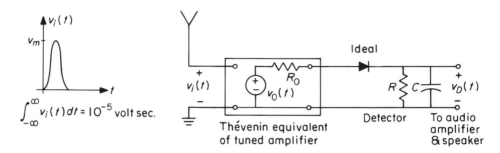

$$\int_{-\infty}^{\infty} v_i(t)\,dt = 10^{-5} \text{ volt sec.}$$

a) Assume that the frequency response of the tuned amplifier on open circuit,

$$H(j\omega) = V_0(j\omega)/V_i(j\omega)$$

is as shown below. Determine an analytic expression for $H(s)$ if it is known to consist effectively of a single pole-pair and zero as shown.

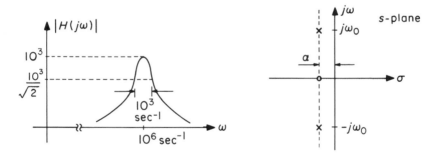

b) If $v_i(t)$ can be approximated as an impulse,

$$v_i(t) \approx 10^{-5}\delta(t)$$

find and sketch $v_0(t)$. Label your sketch.

c) Define the duration of the lightning pulse as

$$\Delta T = \frac{\int_0^\infty v_i(t)\,dt}{v_m}.$$

What is an approximate upper bound on ΔT if the approximation of (b) is to be a reasonable one for this receiver?

d) If the detector functioned as an ideal envelope detector (in the sense described in Section 17.3) for the lightning response $v_0(t)$, sketch $v_D(t)$. Label your sketch.

e) If $R_0 = 100\Omega$, $R = 10^5\Omega$, select a value for C so that the detector will approach the ideal assumed in (d).

Problem 17.12

The *phase-lock loop* is an extremely useful system for a number of purposes, including detection of FM, PM, and SC-AM signals. A simple form of the system is shown below.

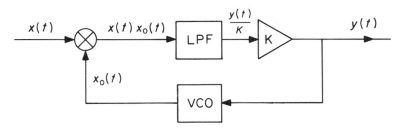

The heart of the device is the *voltage-controlled oscillator* (VCO) whose output (ideally) is a sinusoid,

$$x_0(t) = 2\sin(2\pi f_c t + \psi(t))$$

whose rate of change of phase is proportional to its input:

$$\frac{d\psi(t)}{dt} = \alpha y(t).$$

That is, the VCO is an ideal frequency modulator. Suppose that

$$x(t) = A\cos(2\pi f_c t + \theta(t))$$

and suppose further that $\psi(t)$ and $\theta(t)$ are slowly varying compared with $2\pi f_c t$. Assume that the lowpass filter (LPF) is ideal with unity gain in the pass band and a bandwidth wide enough to pass the spectral components of $x(t)x_0(t)$ around $f = 0$ without distortion while rejecting completely the spectral components around $|f| = 2f_c$.

a) Show that $\psi(t)$ satisfies the non-linear differential equation

$$\frac{d\psi(t)}{dt} = \alpha K A \sin(\psi(t) - \theta(t)).$$

b) Sketch $d\psi(t)/dt$ vs. $\psi(t)$ (this is called a *phase-plane* plot) for $\theta(t) = \theta_0 = $ constant. Argue that $\psi(t) = \theta_0$ is a stable solution of this differential equation if the sign of $\alpha K A$ is properly chosen. (HINT: Consider a small perturbation in $\psi(t)$ to, say, $\theta_0 + \Delta\theta$; if the sign of $d\psi(t)/dt$ is appropriate, then the perturbation should be reduced in a stable system.)

c) The system is said to be in *lock* if the *error* $\psi(t) - \theta(t)$ is small. Under these conditions,

$$\sin(\psi(t) - \theta(t)) \approx \psi(t) - \theta(t)$$

and the dynamic equation characterizing the system becomes linear. Show that, under these conditions and assuming stability, the system can be described by a frequency response of the form

$$\frac{Y(f)}{\Theta(f)} = K_0 \frac{j2\pi f}{j2\pi f + \beta}.$$

Discuss the conditions on $d\theta(t)/dt$ (or its transform) under which $y(t) \sim d\theta(t)/dt$, so that the phase-lock loop behaves as an ideal FM detector.

d) What effect will variations in the amplitude A of the input signal have on the results of (c)?

e) What effect will too large a magnitude of $|d\theta(t)/dt|$ have on the operation of the circuit? (HINT: Consider the situations under which the non-linear differential equation of (a) can have a solution that will approach the steady-state solution $\psi(t) = \theta(t)$ if $d\theta(t)/dt = \Delta\omega u(t)$.)

f) Voltage-controlled oscillators and the phase-lock loop principle have many applications. For example, the system shown below (due to J. P. Costas) is an effective way of detecting SC-AM signals. To analyze its performance, assume that the output of the VCO is

$$x_c(t) = 2\cos(2\pi f_c t + \psi(t))$$

so that

$$x_s(t) = 2\sin(2\pi f_c t + \psi(t))$$

if the box labelled "90°" is described, at least for $|f|$ near f_c, by the frequency response

$$H(f) = \begin{cases} -j, & f > 0 \\ j, & f < 0. \end{cases}$$

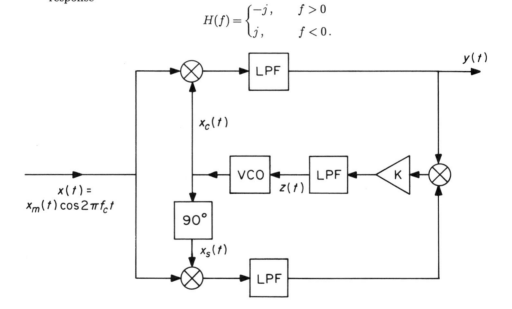

(The output of this box is thus the Hilbert transform of its input—see Problem 15.3.) Assume further that

$$\frac{d\psi(t)}{dt} = \alpha z(t)$$

and that the lowpass filters in the top and bottom branches pass the spectral band near $f = 0$ without distortion while rejecting frequencies near $|f| = f_c$. The lowpass filter in the middle branch has an even narrower pass band, letting through only a very small band of frequencies near $f = 0$.

Analyze the behavior of this system, showing how the output of the VCO locks in phase on the phase of the input, so that $y(t) = x_m(t)$ if the sign of α is properly chosen. Might it be possible for the VCO phase to lock in such a way that $y(t) = -x_m(t)$?

Problem 17.13

The stereo-multiplex system for FM broadcasting works as follows. A stereo pair of audio signals, $v_L(t)$ and $v_R(t)$ (left and right), are combined to form sum and difference signals:

$$v_S(t) = v_L(t) + v_R(t)$$

$$v_D(t) = v_L(t) - v_R(t).$$

Assume that both $v_L(t)$ and $v_R(t)$ are bandlimited to $|f| < W \approx 15$ kHz. The difference signal $v_D(t)$ is suppressed-carrier-modulated on a sinusoid of frequency $2f_0 = 38$ kHz and added to the sum signal, $v_S(t)$, along with a sinusoid of frequency $f_0 > W$. Thus the entire signal to be transmitted is

$$v_T(t) = v_S(t) + \cos 2\pi f_0 t + 2v_D(t)\cos 4\pi f_0 t.$$

This total signal is transmitted as frequency modulation and hence is recovered at the output of the discriminator in the receiver. (The frequency $f_0 = 19$ kHz is beyond the audio pass band of old monaural FM receivers, which can be modelled as ideal lowpass filters with

$$H(f) = \begin{cases} 1, & |f| < W \\ 0, & \text{elsewhere.} \end{cases}$$

Thus listeners using old receivers hear just $v_S(t)$. One of the requirements on the design of the stereo-multiplex system was that it be "compatible" with existing receivers.)

a) Assume "typical" spectra for $v_S(t)$ and $v_D(t)$ and sketch the spectrum of $v_T(t)$.

b) Discuss the bandwidth required for the transmitted WBFM waveform assuming a maximum frequency deviation of 75 kHz. What kind of price is paid to transmit a stereo pair over a system originally intended for a single monaural waveform?

c) One method of retrieving the stereo signals is as follows: First, $\cos 2\pi f_0 t$ is found from $v_T(t)$ (e.g., using the phase-locked loop of Problem 17.12; most of the electronics required in the "decoder" are now available on a single IC chip). Then $\cos^2 2\pi f_0 t$ and $\sin^2 2\pi f_0 t$ are formed, separately multiplied by $v_T(t)$, and lowpass-filtered as shown on the next page. Find $v_a(t)$ and $v_b(t)$ in terms of $v_L(t)$ and $v_R(t)$.

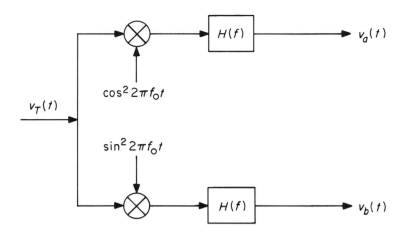

Problem 17.14

For cryptographic reasons, it is proposed to break up the speech spectrum into several distinct frequency bands and to transmit these through separate spectral channels. A scheme for accomplishing this is shown in the figure below.

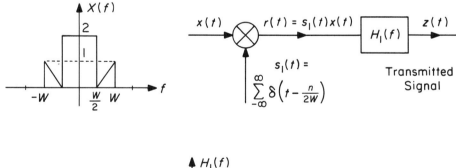

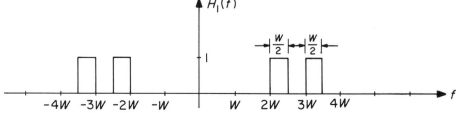

a) Assuming that the spectrum of $x(t)$ has the shape shown, sketch the spectrum of $r(t)$, labelling carefully all amplitudes and frequencies.

b) Sketch the spectrum of $z(t)$, labelling carefully all amplitudes and frequencies.

c) It is proposed to recover $x(t)$ from $z(t)$ by the process shown below. Either show that this scheme will work for appropriate choices of f_1, f_2, and A (give appropriate values), or explain why it will not work and give details of a process that will.

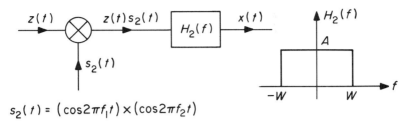

$$s_2(t) = (\cos 2\pi f_1 t) \times (\cos 2\pi f_2 t)$$

Problem 17.15

This problem is intended to illustrate the difference between *group delay* and *phase delay*. Assume that $\xi(t)$ is a lowpass bandlimited waveform; that is, $\Xi(f)$, the Fourier transform of $\xi(t)$, is zero for all f such that $|f| > f_1$. Let $x(t) = \xi(t) \cos 2\pi f_0 t$ and assume $f_1 \ll f_0$.

a) Suppose that $x(t)$ is the input to an LTI system with $H_1(f)$ as shown below. Find a formula (not involving any integrals) for the output $y(t)$ in terms of $\xi(t)$. (Your result can be interpreted as saying that a system such as $H_1(f)$ delays the carrier, $\cos 2\pi f_0 t$, but does not affect the modulating envelope.)

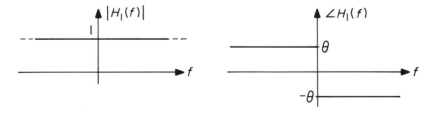

b) Suppose that $x(t)$ is the input to an LTI system with $H(f)$ as shown on the next page. Argue that this system can (at least for waveforms such as $x(t)$) be considered as the cascade of an ideal delay with a system such as $H_1(f)$. Thus conclude that the output $y(t)$ can be written in the form

$$y(t) = A\xi(t - T_1) \cos 2\pi f_0(t - T_2)$$

where

$$T_1 = k_1 \left.\frac{d[\angle H(f)]}{df}\right|_{f=f_0} = \text{group delay} = \text{envelope delay}$$

$$T_2 = k_2[\angle H(f_0)] = \text{phase delay} = \text{carrier delay}$$

and find A, k_1, and k_2.

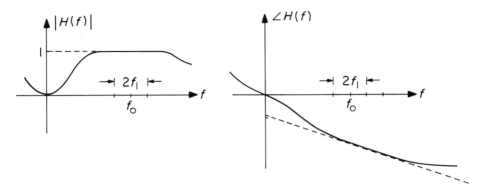

c) As a particular example, consider the circuit below, where $x(t) = \xi(t)\cos 2\pi f_0 t$ is a voltage source and $y(t)$ is the voltage across the capacitor. Let $\sqrt{LC} = 1/2\pi f_0$ and assume that $2\pi f_1 \ll \alpha = R/2L \ll 2\pi f_0$, that is, $x(t)$ is centered on the resonant frequency and is sufficiently narrowband that the gain of the circuit is nearly constant over the band of frequencies contained in $x(t)$. Then $y(t)$ can be approximately represented by an equation such as that in part (b). Find A, T_1, and T_2.

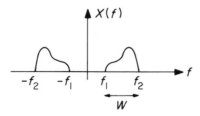

Problem 17.16

Consider a strictly bandlimited bandpass signal whose spectrum is zero except for $f_1 < |f| < f_2 = f_1 + W$ as shown:

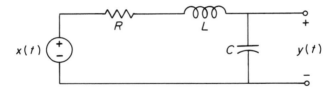

Since this signal is a special case of a lowpass bandlimited signal, with $X(f) = 0$ for $|f| > f_2$, it follows at once from the ordinary Sampling Theorem that $x(t)$ is completely characterized by $2f_2$ periodic samples per second. Usually, however, a much smaller sampling rate will suffice, as this problem illustrates.

a) Let $f_2 = 3W$. By sketching the spectrum of the sampled signal show that periodic samples at a rate $2f_2/3$ preserve all of the information in $x(t)$. Show a block diagram of a system that will recover $x(t)$ from the samples.

b) Generalizing on the results of (a), show that periodic samples at the rate $2f_2/m$, where m is the largest integer such that $m \leq f_2/W$, preserve all of the information in $x(t)$.

c) Show with an example that not all periodic sampling rates greater than the minimum permit $x(t)$ to be recovered from the samples.

d) The formula in (b) gives a periodic sampling rate greater than or equal to $2W$, with equality whenever f_2/W is an integer. By considering samples of the complex envelope of $x(t)$, show that $2W$ numbers per second (not necessarily samples of $x(t)$) are always sufficient to define any bandpass waveform.

Problem 17.17

Lord Rayleigh's classic treatise *The Theory of Sound* (first published in 1877) contains the following paragraph and unlabelled figure.

"The theory of intermittent vibrations is well illustrated by electrically driven forks. A fork interrupter of frequency 128 gave a periodic current, by the passage of which through an electro-magnet a second fork of like pitch could be excited. The action of this current on the second fork could be rendered intermittent by short-circuiting the electro-magnet. This was effected by another interrupter of frequency 4, worked by an *independent* current from a Smee cell. To excite the main current a Grove cell was employed. When the contact of the second interrupter was permanently broken, so that the main current passed continuously through the electro-magnet, the fork was, of course, most powerfully affected when tuned to 128. Scarcely any response was observable when the pitch was changed to 124 or 132. But if the second interrupter were allowed to operate, so as to render the periodic current through the electro-magnet intermittent, then the fork would respond powerfully when tuned to 124 or 132 as well as when tuned to 128, but not when tuned to intermediate pitches, such as 126 or 130."

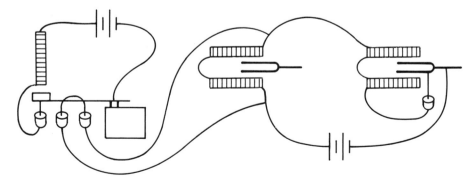

In spite of the fact that you have probably never heard of a Smee or Grove cell, you should have little difficulty in explaining Rayleigh's observations. Describe the operation of the circuit in a few sentences, including perhaps a few equations and sketches of waveforms. Are 124, 128, and 132 Hz the only frequencies for which you would expect the fork to respond? Upon what parameters does the relative strength of these responses depend? Accompany your answer with a copy of Lord Rayleigh's figure in which the various elements—such as batteries, electro-magnets, mercury contacts, and vibrating elements (including their frequency or range of tuning)—are properly labelled.

Problem 17.18

A time-varying capacitor described by the constitutive relation

$$i(t) = \frac{d}{dt}[C(t)v(t)]$$

is still a linear element, so that superposition applies, but—unlike a time-invariant capacitor—it is not usually an ideal energy-storage element. Instead, a time-varying capacitor may supply power to or absorb power from the circuit. If power is absorbed, it is not in general dissipated but rather coupled or *transduced* into whatever form is producing the time variations of the capacitance, such as a mechanical system changing the spacing of the capacitor plates. Properly employed, a periodically varying capacitor can produce *parametric amplification*. A simple circuit illustrating this effect is shown below.

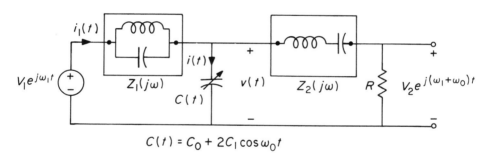

$$C(t) = C_0 + 2C_1 \cos \omega_0 t$$

Exploiting linearity, we shall represent the steady-state voltages and currents as complex exponentials; the actual sinusoidal waveforms can be found by taking real parts. (The sinusoidal variations of the capacitor must not, however, be represented as a complex exponential; circuit parameters are different from dynamic variables.) The circuits $Z_1(j\omega)$ and $Z_2(j\omega)$ are tuned to the output frequency $\omega_1 + \omega_0$. For simplicity we shall idealize these impedances to be either open or short circuits as follows:

$$Z_1(j\omega) = \begin{cases} \infty, & \omega = \omega_1 + \omega_0 \\ 0, & \text{otherwise} \end{cases} \qquad Z_2(j\omega) = \begin{cases} 0, & \omega = \omega_1 + \omega_0 \\ \infty, & \text{otherwise.} \end{cases}$$

a) Exploiting the properties of $Z_1(j\omega)$ and $Z_2(j\omega)$, argue that

$$v(t) = V_1 e^{j\omega_1 t} + V_2 e^{j(\omega_1 + \omega_0)t} .$$

Explain in particular why $v(t)$ does not contain terms at the frequencies $\omega_1 - \omega_0$ and $2\omega_0 + \omega_1$, which (as we shall see) are contained in the capacitor current $i(t)$.

b) Show that

$$i(t) = j\omega_1(C_0 V_1 + C_1 V_2)e^{j\omega_1 t}$$
$$+ j(\omega_1 + \omega_0)(C_0 V_2 + C_1 V_1)e^{j(\omega_1 + \omega_0)t}$$
$$+ j(\omega_1 - \omega_0)C_1 V_1 e^{j(\omega_1 - \omega_0)t}$$
$$+ j(2\omega_0 + \omega_1)C_1 V_2 e^{j(2\omega_0 + \omega_1)t}.$$

c) Argue that the component of $i(t)$ at the frequency $(\omega_1 + \omega_0)$ flows entirely through the load R. From this, solve for V_2 to obtain

$$V_2 = \frac{-j(\omega_1 + \omega_0)C_1 V_1 R}{1 + j(\omega_1 + \omega_0)C_0 R}.$$

d) The power supplied by the sinusoidal voltage source at frequency ω_1 depends only on the component of $i_1(t)$ at the frequency ω_1. Writing this component in complex form as $I_1 e^{j\omega_1 t}$, and recalling that the average input power is $P_{\mathrm{IN}} = \frac{1}{2}\Re e[V_1 I_1^*]$ and the average output power is $P_{\mathrm{OUT}} = \frac{1}{2}|V_2|^2/R$, show that

$$P_{\mathrm{OUT}} = \frac{\omega_1 + \omega_0}{\omega_1} P_{\mathrm{IN}}.$$

Taken together, (c) and (d) imply that V_2 is proportional to V_1; if V_1 varies slowly, V_2 will vary slowly in the same pattern. On the other hand, the output power is greater than the input power. The time-varying capacitor thus functions as an amplifier.

In practice, the effect of a time-varying capacitor would probably be achieved by using a large electrical voltage at frequency ω_0 (called the *pump* voltage) to vary the operating point of a non-linear capacitor. For generalizations of the parametric amplifier principle, see P. E. Penfield, *Frequency-Power Formulas* (Cambridge, MA: Technology Press, 1960) and the fundamental paper by J. M. Manley and H. E. Rowe, *Proc IRE*, 47, 7 (1956).

18

FOURIER TRANSFORMS IN DISCRETE-TIME SYSTEMS

18.0 Introduction

In Chapter 9, we demonstrated that the most general DT LTI system is characterized functionally in the time domain by the convolution formula

$$y[n] = x[n] * h[n] = \sum_{m=-\infty}^{\infty} x[m]h[n-m]$$

$$= h[n] * x[n] = \sum_{m=-\infty}^{\infty} h[m]x[n-m]. \tag{18.0-1}$$

In effect, (18.0−1) states that if we think of $x[n]$ as a sum of weighted delayed unit sample functions, then we can describe $y[n]$ for an LTI system as the sum of the correspondingly weighted unit sample responses.

Just as with CT LTI systems, however, an alternative functional description is possible in the frequency domain. The eigenfunctions of a DT LTI system are DT exponentials,

$$x[n] = z^n \tag{18.0-2}$$

where z is any complex number in an appropriate region (usually an annulus) of the z-plane. To show that DT exponentials have this property, it is only necessary to substitute (18.0−2) into (18.0−1) to obtain

$$y[n] = \sum_{m=-\infty}^{\infty} h[m]z^{n-m} = \tilde{H}(z)z^n \tag{18.0-3}$$

where

$$\tilde{H}(z) = \sum_{m=-\infty}^{\infty} h[m]z^{-m}. \tag{18.0-4}$$

Thus the output $y[n]$ is the same as the input $x[n]$ except for the amplitude factor (eigenvalue) $\tilde{H}(z)$; this is the defining property of an eigenfunction. Finally, any $x[n]$ that can be written as a weighted sum of eigenfunctions,

$$x[n] = \sum_{i} \tilde{X}(z_i)z_i^n \tag{18.0-5}$$

(where the z_i must lie in the convergence region for (18.0–4)), will yield the response

$$y[n] = \sum_i \tilde{Y}(z_i) z_i^n \tag{18.0–6}$$

where

$$\tilde{Y}(z_i) = \tilde{X}(z_i)\tilde{H}(z_i). \tag{18.0–7}$$

In Chapter 14, we derived a general DT signal representation formula similar to (18.0–5):

$$x[n] = \frac{1}{2W} \int_{-W}^{W} X(f) e^{j2\pi n f/2W} \, df \tag{18.0–8}$$

where

$$X(f) = \sum_{n=-\infty}^{\infty} x[n] e^{-j2\pi n f/2W}. \tag{18.0–9}$$

These formulas came from considering the Fourier transform of the impulse train

$$x(t) = \sum_{n=-\infty}^{\infty} x[n] \delta\left(t - \frac{n}{2W}\right) \tag{18.0–10}$$

and are the time-frequency duals of the Fourier series formulas (because $X(f)$ is periodic in f with period $2W$). In the context of DT systems, the impulse spacing $1/2W$ in (18.0–10) has no particular relevance. It is, therefore, convenient to set $2W = 1$, or equivalently to introduce the frequency variable $\phi = f/2W$, so that (18.0–8) and (18.0–9) become

$$x[n] = \int_{-1/2}^{1/2} X(\phi) e^{j2\pi n\phi} \, d\phi \tag{18.0–11}$$

$$X(\phi) = \sum_{n=-\infty}^{\infty} x[n] e^{-j2\pi n\phi}. \tag{18.0–12}$$

$X(\phi)$ is called the *discrete-time Fourier transform* (DTFT) of $x[n]$. $X(\phi)$ is periodic with period 1 in the frequency variable ϕ.

Equation (18.0–11) is a representation of the DT function $x[n]$ as a weighted "sum" of DT exponentials, $z^n = e^{j2\pi n\phi}$. Thus all of the values of z used in this representation lie on the unit circle in the z-plane. Indeed, (18.0–12) with $z = e^{j2\pi\phi}$ becomes

$$\tilde{X}(z) = \sum_{n=-\infty}^{\infty} x[n] z^{-n} \tag{18.0–13}$$

which has exactly the same form as the (unilateral) Z-transform studied in Chapter 8 except that the lower limit has been extended to $-\infty$. Equation (18.0–13) thus describes the *bilateral Z-transform* of $x[n]$, and we can begin to see why the Z-transform was so successful in simplifying the analysis of DT

LTI systems. The Z-transform $\tilde{X}(z)$ describes the weights of the eigenfunctions z^n, which added up appropriately (specifically, by a contour integral along an appropriate circle in the z-plane) yield $x[n]$.

Formally, we see that, provided the unit circle in the z-plane lies in the domain of convergence of $\tilde{X}(z)$ (which will be true if, for example, $x[n]$ is absolutely summable), the DTFT and the Z-transform are related by

$$X(\phi) = \tilde{X}(e^{j2\pi\phi}). \tag{18.0-14}$$

This relationship is entirely analogous to the fact that for CT signals the Fourier transform is identical to the (bilateral) Laplace transform evaluated for $s = j2\pi f$ (assuming that the $j\omega$-axis lies inside the domain of convergence of the Laplace transform).

The purpose of this chapter is to illustrate the important features of the DTFT, with special emphasis on the ways in which it differs from the CT Fourier transform. We shall also consider various DT systems—primarily DT filters—that are effectively described by their frequency response

$$H(\phi) = \tilde{H}(e^{j2\pi\phi}) = \sum_{n=-\infty}^{\infty} h[n]e^{-j2\pi n\phi} \tag{18.0-15}$$

and shall study some of the problems that arise in the design of such systems for typical digital signal processing applications.

18.1 Properties of the Discrete-Time Fourier Transform

Most of the usual theorems and properties of Fourier analysis apply to the DTFT. However, because $X(\phi)$ is periodic and $x[n]$ is a sequence, certain theorems have an altered appearance and others are inapplicable. The most important theorems and several of the more useful DTFT pairs are listed in Tables XVIII.1 and XVIII.2 in the appendix to this chapter. Derivations of the less familiar entries in these tables appear in the following examples.

Example 18.1–1

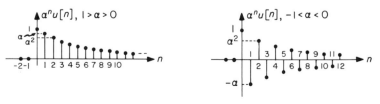

Figure 18.1–1. The sequence $x[n] = \alpha^n u[n]$, $|\alpha| < 1$.

The DTFT of the DT exponential function

$$x[n] = \alpha^n u[n] \tag{18.1-1}$$

shown in Figure 18.1-1 is, from (18.0-12),

$$X(\phi) = \sum_{n=-\infty}^{\infty} x[n] e^{-j2\pi n\phi} = \sum_{n=0}^{\infty} \alpha^n e^{-j2\pi n\phi} = \sum_{n=0}^{\infty} \left(\alpha e^{-j2\pi\phi} \right)^n$$

$$= \frac{1}{1 - \alpha e^{-j2\pi\phi}} \quad \text{if } |\alpha| < 1 \tag{18.1-2}$$

where to sum the series we have once again used the formula

$$1 + x + x^2 + x^3 + \cdots = \frac{1}{1-x} \tag{18.1-3}$$

which is valid provided $|x|$ (and hence $|\alpha|$) is less than 1.

The real and imaginary parts and the magnitude and angle of $X(\phi)$ for the DT exponential function are shown in Figure 18.1-2 for three values of α. Note that $X(\phi)$ is periodic with period 1, as must be the case given the structure of (18.0-12). For small α, it should be evident that $\alpha^n u[n] \to \delta[n]$ and $X(\phi) \to 1$. Indeed, the DTFT pair

$$x[n] = \delta[n] \quad \Longleftrightarrow \quad X(\phi) = 1$$

follows at once from (18.0-12) and is listed in Table XVIII.1. For larger α, the spectrum at lower frequencies approaches in shape the spectrum of the CT signal $e^{-(\ln \alpha)t} u(t) = \alpha^t u(t)$ (as shown by the dotted lines), but deviates at frequencies near 0.5. Samples of this CT signal at integer times have values equal to the DT exponential signal $\alpha^n u[n]$; the deviations for ϕ near 0.5 are thus the result of aliasing as discussed in Section 14.3.

As $\alpha \to 1$ the DT function $\alpha^n u[n] \to u[n]$, and the condition for the validity of the geometric series formula (18.1-3) is no longer met. As in the case of the Fourier transform of the CT unit step function $u(t)$, the DTFT for $u[n]$ must be approached with care. It can be shown that the real part of $X(\phi)$ as $\alpha \to 1$ includes a chain of impulses of area 0.5 at integer values of ϕ, so that we have the DTFT pair

$$x[n] = u[n] \quad \Longleftrightarrow \quad X(\phi) = \frac{1}{1 - e^{-j2\pi\phi}} + \frac{1}{2} \sum_{n=-\infty}^{\infty} \delta(\phi - n) \tag{18.1-4}$$

which is included in Table XVIII.1.

The behavior of the spectrum of the DT exponential function $\alpha^n u[n]$ for negative values of α is interesting and has no direct CT counterpart (see Exercise 18.4). For $|\alpha| > 1$, the DT exponential function blows up rapidly and the DTFT does not exist.

▶ ▶ ▶

Example 18.1-2

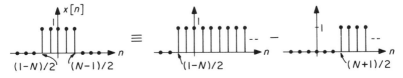

Figure 18.1-3. A DT pulse function. (Drawn for $N = 5$.)

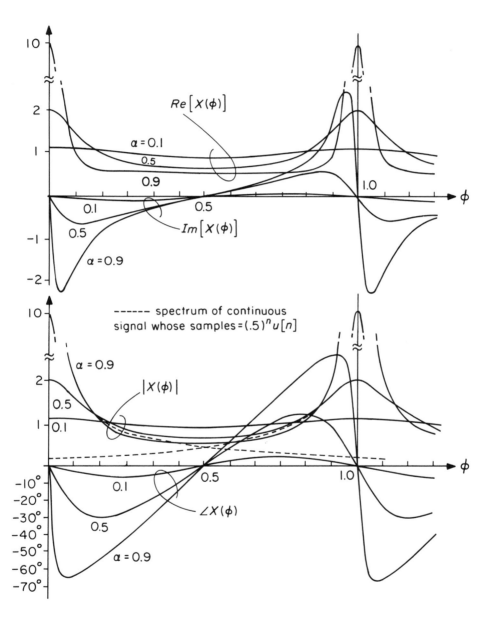

Figure 18.1–2. Sketches of $X(\phi) = \dfrac{1}{1 - \alpha e^{-j2\pi\phi}}$.

The DT pulse function

$$x[n] = \begin{cases} 1, & \frac{1-N}{2} \le n \le \frac{N-1}{2} \\ 0, & \text{elsewhere} \end{cases} \tag{18.1-5}$$

where we assume that N is an odd number so that $x[n]$ is an even function, can be described as the difference of two unit step functions,

$$x[n] = u\left[n + \frac{N-1}{2}\right] - u\left[n - \frac{N+1}{2}\right] \tag{18.1-6}$$

as shown in Figure 18.1–3. Thus by linearity and the Delay Theorem (both of which follow trivially from (18.0−12)), the DTFT of the pulse function is

$$X(\phi) = e^{j2\pi\phi\frac{N-1}{2}}\left[\frac{1}{1 - e^{-j2\pi\phi}} + \frac{1}{2}\sum_{n=-\infty}^{\infty}\delta(\phi - n)\right]$$
$$- e^{-j2\pi\phi\frac{N+1}{2}}\left[\frac{1}{1 - e^{-j2\pi\phi}} + \frac{1}{2}\sum_{n=-\infty}^{\infty}\delta(\phi - n)\right]. \tag{18.1-7}$$

The impulse terms cancel since $e^{j2\pi\phi\frac{N-1}{2}} = e^{-j2\pi\phi\frac{N+1}{2}}$ for all integer values of ϕ, leaving

$$X(\phi) = \frac{e^{j\pi\phi(N-1)} - e^{-j\pi\phi(N+1)}}{1 - e^{-j2\pi\phi}} = \frac{\sin N\pi\phi}{\sin \pi\phi}. \tag{18.1-8}$$

The same result can, of course, be obtained directly from (18.0−12) and the geometric series formula.

   $X(\phi)$ as given by (18.1−8) is real because $x[n]$ is an even function; it is sketched in Figure 18.1–4 for $N = 5$. The ratio of sines is the periodic version of the $\frac{\sin f}{f}$-shaped transform of a CT square pulse. Note that the value at $\phi = 0$ is N and the first zero occurs at $\phi = 1/N$.

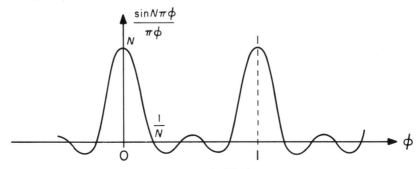

Figure 18.1–4. Sketch of $\dfrac{\sin N\pi\phi}{\sin \pi\phi}$. (Drawn for $N = 5$.)

▶ ▶ ▶

Example 18.1–3

Evaluating the DTFT of the summation operation $\sum_{m=-\infty}^{n} x[m]$ provides another opportunity to combine basic DTFT properties with the DTFT of the step function $u[n]$. This is because the summation operation can be written as a convolution with a step:

$$\sum_{m=-\infty}^{n} x[m] = \sum_{m=-\infty}^{\infty} x[m]u[n-m] = x[n] * u[n]. \qquad (18.1-9)$$

From Table XVIII.2, the DTFT of the convolution of DT time functions is the product of their transforms. Hence the transform of the summation operation is

$$X(\phi)\left[\frac{1}{1-e^{-j2\pi\phi}} + \frac{1}{2}\sum_{n=-\infty}^{\infty} \delta(\phi-n)\right] = \frac{X(\phi)}{1-e^{-j2\pi\phi}} + \frac{X(0)}{2}\sum_{n=-\infty}^{\infty} \delta(\phi-n) \qquad (18.1-10)$$

as listed in Table XVIII.2.

▶ ▶ ▶

Example 18.1–4

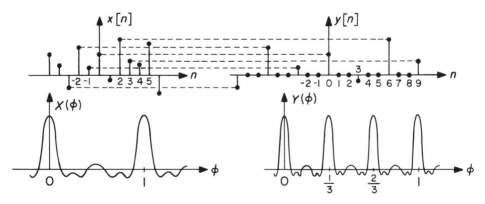

Figure 18.1–5. The DT scale-change property. (Drawn for $N = 3$.)

Sometimes it is necessary to generate a new sequence $y[n]$ from a given sequence $x[n]$ by inserting $N-1$ zeros between each successive pair of values of $x[n]$. This is the DT equivalent of a change of scale. What we have in mind is illustrated in Figure 18.1–5 and described formally by the equation

$$y[n] = \sum_{k=-\infty}^{\infty} x[k]\delta[n-kN]. \qquad (18.1-11)$$

The corresponding DTFT is

$$Y(\phi) = \sum_{n=-\infty}^{\infty} y[n]e^{-j2\pi\phi n} = \sum_{n=-\infty}^{\infty} \sum_{k=-\infty}^{\infty} x[k]\delta[n-kN]e^{-j2\pi\phi n}$$

$$= \sum_{k=-\infty}^{\infty} x[k] \sum_{n=-\infty}^{\infty} \delta[n-kN]e^{-j2\pi\phi n}$$

$$= \sum_{k=-\infty}^{\infty} x[k]e^{-j2\pi\phi Nk}$$

$$= X(\phi N) \tag{18.1-12}$$

which is the result given in Table XVIII.2. The fundamental period of $X(\phi N)$ is $1/N$ as shown in Figure 18.1–5; of course, $X(\phi N)$ is still periodic with period 1 since it is the transform of the DT sequence $y[n]$.

As an example of the scale-change property, the doubly infinite chain of unit samples described by $x[n] = 1$, $-\infty < n < \infty$, has the DTFT $X(\phi) = \sum_{k=-\infty}^{\infty} \delta(\phi - k)$. (This follows directly from (18.0–11).) Hence, the chain of unit samples separated by $N-1$ zeros, that is, the DT function

$$x'[n] = \sum_{k=-\infty}^{\infty} \delta[n-kN] \tag{18.1-13}$$

has the DTFT

$$X'(\phi) = \sum_{k=-\infty}^{\infty} \delta(\phi N - k) = \frac{1}{N} \sum_{k=-\infty}^{\infty} \delta\left(\phi - \frac{k}{N}\right).$$

This result is shown graphically in Figure 18.1–6 and is listed in Table XVIII.1.

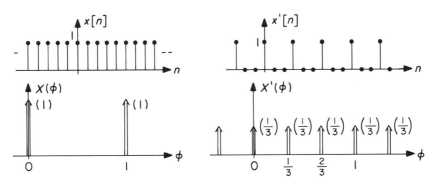

Figure 18.1–6. An example of the scale-change property. (Drawn for $N = 3$.)

▶ ▶ ▶

18.2 Discrete-Time Filters

Our principal reason for being interested in the DTFT is that certain DT systems or operations are easier to describe functionally (and thus easier to design and understand) in terms of their frequency responses $H(\phi)$ and input-output relationships of the form

$$Y(\phi) = H(\phi)X(\phi)$$

than in terms of their unit sample responses $h[n]$ and input-output relationships of the form

$$y[n] = h[n] * x[n].$$

Filters that pass or reject various frequencies are the prototypical examples of systems that are appropriately described in the frequency domain.

For most of the rest of this chapter, we shall focus on the following particular DT filter design problem. Suppose we are given a CT signal $x(t)$ that contains no significant energy at frequencies above 5 kHz. Suppose further that the information-bearing part of this signal lies in the band $0 < f < 1$ kHz; components in $x(t)$ above 1 kHz are thus primarily "noise" corrupting or obscuring the desired signal. Consequently, we would like to pass $x(t)$ through a lowpass filter with a cutoff frequency of 1 kHz. To this end, we first sample the given signal $x(t)$ at the Nyquist interval $T = 1/(2 \times 5 \text{ kHz}) = 100$ μsec to avoid aliasing. The amplitudes of these samples determine a DT signal $\hat{x}[n]$, where the "hat" is introduced to minimize later confusions. We then propose to carry out a DT LTI operation on $\hat{x}[n]$ (using, perhaps, a digital computer) yielding $\hat{y}[n]$. Finally, we intend to construct from $\hat{y}[n]$ a CT signal $y(t)$ that we hope will approximate the output of an ideal lowpass CT filter with cutoff frequency 1 kHz operating on $x(t)$.

The various stages in this process are illustrated in Figure 18.2–1. The boxes labelled "C/D" and "D/C" require further discussion. The first includes the sampling stage (which is why it needs a clock input), but it does more because its output is not the sample impulse train but rather the DT sequence of sample heights. Its effect is completely described by the formula

$$\hat{x}[n] = x(nT). \tag{18.2–1}$$

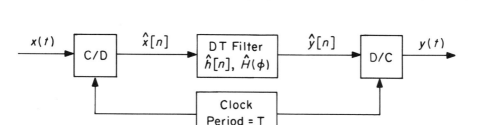

Figure 18.2–1. DT filtering of a CT signal.

The box labelled "D/C" is more complex. In practice, it might contain a zero-order-hold circuit of the type discussed in Problems 7.4 and 8.9. An analytically simpler scheme is to suppose that the D/C box first constructs an impulse train spaced T seconds apart with amplitudes equal to $y[n]$ and then passes this train through an ideal lowpass filter with cutoff frequency $1/2T$. The effect is described by the equation

$$y(t) = \sum_{n=-\infty}^{\infty} \hat{y}[n] \frac{\sin \dfrac{\pi(t-nT)}{T}}{\dfrac{\pi(t-nT)}{T}}. \tag{18.2-2}$$

Obviously, the D/C box also requires clock input.

The overall goal we seek to accomplish with the system of Figure 18.2–1 is described by the first and last spectra in Figure 18.2–2. (As is our custom, we have for diagrammatic purposes treated complex quantities in Figure 18.2–2 as if they were real.) The effect of the C/D box in Figure 18.2–1 is to yield a DT function with DTFT $\hat{X}(\phi)$, as shown on the second line in Figure 18.2–2. $\hat{X}(\phi)$ is proportional to $X(f)$ repeated periodically with period 1 on a frequency scale such that $\phi = 1$ corresponds to $f = 1/T = 10$ kHz. (The amplitude of $\hat{X}(\phi)$ is also scaled by a factor $1/T$ so that $\hat{X}(0) = X(0)/T$. But this scaling is precisely undone by the D/C box if it functions as described; this is easy to show by leaving out the DT filter and thinking about the overall effect of the system in light of the Sampling Theorem.) The desired DT filter thus should have for

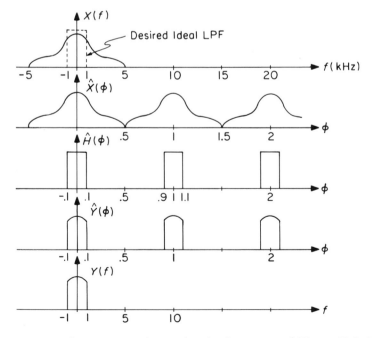

Figure 18.2–2. Spectra at various points in the system of Figure 18.2–1.

$-0.5 < \phi < 0.5$ the frequency response

$$\hat{H}(\phi) = \begin{cases} 1, & |\phi| < 0.1 \\ 0, & \text{elsewhere} \end{cases} \qquad (18.2-3)$$

as shown in the middle line of Figure 18.2–2.

The unit sample response that corresponds to the DTFT of (18.2–3) is readily derived from (18.0–11):

$$\hat{h}[n] = \int_{-1/2}^{1/2} \hat{H}(\phi) e^{j2\pi n\phi}\, d\phi = \int_{-0.1}^{0.1} 1\, e^{j2\pi n\phi}\, d\phi$$

$$= \frac{e^{j2\pi n(0.1)} - e^{j2\pi n(-0.1)}}{j2\pi n} = \frac{\sin(2\pi n/10)}{\pi n}. \qquad (18.2-4)$$

$\hat{h}[n]$ is a special case of a general formula given in Table XVIII.1 and is plotted in Figure 18.2–3.

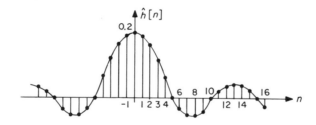

Figure 18.2–3. The unit sample response defined by (18.2–4).

Precisely as in the case of the CT ideal lowpass filter, the DT unit sample response of Figure 18.2–3 extends from $-\infty$ to ∞ and is hence fundamentally non-causal. Also as in the CT case, there are various ways to overcome this difficulty by accepting compromises or approximations to the desired ideal filter characteristic. In the following examples, we consider several schemes that are similar or identical to those we explored in Chapter 15 in connection with the analogous CT problem. In the next section, we shall present a rather different approach that is uniquely applicable in discrete time.

Example 18.2–1

The most straightforward way to achieve a causal sample response is to multiply the ideal $\hat{h}[n]$ of (18.2–4) by a *window function* $\hat{w}[n]$ that is zero for all n less than some negative value. Shifting (delaying) the result appropriately will then yield a causal sample response. In general, simpler algorithms and improved performance result if $\hat{w}[n]$ is chosen to be symmetric about $n = 0$. The product unit sample response is then of finite duration, that is, of FIR type as discussed in Section 9.1. Moreover, the product unit sample response is also symmetric about $n = 0$. The DTFT of the product sample response is thus real, and the frequency response of the causal (shifted) filter has a linear phase characteristic (a pure delay). This turns out to be an important feature, particularly for systems handling pulse signals. (See Exercise 18.1.)

The DTFT of the product of two DT time functions is the convolution of their individual DTFTs, but the convolution formula is a little different from those to which we have become accustomed. The reason for the difference is that the individual DTFTs are both periodic with the same period, and their product is thus periodic with the same period. The infinite integral of a periodic function is in general unbounded or ill-defined. The appropriate convolution formula is easily shown to be

$$x[n]y[n] \quad \Longleftrightarrow \quad \int_{-1/2}^{1/2} X(\nu)Y(\phi-\nu)\,d\nu \qquad (18.2-5)$$

as listed in Table XVIII.2. The right-hand side of (18.2−5) is called *cyclic convolution* because as ϕ is increased a section of $Y(\phi-\nu)$ moves outside the integration interval and (because $Y(\phi)$ is periodic) an equivalent section moves in at the other end of the interval. To distinguish cyclic convolution from the ordinary kind, we shall use the shorthand notation $X(\phi) \circledast Y(\phi)$ for the right-hand side of (18.2−5).

The requirements that a satisfactory window function $\hat{w}[n]$ should meet involve the same conflicting set of features discussed in Section 16.2. Specifically, the DTFT of $\hat{w}[n]$ should be as sharply peaked around integer values of ϕ as possible, and the "tails" of the spectrum of $\hat{w}[n]$ away from the peaks should be as low as possible. The pulse shapes analyzed in Example 16.2–4 are common examples of CT window functions. For purposes of illustration, let us choose $\hat{w}[n]$ to be the triangular pulse

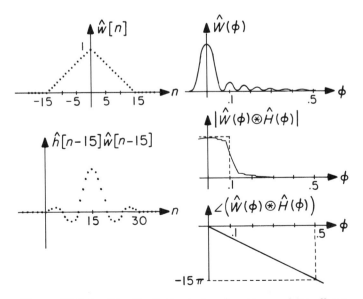

Figure 18.2–4. The Bartlett window function and its effects.

(sometimes called the *Bartlett window function*) shown in Figure 18.2–4. The resulting product $\hat{h}[n]\hat{w}[n]$, shifted to be causal, is also shown together with its DTFT. Note that the phase of the filter frequency response is linear as expected. To reduce the width of the transition region in the magnitude of the filter frequency response, the duration or length of $\hat{w}[n]$ should obviously be made as long as possible. The limits on this duration are ultimately set by the method we choose to realize the DT filter. If we plan to use a computer to evaluate $\hat{x}[n] * (\hat{w}[n]\hat{h}[n])$ directly in real time, then one limit is set by the available computer memory capacity: The length of $\hat{w}[n]\hat{h}[n]$ determines the number of past values of the input $\hat{x}[n]$ that must be retained at each moment to enable computation of future outputs. But a more serious limit may be set by the computer multiply-and-add time: If the length of $\hat{w}[n]\hat{h}[n]$ is N samples, then N multiply-and-adds must be carried out to compute the current output value in the time between input sample arrivals. For the example, $N = 30$ and $T = 100$ μsec. Thus 3.3 μsec are allowed for a multiply-and-add—within the current state of the art, although another factor of ten might begin to present difficulties.

▶ ▶ ▶

Example 18.2–2

Another technique for selecting a causal FIR $\hat{h}[n]$ to realize a DT lowpass filter with cutoff frequency 0.1 is to sample a finite segment of the impulse response of a CT LPF with the desired cutoff. For example, the impulse response of the 3rd-order Butterworth lowpass filter, extensively considered in earlier chapters, can readily be found from the system function

$$H(s) = \frac{1}{(\tau s)^3 + 2(\tau s)^2 + 2(\tau s) + 1} \tag{18.2–6}$$

and shown to be

$$h(t) = \left[\frac{1}{\tau}e^{-t/\tau} + \frac{2}{\sqrt{3}}\frac{1}{\tau}e^{-t/2\tau}\cos\left(\frac{\sqrt{3}t}{2\tau} - \frac{5\pi}{6}\right)\right]u(t). \tag{18.2–7}$$

To obtain a cutoff frequency of 0.1, set $1/2\pi\tau = 0.1$ or $\tau = 5/\pi$. The resulting $h(t)$ is plotted in Figure 18.2–5. Setting

$$\hat{h}[n] = \begin{cases} h(n), & 0 \le n < 30 \\ 0, & \text{elsewhere} \end{cases} \tag{18.2–8}$$

gives the DT frequency response $\hat{H}(\phi)$, which differs from the CT frequency response $H(j2\pi f)$ (shown dotted) because of aliasing effects as well as the effects of truncating $\hat{h}[n]$ to a finite number of nonzero terms.

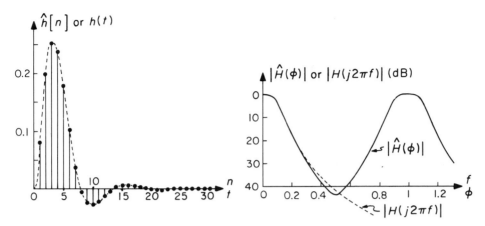

Figure 18.2–5. Comparison of the DT and CT behavior of a Butterworth filter.

▶ ▶ ▶

Example 18.2–3

Still another technique for DT filter design uses a known CT causal lumped filter characterization to determine appropriate pole locations for a DT system function $\tilde{H}(z)$ instead of determining values for $\hat{h}[n]$ directly. The resulting unit sample responses are of infinite duration (IIR) but can be realized recursively with finite structures or algorithms. For example, the 3rd-order Butterworth filter with cutoff frequency 0.1 Hz discussed in the preceding example has poles in the s-plane as shown on the left in Figure 18.2–6.

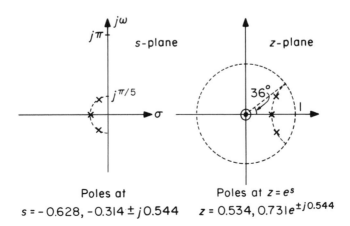

Figure 18.2–6. Butterworth pole locations in the s-plane and the z-plane.

Carrying out the transformation $z = e^s$ maps these poles into the locations in the z-plane shown on the right in Figure 18.2–6. This transformation also maps the piece of the $j\omega$-axis between $-j\pi$ and $+j\pi$ in the s-plane into the unit circle in the z-plane traversed once counterclockwise starting at $z = -1$. Hence, except for the small effect of aliasing,* $\tilde{H}(e^{j2\pi\phi}) = \hat{H}(\phi)$ should have Butterworth behavior with cutoff frequency $|\phi| = 0.1$. The actual response is shown in Figure 18.2–7 for

$$\tilde{H}(z) = \frac{0.1317}{(1 - 0.534z^{-1})(1 - 0.731e^{j0.544}z^{-1})(1 - 0.731e^{-j0.544}z^{-1})}$$

$$= \frac{0.1317}{1 - 1.785z^{-1} + 1.202z^{-2} - 0.2853z^{-3}} \qquad (18.2\text{–}9)$$

where the constant multiplier has been chosen so that $\hat{H}(0) = \tilde{H}(1) = 1$. This Butterworth filter can be synthesized with the block diagram shown in Figure 18.2–8. Note that only three delay elements (or memory registers) are required.

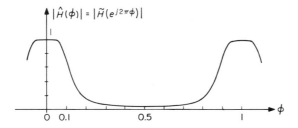

Figure 18.2–7. Frequency response magnitude for (18.2–9).

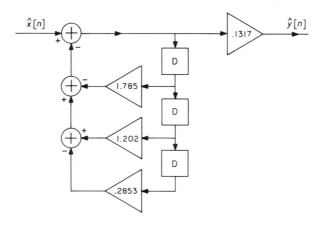

Figure 18.2–8. Gain-adder-delay block diagram realizing (18.2–9).

▶ ▶ ▶

*The effect of aliasing is small in this example. To avoid aliasing effects in general, we can replace the exponential transformation $z = e^s$ used here to relate the s-plane and the z-plane by some other transformation. One example is discussed in Problem 18.6. For further details on this as well as many other aspects of the design of DT filters, see the classic text of A. V. Oppenheim and R. W. Schafer, *Digital Signal Processing* (Englewood Cliffs, NJ: Prentice-Hall, 1975).

18.3 The DT Fourier Series and the Discrete Fourier Transform (DFT)

In Chapter 14, we observed that a periodic time function has a transform that is an impulse train in frequency, the areas of the impulses being the Fourier series coefficients. On the other hand, an impulse train in time has a transform that is a periodic frequency function—the DTFT of the DT sequence of the impulse areas. Hence, a time function that is both an impulse train and periodic (in time) should have a transform that is both an impulse train and periodic (in frequency). It should then be possible to write both the analysis and the synthesis formulas in forms that involve sums of impulse areas rather than integrals. And both of the sums should be finite because a periodic impulse train contains only a finite number of different impulse areas.

There are many ways to derive the desired formulas. Perhaps the most illuminating starts from the observation that we can write a periodic DT function $x[n]$ as the convolution of the sequence $x_N[n]$ describing a single period,

$$x_N[n] = \begin{cases} x[n], & 0 \le n < N \\ 0, & \text{elsewhere} \end{cases} \tag{18.3-1}$$

with the DT unit sample chain $\sum_{k=-\infty}^{\infty} \delta[n - kN]$:

$$x[n] = x_N[n] * \left(\sum_{k=-\infty}^{\infty} \delta[n - kN] \right). \tag{18.3-2}$$

Hence, from the DTFT Convolution Theorem of Table XVIII.2 and (18.1–14) we should have

$$X(\phi) = X_N(\phi)\left(\frac{1}{N} \sum_{k=-\infty}^{\infty} \delta\left(\phi - \frac{k}{N} \right) \right). \tag{18.3-3}$$

These formulas are illustrated in Figure 18.3–1.

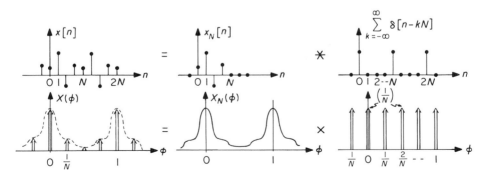

Figure 18.3–1. Construction of a periodic $x[n]$. (Drawn for $N = 4$.)

Let $X[m]$ be the area of the impulse in $X(\phi)$ at $\phi = m/N$. Then, from (18.3–3) and (18.0–12),

$$X[m] = \frac{1}{N}X_N(m/N) = \frac{1}{N}\sum_{n=0}^{N-1} x[n]e^{-j2\pi nm/N} \tag{18.3–4}$$

which is the analysis formula sought. To obtain the synthesis formula, combine (18.0–11) and (18.3–3) to give

$$x[n] = \int_{-1/2}^{1/2} X(\phi)e^{j2\pi n\phi}\,d\phi = \sum_{m=0}^{N-1} X[m]e^{j2\pi nm/N}. \tag{18.3–5}$$

Equation (18.3–5) represents the periodic sequence $x[n]$ as a *discrete-time Fourier series* with coefficients given by (18.3–4). Together, (18.3–4) and (18.3–5) constitute a transform pair:

$$\boxed{\begin{aligned} x[n] &= \sum_{m=0}^{N-1} X[m]e^{j2\pi nm/N} \\ X[m] &= \frac{1}{N}\sum_{n=0}^{N-1} x[n]e^{-j2\pi nm/N}. \end{aligned}} \tag{18.3–6}$$

Although our derivation of the formula pair (18.3–6) emphasized the periodic character of both $x[n]$ and $X[m]$, the resulting equations have an alternative interpretation that is independent of periodicity. Each equation has the form of a set of N linear algebraic equations relating one set of N quantities $x[n]$, $0 \le n \le N-1$, to another set of N quantities $X[m]$, $0 \le m \le N-1$. The two sets of equations are inverses of one another—the first gives the set $x[n]$ if the set $X[m]$ is known, the second gives the set $X[m]$ if the set $x[n]$ is known. When interpreted this way, the set $X[m]$ is said to be the *discrete Fourier transform* (DFT) of the set $x[n]$. For a direct derivation of the pair (18.3–6) from this point of view, see Problem 18.10.

Example 18.3–1

The DFT formulas (18.3–6) have an immediate application to the filtering problem considered in the preceding section, namely to specify a causal unit sample response $\hat{h}[n]$ whose DTFT approximates an ideal lowpass filter with cutoff frequency $\phi = 0.1$. Suppose we also desire $\hat{h}[n]$ to be of finite length, that is, $\hat{h}[n] = 0$ except for $0 \le n \le N-1$. Then the values of the DTFT $\hat{H}(\phi)$ at the N frequencies $\phi = m/N$, $0 \le m \le N-1$, are given by

$$\hat{H}(m/N) = \sum_{n=-\infty}^{\infty} \hat{h}[n]e^{-j2\pi\phi n}\bigg|_{\phi=m/N}$$

$$= \sum_{n=0}^{N-1} \hat{h}[n]e^{-j2\pi nm/N}. \tag{18.3–7}$$

If we identify $\hat{H}(m/N) = N\hat{H}[m]$, then (18.3–7) has precisely the same form as the second of the DFT equations (18.3–6). Hence presumably we may write

$$\hat{h}[n] = \frac{1}{N}\sum_{m=0}^{N-1}\hat{H}(m/N)e^{j2\pi nm/N}, \quad 0 \le n \le N-1. \qquad (18.3\text{–}8)$$

In other words, an FIR $\hat{h}[n]$ and its DTFT $\hat{H}(\phi)$ are completely specified by N samples of $\hat{H}(\phi)$ at the frequencies $\phi = m/N$, $0 \le m \le N-1$. On reflection this should hardly be surprising: An FIR $\hat{h}[n]$ is described by N numbers and thus must be vastly overspecified by the degrees of freedom available in general to $\hat{H}(\phi)$, a function of the continuous variable ϕ, $0 < \phi < 1$. Moreover, we know from the Sampling Theorem that the transform of a function of finite duration is uniquely specified by appropriately chosen samples. Either of these observations could also have led us to (18.3–8).

Equation (18.3–8) permits us to pick directly and arbitrarily any desired complex amplitudes for the DTFT of the FIR function $\hat{h}[n]$ at the N frequencies $\phi = m/N$, $0 \le m \le N-1$. These amplitudes then determine the N nonzero values of $\hat{h}[n]$. The actual DTFT of the resulting FIR filter is, of course, defined for all ϕ, not just the N points whose amplitudes we selected. Thus even though the DTFT of the FIR filter is guaranteed to have the selected values at the specified points, its values in between may not provide a satisfactory approximation to the desired response. Hence some iteration may be required.

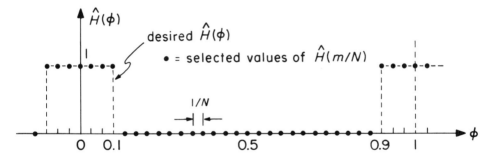

Figure 18.3–2. Selected values for $\hat{H}(m/N)$. (Drawn for $N = 30$.)

To apply this idea to the filter problem of the previous section, suppose we seek the $\hat{H}(\phi)$ shown in Figure 18.3–2 and choose $N = 30$. Initially, let us try selecting the values of $\hat{H}(m/N)$ as shown, that is, equal to 1 in the pass band and zero in the stop band—with the band-edge point $m = 3$ treated as if it were inside the pass band. If we choose $\angle\hat{H}(m/N) = 0$, the resulting $\hat{h}[n]$ will have peaks at the beginning and end of the interval $0 \le n \le N-1$ in which it is nonzero; $\hat{H}(\phi)$ will have the specified values at $\phi = m/N$ but will deviate dramatically from the desired behavior for values of ϕ in between. We get much better results if we choose $\angle\hat{H}(m/N) = -2\pi\phi(N/2)|_{\phi=m/N}$, which corresponds to a delay of $N/2$. The peak of $\hat{h}[n]$ will then occur in the middle of its nonzero interval. Such a phase choice is in fact completely equivalent to requiring that $\hat{H}(m/N) = (-1)^m$ instead of 1 in the pass band. From (18.3–8) it then follows*

---

*Since $\hat{H}(\phi)$ is periodic, the summation in (18.3–8) can be taken over any N consecutive samples, instead of $0 \le m \le N-1$, if convenient. The closed form in (18.3–9) derives once again from the formula for the partial sum of a geometric series.

that

$$\hat{h}[n] = \frac{1}{30}\left[-e^{-j2\pi 3n/30} + e^{-j2\pi 2n/30} - e^{-j2\pi n/30} + 1\right.$$
$$\left. - e^{j2\pi n/30} + e^{j2\pi 2n/30} - e^{j2\pi 3n/30}\right]$$

$$= \frac{\sin\dfrac{7\pi(n-15)}{30}}{30\sin\dfrac{\pi(n-15)}{30}}, \quad 0 \le n \le 29. \tag{18.3-9}$$

The corresponding FIR unit sample response $\hat{h}[n]$ and the magnitude of its actual DTFT are shown in Figure 18.3-3. Note that to preserve symmetry (and hence achieve a linear phase for the DTFT) the value of the FIR $\hat{h}[n]$ shown at $n = 0$ is in fact one-half the value given by (18.3-9); an equal term has been added at $n = 30$.

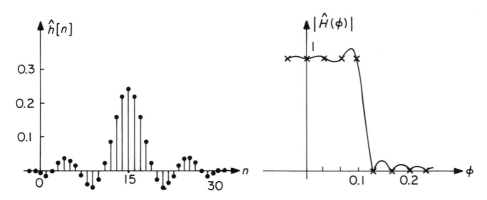

Figure 18.3-3. The unit sample response of (18.3-9) and its DTFT.

$|\hat{H}(\phi)|$ clearly goes through the specified values, but in between it deviates substantially from the desired ideal response. In particular, the filter of (18.3-9) lets through significant signal energy just above the cutoff frequency; the attenuation in this region would probably be unacceptable for many purposes.

An improved attenuation at the cost of a somewhat more gradual transition band can be obtained by choosing the selected values of $\hat{H}(m/N)$ to jump less abruptly from 1 to 0. For example, we might pick the magnitude of the sample at the band edge to be 1/2 instead of 1. The resulting FIR unit sample response is then readily shown to be

$$\hat{h}[n] = \frac{\sin\dfrac{\pi(n-15)}{5}}{30\tan\dfrac{\pi(n-15)}{30}}, \quad 0 \le n \le 30. \tag{18.3-10}$$

This choice of values for $\hat{H}(m/N)$ is equivalent to the truncation scheme of Example 18.2-1 with a window function

$$\hat{w}[n] = \frac{\pi(n-15)}{30\tan\dfrac{\pi(n-15)}{30}}. \tag{18.3-11}$$

Figure 18.3–4 shows $\hat{w}[n]$, $\hat{h}[n]$, and the magnitude of the DTFT of $\hat{h}[n]$. $|\hat{H}(\phi)|$ has the expected behavior—a wider transition but substantially greater attenuation in the stop band, and a smaller ripple in the pass band. Even better performance might be obtained by describing explicitly the relative costs of transition and attenuation and then optimally choosing the transition value rather than arbitrarily setting it equal to 1/2. Extensive studies of this and other techniques have been carried out.*

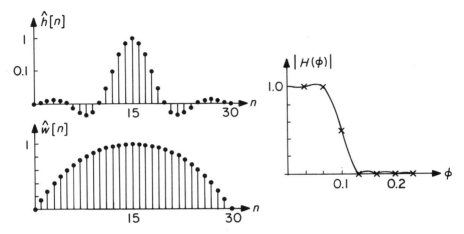

Figure 18.3–4. The unit sample response of (18.3–3) and its DTFT.

▶ ▶ ▶

18.4 Properties of the DT Fourier Series and the DFT

The most basic property of any Fourier transform is that convolution in one domain corresponds to multiplication in the other. Sequences related by the DT Fourier series or DFT pair (18.3–6) are periodic in both domains: $x[n] = x[n+N]$ and $X[m] = X[m+N]$. Hence, the appropriate form of convolution is the DT equivalent of cyclic convolution introduced in Example 18.2–1:

$$x[n] \circledast y[n] = \sum_{k=0}^{N-1} x[k]y[n-k]. \qquad (18.4\text{–}1)$$

It is easy to show (see Problem 18.10) that the DFT of the cyclic convolution of two periodic sequences (with the same period N) is N times the product of their individual DFTs:

$$x[n] \circledast y[n] \quad \Longleftrightarrow \quad NX[m]Y[m]. \qquad (18.4\text{–}2)$$

By symmetry, it follows that the DFT of the product of two periodic sequences (with the same period N) is the cyclic convolution of their individual DFTs:

$$x[n]y[n] \quad \Longleftrightarrow \quad X[m] \circledast Y[m]. \qquad (18.4\text{–}3)$$

---

*See, for example, L. R. Rabiner, B. Gold, and C. H. McGonegal, *IEEE Trans. Audio Electro-acoust.* AU-18, 2 (June 1970): 83–106.

An algorithm for computing the DFT $X[m]$ of a set of N quantities $x[n]$ would seem to require computation time proportional to N^2; that is, N multiply-and-adds are needed for each of N values of m. However, schemes have been discovered that require time proportional to only $N \log_2 N$. This has major implications. For example, a straightforward algorithm for computing the cyclic convolution of two sequences of period N requires time proportional to N^2. But suppose instead we first compute the DFT of each sequence and then take the inverse transform of the product of the results. If we use a "fast" algorithm, the computation time is proportional to about $3N \log_2 N + N$, which is very much less than N^2 for large N. For $N = 10^3$, for example, the numbers are about 3.1×10^4 and 10^6 respectively—a saving of a factor greater than 30 in computation time. Such savings make it very attractive to try to use "fast" algorithms even when the sequences being convolved are not periodic—as, for example, in the filtering problem studied earlier in this chapter. One way of doing this is considered in Problem 18.13.

The fast DFT algorithms exploit various symmetries and periodicities of the factors $e^{\pm j 2\pi mn/N}$ that appear in the DFT formulas (18.3–6). Related techniques were known long before high-speed computers became widely available. But in the era of hand computation it was feasible to consider only small values of N, for which the advantages of being clever are real but not overwhelming. The truly spectacular improvements that these techniques offer for large N were generally overlooked until the publication of a famous paper by Cooley and Tukey in 1965.* Since that date, many variations, refinements, and extensions of the basic idea have been developed. We shall restrict ourselves here to discussing a simple example of the *fast Fourier transform* (FFT).

Suppose that $N = 8$ and we want to compute the set of eight numbers $\{x[0], x[1], \ldots, x[7]\}$ from a given set of eight numbers $\{X[0], X[1], \ldots, X[7]\}$ according to the formula

$$x[n] = \sum_{m=0}^{7} X[m] e^{j 2\pi mn/8} \tag{18.4–4}$$

which is the first of the DFT formulas (18.3–6). Rearrange the order of the terms in (18.4–4) and group them to give

$$x[n] = x_e[n] + x_o[n] e^{j 2\pi n/8}, \quad 0 \le n < 8 \tag{18.4–5}$$

where

$$\begin{aligned} x_e[n] &= \left[X[0] + X[2] e^{j 2\pi n/4} + X[4] e^{j 2\pi 2n/4} + X[6] e^{j 2\pi 3n/4} \right] \\ x_o[n] &= \left[X[1] + X[3] e^{j 2\pi n/4} + X[5] e^{j 2\pi 2n/4} + X[7] e^{j 2\pi 3n/4} \right]. \end{aligned} \tag{18.4–6}$$

*J. W. Cooley and J. W. Tukey, *Math. Computation, 19* (1965): 297–301. Somewhat similar schemes were suggested earlier by I. J. Good, L. H. Thomas, and (as early as 1905) R. Runge. For a modern discussion, see *Programs for Digital Signal Processing* (New York, NY: IEEE Press, 1979).

$x_e[n]$ and $x_o[n]$ are, from (18.3–6), the DFTs of the sequences of $N/2 = 4$ terms corresponding, respectively, to the even-positioned and odd-positioned terms of the original sequence of $N = 8$ numbers $\{X[0], X[1], \ldots, X[7]\}$. If $x_e[n]$ and $x_o[n]$ were known, then, by (18.4–5), $N = 8$ multiply-and-adds would suffice to compute $x[n]$.*

Generalizing, the DFT of N numbers (if N is even) can be reduced to the computation of two DFTs of $N/2$ numbers, together with N multiply-and-adds. But, using the same strategy, we can reduce each of the DFTs of $N/2$ numbers (if $N/2$ is even) to the computation of two DFTs of $N/4$ numbers, together with $N/2$ multiply-and-adds. If N is a power of 2, this process can be continued through $\log_2 N$ stages, until we finally arrive at the need to compute N DFTs of single numbers—which from (18.3–6) are simply the numbers themselves. At each stage a total of N multiply-and-adds must be executed, for an overall number of operations equal to $N \log_2 N$ as promised. A specific program for executing this process is given in Problem 18.14.

Some further examples and properties of the DFT are discussed in Problems 18.9–12; for a more complete development, see the book by Oppenheim and Schafer cited above.

18.5 Summary

The frequency domain is as useful for the analysis and synthesis of DT systems as it is for CT systems. The sequence $x[n] = z^n$ is an eigenfunction for LTI DT systems provided that $|z|$ is appropriately constrained. Hence, if $x[n] = z^n$ is the input to an LTI DT system, the output has the form $y[n] = \tilde{H}(z)z^n$, and any input that can be expressed as a weighted sum of eigenfunctions will yield an output of the same form, except that the weights will be multiplied by the appropriate value of $\tilde{H}(z)$. In particular, inputs that grow no faster than a finite power of n can be represented by

$$x[n] = \int_{-1/2}^{1/2} X(\phi)e^{j2\pi n\phi}\, d\phi$$

where the DTFT $X(\phi)$ is given by

$$X(\phi) = \sum_{n=-\infty}^{\infty} x[n]e^{-j2\pi n\phi}.$$

If $x[n]$ is the input to an LTI DT system, the DTFT of the response will be

$$Y(\phi) = X(\phi)H(\phi)$$

---

*Note that $e^{j2\pi n/N} = -e^{j2\pi(n+N/2)/N}$ and that $x_e[n]$ and $x_o[n]$ are periodic with period $N/2$. The number of multiplications needed can thus be reduced by a factor of 2 from the number implied by (18.4–5).

where the frequency response $H(\phi)$ is given by

$$H(\phi) = \tilde{H}(e^{j2\pi\phi}) = \sum_{n=-\infty}^{\infty} h[n]e^{-j2\pi n\phi}.$$

The properties of $X(\phi)$ are similar to those of Fourier transforms of CT functions, except for modifications required by the fact that $X(\phi)$ is periodic with period 1. Thus the DTFT of a product of DT functions is the cyclic convolution of their transforms:

$$x[n]y[n] \quad \Longleftrightarrow \quad X(\phi) \circledast Y(\phi) = \int_{-1/2}^{1/2} X(\nu)Y(\phi-\nu)\,d\nu.$$

A major focus of the chapter was on various schemes for designing a causal approximation to an ideal lowpass DT filter. The most direct technique was to multiply the non-causal ideal filter unit sample response by a window function of finite duration. Two other schemes exploited known solutions to the corresponding CT problem; one specified the unit sample response $h[n]$ in terms of the CT impulse response

$$h[n] = h(n)$$

and the other specified the system function $\tilde{H}(z)$ in terms of the CT system function through the transformation $z = e^s$. The fourth, and perhaps most novel, approach was to use the DFT formulas to derive $h[n]$ from selected samples of the desired frequency response $H(\phi)$. The DFT implemented in the form of the FFT also provides an efficient means for realizing the DT filtering operation.

We have already pointed out that our goal in this book is more to emphasize the power of our tools as a *language* for describing and thinking about complex system behavior than to provide specific analytical methods for particular problems. Our treatment of DT transforms in this chapter has been consistent with that objective—we have stressed their continuity with the CT transform techniques studied earlier, and have used them as a topic for illustrating how effective our language is and how facile we have become in speaking it. The DTFT, DFT, FFT, etc., are themselves pre-eminently tools for *doing* things—not talking about them. They are thus in a sense more similar to such techniques as feedback or modulation than to CT transforms, which they resemble mostly because they are abstractions with similar forms. As practical engineering tools for digital signal processing, the DT transforms have received extensive fine-tuning and development. This is an important topic that you will want to consider as part of your later professional studies.

APPENDIX TO CHAPTER 18
Table XVIII.1—Short Table of Discrete-Time Fourier Transforms

$$x[n] = \int_{-1/2}^{1/2} X(\phi) e^{j2\pi\phi n}\, d\phi \qquad\qquad X(\phi) = \sum_{n=-\infty}^{\infty} x[n] e^{-j2\pi\phi n}$$

$$\delta[n] = \begin{cases} 1, & n = 0 \\ 0, & n \neq 0 \end{cases} \qquad\Longleftrightarrow\qquad 1$$

$$1 \qquad\Longleftrightarrow\qquad \sum_{m=-\infty}^{\infty} \delta(\phi - m)$$

$$\alpha^n u[n], \quad |\alpha| < 1 \qquad\Longleftrightarrow\qquad \frac{1}{1 - \alpha e^{-j2\pi\phi}}$$

$$u[n] \qquad\Longleftrightarrow\qquad \frac{1}{1 - e^{-j2\pi\phi}} + \frac{1}{2}\sum_{m=-\infty}^{\infty} \delta(\phi - m)$$

$$e^{j2\pi\phi_0 n} \qquad\Longleftrightarrow\qquad \sum_{m=-\infty}^{\infty} \delta(\phi - \phi_0 - m)$$

$$\sum_{m=(1-N)/2}^{(N-1)/2} \delta[n-m], \quad N \text{ odd} \qquad\Longleftrightarrow\qquad \frac{\sin N\pi\phi}{\sin \pi\phi}$$

$$\frac{\sin 2\pi\phi_0 n}{\pi n} \qquad\Longleftrightarrow\qquad \begin{cases} 1, & |\phi - m| < \phi_0, \quad -\infty < m < \infty \\ 0, & \text{elsewhere} \end{cases}$$

$$\sum_{m=-\infty}^{\infty} \delta[n-mN] \qquad\Longleftrightarrow\qquad \frac{1}{N}\sum_{m=-\infty}^{\infty} \delta\left(\phi - \frac{m}{N}\right)$$

Table XVIII.2—Important Discrete-Time Fourier Transform Theorems

Delay	$x[n-N]$	$\Longleftrightarrow$	$e^{-j2\pi\phi N} X(\phi)$
Multiply-by-$e^{j2\pi\phi_0 n}$	$e^{j2\pi\phi_0 n} x[n]$	$\Longleftrightarrow$	$X(\phi - \phi_0)$
Multiply-by-n	$n x[n]$	$\Longleftrightarrow$	$\dfrac{1}{-j2\pi}\dfrac{dX(\phi)}{d\phi}$
Convolution	$\displaystyle\sum_{m=-\infty}^{\infty} x[m] y[n-m]$	$\Longleftrightarrow$	$X(\phi) Y(\phi)$
Product	$x[n] y[n]$	$\Longleftrightarrow$	$\displaystyle\int_{-1/2}^{1/2} X(\nu) Y(\phi - \nu)\, d\nu$
Summation	$\displaystyle\sum_{m=-\infty}^{n} x[m]$	$\Longleftrightarrow$	$\dfrac{X(\phi)}{1 - e^{-j2\pi\phi}} + \dfrac{X(0)}{2}\displaystyle\sum_{m=-\infty}^{\infty} \delta(\phi - m)$
Scale-Change	$\displaystyle\sum_{m=-\infty}^{\infty} x[m]\delta[n-mN]$	$\Longleftrightarrow$	$X(\phi N)$

The symmetry and linearity properties are the same as for CT transforms. The duality and differentiation properties are essentially meaningless for DT transforms.

$$\sum_{n=-\infty}^{\infty} |x[n]|^2 = \int_{-1/2}^{1/2} |X(\phi)|^2\, d\phi; \quad x[0] = \int_{-1/2}^{1/2} X(\phi)\, d\phi; \quad X(0) = \sum_{n=-\infty}^{\infty} x[n]$$

EXERCISES FOR CHAPTER 18

Exercise 18.1

a) Show directly that the DTFT formulas (18.0–11) and (18.0–12) are inverses of one another by the following procedure:

  i) Substitute $X(\phi)$ from (18.0–12) into (18.0–11). (Be careful to distinguish by an appropriate change of notation between n as a variable of summation and n as the value of the argument of $x[n]$.)

  ii) Interchange the order of integration and summation. (This would require careful investigation in a rigorous discussion.)

  iii) Carry out the integration to obtain the identity $x[n] = x[n]$.

b) Similarly, derive the DTFT version of Parseval's Theorem

$$\int_{-1/2}^{1/2} |X(\phi)|^2 \, d\phi = \sum_{n=-\infty}^{\infty} |x[n]|^2$$

by the following procedure:

  i) Think of $|X(\phi)|^2$ as $X(\phi)X^*(\phi)$ and substitute (18.0–12) for one of these terms on the left. (Allow for the possibility that $x[n]$ might be complex.)

  ii) Interchange the order of integration and summation.

  iii) Carry out the integration using (18.0–11).

Exercise 18.2

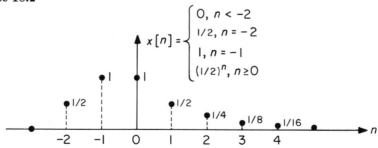

$$x[n] = \begin{cases} 0, & n < -2 \\ 1/2, & n = -2 \\ 1, & n = -1 \\ (1/2)^n, & n \geq 0 \end{cases}$$

The DTFT of the DT signal $x[n]$ shown above is $X(\phi) = \displaystyle\sum_{n=-\infty}^{\infty} x[n] e^{-j2\pi\phi n}$.

a) Argue that $\displaystyle\int_{-1}^{1} X(\phi) \, d\phi = 2$.

b) Argue that $\displaystyle\int_{0}^{1/2} |X(\phi)|^2 \, d\phi = \frac{31}{24} \approx 1.29$.

Exercise 18.3

Argue that a finite-duration unit sample response $h[n]$ is a necessary condition for a causal stable DT LTI filter to have a strictly linear phase

$$\angle H(\phi) \sim \phi, \quad -1/2 < \phi \leq 1/2.$$

Exercise 18.4

Example 18.1–1 explored the properties of

$$X(\phi) = \frac{1}{1 - \alpha e^{-j2\pi\phi}}$$

which is the DTFT of

$$x[n] = \alpha^n u[n], \quad |\alpha| < 1.$$

In particular, the magnitude and angle of $X(\phi)$ were sketched for three positive values of α.

a) Argue that $x[n]$ for α negative is equal to $x[n]$ for α positive multiplied by the DT function $r[n] = e^{j2\pi\phi_0 n}$, $-\infty < n < \infty$, with an appropriately chosen value of ϕ_0.

b) From the appropriate entry in Table XVIII.2, conclude that the DTFT of $x[n]$ for α negative is the DTFT of $x[n]$ for the corresponding positive value of α but shifted along the frequency axis. Sketch $|X(\phi)|$ for $\alpha = -0.5$ using the results of Example 18.1–1 and the above argument.

c) Similarly, show that

$$(-1)^n u[n] \quad \Longleftrightarrow \quad \frac{1}{1 + e^{-j2\pi\phi}} + \frac{1}{2} \sum_{n=-\infty}^{\infty} \delta(\phi - n - 0.5).$$

Sketch the real and imaginary parts of the DTFT of $(-1)^n u[n]$.

Exercise 18.5

A computer program is available that computes the DFT $X[m]$ of a sequence of N complex numbers $x[n]$ according to the second equation of the pair (18.3–6). The resulting sequence of N numbers is then operated on again by the same program. What is the relationship between the output of this second operation and the original set of N complex numbers?

PROBLEMS FOR CHAPTER 18

Problem 18.1

a) Compute the DTFT

$$X(\phi) = \sum_{n=-\infty}^{\infty} x[n]e^{-j2\pi\phi n}$$

for each of the following DT functions:

i)

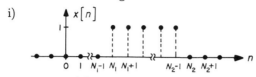

ii) $x[n] = \alpha^{|n|}, \quad 0 < \alpha < 1, \quad -\infty < n < \infty$

iii) $x[n] = n, \quad -\infty < n < \infty$

iv) $x[n] = \cos\dfrac{\pi n}{4}, \quad -\infty < n < \infty$

v) $x[n] = \cos\dfrac{9\pi n}{4}, \quad -\infty < n < \infty$

vi) $x[n] = \cos\dfrac{\pi n}{4}u[n].$

b) Compute the inverse DTFT

$$x[n] = \int_{-1/2}^{1/2} X(\phi)e^{j2\pi\phi n}\,d\phi$$

for each of the following periodic functions:

i)

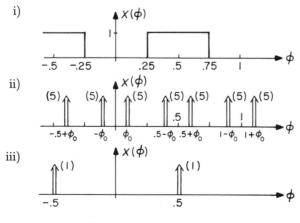

iv) $X(\phi) = \displaystyle\sum_{n=-\infty}^{\infty} \dot{\delta}(\phi - n)$

v) $X(\phi) = je^{-j3\pi\phi}\dfrac{\sin 2\pi\phi}{\cos \pi\phi}.$

Problem 18.2

A DT signal $x[n]$ with DTFT

$$X(\phi) = \sum_{n=-\infty}^{\infty} x[n]e^{-j2\pi\phi n} = e^{-j\pi\phi}\frac{\sin 4\pi\phi}{\sin \pi\phi}$$

is the input to a DT LTI system with unit sample response

$$h[n] = 2^n u[n].$$

Find and sketch the output

$$y[n] = x[n] * h[n].$$

Problem 18.3

A DT sequence $y[n]$ is constructed from another DT sequence $x[n]$ according to the formula

$$y[n] = x[nN]$$

where N is a constant positive integer greater than one. (This process is usually called *decimation*, although this name would strictly be appropriate only if $N = 10$.)

a) Sketch a typical $x[n]$ and the corresponding $y[n]$ for, say, $N = 3$.

b) Suggest a set of conditions to be imposed on the DTFT of $x[n]$ such that it will be possible to reconstruct $x[n]$ for *all* n from $y[n]$.

c) Describe a specific scheme for carrying out the reconstruction if the conditions of (b) apply.

d) The conditions on $x[n]$ that are sufficient to permit reconstruction of $x[n]$ for all n given $y[n]$ are not unique. Describe at least one set of conditions different from those in (b) that will also suffice.

Problem 18.4

In the diagram on the next page, $X(\phi) = \sum_{n=-\infty}^{\infty} x[n]e^{-j2\pi\phi n}$ in the period $-1/2 \leq \phi < 1/2$ is bandlimited to the band $|\phi| < 1/4$, and $H(\phi) = \sum_{n=-\infty}^{\infty} h[n]e^{-j2\pi\phi n}$ in the period $-1/2 \leq \phi < 1/2$ is a highpass filter as shown.

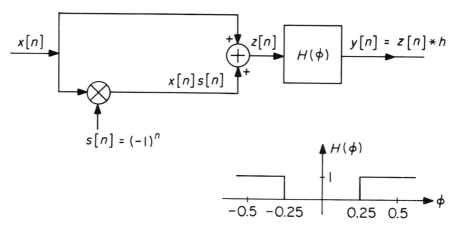

a) Determine $S[m]$ in the formula

$$s[n] = \sum_{m=0}^{N-1} S[m]e^{j2\pi nm/N}, \quad N = 2.$$

b) Determine and sketch $S(\phi)$ in the formula

$$s[n] = \int_{-1/2}^{1/2} S(\phi)e^{j2\pi\phi n} \, d\phi.$$

c) Show that $y[n]$ can be written $y[n] = r[n]x[n]$. Find and sketch $r[n]$.

Problem 18.5

The figure below depicts a DT system consisting of a parallel combination of N DT LTI filters with unit sample responses $h_k[n]$, $k = 0, 1, \ldots, N-1$. $h_k[n]$ for any k is related to $h_0[n]$ by

$$h_k[n] = e^{j2\pi nk/N}h_0[n].$$

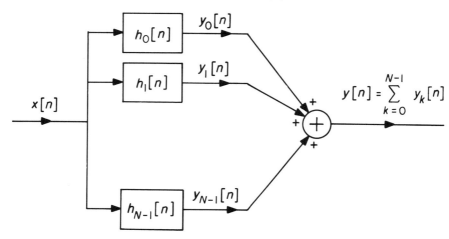

a) If $h_0[n]$ is an ideal lowpass filter with frequency response $H_0(\phi)$ as shown for $-1/2 \le \phi < 1/2$ in the figure below, sketch the DTFTs of $h_1[n]$ and $h_{N-1}[n]$ for ϕ in the range $-1/2 \le \phi < 1/2$.

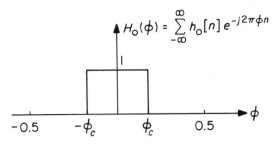

$$H_0(\phi) = \sum_{-\infty}^{\infty} h_0[n]\, e^{-j2\pi\phi n}$$

b) Determine, in terms of N, the value of ϕ_c, $0 < \phi_c < 1/2$, such that the system shown on the preceding page is an identity system, $y[n] = x[n]$, for all n and for any input $x[n]$.

c) Now suppose that $h_0[n]$ is no longer restricted to be an ideal lowpass filter. If $h[n]$ denotes the unit sample response of the overall system with input $x[n]$ and output $y[n]$, show that $h[n]$ can be expressed in the form

$$h[n] = r[n] h_0[n].$$

Determine and sketch $r[n]$. From this result determine a necessary and sufficient condition on $h_0[n]$ to ensure that the overall system shown on the preceding page will be an identity system.

Problem 18.6

Another transformation that maps the left half of the s-plane into the inside of the unit circle in the z-plane is the *bilinear transformation*

$$z = \frac{1+s}{1-s}.$$

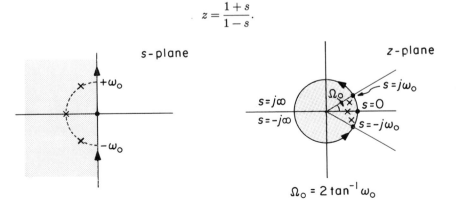

Compared with the exponential transformation $z = e^s$ discussed in the chapter, the bilinear transformation has several advantages, of which the most obvious is that aliasing effects are eliminated since the entire $j\omega$-axis, $-\infty < \omega < \infty$, is mapped only once onto the unit circle.

a) Suppose that the bilinear transformation had been used in Example 18.2–3 in place of the exponential. Determine Ω_0 and ω_0 such that the cutoff frequency of the DT filter will be $\phi_0 = 0.1$.

b) The poles of the 3rd-order Butterworth filter with system function

$$H(s) = \frac{1}{(s/\omega_0)^3 + 2(s/\omega_0)^2 + 2(s/\omega_0) + 1}$$

are located at

$$(s/\omega_0) = -1, \frac{-1 \pm j\sqrt{3}}{2}.$$

Find the corresponding pole locations for $\tilde{H}(z)$.

c) In accordance with the bilinear transformation, where are the zeros of $\tilde{H}(z)$? Compare these zero locations with those for the exponential transformation as given in Example 18.2–3. Discuss the effect of this difference on $H(\phi)$.

d) Determine a formula for $\tilde{H}(z)$ and devise a gain-adder-delay realization of this DT filter.

Problem 18.7

Reactionary Systems, Inc., has proposed the following hybrid scheme to overcome certain computer speed limitations affecting the realization of DT LTI filters. They suggest (see the diagram below) converting the incoming DT sequence $\hat{x}[n]$ into a CT impulse train

$$x(t) = \sum_{n=-\infty}^{\infty} \hat{x}[n]\delta(t - nT)$$

which is then filtered by a CT system with impulse response $h(t)$ to yield

$$y(t) = h(t) * x(t).$$

The output is finally sampled and reconverted to a DT sequence

$$\hat{y}[n] = y(nT).$$

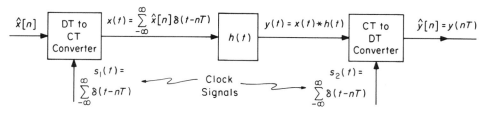

a) Find the overall response of this system to the input $\hat{x}[n] = \delta[n]$. Assume some arbitrary shape for $h(t)$, and illustrate your argument with sketches of $x(t)$, $y(t)$, and $\hat{y}[n]$.

b) This system contains non-constant inputs in addition to $\hat{x}[n]$ (the clock signals) and includes subsystems that are not exactly LTI (the DT/CT converters). Nevertheless, it functions as a linear time-(or shift-)invariant DT system. Justify this statement, and give a formula for the DT unit sample response $\hat{h}[n]$.

c) Does the system remain LTI if the phases of the clock signals are different, for example, if $s_1(t) = \sum_{n=-\infty}^{\infty} \delta(t - nT)$, $s_2(t) = \sum_{n=-\infty}^{\infty} \delta(t - nT - \Delta)$, $\Delta \neq 0$? If it does remain LTI, determine the unit sample response (if different from (b)).

d) Repeat part (c) for the case where the periods of the clock signals are different, for example, for $s_1(t) = \sum_{n=-\infty}^{\infty} \delta(t - nT_1)$, $s_2(t) = \sum_{n=-\infty}^{\infty} \delta(t - nT_2)$, $T_1 \neq T_2$.

e) For the original clock signals, find the DT frequency response

$$\hat{H}(\phi) = \sum_{n=-\infty}^{\infty} \hat{h}[n] e^{-j2\pi\phi n}$$

in terms of

$$H(f) = \int_{-\infty}^{\infty} h(t) e^{-j2\pi ft} \, dt.$$

HINT: Note that

$$\sum_{n=-\infty}^{\infty} \hat{h}[n] e^{-j2\pi\phi n} = \int_{-\infty}^{\infty} \left[\sum_{n=-\infty}^{\infty} \hat{h}[n] \delta(t - nT) \right] e^{-j2\pi\phi t/T} \, dt$$

and that

$$\sum_{n=-\infty}^{\infty} \hat{h}[n] \delta(t - nT) = h(t) \sum_{n=-\infty}^{\infty} \delta(t - nT).$$

Problem 18.8

In many practical situations, an interesting but unknown signal $s(t)$ is corrupted by an echo; the observed signal $x(t)$ then has the form

$$x(t) = s(t) + ks(t - T_0).$$

Frequently, the echo amplitude k and the delay T_0 can be estimated from previous observations or knowledge of the physical situation. We propose to study various ways to recover $s(t)$ from $x(t)$ if k and T_0 are known.

a) Find (in terms of k and T_0) the frequency response $H(f)$ of an LTI CT system that will yield $s(t)$ as its output if $x(t)$ having the form above is its input.

b) Show that $H(f)$ in (a) can be realized by the block diagram below. Find the values of k_0, k_1, and T_1 in terms of k and T_0. Sketch the impulse response $h(t)$ of this system.

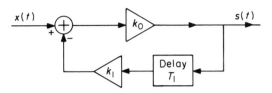

c) Assume that we know, in addition to k and T_0, that $s(t)$ (and hence $x(t)$) is band-limited with $S(f) = 0$ for $|f| > W$. Moreover, $T_0 \leq 1/2W$. Determine the difference equation describing the DT LTI system in the block diagram below such that the overall output is $s(t)$ when $x(t)$ is the input. Note that the sampling period T has been chosen equal to the echo delay T_0.

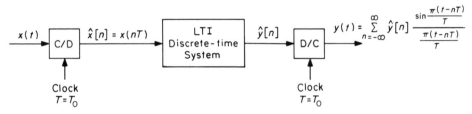

d) Explain why the system in (c) would fail if $1/2W < T_0 \leq 1/W$. Argue, however, that doubling the clock rate ($T = T_0/2$) and making appropriate changes in the DT system will restore the desired performance. Specify the frequency response (DTFT) of the new DT system.

Problem 18.9

a) Find explicitly the DFT

$$X[m] = \frac{1}{N} \sum_{n=0}^{N-1} x[n] e^{-j2\pi mn/N}$$

for the DT function $x[n] = \sin(2\pi n/8)$, $N = 4$.

b) Demonstrate explicitly that $x[n]$, $0 \leq n < 4$, can be recovered from $X[m]$ through the formula

$$x[n] = \sum_{m=0}^{N-1} X[m] e^{j2\pi mn/N}.$$

Problem 18.10

a) Use the formula for the partial sum of a geometric series to prove that

$$\sum_{n=0}^{N-1} e^{j2\pi mn/N} = \begin{cases} N, & m = 0 \\ 0, & 0 < m < N. \end{cases}$$

b) Use the result of (a) to show that the set of linear equations

$$x[n] = \sum_{m=0}^{N-1} X[m]e^{j2\pi mn/N}$$

are in fact the solution to the set of linear equations

$$X[m] = \frac{1}{N} \sum_{n=0}^{N-1} x[n]e^{-j2\pi mn/N}.$$

c) Use the result of (a) to prove Parseval's Theorem for the DFT:

$$\frac{1}{N} \sum_{n=0}^{N-1} |x[n]|^2 = \sum_{m=0}^{N-1} |X[m]|^2.$$

d) Use the result of (a) or (b) to prove the convolution theorems, equations (18.4–2) and (18.4–3), for the DFT.

Problem 18.11

The CT input function $x(t)$ to the C/D box described in Section 18.2 yields the DT output function $\hat{x}[n] = x(nT)$. Suppose $x(t)$ is periodic with period T_0 so that it can be expanded in a Fourier series

$$x(t) = \sum_{m=-\infty}^{\infty} X[m]e^{j2\pi mt/T_0}.$$

a) Argue that the DT function $\hat{x}[n]$ is periodic if and only if T_0/T is an integer.

b) If $T_0/T = N$ is an integer, then N is the period of $\hat{x}[n]$ and we can expand

$$\hat{x}[n] = \sum_{m=0}^{N-1} \hat{X}[m]e^{j2\pi mn/N}.$$

What is the relationship between the two sets of Fourier series coefficients $X[m]$ and $\hat{X}[m]$? (Note that one of these sets is finite, the other infinite.)

Problem 18.12

a) If $x[n]$ is zero outside the interval $0 \le n < N$, show that the output $y[n]$ of the system below is $x[n]$.

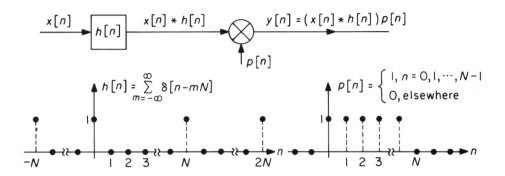

b) By expressing the operation of this system in the frequency domain using the DTFT, show that equating $y[n] = x[n]$ gives another way of deriving (18.3–5)

$$x[n] = \sum_{m=0}^{N-1} X[m]e^{j2\pi mn/N}, \qquad 0 \le n < N.$$

Problem 18.13

Suppose we want to convolve a long sequence $x[n]$ with an FIR $h[n]$ that is zero outside the interval $0 \le n < N$. Show by means of pictures and formulas that it is possible to carry out this convolution using the DFT and FFT by the following procedure:

1. Break up $x[n]$ into subsequences of length N.

2. Construct new finite sequences of length $2N$ by appending to each subsequence of $x[n]$ and to $h[n]$ a sequence of N zeros.

3. DFT these extended sequences of length $2N$.

4. Multiply each DFT of the extended $x[n]$ subsequences by the DFT of $h[n]$ extended.

5. Inverse DFT each product.

6. Combine the resulting sequences of length $2N$, overlapping successive sequences by N samples.

If the length of the original sequence $x[n]$ is $N_0 \gg N$, compare the number of multiply-and-adds required to evaluate $x[n] * h[n]$ by this approach with the number needed for a direct evaluation of the convolution sum.

Problem 18.14

The following program is the core of an FFT procedure evaluating

$$x[n] = \sum_{m=0}^{N-1} X[m]e^{j2\pi mn/N}$$

for a given set of N complex numbers $X[m]$, assuming that N is any power of 2. It is written in the SCHEME dialect of LISP,* which has the advantage of making the underlying mechanism of the FFT procedure readily apparent (although at some sacrifice in speed compared with, say, FORTRAN[†]):

```
(define (fft list)
   (let ((n (length list)))
      (cond ((= n 1) list)
            ((odd? n) (error "FFT input length not 2**N"))
            (else (let ((even-terms (fft (evens list)))
                        (odd-terms (w* (fft (odds list)))))
                     (append (add even-terms odd-terms)
                             (sub even-terms odd-terms)))))))
```

The procedure FFT takes a list of N complex numbers $(X[0]\,X[1]\ldots X[N-1])$ and returns the list of N (possibly complex) numbers $(x[0]\,x[1]\ldots x[N-1])$. The procedure is built upon two recursive calls to evaluate the FFTs of the two lists of $N/2$ terms provided by the subprocedures EVENS and ODDS:

$$(\text{evens list}) = (X[0]\,X[2]\ldots X[N-2])$$
$$(\text{odds list}) = (X[1]\,X[3]\ldots X[N-1]).$$

The subprocedure W* takes a list of $N/2$ complex numbers $(x_0\,x_1\,x_2\ldots x_{N/2-1})$ and returns the list of $N/2$ complex numbers $(x_0\,x_1 W\,x_2 W^2\ldots x_{N/2-1}W^{N/2-1})$, where $W = e^{j2\pi/N}$. The APPEND operation makes a single list of N elements, of which the first $N/2$ are obtained by ADDing term by term the elements of EVEN-TERMS and ODD-TERMS, and the last $N/2$ are obtained by SUBtracting term by term the elements of EVEN-TERMS and ODD-TERMS. As implied by (18.4-5) and the footnote on that same page, the result is the FFT sought.

   a) Devise procedures EVENS, ODDS, ADD, SUB, and W* to carry out the manipulations desired. Since all lists may contain complex numbers, you will also have to devise a representation for complex numbers (e.g., as pairs) and appropriate procedures for adding, subtracting, and multiplying complex numbers.

   b) Demonstrate that the complete FFT package works by testing it with various simple sets of $X[m]$ for which you can easily compute the set $x[n]$.

   c) Explore with examples how the computation time depends on N.

---

*H. Abelson and G. J. Sussman with J. Sussman, *Structure and Interpretation of Computer Programs* (Cambridge, MA: MIT Press and McGraw-Hill, 1985).

[†]For a collection of programs optimized for speed, see *Programs for Digital Signal Processing* (New York, NY: IEEE Press, 1979).

d) Devise a modified procedure to recover the set $X[m]$ from the set $x[n]$, based upon

$$X[m] = \frac{1}{N} \sum_{n=0}^{N-1} x[n]e^{-j2\pi nm/N}.$$

Demonstrate that your new procedure works by first taking the FFT of some "random" set $X[m]$ and then operating on the result with your modified procedure to recover $X[m]$.

19

AVERAGES AND RANDOM SIGNALS

19.0 Introduction

In the preceding chapters, we have typically assumed that certain of the signals in a system (such as the sources) are completely specified, either as ordinary functions in terms of what they "are" or as generalized functions in terms of what they "do." Our problem has usually been to find complete specifications of some or all of the remaining signals in the system (such as the responses). Often, however, it is desirable or necessary to study system behavior in response to source signals that are only partially specified, that is, signals for which we know only a limited, incomplete set of *properties* or *features*. We might, for example, ask what we can say about the system response if we know only that the source magnitude is less than some maximum value, or only that the average power of the source has a specified value, or only that the spectrum of the source waveform contains no frequencies greater than W.

In such cases, there generally exists a large class or *ensemble* of source signal waveforms sharing the specified properties. Thus, in our discussion of the Sampling Theorem in Chapter 14 we showed that every waveform of the form

$$x(t) = \sum_{n=-\infty}^{\infty} x[n] \frac{\sin 2\pi W(t - n/2W)}{2\pi W(t - n/2W)} \qquad (19.0-1)$$

has its spectrum restricted to the band $|f| < W$ for any choice of the numbers $x[n]$; each distinct choice of the sequence $x[n]$, $-\infty < n < \infty$, determines a distinct member of the ensemble of waveforms whose spectra contain no frequencies greater than W.

Each member of an ensemble of source signals sharing some common features determines a member of the ensemble of response signals. Our objective in this chapter is to find descriptions of the properties or features that characterize the response ensemble as functions of the specified properties or features that determine the source ensemble. We cannot expect that this will be possible in general for arbitrary source features and arbitrary systems. But if the properties characterizing the source ensembles are long-term averages of various kinds, then the response ensembles are typically also described by long-term averages. If the system is LTI, moreover, then knowledge of a small set of appropriate averages of the source signals may permit us to compute corresponding averages of the response signals. As we shall illustrate, this relationship between certain averages of the inputs and outputs of LTI systems is widely useful.

Signals described in terms of sets of averages are sometimes called *random signals*. It is not hard to see why. For example, signals as regular as sinusoids will, of course, be members of the ensemble of bandlimited waveforms,* but most[†] of the waveforms specified by (19.0–1) will look haphazard, disordered, unpredictable, unstructured—as the word "random" implies. Indeed, in many cases apparent chaos is the most obvious characteristic of signals that we might wish to describe by averages. One example is the "noise" produced by the thermal agitation of charged particles in a resistor or semiconductor or any dissipative system. But even meaningful speech and music waveforms may seem chaotic and meaningless if displayed on an oscilloscope rather than presented acoustically to the ear. Moreover, for signals in communication systems it is precisely our inability to predict the exact future of a waveform that permits the waveform to tell us something new, that is, to convey information.

We cannot, however, base mathematics on the absence of order; what is important about random signals in the sense to be discussed in this chapter is not that they are or may be unpredictable in detail, but rather that they possess certain specific and predictable averages. Thus we can partially characterize thermal agitation noise by saying that it has a certain average *amplitude distribution*. And similarly we can claim that the *power density spectrum* (which as we shall see is related to an average) of a speech wave in some communication system has a certain shape that is largely restricted to the frequency band 300 to 3000 Hz.

Instead of emphasizing the "unpredictable" aspect of random signals, it is perhaps more to the point to observe that signals that are usefully (or necessarily) characterized by averages (instead of by specific time functions) are nearly always complex. Thus it is difficult to know what we would do (if we had it) with a complete mathematical specification of the response of a telephone line to a particular speaker giving a particular speech. We would be overwhelmed with details. From appropriate average properties of the response to many speakers and many speeches, however, we may be able to describe the quality of the phone line as a speech communication link, suggest ways to improve its performance, or provide information vital to the design of the terminal equipment.

We must emphasize that applications of the methods to be introduced in this

---

*As we shall show, the feature of band limitation is derived from a condition on long-term averages.

[†] "Most" sounds intuitively correct, but we have not provided (and will not provide) any measure or weighting for the relative likelihood or density of various waveforms in our ensembles and hence cannot rigorously conclude that "most" ensemble members will appear unstructured. Indeed, by suitably arranging the procedure for constructing the ensemble of bandlimited waveforms, we could make the fraction that are pure sinusoids as large as we wish. This is precisely where the mathematically rigorous theory of *random processes* (which we do not intend to discuss) begins—by assigning measures or probabilities to each member of the ensemble.

chapter can be controversial. Not all "random-looking" waveforms are regular or uniform or homogeneous enough to be usefully or effectively characterized by averages. For example, the strength of a radio wave reflected from the ionosphere fluctuates irregularly from moment to moment. The average signal strength over successive short periods such as an hour will usually be nearly the same, but one-hour averages from day and night or from successive days will often be markedly different. A similar short-term homogeneity and long-term variability is characteristic of many other phenomena, such as weather data, sunspot numbers, biological and medical records, geological and paleological observations, and most socioeconomic activity such as stock-market prices. If the fluctuations in the averages are regular and predictable, then (by more complicated methods than we shall discuss) the data can often be handled as a *non-stationary* random signal—price indices, for example, are usually "corrected for seasonal variations," and sunspots tend to occur in "11-year cycles." But if the fluctuations in the averages are themselves irregular, then either we must have a very long data record so that we can estimate the "averages of the averages," or we must try to design systems that will continually measure the short-time averages and "adapt" to them, or perhaps we must simply admit that it is not very helpful to try to characterize some signals by averages. The history of nuclear power plants, for example, is so limited, and the rate of change of their technology so rapid, that it is simply not feasible to determine with any precision a credible value for such an average as the probability of a "meltdown" per reactor per year. Those who attempt to justify public policy by relying on the accuracy of such probability estimates are not being "rational" or "scientific"—they are just being naive.

The important things about any signal are the things we can say about it, not the things we cannot. When we consider a signal as random we assume that what we do know about it—indeed, the only things we know about it—are certain averages. The goal in preceding chapters has usually been to develop methods for describing in detail the output of some system given a detailed description of the input; in this chapter the analogous goal is to develop methods for describing certain averages of the output given only averages of the input.

19.1 Averages of Periodic Functions

It is helpful to begin our study of averages and their properties by reviewing and extending the discussion in Section 12.5 of averages of periodic functions. The infinite-time average of a function $x(t)$, defined by

$$\langle x(t) \rangle = \lim_{T_0 \to \infty} \frac{1}{2T_0} \int_{-T_0}^{T_0} x(t)\,dt \qquad (19.1\text{--}1)$$

can be computed explicitly if $x(t)$ is periodic. As argued in Section 12.5, it is equal to the average of $x(t)$ over a single period T:

$$\langle x(t) \rangle = \frac{1}{T} \int_0^T x(t)\,dt. \qquad (19.1\text{--}2)$$

Hence, for periodic functions we can directly extend our methods for analyzing input-output behavior to explore the relationships between averages of inputs and averages of outputs. On the other hand, if $x(t)$ is not periodic, then (19.1–1) may still describe what we intend to mean by the infinite-time average of $x(t)$, but it is not very satisfactory mathematically because (unless perhaps $\langle x(t) \rangle = 0$) it is hard to imagine how we could provide the huge amount of information necessary to specify $x(t)$ in such detail as to allow the limiting operation to be carried out. Studying the periodic case—where we can carry out the limit—will yield examples, features, and properties that will guide us in the generalization we seek.

In Section 12.5, we extended (19.1–2) to show that the Fourier series coefficients $X[n]$ can be derived as averages,

$$X[n] = \langle x(t)e^{-j2\pi nt/T} \rangle \tag{19.1–3}$$

and that Parseval's Theorem for real periodic waveforms can be written

$$\langle x^2(t) \rangle = \frac{1}{T} \int_0^T x^2(t)\,dt = \sum_{n=-\infty}^{\infty} |X[n]|^2 \tag{19.1–4}$$

which can be interpreted as stating that the average power in a periodic function $x(t)$ is the sum* of the average powers in its harmonic components. Since the powers in individual Fourier components can be added to determine the total average power, it is possible to define a *power density spectrum* $S_x(f)$ for periodic functions,

$$S_x(f) \overset{\Delta}{=} \sum_{n=-\infty}^{\infty} |X[n]|^2 \delta(f - n/T) \tag{19.1–5}$$

which measures the distribution of power in $x(t)$ as a function of frequency. The total average power in a real periodic waveform $x(t)$ is then

$$\int_{-\infty}^{\infty} S_x(f)\,df = \sum_{n=-\infty}^{\infty} |X[n]|^2 = \langle x^2(t) \rangle. \tag{19.1–6}$$

Note that the impulses comprising $S_x(f)$ always have non-negative real areas for any $x(t)$. If $x(t)$ is real (as will be assumed throughout this chapter), then $|X[-n]| = |X[n]|$ and $S_x(f)$ will be an even function of f.

Since in general the Fourier components of the output $y(t)$ of an LTI system with frequency response $H(f)$ in response to a periodic input $x(t)$ are

$$Y[n] = X[n]H(n/T) \quad \text{or} \quad |Y[n]|^2 = |X[n]|^2 |H(n/T)|^2 \tag{19.1–7}$$

---

*It is not, of course, generally true that the average power in a sum of waveforms is the sum of the average powers; that is, in general $\langle [x(t) + y(t)]^2 \rangle \neq \langle x^2(t) \rangle + \langle y^2(t) \rangle$. The fact that the power in the sum is the sum of the powers for a Fourier series follows from the orthogonality of harmonically related sinusoids, as discussed in the appendix to Chapter 14. See, however, Problem 19.8.

the output power density spectrum is

$$S_y(f) = S_x(f)|H(f)|^2 \qquad (19.1\text{--}8)$$

and the total average output power is

$$\langle y^2(t) \rangle = \int_{-\infty}^{\infty} S_y(f)\,df = \int_{-\infty}^{\infty} S_x(f)|H(f)|^2\,df. \qquad (19.1\text{--}9)$$

If, in particular, the LTI system is an ideal filter passing with unit gain a limited band of frequencies, then the average output power is simply the sum of the powers in just those input components lying in that band, which is another way of understanding the sense in which $S_x(f)$ measures the distribution of power in $x(t)$ as a function of frequency.

The power density spectrum of a periodic signal consists of impulses at harmonically related frequencies and hence is the transform of some periodic time function. Indeed, it is easy to show that for real $x(t)$ this time function is

$$R_x(\tau) \overset{\Delta}{=} \langle x(t)x(t-\tau) \rangle = \frac{1}{T} \int_0^T x(t)x(t-\tau)\,dt$$

$$= \frac{1}{T} x(t) \circledast x(-t) = \sum_{n=-\infty}^{\infty} |X[n]|^2\, e^{j2\pi n\tau/T}$$

$$= \int_{-\infty}^{\infty} S_x(f)e^{j2\pi f\tau}\,df. \qquad (19.1\text{--}10)$$

$R_x(\tau)$ is called the *autocorrelation function* of the real periodic function $x(t)$; it has a number of characteristic properties:

i) $R_x(\tau) = \langle x(t)x(t-\tau) \rangle$ is periodic with the same period as $x(t)$. The Fourier coefficients of $R_x(\tau)$ are $|X[n]|^2$.

ii) The Fourier transform of $R_x(\tau)$ is $S_x(f)$, which is real and non-negative. This implies among other things that $R_x(\tau)$ is an even function for real $x(t)$.

iii) $R_x(0) = \langle x^2(t) \rangle \geq |R_x(\tau)|$; that is, $R_x(\tau)$ has its largest magnitude at $\tau = 0, \pm T, \pm 2T$, etc., at which points it is equal to the average power in $x(t)$. The proof of this property follows immediately from the equation above since

$$|R_x(\tau)| = \left| \int_{-\infty}^{\infty} S_x(f)e^{j2\pi f\tau}\,df \right|$$

$$\leq \int_{-\infty}^{\infty} |S_x(f)e^{j2\pi f\tau}|\,df = \int_{-\infty}^{\infty} S_x(f)\,df = R_x(0)$$

where we have used the fact that $S_x(f)$ is real and non-negative (see also Problem 14.17).

$R_x(\tau)$ measures the extent to which $x(t)$ and $x(t-\tau)$ are the same; that is, it measures the rate at which $x(t)$ changes—as the following examples illustrate.

Example 19.1–1

Suppose that

$$x(t) = A\cos(\omega_0 t + \theta), \quad -\infty < t < \infty. \tag{19.1-11}$$

Then

$$R_x(\tau) = \langle x(t)x(t-\tau) \rangle$$

$$= \frac{\omega_0 A^2}{2\pi} \int_0^{2\pi/\omega_0} \cos(\omega_0 t + \theta)\cos(\omega_0 t - \omega_0 \tau + \theta)\, dt$$

$$= \frac{\omega_0 A^2}{4\pi} \int_0^{2\pi/\omega_0} \cos(2\omega_0 t - \omega_0 \tau + 2\theta)\, dt$$

$$+ \frac{A^2}{2}\frac{\omega_0}{2\pi} \int_0^{2\pi/\omega_0} \cos\omega_0\tau\, dt. \tag{19.1-12}$$

The first integral in the last term is zero (since it is the integral of a sinusoid over two full periods), and the second integral is the average of a constant. Hence

$$R_x(\tau) = \frac{A^2}{2}\cos\omega_0\tau. \tag{19.1-13}$$

As expected, $R_x(\tau)$ is periodic with period $2\pi/\omega_0$ and is an even function. $R_x(0)$ is the average power $A^2/2$ in $x(t)$ and is greater than or equal to $|R_x(\tau)|$ at any other point. The Fourier transform of $R_x(\tau)$ is the power density spectrum

$$S_x(f) = \frac{A^2}{4}\delta(f - \frac{\omega_0}{2\pi}) + \frac{A^2}{4}\delta(f + \frac{\omega_0}{2\pi}). \tag{19.1-14}$$

The amplitudes of the impulses in $S_x(f)$ are the squared magnitudes of the coefficients in the Fourier series expansion of $x(t)$, which can be written

$$x(t) = \frac{Ae^{j\theta}}{2}e^{j\omega_0 t} + \frac{Ae^{-j\theta}}{2}e^{-j\omega_0 t}. \tag{19.1-15}$$

Note that the phase angle θ disappears in the evaluation of $R_x(\tau)$; there are thus many waveforms having the same autocorrelation function. Specifying the set of averages $R_x(\tau)$ (one average for each value of τ) does not define a unique $x(t)$ but rather an ensemble of possible $x(t)$'s.

▶ ▶ ▶

Example 19.1–2

The autocorrelation function $R_p(\tau)$ of the impulse train $p(t)$ shown in Figure 19.1–1 is also an impulse train. This can be derived directly by cyclic convolution (19.1–10) or from the fact that the power density spectrum $S_p(f)$ is, by (19.1–5), an impulse train in frequency with impulse areas equal to the squares of the Fourier series coefficients for $p(t)$, as shown.

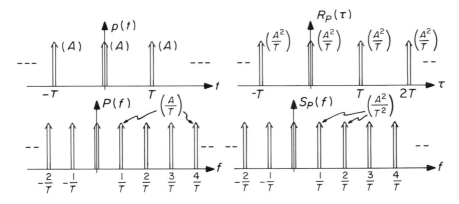

Figure 19.1–1. An impulse train and its autocorrelation function, Fourier transform, and power density spectrum.

▶ ▶ ▶

Example 19.1–3

Any periodic waveform $x(t)$ can be represented as the convolution of the impulse train $p(t)$ of Figure 19.1–1 (with $A = 1$) with the waveform

$$x_T(t) = \begin{cases} x(t), & 0 < t < T \\ 0, & \text{elsewhere} \end{cases} \tag{19.1–16}$$

which describes a single period of $x(t)$. That is (with $A = 1$),

$$x(t) = p(t) * x_T(t) \tag{19.1–17}$$

or in frequency

$$X(f) = P(f)X_T(f). \tag{19.1–18}$$

From (19.1–8) it then follows that

$$S_x(f) = S_p(f)|X_T(f)|^2 \tag{19.1–19}$$

or in time

$$R_x(\tau) = R_p(\tau) * [x_T(\tau) * x_T(-\tau)]. \tag{19.1–20}$$

These formulas are illustrated in Figure 19.1–2 for a periodic rectangular pulse. Note how the various properties of the autocorrelation function are exemplified in this case.

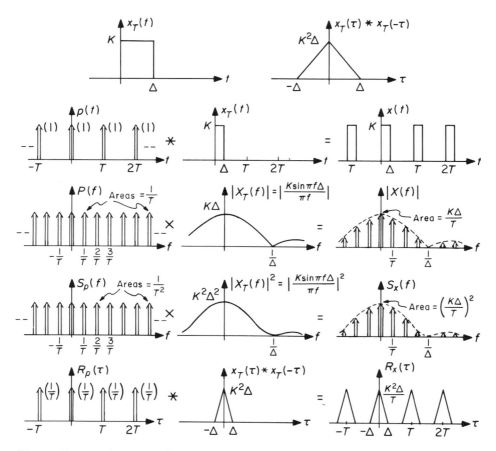

Figure 19.1–2. A rectangular pulse and its transform, power density spectrum, and autocorrelation function.

▶ ▶ ▶

Example 19.1–4

The previous example illustrates how convolving a periodic waveform (in this case $p(t)$) with another waveform (in this case $x_T(t)$) produces a new periodic waveform that in general has a different autocorrelation function. However, the new waveform does not have to have a different autocorrelation function. The autocorrelation function of $x(t) = p(t) * h(t)$ will be the same as that of $p(t)$ if the magnitude of the Fourier transform of $h(t)$ is unity, that is, if $h(t)$ is the impulse response of an all-pass network. A simple specific case is

$$H(f) = \frac{1 - j2\pi f}{1 + j2\pi f} \quad \Longleftrightarrow \quad h(t) = 2e^{-t}u(t) - \delta(t) \tag{19.1–21}$$

for which $x(t) = p(t) * h(t)$ is shown in Figure 19.1–3. That the autocorrelation functions for $x(t)$ and $p(t)$ should be the same is certainly not obvious. This example illustrates once again that the relationship between functions and their autocorrelation functions is many-to-one.

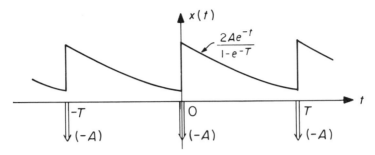

Figure 19.1–3. Another waveform having the same autocorrelation function as $p(t)$ in Figure 19.1–1.

▶ ▶ ▶

This brief survey of the properties of certain averages of periodic functions should help us to appreciate two important principles that apply to averages of other types of signals as well. First, knowledge of the values of even a large number of averages of operations on $x(t)$—such as $\langle x(t) \rangle$, $\langle x^2(t) \rangle$, and even $\langle x(t)x(t-\tau) \rangle = R_x(\tau)$ for all τ—does not in general provide enough information to determine $x(t)$ unambiguously.* Specifically, only the amplitudes, but not the phases, of the Fourier coefficients of $x(t)$ can be determined from $R_x(\tau)$. In general, there is a large ensemble of waveforms that share a common set of averages. When we agree to characterize a waveform in terms of its averages— because no other knowledge is available, or because averages are sufficient for our purposes—we are implying that we do not know or do not care which specific member of this ensemble is in fact present. It is in this sense, and in this sense only, that the methods of this chapter will make any connection with the usual connotations of the word "random."

Second, even though a knowledge of averages is insufficient to specify the input to some system completely, it may be adequate if all we want to compute are certain averages of the output. Thus, as shown by (19.1–8), specification of the autocorrelation function or power density spectrum of the input to an LTI system is sufficient to determine the autocorrelation function and power density spectrum of the output. As we shall now illustrate, such limited means and limited ends can take us a long way.

---

*To be sure, if we know $\langle x(t)e^{-j2\pi nt/T} \rangle = X[n]$ for all n and if we know that $x(t)$ is periodic with period T, then we can reconstruct $x(t)$ exactly. But the knowledge that $x(t)$ is periodic is not an average.

19.2 Properties of Infinite-Time Averages

Most of the ideas applied in the preceding section to periodic waveforms can be immediately extended to non-periodic situations. Thus, suppose that the input to some system of interest is an (unknown) member of an ensemble of non-periodic waveforms on $-\infty < t < \infty$, all of which are characterized by the same given values of some set of infinite-time averages such as $\langle x(t) \rangle$ and $R_x(\tau) = \langle x(t)x(t + \tau) \rangle$. How to determine an appropriate set of averages to represent some situation of interest is an extremely important and difficult problem whose features we can do little more than hint at in this introductory discussion. Usually, of course, we are not given enough information to describe any or all of the ensemble of waveforms $x(t)$ completely as time functions, $-\infty < t < \infty$, and thus we cannot really expect to be able to carry out the limiting operations corresponding to infinite-time averages. Instead, we shall likely be faced with a physical situation in which we can make certain measurements. We shall have to make these measurements under conditions that we believe to be "typical," and we shall have to convince ourselves somehow that the situation shows sufficient *statistical regularity* to justify representing it in terms of infinite-time averages. Taking into account the kind of operations we expect to carry out on the waveforms and the kind of conclusions we hope to reach, we shall then have to select and compute a specific finite set of finite-time averages as a summary of our observations. Finally, we must model our experimental results by an idealized set of postulated infinite-time measurements that meet various consistency and realizability conditions (as discussed briefly below). Our mathematical model will bear much the same relationship to reality that the function $A \cos 2\pi ft$ does to the output of an oscillator; whether a certain physical situation can be usefully described in this way, and if so what values to assign to the parameters, are not mathematical questions. In all of this, mathematics can provide suggestions and guides, but sophisticated theory cannot be substituted for the wisdom, experience, and judgment of the investigator. *There is probably no aspect of science and technology that is less understood—by laymen and professionals alike—or more likely to cause mischief and trauma than the process of making appropriate statistical models.*

Blithely ignoring all of the above, let us suppose that we have somehow arrived at an appropriate statistical model—a set of averages—describing what we believe we know about the input waveform (or more correctly about the ensemble of input waveforms) to some system. We now wish to compute certain averages of the output waveform. Since the output is a functional of the input, this means we want to find the values of averages of certain operations on a waveform given the values of the averages of other operations on that waveform. To proceed we need to define some rules for manipulating averages. Of course, we shall choose these rules to be consistent with the basic idea of an infinite-time average implied by the formula

$$\langle x(t) \rangle = \lim_{T_0 \to \infty} \frac{1}{2T_0} \int_{-T_0}^{T_0} x(t)\,dt\,. \tag{19.2–1}$$

For example, several simple rules are:

1. *TIME-INVARIANCE:*

    For every T, $-\infty < T < \infty$,

    $$\langle x(t+T) \rangle = \langle x(t) \rangle . \tag{19.2-2}$$

2. *LINEARITY:*

    For any waveforms $x(t)$ and $y(t)$ and constants A and B,

    $$\langle Ax(t) + By(t) \rangle = A\langle x(t) \rangle + B\langle y(t) \rangle . \tag{19.2-3}$$

3. *LIMITS ON THE MEAN:*

    If $x_{min} < x(t) < x_{max}$, then $x_{min} < \langle x(t) \rangle < x_{max}$. As a special case, if $x(t) \geq 0$ for all t, then $\langle x(t) \rangle \geq 0$.

To illustrate the application of these rules, consider the following example.

Example 19.2–1

Suppose we seek the average of the output $y(t)$ of a linear system described by

$$y(t) = \int_{-\infty}^{\infty} x(t-\tau)h(\tau)\, d\tau .$$

Applying the rules above, we conclude that

$$\langle y(t) \rangle = \left\langle \int_{-\infty}^{\infty} x(t-\tau)h(\tau)\, d\tau \right\rangle = \int_{-\infty}^{\infty} \langle x(t-\tau) \rangle h(\tau)\, d\tau = \int_{-\infty}^{\infty} \langle x(t) \rangle h(\tau)\, d\tau$$

$$= \langle x(t) \rangle \int_{-\infty}^{\infty} h(\tau)\, dt = \langle x(t) \rangle H(0) . \tag{19.2-4}$$

The second equality interprets the integral as a sum and applies the linearity rule; the third applies the time-invariance rule. In words, (19.2–4) states that the average value of the output of a linear system is equal to the average value of the input multiplied by the gain or frequency response of the system for zero frequency.*

▶ ▶ ▶

---

*Actually our derivation of this reasonable-sounding result has been more graphic than rigorous. A more careful treatment would show the necessity both for extending the linearity rule explicitly to infinite sums and for restricting the system to be stable.

Equation (19.2–4) is an example of the sort of results we shall be seeking in this chapter. Here a large number of averages of functionals of $x(t)$ (specifically the average of any LTI operation on $x(t)$) turn out to be completely specified by giving only one quantity, $\langle x(t) \rangle$, and two rules for manipulating averages. But, of course, we shall need to know more about $x(t)$ than simply $\langle x(t) \rangle$ if we seek the average of some more general functional of $x(t)$. We cannot, for example, determine the average of even such a simple function as the square of $x(t)$ if all we know is $\langle x(t) \rangle$. This conclusion should be immediately obvious from our knowledge of the properties of periodic functions. But it should also be obvious that we cannot specify $\langle x(t) \rangle$ and $\langle x^2(t) \rangle$ completely independently; we must at least require that

$$\langle x^2(t) \rangle \geq \langle x(t) \rangle^2 \tag{19.2–5}$$

that is, the total average power must be greater than or equal to the power in the average or d-c component. For periodic waveforms, (19.2–5) is a direct consequence of Parseval's Theorem (see Problem 14.17), but it also follows from the rules above. Thus consider

$$\left\langle [x(t) - \langle x(t) \rangle]^2 \right\rangle = \left\langle x^2(t) - 2x(t)\langle x(t) \rangle + \langle x(t) \rangle^2 \right\rangle$$

$$= \langle x^2(t) \rangle - 2\langle x(t) \rangle\langle x(t) \rangle + \langle x(t) \rangle^2$$

$$= \langle x^2(t) \rangle - \langle x(t) \rangle^2 . \tag{19.2–6}$$

The left-hand side of (19.2–6) is sometimes called the a-c *power* in $x(t)$, that is, the power in the fluctuations of $x(t)$ about its mean (average) value. In words, (19.2–6) states that the total power $\langle x^2(t) \rangle$ is the sum of the a-c power and the d-c power, $\langle x(t) \rangle^2$. The a-c power is the average of a positive quantity (a square) and hence, by the special case of the limits-on-the-mean rule, must be positive; thus (19.2–6) implies (19.2–5).

In general, we must simultaneously specify a number of average properties of $x(t)$, such as $\langle x(t) \rangle$ and $\langle x^2(t) \rangle$, if we wish to compute the average properties of an interesting set of operations on $x(t)$, that is, averages of the outputs of interesting systems with $x(t)$ as their input. In general also, as we have just seen, the specified properties of $x(t)$ may not be independent. If, at least in principle, an $x(t)$ exists having the specified average properties, we shall call these properties *consistent*. Specifying an inconsistent set of average properties would be equivalent to trying to specify both a time function and its Fourier transform, ignoring the linkages between them. Sometimes it is easy to show by means of an example that certain averages are consistent; you should have no difficulty, for example, constructing a periodic waveform to prove that any values assumed for $\langle x(t) \rangle$ and $\langle x^2(t) \rangle$ that satisfy (19.2–5) are consistent. But more often it is tedious: How would you go about showing that, say, certain assumed values for $\langle x(t) \rangle$ and $\langle \ln |x(t)| \rangle$ were consistent? (It can in fact be shown that these averages can be assigned values entirely independently.)

The sets of averages for which the consistency issue is easiest to resolve are various *probability distribution functions*. These are also a powerful set of averages in that knowledge of a modest set of appropriate probability distribution functions for $x(t)$ permits calculation of the averages of a large set of operations on $x(t)$. Indeed, in certain cases it is possible to specify (algorithmically) a set of probability distribution functions that provide a *complete statistical description* of $x(t)$, that is, sufficient information to determine (at least in principle) the average of *any* desired operation on $x(t)$. A full formal discussion of such matters is well beyond our scope—it is a topic in advanced courses in the *theory of random processes*. In the next section, however, we shall give several examples in which simple intuitive notions of probability provide useful models for real waveforms in communication systems.

Before we proceed to these examples, it will be helpful to state our remaining objectives in this chapter more narrowly and precisely. We shall restrict ourselves to situations in which a random signal $x(t)$ is the input to a stable LTI system with impulse response $h(t)$. We shall assume that the input is present throughout all time, $-\infty < t < \infty$, and that we are interested at most in averages of quadratic functions of the output $y(t)$, such as the autocorrelation function $R_y(\tau)$ defined by

$$R_y(\tau) = \langle y(t)y(t+\tau)\rangle. \tag{19.2--7}$$

Substituting

$$y(t) = \int_{-\infty}^{\infty} h(\mu)x(t-\mu)\,d\mu$$

and

$$y(t+\tau) = \int_{-\infty}^{\infty} h(\nu)x(t+\tau-\nu)\,d\nu$$

and invoking the linearity rule (19.2–3), we obtain from (19.2–7)

$$R_y(\tau) = \int_{-\infty}^{\infty}\int_{-\infty}^{\infty} h(\mu)h(\nu)\langle x(t-\mu)x(t+\tau-\nu)\rangle\,d\mu\,d\nu. \tag{19.2--8}$$

Defining

$$R_x(\tau) = \langle x(t)x(t+\tau)\rangle \tag{19.2--9}$$

and using the time-invariance rule (19.2–2), it is easy to show that (19.2–8) becomes

$$\boxed{R_y(\tau) = \int_{-\infty}^{\infty}\int_{-\infty}^{\infty} h(\mu)h(\nu)R_x(\tau+\mu-\nu)\,d\mu\,d\nu.} \tag{19.2--10}$$

Equation (19.2–10) is the time-domain equivalent of (19.1–8), now extended to non-periodic $x(t)$. In words, if we wish to know the autocorrelation function $R_y(\tau)$ of the output of a linear system, then the set of averages of the input that we must know is only the input autocorrelation function $R_x(\tau)$. Fortunately for the utility of this simple result, linear systems are common and autocorrelation functions convey a great deal of interesting information about the character of random waveforms.

19.3 Probabilistic Models of Simple Random Processes

A simple illustration of the way in which probabilistic information can be utilized to determine average properties of waveforms is provided by the following example.

Example 19.3-1

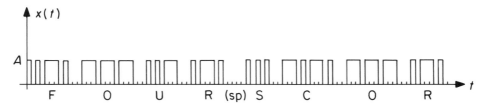

Figure 19.3-1. A segment of a telegraph waveform.

A section of the electrical waveform on a telegraph wire carrying a Morse code signal might appear as shown in Figure 19.3-1. What is the average power in such a waveform? In a message represented in Morse code, five types of symbols appear—dots (1 elementary space long), dashes (3 elem. sp.), symbol spaces (1 elem. sp.), letter spaces (3 elem. sp.), and word spaces (5 elem. sp.). Suppose that on examining a long string of text it is observed that approximately

28% of the symbols are dots
22% of the symbols are dashes
31% of the symbols are symbol spaces
16% of the symbols are letter spaces
3% of the symbols are word spaces.

These numbers can be interpreted as probabilities that a symbol picked "at random" is a dot, dash, etc.

To use this probability information to compute the average power in $x(t)$, note that 100 symbols will require about

$$(28 \times 1) + (22 \times 3) + (31 \times 1) + (16 \times 3) + (3 \times 5) = 188$$

elementary spaces on the average, of which

$$(28 \times 1) + (22 \times 3) = 94$$

or precisely half on the average will contain a pulse of height A, the remainder being zero. Hence

$$\langle x^2(t) \rangle = \frac{1}{2}(A^2) + \frac{1}{2}(0) = \frac{A^2}{2}$$

exactly as for an ordinary square wave.

▶ ▶ ▶

If we sought the autocorrelation function $R_x(\tau) = \langle x(t)x(t+\tau) \rangle$ for the Morse code waveform of Example 19.3–1, the information available would be insufficient to find it. We would need to know the relative frequencies with which various combinations of symbols occur. This rapidly gets complex; we shall turn instead to a simpler example.

Example 19.3–2

Figure 19.3–2. A segment of a random pulse waveform.

Suppose that a waveform, representing perhaps some idealization of the pulse-code-modulation signal discussed in Section 14.4, can be described as follows. Time is divided into intervals of length T seconds; $x(t)$ is zero in the last half of each interval, and in the first half it is of height A in half the intervals and zero in the remainder. Suppose further that whether or not a particular interval contains a pulse is independent of whether or not any other intervals contain pulses. Informally, we can consider the waveform to be generated by tossing a "fair" coin successively, once for each interval, inserting a pulse if the coin comes up "heads" and otherwise a zero. A segment of a typical waveform might appear as in Figure 19.3–2.

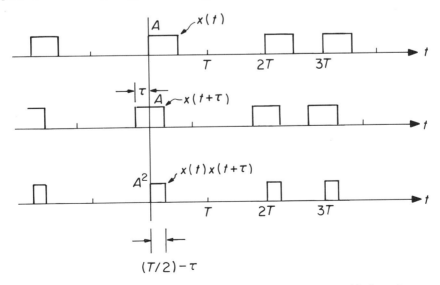

Figure 19.3–3. Construction of the product waveform $x(t)x(t+\tau)$.

The information given provides a complete statistical description of an ensemble of waveforms $x(t)$, allowing us in principle to compute the average of any functional of $x(t)$. As an illustration we can compute the autocorrelation function $R_x(\tau) = \langle x(t)x(t+\tau)\rangle$. This is done informally as follows. For $0 \le \tau \le T/2$, the product waveform appears as in Figure 19.3–3. From the properties of the sequence, we know that on average half the intervals in the product waveform will contain an overlap pulse, and hence the average value of the product will be

$$R_x(\tau) = \frac{1}{2}\left(\frac{1}{T}\right)A^2\left(\frac{T}{2}-\tau\right) = \frac{A^2}{4}\left(1-\frac{2\tau}{T}\right), \quad 0 \le \tau \le T/2$$

that is, a straight line from $A^2/4$ at $\tau = 0$ to 0 at $\tau = T/2$. As τ increases further, $R_x(\tau)$ will increase again as the pulses begin to overlap with their neighbors, but only one-fourth of the intervals contain such overlaps, and hence the value of $R_x(T)$ is only half the value of $R_x(0)$. Continuing, $R_x(\tau)$ will appear as in Figure 19.3–4—a central triangle with a periodic train of smaller triangles on each side.

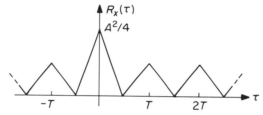

Figure 19.3–4. The autocorrelation function of the waveform of Figure 19.3–2.

▶ ▶ ▶

We can use the result of Example 19.3–2 to explore several general properties of correlation functions. First, consider $R_x(\tau) = \langle x(t)x(t+\tau)\rangle$ for a general waveform $x(t)$ and for large values of τ. Many random waveforms appear to have a "limited memory" in that knowledge of the value $x(t)$ had at some moment in the remote past tells us little or nothing about its present value. In such a case $x(t)$ and $x(t+\tau)$ become independent waveforms for large τ, so that $R_x(\tau) \to \langle x(t)\rangle^2$ as $\tau \to \infty$. Hence it is often possible to estimate the mean value of a waveform from the asymptotic tail of its autocorrelation function. Clearly, however, the waveform of Figure 19.3–2 is not of this "limited memory" type since the tails of $R_x(\tau)$ do not approach an asymptote. For this waveform we require a similar but slightly more complicated notion.

Let us next consider a general waveform constructed as the sum of two waveforms, $z(t) = x(t) + y(t)$. The autocorrelation function of the sum is

$$R_z(\tau) = \langle z(t)z(t+\tau)\rangle = \langle [x(t)+y(t)][x(t+\tau)+y(t+\tau)]\rangle$$
$$= R_x(\tau) + R_y(\tau) + \langle x(t)y(t+\tau)\rangle + \langle x(t+\tau)y(t)\rangle.$$

In many cases of interest, the last two terms (which are called *cross*-correlation functions) will vanish and the autocorrelation function of the sum will be the sum of the autocorrelation functions. Sufficient conditions for the cross-correlation functions to vanish are that $x(t)$ and $y(t)$ be completely independent processes and that at least one of the processes have zero average value. (These conditions are far from necessary, however.)

Finally, consider the autocorrelation function of the sum of a "limited memory" waveform with zero average value and an independent periodic waveform. Typical autocorrelation functions might appear as in Figure 19.3–5. Note that, even if the power in the periodic waveform is much less than that in the "limited memory" waveform (as evidenced by comparing their autocorrelation functions at $\tau = 0$), the tail of the autocorrelation function for the sum will still become periodic for sufficiently large τ and thus betray the presence of the periodic component. This result provides the basis for several common methods for detecting "hidden periodicities" or "cycles" in experimental data.

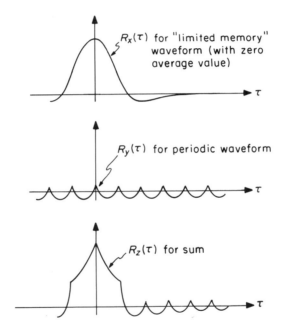

Figure 19.3–5. Detecting hidden periodicities.

In these terms, then, the periodic tail of Figure 19.3–4 suggests that the corresponding waveform contains a periodic part. To be sure, there is no guarantee that we can actually split the waveform of Figure 19.3–2 into two components as above. There are many periodic waveforms having the same autocorrelation function, and we have no way of deducing which one (if there even exists one) to subtract out to leave an independent "limited memory" waveform. However, Example 19.1–3 with $\Delta = T/2$ shows that the autocorrelation function of a square wave is a periodic triangular wave, and this suggests a decomposition as shown in Figure 19.3–6. It is easy to show that $R_{x_1}(\tau)$ and $R_{x_2}(\tau)$ have the form illustrated. Moreover, $\langle x_1(t)x_2(t+\tau) \rangle$ is zero for all τ, so that the autocorrelation function of the sum is the sum of the autocorrelation functions as desired. But

$x_1(t)$ and $x_2(t)$ are not really entirely independent processes—indeed, $x_1(t) = (2/A)x_2^2(t)$—so this decomposition is not quite the sort envisioned earlier.

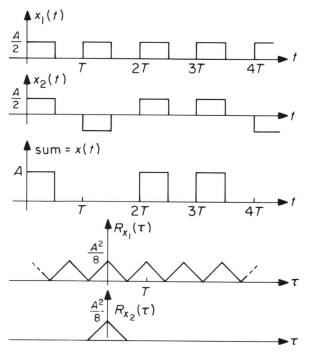

Figure 19.3–6. Decomposition of the waveform of Figure 19.3–2.

Example 19.3–3

Many other useful examples of random processes can be described as in Example 19.3–2. One of these—called a *white shot noise* process for reasons we shall discuss shortly—requires only a minor extension of the methods introduced in the last example. Informally, we can describe this process as follows. Toss a coin every δ seconds. In the corresponding time interval, erect a pulse of height $1/\delta$ if the coin comes up heads; otherwise, let the waveform height be zero. A typical result might look like Figure 19.3–7. Assume that the coin is heavily unbalanced so that only a small fraction of the tosses come up heads. Intervals containing pulses are thus rare; the average number of pulses per second is $\bar{n} = p/\delta$. We ultimately intend to let both p and δ tend to zero holding $\bar{n}$ fixed, so that a typical sample function consists of impulses occurring "at random" with an average rate of $\bar{n}$ per second. But first we compute the autocorrelation function for this waveform with nonzero p and δ. For $0 < \tau < \delta$ the average of the product of $x(t)$ and $x(t+\tau)$ is made up of two parts. First, there is the overlap of each pulse with itself, which creates a pulse in the product waveform of height $1/\delta^2$ and width $\delta - \tau$; there are $\bar{n}$ such product pulses per second. Second, there is the overlap of each pulse with its neighbor (if any), producing a product pulse of height $1/\delta^2$ and width τ; there are $\bar{n}p$ such pulses per second. Hence

$$R_x(\tau) = (1/\delta^2)(\delta - \tau)\bar{n} + (1/\delta^2)(\tau)\bar{n}p$$
$$= (\bar{n}/\delta)(1 - \tau/\delta) + (\bar{n})^2\tau/\delta, \quad 0 < \tau < \delta.$$

As a function of τ, the first term (from the overlap of a pulse with itself) contributes the solid line in the interval $0 < \tau < \delta$ in Figure 19.3–8; the second term (from the overlap of a pulse with its nearest neighbor) contributes the dashed line. For larger values of τ the analysis proceeds similarly; the overall result for $R_x(\tau)$ for finite δ is the sum of all the individual triangular pulses in Figure 19.3–8.

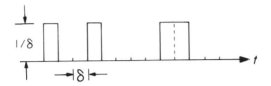

Figure 19.3–7. A typical waveform describing a shot noise process.

Passing to the limit as $\delta \to 0$ with $\bar{n}$ fixed, the autocorrelation function for the white shot noise process composed of impulses occurring at random consists (Figure 19.3–9) of an impulse and a constant (which we note has a value $\bar{n}^2 = \langle x(t)\rangle^2$ as our preceding discussion would suggest).

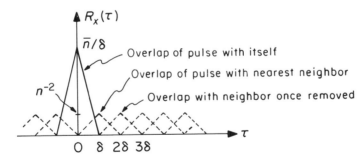

Figure 19.3–8. $R_x(\tau)$ for an approximation to a white shot noise process.

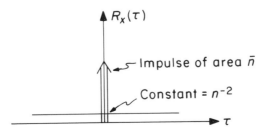

Figure 19.3–9. $R_x(\tau)$ in the limit as $\delta \to 0$.

▶ ▶ ▶

One of the most important applications of correlation functions—as illustrated in the last section and summarized in (19.2–10)—is to the study of the effect of LTI systems on random processes. As an example, suppose that the input to an LTI system is the white noise process of Example (19.3–3) with correlation function described by Figure 19.3–9, that is,

$$R_x(\tau) = \bar{n}^2 + \bar{n}\delta(\tau). \qquad (19.3\text{–}1)$$

Then the output $y(t)$ will consist of a succession of pulses shaped like $h(t)$, distributed "uniformly at random" in time with average rate $\bar{n}$ (see Figure 19.3–10). The autocorrelation function for $y(t)$ is given by (19.2–10):

$$
\begin{aligned}
R_y(\tau) &= \int_{-\infty}^{\infty}\int_{-\infty}^{\infty} h(\mu)h(\nu)\left[\bar{n}^2 + \bar{n}\delta(\tau - \nu + \mu)\right] d\mu\, d\nu \\
&= \bar{n}\int_{-\infty}^{\infty} h(\mu)h(\mu+\tau)\, d\mu + \left(\bar{n}\int_{-\infty}^{\infty} h(\mu)\, d\mu\right)^2. \qquad (19.3\text{–}2)
\end{aligned}
$$

Equation (19.3–2), called *Campbell's Theorem*, is important in applications, since there are a variety of phenomena—from the patter of rain on a tin roof to the inhomogeneities of current in a resistor resulting from the quantized nature of electric charge—that can be successfully modelled by the sort of waveform shown in Figure 19.3–10. Such applications may also suggest why this process is called "shot" noise.

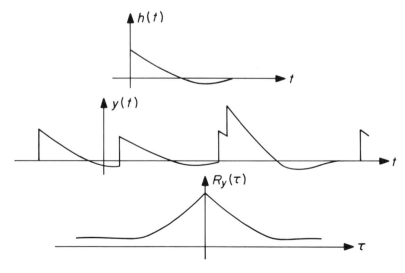

Figure 19.3-10. The result of passing white shot noise through a filter with impulse response $h(t)$.

Our previous experiences with averages of periodic functions suggest that (19.2–10) might be considerably simplified if the autocorrelation functions were replaced by their Fourier transforms. Thus let

$$S_x(f) = \int_{-\infty}^{\infty} R_x(\tau)e^{-j2\pi f\tau}\, d\tau, \quad S_y(f) = \int_{-\infty}^{\infty} R_y(\tau)e^{-j2\pi f\tau}\, d\tau \qquad (19.3–3)$$

and

$$H(f) = \int_{-\infty}^{\infty} h(t)e^{-j2\pi ft}\, dt.$$

Then (19.2–10) becomes

$$\boxed{S_y(f) = |H(f)|^2 S_x(f)} \qquad (19.3–4)$$

which is a most important formula.

In Section 19.1 we arrived at this equation by an entirely different route— by exploring how the spectral distribution of power was transformed by an LTI system. In that approach $S_x(f)$ was the power density spectrum. This remains an appropriate name for the transform of the autocorrelation function for the following reasons. First, the total average power in $x(t)$ is simply

$$\langle x^2(t)\rangle = R_x(0) = \int_{-\infty}^{\infty} S_x(f)\, df \qquad (19.3–5)$$

from the Fourier inversion formula. Equation (19.3–5) is clearly consistent with the notion of $S_x(f)$ describing power *density* (power per differential bandwidth). Next, suppose that $x(t)$ is passed through the special linear filter described by Figure 19.3–11. The average power in the output of the filter of Figure 19.3–11 is, from (19.3–5) and (19.3–4),

$$\int_{-\infty}^{\infty} |H(f)|^2 S_x(f)\, df \approx S_x(f_0)\Delta f. \qquad (19.3–6)$$

That is, $S_x(f_0)\Delta f$ is the amount of average power in $x(t)$ "contained" in the narrow band of frequencies of width Δf at f_0. Hence the name "power density spectrum."

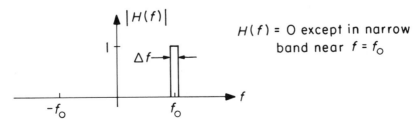

Figure 19.3-11. A special filter for probing the power density spectrum.

As a special consequence of (19.3–6), note that the average output power—and hence $S_x(f_0)$—must be a non-negative quantity. Since f_0 is an arbitrary frequency and $x(t)$ is an arbitrary process, we conclude that every autocorrelation function $R_x(\tau)$ must be so structured that its Fourier transform, the power density spectrum $S_x(f)$, is real and non-negative. Since not every function has this property, we have apparently derived a constraint that autocorrelation functions must satisfy. For example, it follows at once that $R_x(\tau)$ for a real waveform must be an even function, and in general

$$|R_x(\tau)| = \left| \int_{-\infty}^{\infty} S_x(f) e^{j2\pi f \tau} \, df \right|$$
$$\leq \int_{-\infty}^{\infty} |S_x(f) e^{j2\pi f \tau}| \, df = \int_{-\infty}^{\infty} S_x(f) \, df = R_x(0).$$

That is, an autocorrelation function has its largest magnitude at $\tau = 0$. These particular properties of $R_x(\tau)$ can also be derived directly (see Problem 19.2).

The fact that $S_x(f)$ must be non-negative and real is a much stronger condition, however. Indeed, it is easy to prove that this condition is also sufficient; that is, a necessary and sufficient condition for a function $R_x(\tau)$ to be the autocorrelation function of some waveform is that its Fourier transform $S_x(f)$ be real and non-negative. To prove sufficiency, we must demonstrate the existance of at least one waveform having the prescribed autocorrelation function, or equivalently having the corresponding power density spectrum. This we can do in terms of the white shot noise process, whose autocorrelation function is given by

$$R_x(\tau) = \bar{n}^2 + \bar{n}\delta(\tau). \tag{19.3–7}$$

The corresponding power density spectrum is

$$S_x(f) = \bar{n}^2 \delta(f) + \bar{n}. \tag{19.3–8}$$

The impulse at zero frequency corresponds to the average value of the process and can be removed simply by subtracting a constant $\bar{n}$ from $x(t)$. The remaining spectral density is a constant for all frequencies. Such a spectrum is called *white* (by analogy with white light, which is composed of a superposition of all colors). If $(x(t) - \bar{n})$ is now passed through a filter with frequency response $H(f)$, the resulting power density spectrum is $\bar{n}|H(f)|^2$. An arbitrary non-negative real spectrum $S_x(f)$ can thus be realized simply by passing a white shot noise process (with zero mean) through a filter with $|H(f)|$ proportional to $\sqrt{S_x(f)}$. The effect of such a filter can always be approximated as closely as desired (although not necessarily in "real time"—that is, by a causal filter—unless $\sqrt{S_x(f)}$ satisfies the Paley-Wiener condition of Chapter 15).

Example 19.3–4

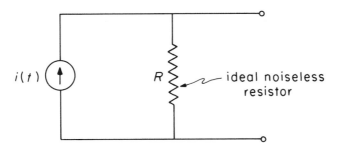

Figure 19.3–12. Equivalent circuit for thermal noise in a resistor.

An important result in random signal theory concerns the effect of the thermal agitation of charges in a resistor and is known as *thermal noise* or *Nyquist noise* (after the man who analyzed it theoretically) or *Johnson noise* (after the man who first studied it experimentally). It can be shown* under rather general conditions and in quite fundamental ways that a resistor R in thermal equilibrium at an absolute temperature T can be represented by the equivalent circuit in Figure 19.3–12; here $i(t)$ is a white noise source (at least up to frequencies at which the whole concept of a lumped resistor breaks down) with power density spectrum

$$S_i(f) = 2kT/R \qquad (19.3–9)$$

where $k = 1.38 \times 10^{-23}$ watt-sec/deg Kelvin is Boltzmann's constant. Alternatively, we can use Thévenin's Theorem to deduce an equivalent circuit consisting of a noiseless resistor R in series with a voltage source with power density spectrum $S_v(f) = 2kTR$.

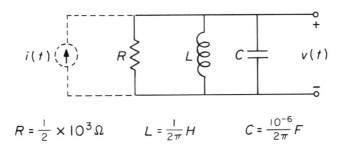

$$R = \frac{1}{2} \times 10^3\,\Omega \qquad L = \frac{1}{2\pi}\,H \qquad C = \frac{10^{-6}}{2\pi}\,F$$

Figure 19.3–13. A tuned circuit.

As an application of (19.3–9), we can compute the rms open-circuit voltage that would appear due to thermal agitation across the tuned circuit of Figure 19.3–13 at room temperature. Adding the source $i(t)$ (shown dotted) to represent the thermal noise, $v(t)$ is simply the output of an LTI system with input $i(t)$ and frequency response

---

*See, for example, J. L. Lawson and G. E. Uhlenbeck, *Threshold Signals* (New York, NY: McGraw-Hill, 1950) Chapter 4.

equal to the complex impedance of the circuit. That is,

$$H(f) = \cfrac{1}{\cfrac{1}{R} + j2\pi fC + \cfrac{1}{j2\pi fL}} = \cfrac{1}{2 \times 10^{-3} + j10^{-6}f + \cfrac{1}{jf}}$$

$$= \frac{jf}{(j10^{-3}f + 1)^2}\, \Omega. \qquad (19.3\text{--}10)$$

For simplicity the numbers have been chosen so that the poles coincide and the circuit is critically damped. The power density spectrum of the voltage is then

$$S_v(f) = |H(f)|^2 S_i(f) = |H(f)|^2 \frac{2kt}{R}$$

$$= \frac{f^2}{(1 + 10^{-6}f^2)^2} \cdot \frac{2 \times 1.38 \times 10^{-23} \times 293}{0.5 \times 10^3}$$

$$= \frac{16.2 \times 10^{-24} f^2}{(1 + 10^{-6}f^2)^2} \quad \text{volt}^2 \text{ sec} \qquad (19.3\text{--}11)$$

(where room temperature has been chosen as 20°C or 293°K). Finally,

$$\langle v^2(t) \rangle = \int_{-\infty}^{\infty} S_v(f)\,df = 25.5 \times 10^{-15}\ \text{volt}^2.$$

(The integral can be reduced to an elementary one by using Parseval's Theorem.) Thus

$$\sqrt{\langle v^2(t) \rangle} \approx 1.60 \times 10^{-7}\ \text{volts}$$

which is by no means negligible in many applications. It is interesting here (and can be shown in general) that the average energy stored in the capacitor is

$$\frac{1}{2}C\langle v^2(t) \rangle = \frac{1}{2}kT \qquad (19.3\text{--}12)$$

independent of the values of the circuit parameters. The average energy stored in the inductor has identically the same value; this is an example of the Equipartition Theorem of statistical physics.

▶ ▶ ▶

19.4 Summary

If all that is known about the source signals in some system is a certain set of long-term averages rather than a detailed description of the signals as waveforms or time functions, then all that can be determined about the responses is also a certain set of averages. Sometimes, however, such average information is quite sufficient for the system design or understanding that we seek. In particular, if the system is LTI, then knowledge of the input autocorrelation function

$R_x(\tau)$—or its Fourier transform, the power density spectrum $S_x(f)$—determines the autocorrelation function and the power density spectrum of the output. Autocorrelation functions describe the temporal rates at which signals take on new values; power density spectra specify the relative importance of various spectral regions in the frequency syntheses of signals. For signals characterized by averages, these functions play much the same roles that time and frequency specifications of ordinary signals played in earlier chapters. Most of the chapter was devoted to examples illustrating the properties of correlation functions and power spectra, starting from periodic waveforms, for which long-term averages can be directly computed, and proceeding to more interesting cases in which various other kinds of average information, such as probability distributions, can be used to determine the desired correlation functions. In the final chapter, we shall study how averaging notions such as those discussed here can be combined with the ideas of filtering, sampling, modulation, and the like to help us understand the design of modern communication systems.

PROBLEMS FOR CHAPTER 19

Problem 19.1

A waveform can be built up from the decimal representation for $\pi = 3.141592653\ldots$ as shown below:

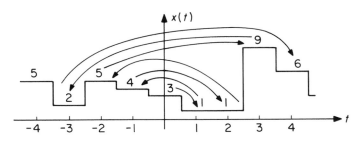

a) A waveform $x_T(t)$ is constructed from $x(t)$ by the rule

$$x_T(t) = \begin{cases} x(t), & |t| < T \\ 0, & |t| > T. \end{cases}$$

Discuss the general character of $X_T(f)$ as $T \to \infty$. (Do not attempt to compute $X_T(f)$.)

b) Would it be appropriate to consider $x(t)$ a "random" waveform? For what purposes? Suggest an application for which it would not be appropriate to consider $x(t)$ a "random" waveform. In formulating your answer you may find the following "data" helpful (results for integers other than 4 are similar):

Relative frequency of digit 4 in digits of $\pi = \dfrac{\text{Number of 4's in } N \text{ places}}{N}$

Places	Rel. Freq.	Places	Rel. Freq.
1–50	0.080	251–300	0.180
51–100	0.120	301–350	0.080
101–150	0.100	351–400	0.120
151–200	0.140	401–450	0.020
201–250	0.120	451–500	0.100

Problem 19.2

a) Use the time-invariance rule to prove that $R_x(\tau) = \langle x(t)x(t-\tau) \rangle$ is an even function for $x(t)$ real, that is, $R_x(-\tau) = R_x(\tau)$.

b) Use various rules applied to the average $\langle [x(t) \pm x(t-\tau)]^2 \rangle$ to prove that $R_x(0) \geq \pm R_x(\tau)$, that is, $R_x(0) \geq |R_x(\tau)|$. Assume that $x(t)$ is real.

Problem 19.3

A real random signal $x(t)$ has an autocorrelation function

$$R_x(\tau) = \langle x(t)x(t-\tau) \rangle = \frac{\sin 2\pi W \tau}{2\pi W \tau}, \quad -\infty < \tau < \infty.$$

a) What is the d-c value $\langle x(t) \rangle$ of $x(t)$?

b) What is the total average power $\langle x^2(t) \rangle$ of $x(t)$?

c) If $x(t)$ is passed through an LTI filter whose frequency response is

$$H(f) = \begin{cases} -j, & 0 < f < W/4 \\ j, & -W/4 < f < 0 \\ 0, & |f| > W/4 \end{cases}$$

what is the total average power in the output?

d) What is the output power density spectrum?

Problem 19.4

Assume that $x(t)$ is a real random signal characterized by

$$\langle x(t) \rangle = 0$$

$$\langle x(t)x(t-\tau) \rangle = R_x(\tau) = \int_{-\infty}^{\infty} S_x(f)e^{j2\pi f\tau} \, df.$$

Assume also that $H_1(f)$ and $H_2(f)$ are the frequency responses of LTI systems, as shown below.

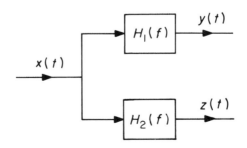

a) Find an expression for $S_{yz}(f)$, the Fourier transform of the cross-correlation function $R_{yz}(\tau) = \langle y(\tau)z(t-\tau) \rangle$, in terms of $S_x(f)$, $H_1(f)$, and $H_2(f)$.

b) Suppose that $x(t)$ is a white noise, that is, $S_x(f)$ is a constant. Find nontrivial impulse responses $h_1(t)$ and $h_2(t)$ such that $R_{yz}(\tau) = 0$ for all τ.

Problem 19.5

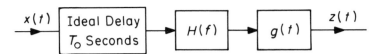

Three LTI systems are cascaded as above. If

$$H(f) = \frac{j2\pi f - 1}{j2\pi f + 2}$$

find the impulse response $g(t)$ of a (not necessarily unique) *stable causal* system such that $\langle z(t)z(t-\tau)\rangle = \langle x(t)x(t-\tau)\rangle$ for all τ.

Problem 19.6

Which if any of the following can be autocorrelation functions?

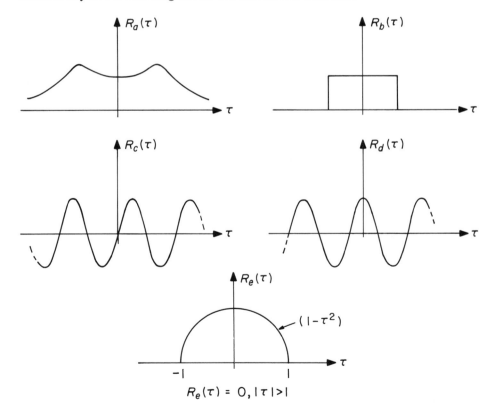

Problem 19.7

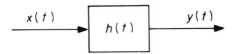

The impulse response of the LTI system above is

$$h(t) = \delta(t) - 2e^{-t}u(t).$$

Determine $\langle y(t)\rangle$ and $R_y(\tau) = \langle y(t)y(t-\tau)\rangle$ if

$$R_x(\tau) = \langle x(t)x(t-\tau)\rangle = e^{-|\tau|}.$$

Problem 19.8

a) Show that a necessary and sufficient condition for the power in the sum of two random signals to be the sum of the powers, that is, for

$$\langle [x(t) + y(t)]^2\rangle = \langle x^2(t)\rangle + \langle y^2(t)\rangle$$

is that $\langle x(t)y(t)\rangle = 0$.

b) Let $x(t)$ and $y(t)$ be periodic waveforms with incommensurable periods (so that T_x/T_y is not a rational number).

   i) Argue that

$$\langle x(t)y(t)\rangle = \langle x(t)\rangle\langle y(t)\rangle.$$

   ii) If either $\langle x(t)\rangle = 0$ or $\langle y(t)\rangle = 0$ (or both), show that

$$\langle [x(t) + y(t)]^2\rangle = \langle x^2(t)\rangle + \langle y^2(t)\rangle.$$

Problem 19.9

a) Show that $x(t)$ and $x(t-a)$ have the same autocorrelation function, independent of a.

b) Describe with a formula or sketch at least one other specific waveform—not simply a delayed version—that has the same autocorrelation function as the square wave of Figure 19.1–2 with $\Delta = T/2$.

Problem 19.10

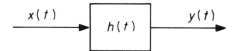

Let the input to the LTI system above be white noise, $R_x(\tau) = N_0 \delta(\tau)$. Show that

$$R_{xy}(\tau) = \langle x(t)y(t+\tau)\rangle = N_0 h(\tau).$$

This scheme is sometimes proposed as a method for experimentally determining $h(t)$. It is particularly useful if the normal input to the system being studied can be observed but not disconnected and if this input can be approximately described as white noise over the frequency range that is important for the system.

Problem 19.11

The received waveform $r(t)$ in a communication system consists of a complicated signal $s(t)$ added to an independent noise $n(t)$:

$$r(t) = s(t) + n(t).$$

The known autocorrelation functions of $s(t)$ and $n(t)$ are $R_s(\tau)$ and $R_n(\tau)$ respectively. To reduce the effect of the noise, it is proposed to pass $r(t)$ through an LTI system whose impulse response $h(t)$ has been chosen to minimize the average square of the difference between $r(t) * h(t)$ and $s(t)$.

a) Argue that an equivalent way to describe the LTI system is to choose $H(f)$ (the frequency response corresponding to $h(t)$) so as to minimize

$$\mathcal{E} = \int_{-\infty}^{\infty} \left[S_s(f)|H(f) - 1|^2 + S_n(f)|H(f)|^2 \right] df$$

where $S_s(f)$ and $S_n(f)$ are the power density spectra of $s(t)$ and $n(t)$ respectively.

b) If there are no other constraints (such as causality*) on $H(f)$, then it is easy to argue that the minimum value of $\mathcal{E}$ results if the integrand is minimized separately for each frequency. By differentiating

$$S_s(f)[H(f) - 1]^2 + S_n(f)H^2(f)$$

with respect to $H(f)$ and equating to zero, show that the optimum $H(f)$ is

$$H(f) = \frac{S_s(f)}{S_s(f) + S_n(f)}.$$

c) Sketch $H(f)$ for $S_n(f) = N_0$, $S_s(f) = \dfrac{K}{1+f^2}$, and several values of K, such as $N_0/10$, N_0, and $10N_0$.

---

*If a causal filter is required, this problem becomes much more difficult and has several famous solutions—such as the *Wiener* and *Kalman* filters. See, for example, B. D. O. Anderson and J. B. Moore, *Optimal Filtering* (Englewood Cliffs, NJ: Prentice-Hall, 1979).

Problem 19.12

A voltage waveform $x(t)$, observed during the interval $0 < t < T$, is known to consist either of a specific known signal waveform $s(t)$, $0 < t < T$, plus white noise $n(t)$ with power density N_0 volt^2 sec, or of the noise waveform $n(t)$ alone. To decide which condition is more likely to be true for a particular $x(t)$, it is proposed to pass $x(t)$ through an appropriate LTI system and sample the output $y(t)$ at $t = T$. If $y(T)$ is larger than some threshold value, then $s(t)$ will be announced as present; otherwise, absent. It is argued, then, that an appropriate impulse response $h(t)$ for the LTI system should give a large response to the signal $s(t)$ at $t = T$ but a small response to the noise on the average. Specifically, we seek to find $h(t)$ to maximize the signal-to-noise ratio

$$\frac{S}{N} = \frac{\left(\int_0^T s(\tau)h(T-\tau)\,d\tau\right)^2}{N_0 \int_0^T h^2(t)\,dt} = \frac{(\text{Response to } s(t) \text{ at } t = T)^2}{\text{Mean square noise response}}.$$

a) Use the Schwarz Inequality (appendix to Lecture 14) to show that

$$\frac{S}{N} \leq \frac{E}{N_0}$$

where $E = \int_0^T s^2(t)\,dt$. Show also that the $h(t)$ that achieves the maximum is

$$h(t) \sim s(T-t)$$

which is called the *matched filter* for $s(t)$.

b) Sketch the matched filter $h(t) = s(T-t)$ and the signal component $s(t) * h(t)$ of the matched filter output for the following $s(t)$:

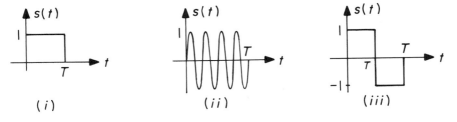

(i)   (ii)   (iii)

Problem 19.13

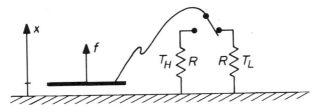

An air-dielectric capacitor with a movable plate can be arranged to behave as a heat engine if resistors in thermal equilibrium with heat reservoirs at two temperatures, T_H and $T_L < T_H$, are available.

a) The force of attraction between the plates of the parallel-plate capacitor when the plates are fixed a distance x apart and carry a fixed charge q is

$$f = \frac{q^2}{2Cx}$$

where C is the capacitance at that spacing. Using methods similar to those of Example 19.3–4, show that

$$\langle f \rangle = \frac{kT}{2x}$$

when the switch in the diagram above is in either position and T is the corresponding resistor temperature.

b) Suppose that the switch is connected to the hotter resistor and the plates are allowed to move slowly together, from an initial spacing x_1 to a final spacing $x_2 < x_1$. Find the mechanical work done on the external world during this contraction, which is approximately $W_{\text{OUT}} = \int_{x_2}^{x_1} \langle f \rangle \, dx$.

c) Show that the average energy stored in the capacitor remains fixed at $kT_H/2$ throughout this contraction, so that by conservation of energy the work done, W_{OUT}, must equal the energy given up by the heat reservoir.

d) Suppose now that the switch is connected to the colder resistor and external mechanical work is used to pull the plates slowly apart from x_2 to their original spacing x_1. Find the mechanical work done on the system during this expansion, which is approximately $W_{\text{IN}} = \int_{x_2}^{x_1} \langle f \rangle \, dx$.

e) Show that the efficiency of this engine,

$$\eta = \frac{\text{Net work out}}{\text{Energy given up by the hotter heat reservoir}} = \frac{W_{\text{OUT}} - W_{\text{IN}}}{W_{\text{OUT}}}$$

is given by

$$\eta = \frac{T_H - T_L}{T_H}.$$

This *Carnot efficiency* bounds the efficiency of any cyclic heat engine working between reservoirs at these temperatures.

20

MODERN COMMUNICATION SYSTEMS

20.0 Introduction

Modern communication systems differ from traditional systems in two characteristic ways:

1. The basic message to be transmitted is typically taken to be a sequence of discrete symbols selected from a finite alphabet, as contrasted with a waveform that is a continuous-valued function of continuous time. If the actual message is written text, numerical data, or computer code, such a representation is natural. If the original message is speech, music, or pictures, then (as discussed in Chapter 14) both a *sampling* process (to transform from continuous to discrete time) and some sort of *quantizing* operation (for example, replacing the continuous range of amplitudes by a finite set of discrete levels) are required. The advantages of considering the basic message as a discrete sequence of symbols from a finite alphabet are manifold:

   a) It makes possible a standardization of the modulators and demodulators (modems) as well as the channels, independent of the character of the message (speech, pictures, data, etc.).

   b) It permits easy manipulation of the message sequence to provide error-correcting coding, privacy, multiplexing, and the like, while leaving the message in the same basic form.

   c) It allows a reshaping of signals at repeaters so that errors and noise do not accumulate in a long cascade of operations.

2. The modulation process is typically described abstractly as the assignment of specific waveforms to specific sequences of symbols (that is, specific "words"). Even if the original message is an analog waveform, the analog waveform actually transmitted need look nothing like it. Instead, its duration, bandwidth, and other characteristics can be chosen to match the available channel and to reduce the effects of noise, interference, and distortion.

Our objective in this final chapter is to illustrate these characteristic features of modern communication systems with several examples, showing in the process how the language and tools we have developed in earlier chapters can be successfully applied.

20.1 Sampling and Quantizing

A simple example of the process of sampling and quantizing—collectively called *digitizing*—is shown in Figure 20.1–1. Part (a) of the figure shows the original waveform $x(t)$. If $x(t)$ is bandlimited, that is, if $X(f) = 0$, $|f| > W$, then samples taken at a rate $2W$ per second completely characterize $x(t)$, as in (b). The samples in (c) are obtained from those in (b) by replacing each actual sample height by the nearest integer multiple, n, of a uniform *sampling interval* Δ. The error in the quantized samples is shown in (d); the amplitude of the error fluctuates erratically between limits $\pm\Delta/2$. If, finally, a continuous signal $\hat{x}(t)$ is reconstructed from the quantized samples, it will differ from the original signal $x(t)$ by the *quantizing error*, which is the signal $\epsilon(t)$ in (f) reconstructed from the sequence of quantizing error samples in (d). If the number of levels used is large enough (greater than, say, 10), then the properties of $\epsilon(t)$ will be essentially independent of $x(t)$; successive error samples will be nearly independent and uniformly distributed over the interval $-\Delta/2$ to $+\Delta/2$. If $\epsilon(t)$ is reproduced acoustically, it will sound like a hiss or noise, and $\hat{x}(t)$ will sound like the original signal $x(t)$ but with a background hiss added (or rather subtracted, but $\epsilon(t)$ and $-\epsilon(t)$ have similar statistical properties) because

$$\hat{x}(t) = x(t) - \epsilon(t). \qquad (20.1-1)$$

Example 20.1–1

For later use, it will be helpful to have an estimate of $\langle \epsilon^2(t) \rangle$, the average power in the quantizing noise. It seems reasonable (and can be shown formally if necessary*) that $\langle \epsilon^2(t) \rangle$ should be equal to the average of the square of the quantizing error samples:

$$\langle \epsilon^2(t) \rangle = \lim_{N \to \infty} \frac{1}{2N+1} \sum_{n=-N}^{N} \epsilon^2\left(\frac{n}{2W}\right). \qquad (20.1-2)$$

If $\epsilon(n/2W)$ is assumed to be uniformly distributed over the range $-\Delta/2$ to $+\Delta/2$, then for large N we can expect $(2N+1)\delta\epsilon/\Delta$ of the error samples to lie in the range $-\Delta/2 < \epsilon(n/2W) < -\Delta/2 + \delta\epsilon$, the same number to lie in the range $-\Delta/2 + \delta\epsilon < \epsilon(n/2W) < -\Delta/2 + 2\delta\epsilon$, etc. Hence

---

*The simplest argument requires expanding the function $\epsilon(t)$, which is bandlimited, in a sum of $\dfrac{\sin 2\pi W(t - n/2W)}{2\pi W(t - n/2W)}$ functions (as in the Sampling Theorem of Chapter 14), and exploiting the orthogonality of these functions (see Problem 13.18).

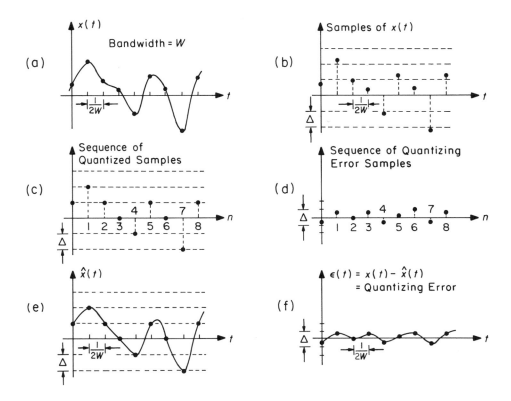

Figure 20.1–1. Sampling and quantizing.

$$\frac{1}{2N+1}\sum_{n=-N}^{N}\epsilon^2\left(\frac{n}{2W}\right)\approx\frac{1}{2N+1}\left[\left(-\frac{\Delta}{2}\right)^2(2N+1)\frac{\delta\epsilon}{\Delta}\right.$$

$$+\left(-\frac{\Delta}{2}+\delta\epsilon\right)^2(2N+1)\frac{\delta\epsilon}{\Delta}$$

$$+\left(-\frac{\Delta}{2}+2\delta\epsilon\right)^2(2N+1)\frac{\delta\epsilon}{\Delta}$$

$$\vdots$$

$$\left.+\left(\frac{\Delta}{2}\right)^2(2N+1)\frac{\delta\epsilon}{\Delta}\right]$$

$$\Longrightarrow\int_{-\Delta/2}^{\Delta/2}\epsilon^2\frac{d\epsilon}{\Delta}=\frac{\Delta^2}{12}$$

so that

$$\langle\epsilon^2(t)\rangle=\frac{\Delta^2}{12}. \tag{20.1-3}$$

▶ ▶ ▶

The effects of quantizing noise depend on the character of the original signal and the purposes of the communication system. Thirty-two levels are considered adequate for "commercial-grade" reproduction of speech—indeed, speech remains understandable if only *two* levels (< 0 and > 0) are employed, and such "clipped" speech even has certain advantages with respect to intelligibility in very noisy environments such as taxicabs and airplane cockpits. On the other hand, hi-fi reproduction of music requires 128 levels or more. Quantizing pictures introduces special problems. The eye is extremely sensitive to the edges of areas of equal quantizing level that appear in regions such as the sky where the picture intensity is smoothly changing. One way to reduce the objectionable characteristics of such edges is to add a small random noise—"snow"—to the picture before quantizing. This increases by perhaps a factor of 2 or so the total effective quantizing error, but the visual effect may be improved because the edges are blurred.

It should also be observed that nothing in the quantizing process requires the quantizing levels to be equally spaced; significant improvements in overall quality for a given total number of levels can often be achieved by spacing the levels more closely in those amplitude ranges that are judged to be more important. Even more significantly, nothing requires that the sampling and quantizing process be carried out on the original waveform. The overall goal of minimizing the number of symbols* per second required to describe a waveform with a given fidelity can often be best accomplished by working with transformations of the original

---

*The size of the alphabet from which the symbols are selected is also important. To compare different systems, it is conventional to reduce sequences to equivalent sequences of binary digits—0 or 1. Thus, a quantizer with $2^5 = 32$ levels per sample requires 5 binary digits (*bits*) per sample to describe its output.

waveform, or by using techniques that are not even describable as sampling and quantizing. This process has been most extensively studied for speech. For example, an attractive scheme for digitizing speech is *delta modulation*,* which is illustrated in Figure 20.1–2. At each sample time, the speech waveform is compared with a ramp waveform. If the speech sample is greater than the ramp value, the ramp moves upward during the next sampling interval; otherwise it moves downward. In the simplest scheme, the slopes of the ramp waveform are equal in magnitude and unchanged throughout the process. The binary waveform representing the speech can be considered to be the derivative of the ramp waveform. By making δ small enough, we can make the representation as accurate as desired. The advantages of delta modulation as compared with, say, PCM (see Chapter 14), which can also yield a binary waveform representing speech, are not so much in accuracy for a given digit rate as in simplicity of implementation: The ramp can be recovered from the binary waveform by integration, and a still smoother approximation can be achieved by further lowpass filtering. More elaborate schemes for digitizing speech, such as vocoders and linear predictive coding,† can substantially reduce the digit rate required for a given fidelity.

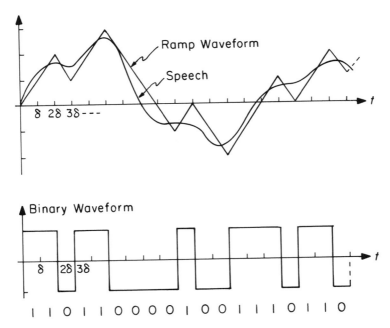

Figure 20.1–2. Delta modulation.

---

*Linear delta modulation was invented by E. M. Deloraine, S. van Miero, and B. Derjavitch (French patent 932140, August 1946). See also an early paper, F. DeJager, *Philips Research Report*, 7 (Dec. 1952): 442–466, and a more recent book, R. Steele, *Delta Modulation Systems* (New York, NY: John Wiley, 1975).

†See, for example, R. W. Schafer and L. R. Rabiner, *Proc. IEEE*, 63, 4 (Apr. 1975): 662–677.

20.2 Error-Correcting Codes

Once we have replaced a continuous-amplitude signal by a sequence of discrete symbols from a finite alphabet, we have lost information and have no way to recover it. The noise or error resulting from the effects of quantizing can be reduced only by selecting more closely spaced quantizing levels (or their equivalent) with a corresponding increase in the length of the symbol sequence or in the size of the alphabet from which the symbols are chosen. With finite sequence lengths and alphabet sizes the quantizing error cannot in general be zero. Quantizing noise thus sets an upper bound on the performance of the communication system; even if the symbol sequence is reproduced at the receiver with absolute accuracy, the effects of quantizing errors will remain.

In general, to be sure, various forms of noise and interference in the communication system will preclude perfect reproduction at the receiver of the transmitted sequence. Unlike quantizing noise, however, transmission errors can often be reduced, and even virtually eliminated, by clever design. And for messages originally in discrete-sequence form, of course, these transmission errors are the only errors there are.

Example 20.2–1

One error-reduction technique involves adding to the message sequence extra symbols derived from the sequence that can serve as check digits to detect or even correct errors. An example is the following simple *parity check* coding scheme. Suppose that the message has been reduced to a sequence of binary values, 0 and 1. Break up the message into blocks of 4 successive digits and insert between each pair of blocks a block of 3 check digits derived from the preceding 4 as shown in Figure 20.2–1. Transmit the entire sequence of message and interleaved check digits. Because of channel noise or interference, the sequence of digits reproduced at the receiver will typically have occasional errors—a 0 turned into a 1 or vice versa. In the example, errors occur in the 4th digit of the first block of 7, and in the 5th digit of the second block of 7. Suppose that errors are known to be so rare that with high probability at most 1 error occurs in any block of 7. If we know that at most 1 error has occurred in each block, we can use the check digits to detect and correct such errors. Thus, in the first block of 7, when we compute the check digits we would expect if there were no errors, we find that the 2nd and 3rd parity checks fail; if there is only 1 error, such a result can occur if and only if the 4th digit is in error. Similarly in the second block of 7, the failure of only the first check shows that the 5th digit (the first check digit) is in error. In the third block of 7, there are no errors, so all parity checks succeed. You should have no difficulty working out the rules for correcting all single errors, as well as describing how the system would behave if there happened to be more than 1 error in a block of 7.

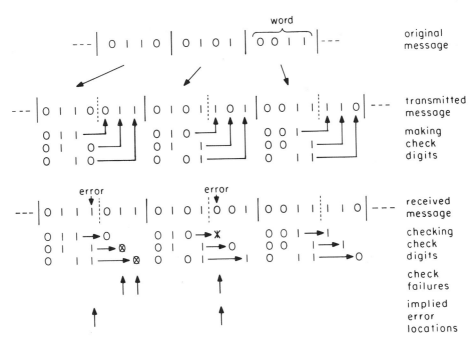

Figure 20.2–1. An error-correcting code. Check digits are 0 if the number of 1's is even, 1 if the number of 1's is odd.

▶ ▶ ▶

In general, the coding problem in information theory is to devise a way to associate with each word of length N in the original message a code word of length $M > N$ to serve as the transmitted message sequence. Ideally, the association is to be done in such a way that the effects of the errors introduced by a specific channel are minimized. A key theorem due to Claude Shannon[*] shows—surprisingly—that, provided the ratio N/M is less than a critical nonzero quantity called the *channel capacity*, the effects of channel errors can be made arbitrarily small, at least if we are willing to let the word lengths M and N become large. Of course, large word lengths impose severe memory, delay, and computational problems, so in practice compromises are necessary. But the surprising thing is that small error rates can be achieved in principle without requiring that the information transmission *rate* also be small. This whole area has been extensively studied in the last several decades, and many ingenious and useful results have been obtained.

---

[*]C. Shannon, *Bell Sys. Tech. J.*, 27 (1948): 379, 623.

20.3 Modulation and Detection

After digitizing and coding, in most communication systems we must convert the symbol sequence to an analog waveform having appropriate characteristics for the channel. In a radio communication system, for example, the transmitted waveform must have a spectrum confined to the assigned frequency channel. In principle, the sequence-to-waveform transformation can be done in quite arbitrary ways. Thus, we could prepare a "code book" listing each possible word sequence of some chosen length M and describing for each a waveform to be transmitted when that sequence occurs. The receiver would compare the received waveform—corrupted by channel disturbances—with the waveforms described in a copy of the "code book," select the closest match, and report the corresponding sequence. The waveforms used in this process need bear no obvious relationship to the sequences they represent, and of course they need not look in any way like the analog waveforms from which the sequence may have been derived initially. Nevertheless, it will surely be easier to build the system if the waveforms can be derived by some simple algorithm from each word sequence. And the performance of the system will obviously be improved if we design the waveforms to be as "different" as possible rather than selecting them arbitrarily.

In this section we shall explore semiquantitatively the performance of three different algorithms for assigning waveforms to code words—specifically the two pulse-modulation systems, PCM and PPM, described in Chapter 14, and a direct conversion of quantized sample heights into pulse amplitudes that we might call *pulse-amplitude modulation* (PAM). As we shall see, the first two require a larger bandwidth to transmit digits at the same rate as PAM, but in compensation their error rate is much lower. An understanding of how bandwidth can be traded for performance is one of the great insights of modern communication theory.

20.3(a) Pulse-Amplitude Modulation (PAM)

In each of the following systems we assume that the basic coded message is a sequence of binary digits to be transmitted at a rate of one digit every T seconds. To provide a basis for comparison, we shall assume that each system treats 3 successive digits as a single word. There are thus $2^3 = 8$ possible words. Each system will assign to each of these 8 words a different waveform of duration $3T$ seconds. In the case of PAM, we assume that these 8 waveforms are square pulses of duration $3T$ with 8 different amplitudes, as shown in Figure 20.3-1. The maximum amplitude is A, the minimum 0, and the remainder equally spaced at multiples of $A/7$. We shall assume that $s(t)$ as shown is the actual transmitted waveform, although in practice it would probably be filtered or shaped to reduce its spectral side lobes and would be modulated on a carrier. The spectrum of the PAM transmitted waveform has a bandwidth on the order of the reciprocal of the pulse length, $1/3T$. If we assume that the 8 levels are equally probable, then the average power in the PAM transmitted waveform is easily shown to be $P_{\text{ave}} = 0.36\,A^2$.

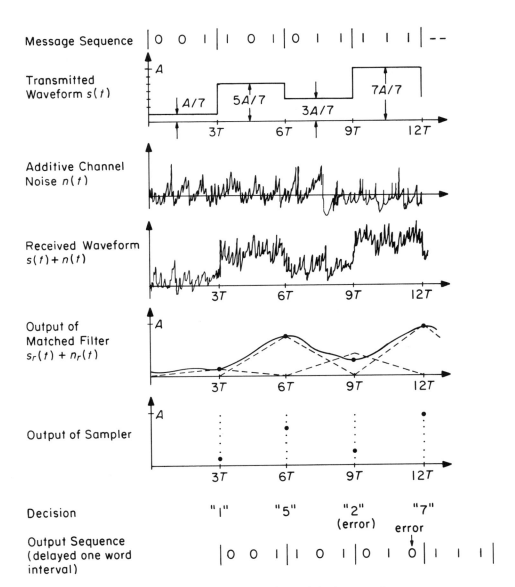

Figure 20.3–1. Pulse-amplitude-modulation waveforms.

The waveform arriving at the receiver will be assumed to be the transmitted waveform $s(t)$, reduced in amplitude (because of channel attenuation) and corrupted by additive disturbances or noise $n(t)$. (Disturbances are due typically to interference from other services as well as to thermal and shot noises introduced in the low-level amplifier stages at the input to the receiver; real channels, of course, need not be linear, and the disturbances need not be additive.) For simplicity, we shall model $n(t)$ as having a white spectrum with power spectral density N_o watts/Hz, and we shall ignore the channel attenuation. (Alternatively we can scale up the noise power by the amount of the attenuation. What counts is the ratio of signal power to noise power in the later stages of the receiver.)

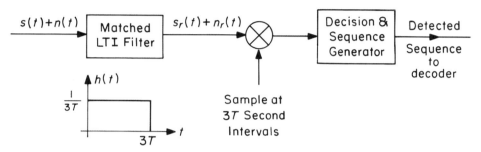

Figure 20.3–2. A matched-filter receiver for the PAM signal.

To recover the message sequence, the receiver should first integrate or average the received waveform, $s(t) + n(t)$, during each $3T$ pulse interval. This minimizes the effect of the noise. Equivalently, the averaging operation can be carried out by the matched filter shown in Figure 20.3–2 (see Problem 19.12). The response $s_r(t)$ of the matched filter to the signal part $s(t)$ of the received waveform is the sum of the dotted triangles shown in the fifth waveform of Figure 20.3–1. The mean square value of the noise response $n_r(t)$ to the noise component $n(t)$ of the received waveform is, by the methods of Chapter 19,

$$\langle n_r^2(t) \rangle = N_o \int_{-\infty}^{\infty} h^2(t)\, dt = \frac{N_o}{3T}. \tag{20.3-1}$$

Each sample height at the output of the sampler in Figure 20.3–2 thus consists of a voltage equal to the signal amplitude during the preceding $3T$ seconds, that is, $A/7, 2A/7, \ldots, 7A/7$, plus a noise voltage whose rms value is $\sqrt{N_o/3T}$. If the decision as to which level was sent during the preceding interval is to be reliable, that is, if errors are to be infrequent, then clearly the rms value of the noise must be small compared with the difference between levels,

$$\sqrt{\frac{N_o}{3T}} \ll \frac{A}{7} \tag{20.3-2}$$

or equivalently, setting $0.36\, A^2 = P_{\text{ave}}$ (the average transmitted power),

$$\frac{P_{\text{ave}} T}{N_o} \gg \frac{49}{3} \times 0.36 = 5.88. \tag{20.3-3}$$

This describes the transmitted power required and/or the rate at which binary digits can be reliably transmitted with this system for a given power.

20.3(b) Pulse-Code Modulation (PCM)

The difference between PCM and PAM in the present context is illustrated in Figure 20.3–4. In effect, each binary digit is transmitted separately—1 corresponding to a pulse of duration T and height B, 0 corresponding to no pulse. If 0 and 1 are equally likely, then the average power in the transmitted waveform is $P_{\text{ave}} = 0.5\,B^2$, and its bandwidth will be about $1/T$. Thus, the bandwidth of the PCM signal is 3 times larger than the PAM signal for the same message sequence rate—a distinct disadvantage.

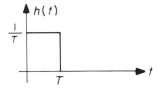

Figure 20.3–3. Matched filter for the PCM signal.

The receiver for the PCM system is similar to that for the PAM signal except that the matched filter or averager should have an impulse response $1/3$ as long and a bandwidth 3 times wider, as shown in Figure 20.3–3. As a result, the mean square value of the output noise is 3 times larger than the value derived in (20.3–1) for the PAM receiver:

$$\langle n_r^2(t)\rangle = \frac{N_o}{T}. \tag{20.3–4}$$

This would also seem to be a distinct disadvantage for the PCM system. However, the difference between signal levels in the sampler output of the PCM receiver is the full peak amplitude of the signal, B, rather than only $1/7$ of the peak amplitude. This more than makes up for the noise disadvantage since a low error rate for the PCM system requires

$$\sqrt{\frac{N_o}{T}} \ll B \tag{20.3–5}$$

or, setting $0.5\,B^2 = P_{\text{ave}}$,

$$\frac{P_{\text{ave}}T}{N_o} \gg 0.5. \tag{20.3–6}$$

The average power in PCM is more than a factor of 10 smaller than that in PAM for the same error rate.* Or, for the same power, performance is vastly

---

*This is not quite fair. Three PCM digits must be received correctly to accomplish the same result as the correct identification of one PAM level. On the other hand, a level error in PAM may mean more than one binary digit in the output sequence will be in error (although code-to-level transformations, called *Grey codes*, can be devised in which adjacent levels differ in only one binary digit). Precise comparisons are not easy to make, but for low error rates the effect of these corrections is small in any event.

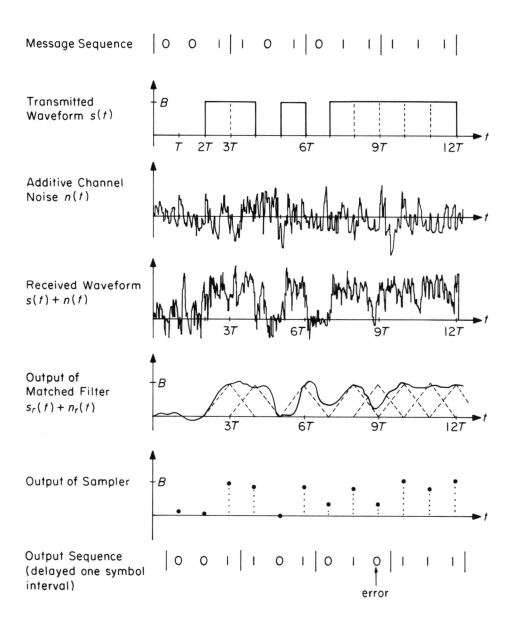

Figure 20.3–4. Pulse-code-modulation waveforms.

improved. Thus we have the remarkable result that bandwidth can be exchanged for signal-to-noise ratio—a basic tenet of modern communication theory that helps explain many puzzling phenomena such as the advantage of FM over AM.

20.3(c) Pulse-Position Modulation (PPM)

The PPM system shown in Figure 20.3–6 has performance superior to the PCM system just considered, but the price in bandwidth is high. One pulse of fixed height is transmitted in each interval of length $3T$, but the pulse has duration $3T/8$ and is transmitted in one of eight positions. The bandwidth of this signal is thus about $8/3T$, which is 8 times the bandwidth of the PAM system and 2.7 times that of the PCM system. The average power is $P_{\text{ave}} = 0.125\,C^2$.

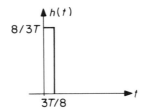

Figure 20.3–5. Matched filter for the PPM signal.

In the receiver, it is again necessary to adjust the duration of the matched filter, as shown in Figure 20.3–5. The mean square value of the output noise is then

$$\langle n_r^2(t)\rangle = \frac{8N_o}{3T} \tag{20.3–7}$$

which is much larger than for either PCM or PAM. But the difference in signal levels is also much larger. For a low error rate we must have*

$$\sqrt{\frac{8N_o}{3T}} \ll C \tag{20.3–8}$$

or, setting $0.125\,C^2 = P_{\text{ave}}$,

$$\frac{P_{\text{ave}}T}{N_o} \gg 0.33\,. \tag{20.3–9}$$

Hence performance equivalent to that of the PCM system can be achieved with about 2/3 the average power, but nearly 3 times the bandwidth is required. PPM is thus less efficient than PCM in trading bandwidth for signal-to-noise ratio.

---

*Again we are not being quite fair, since to avoid error the amplitude at the correct place must be larger than the noise value at any other place, and there are 7 places to check. But again the correction is small.

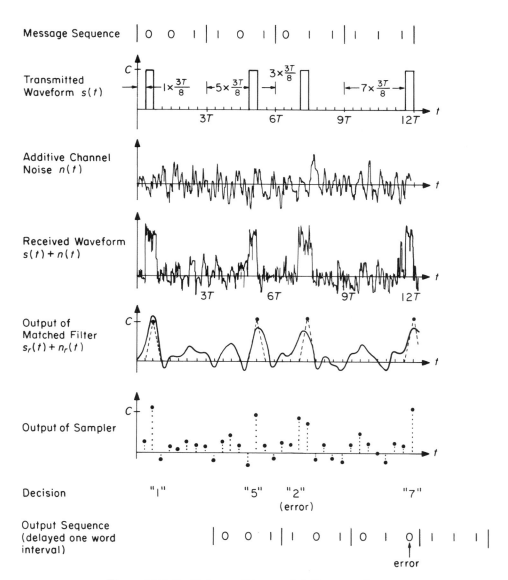

Figure 20.3–6. Pulse-position-modulation waveforms.

20.4 Summary

It is a characteristic of modern communication systems that they start by trans-forming their inputs into a sequence of standardized discrete symbols. If the inputs are analog waveforms with a continuous range of amplitudes, the process of digitization introduces some error. But the advantages gained in flexibility and in the ability to control errors introduced by channel noise or interference are often more important.

One of the techniques used to reduce the effects of noise is to spread out the signal spectrum so that it occupies a wider bandwidth. An extended ex-ample showed how this works by comparing three different pulse communication systems—PAM, PCM, and PPM. The key to understanding why it is possible to trade bandwidth for signal-to-noise ratio is the observation that the number of significantly different waveforms of a given energy and duration goes up directly with the allowed bandwidth. This is one way of interpreting the statement in Chapter 14 that a waveform has $2TW$ "degrees of freedom," that is, there are roughly $2TW$ orthogonal waveforms of duration T and bandwidth W. All three of the sample communication systems in Section 20.3 were required to send one of 8 different waveforms in $3T$ seconds. But if the bandwidth from which they were selected were larger, these 8 waveforms could be designed to be "further apart," or more "unusual," and thus more easily recognized. In a similar way FM (which may in a sense be considered the time-frequency dual of PPM) is a very special waveform in its bandwidth—one that changes frequency *slowly*— and hence can be separated from its background with high reliability, that is, with a high signal-to-noise ratio.

There is, however, another cost—in addition to increased bandwidth and increased system complexity—that must be paid to realize the advantages of PPM, PCM, FM, and similar systems. As the noise level is increased (or the signal-to-noise ratio reduced), the wideband systems characteristically exhibit a *threshold effect*. Little deterioration in performance is observed until a critical signal-to-noise ratio is reached, at which point the behavior of the wideband system suddenly collapses. Most of us have experienced this threshold effect in listening to an FM radio in a car; as the distance from the transmitter increases, a fraction of a mile will often make the difference between high-quality reception and nothing but noise. A PAM system, on the other hand, fails more gracefully; indeed, under poor conditions an AM system may still provide some degree of communication effectiveness when an FM system has failed completely. When wideband systems are good, "they are very, very good; when they are bad, they are horrid!" It would be hard to find a better example of what increasingly seems to be a general principle: Sophisticated technology can provide spectacular benefits under the conditions for which it was designed, but it is often extremely sensitive to these conditions and fails dramatically if circumstances change or have been misjudged.

EPILOGUE

To read a poem (so the King of Hearts told the White Rabbit), "Begin at the beginning and go on till you come to the end: then stop." Books, like poems, are sequentially ordered structures, and thus inevitably have a beginning and an end; this is the final paragraph of this book. But the theory of systems and signals, as we have seen, is not simply a cascaded arrangement of topics. There are multiple loops and branches, many parallel and crossing paths. Most ideas are linked directly and indirectly to many others. There is no simple step-by-step route by which this multidimensional web can be systematically explored and comprehended. There is really no beginning, and no end. We cannot expect to appreciate one topic fully until we have considered others. And so we must continually circle back to examine earlier concepts from a new vantage point. If you have enjoyed your introduction to this fascinating field, I hope you will have many future opportunities to return and extend your knowledge of circuits, signals, and systems.

Index